Proceedings in Life Sciences

Drugs Affecting Lipid Metabolism

Edited by R. Paoletti,
D. Kritchevsky and W. L. Holmes

With 132 Figures

Springer-Verlag
Berlin Heidelberg New York
London Paris Tokyo

Professor Dr. Rodolfo Paoletti
Dean
School of Pharmacy
University of Milan
Via Balzaretti 9
20133 Milan, Italy

Professor Dr. David Kritchevsky
Associate Director
The Wistar Institute
3601 Spruce Street
Philadelphia, PA 19104, USA

Dr. William L. Holmes
238 Chatham Way
West Chester, PA 19380, USA

ISBN-13: 978-3-642-71704-8   e-ISBN-13: 978-3-642-71702-4
DOI: 10.1007/978-3-642-71702-4

Library of Congress Cataloging in Publication Data. Drugs affecting lipid metabolism/
edited by R. Paoletti, D. Kritchevsky, and W.L. Holmes. p. cm. – (Proceedings in life sciences)
Includes index. Hyperlipidemia-Chemotherapy. 2. Antilipemic
agents-Testing. 3. Blood lipoproteins-Metabolism. 4. Blood cholesterol-Metabolism. I. Paoletti,
Rodolfo. II. Kritchevsky, David, 1920-. III. Holmes, Williams L., 1918-. IV. Series. [DNLM:
1. Antilipemic Agents. 2. Arteriosclerosis. 3. Lipids-metabolism. 4. Lipoproteins-metabolism.
5. Metabolism-drug effects.   QU 85 D794]   RC632.H87D78   987   616.3'997061-dc 19

2131/3130-543210

Preface

The recent symposium and the appearance of this new book on *Drugs Affecting Lipid Metabolism* take place at a very unusual time for the development of this area.

After the publication and wide acceptance of the results of the cholestyramine study by the Lipid Clinics in the USA, showing for the first time a direct association between drug induced reduction of plasma levels of total and LDL cholesterol and coronary heart disease in a high risk population, an unparalleled interest in drugs and other procedures able to control plasma cholesterol levels has been activated.

Two other significant events occurred during 1986 and 1987: the availability of compact instruments for the immediate determination of total cholesterol in plasma or total blood and the developments of new agents such as the inhibitors of HMG-CoA (hydroxymethylglutaryl CoA) reductase and ACAT inhibitors, with potentially great effect on plasma lipid levels after oral administration.

These new advances, together with the combined efforts of cell biologists and lipoprotein chemists, have set the pace for an exciting period of research and clinical applications of diets and drugs affecting lipids.

This volume, which includes the work of many of the leading world laboratories, represents an authoritative and up-to-date appraisal of the status of the art and a stimulus to future research at laboratory and clinical level in an area of opportunity for clinical and preventive medicine.

Milan, October 1987   Rodolfo Paoletti

Acknowledgements

We thank the contributors for their appreciable efforts in writing the papers for the Drugs Affecting Lipid Metabolism Meeting (Florence, Italy, October 22–25, 1986) upon which the chapters in this volume are based. We are particularly indebted to the Giovanni Lorenzini Foundation for its help and generous support, which made the meeting possible.

We are also grateful to the scientific secretaries, Drs. A.L. Catapano and C.R. Sirtori, and to the International Advisory Board for their valuable contributions.

The Editors

Contents

Contributors

You will find the addresses at the beginning of the respective contributions

Albarede, J.L. 328
Allievi, L. 131
Apebe, P. 442
Austen, K.F. 167
Avogaro, P. 407, 433
Axthelm, M. 426
Baggio, G. 355
Balestreri, R. 283
Ball, M.J. 56
Baralle, F.E. 56
Barbara, L. 350
Bard, J.M. 333
Barter, P.J. 244
Bass, N.M. 105
Bazzoli, F. 350
Bell, G.D. 223
Bengtsson-Olivecrona, G. 88
Bertolini, S. 283
Beynen, A.C. 122
Bittolo Bon, G. 407, 433
Blane, G.F. 309
Bogaievsky, Y. 343
Bondioli, A. 142
Bonnefous, F. 343
Bosisio, E. 21, 126, 380, 394
Bradley, W.A. 52
Branchi, A. 278
Brewer Jr., H.B. 236
Cacciaguerra, F. 223
Caldwell, J. 324
Canavesi, A. 174
Cao Danh, H. 317, 338
Capaldo, B. 295

Catapano, A.L. 126
Cazzolato, G. 407, 433
Chan, L. 52
Chapman, M.J. 117
Chernick, S.S. 88
Chesne, E. 338
Ciuffetti, G. 291, 372
Cighetti, G. 421
Connor, W.E. 155
Crepaldi, G. 355
Crestani, M. 126, 380
Crook, D. 215
Cuzzolaro, S. 283
Daga, A. 283
De Fabiani, E. 380, 394
Del Puppo, M. 421
Di Santo, C. 358
Dorigo, P. 94
Dostert, P. 324, 338
Douste-Blazy, P. 333
DoVale, H. 1
Drouin, P. 333
Duane, W.C. 150
Dujovne, C.A. 136
Eisenberg, S. 48, 79, 305
Elicio, N. 283
Fantappié, S. 126
Fievet, C. 333
Fragiacomo, C. 394, 410
Franceschini, G. 442
Freeman, M.L. 150
Fruchart, J.C. 333
Fujita, H. 260
Funke, P.T. 255
Gaddi, A. 442

Gaion, R.M. 94
Galli, C. 131, 162
Galli, G. 126, 380, 394
Galli Kienle, M. 421
Gebhard, R.L. 150
Genovese, S. 295
Ghiselli, G. 63
Gianfranceschi, G. 442
Gianturco, S.H. 52
Gibson, D.M. 9
Giudici, G.A. 358
Glatz, J.F.C. 122
Godsland, I.F. 215
Goto, Y. 247, 274
Gotto, A.M., Jr. 52, 63
Graham, K.J. 150
Gregg, R.E. 236
Grundy, S.M. 34, 415
Guichard, J.P. 328
Harris, W.S. 155
Heibig, J. 63
Hibbard, D.M. 150
Hoeg, J.M. 236
Hopkins, G.J. 244
Houin, G. 328
Hunninghake, D.B. 150
Illingworth, D.R. 155
Ito, T. 251
Itoh, H. 260
Javitt, N.B. 29
Jost, G. 343
Kajinami, K. 260
Kamon, N. 260
Kashyap, M.L. 367
Katan, M.B. 122, 158

Relationship of Cholesterol to DNA Synthesis in Normal and Cancerous Cells

M. D. Siperstein, H. DoVale, and J. R. Silber [1]

It has been known since 1950 that there is a striking correlation between the rate at which acetate is converted to cholesterol and the rate of cell growth. Specifically, in tissues such as the kidney, cell replication is very slow and such tissues have low rates of cholesterol synthesis. By contrast, baby brain and intestine replicate at rapid rates and have active cholesterol synthesis. As we showed some years ago, the very rapid cell growth of regenerating liver is accompanied by one of the highest rates of cholesterolgenesis seen in mammalian cells (1).

Our interest in the relation between cell replication and cholesterol biosynthesis developed from two observations. One was the finding that the primary site of feedback control of cholesterol synthesis is located at the synthesis of mevalonate (2,3). Second was our initial observation that this feedback control of mevalonic acid, at least in vivo, is consistently impaired or completed deleted in a series of malignant tumors (1,4,5). In contrast to the liver, where the feeding of cholesterol leads to a marked decrease in the conversion of acetate to cholesterol and a comparable decrease in the activity of HMG CoA reductase, in the slowly growing hepatoma 9121 feeding cholesterol does not inhibit either cholesterogenesis or HMG CoA reductase activity (Table 1) (6,7). Similar data have been reported by Goldfarb and Pitot (7).

Table 1. Absence of feedback control of HMG CoA reductase in hepatoma 9121

Tissue	Diet	$(2\text{-}^{14}C)$ Acetate converted to cholesterol $m\mu mol\ g^{-1}\ h^{-1}$	β-hydroxy-β-methylglutaryl CoA reductase $m\mu mol\ g^{-1}\ h^{-1}$
Liver	Low cholesterol	45	$0.77\ \pm 0.11$
	5% cholesterol	0.8	0.007 ± 0.01
Hepatoma 9121	Low cholesterol	46	$1.34\ \pm 0.14$
	5% cholesterol	96	$1.69\ \pm 0.17$

1 University of California, San Francisco Metabolism Section, Veterans Administration Medical Center, San Francisco, CA 94121, USA

Drugs Affecting Lipid Metabolism
Ed. by R. Paoletti et al.
© Springer-Verlag Berlin Heidelberg 1987

Table 2. Absence of cholesterol feedback control in rat hepatomas

Tissue	Diet	
	Normal Acetate → Chol[a]	Cholesterol 5% Acetate → Chol[a]
Liver	8	0.02
Hepatomas		
9618A	152	126
9633	72	54
7787	33	32
9121	27	25
5121	9	7
7795	6	7
7793	3	4
7794A	7	18
7316A	4	8
7800	7	9
H-35	6	9
7288C	0.9	1
3924A	0.5	0.6
3683	1	3

[a] nmol 2-14 C acetate g^{-1} h^{-1}.

This phenomenon can be demonstrated in a variety of minimal deviation hepatomas. As shown by the data in Table 2, in contrast to the marked decrease in the conversion of acetate to cholesterol characteristic of the livers of animals fed cholesterol, cholesterol feeding consistently has no significant effect upon cholesterol synthesis in hepatomas. Loss of cholesterol feedback control has been shown to occur in non-hepatic tumors (8) and has been confirmed and extended by a number of investigators (9–12). Moreover, loss of the cholesterol feedback system can be demonstrated even in the precancerous state (13,14). For example, feeding aflatoxin to a rat for a period of only 2 days, led to complete loss of feedback control of cholesterol synthesis and of mevalonate synthesis (15). This observation, too, has been confirmed with a wide variety of cancer-producing agents, primarily by Sabine's laboratory (16–18).

These observations raised the question of what might be the role of cholesterol synthesis in both normal and abnormal cell growth. It is known that cholesterol is required for cell membrane synthesis, and it was shown initially by Chen and Kandutsch (19,20) and by Brown and Goldstein (21) that treating cells in tissue culture with hydroxysterols, which inhibit sterol synthesis by blocking HMG CoA reductase, will also inhibit cell growth. This inhibition could be reversed by adding cholesterol in the form of lipoprotein, a finding that logically led to the conclusion that hydroxysterols prevent cell proliferation by depleting the cell of cholesterol.

Because of our earlier interest in the control of mevalonic acid synthesis, we decided to look at the question of whether mevalonic acid, independent of its function

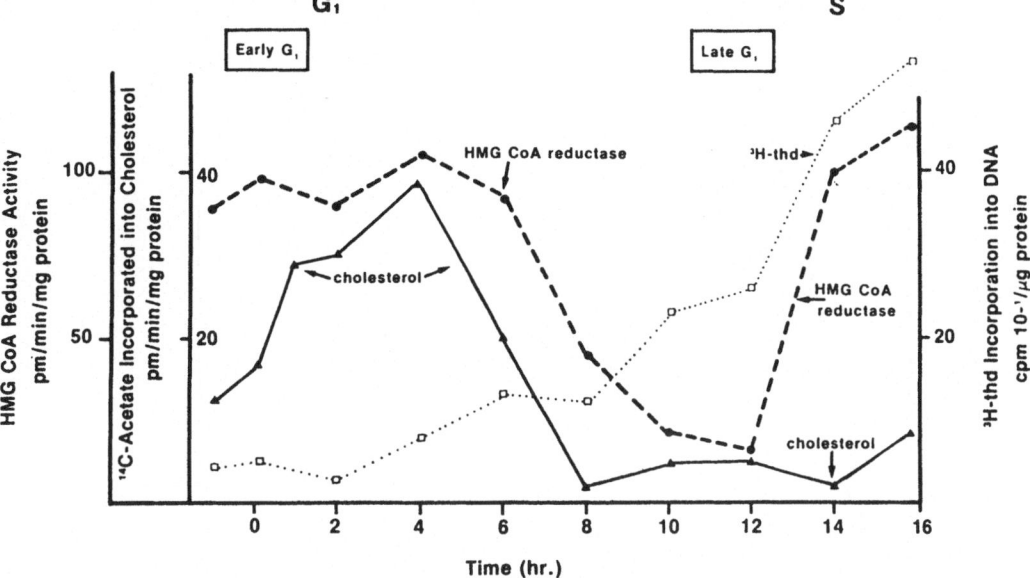

Fig. 1. The interrelationship between HMG CoA reductase activity, cholesterol synthesis, and DNA synthesis in serum-stimulated cells

as a precursor of cholesterol might play a role in cell growth and reproduction. Initially, Dr. Huneeus, in our laboratory, looked at the relationship between the activity of HMG CoA reductase and hence mevalonate synthesis and the stages of the cell cycle (22). As shown in Fig. 1, during the normal cell cycle in BHK cells, the activity of the HMG CoA reductase undergoes a biphasic rise, the first increase occurring during the Gl, growth phase of the cell cycle, and the second simultaneous with the rise of thymidine incorporation into DNA, i.e., during the S-phase of the cell cycle. By contrast, cholesterol synthesis, while showing an increase during the growth phase of the cell cycle, presumably to supply the cholesterol required for membrane synthesis, does not undergo a comparable rise during the S-phase of the cell cycle. A similar dissociation between mevalonate and cholesterol synthesis has been noted by Trentalance et al. in the case of the regenerating liver (23). These findings raised the possibility that mevalonate, independent of its function in cholesterol synthesis, might play a role in the progression of the cell cycle and specifically in DNA synthesis.

Kaneko et al. (24) had previously examined this possibility in tissue culture using compactin, a potent inhibitor of HMG CoA reductase, to deplete the cells of mevalonate and cholesterol. He found, however, that whereas mevalonate, independent of cholesterol, was required for cell growth, no specific requirement for mevalonate in S-phase DNA synthesis could be detected. We were, therefore, somewhat surprised when Dr. Huneeus, in our laboratory (22,25) was able to demonstrate (as shown in Fig. 2), in contrast to the findings of Kaneko et al. (24), that compactin does in fact in-

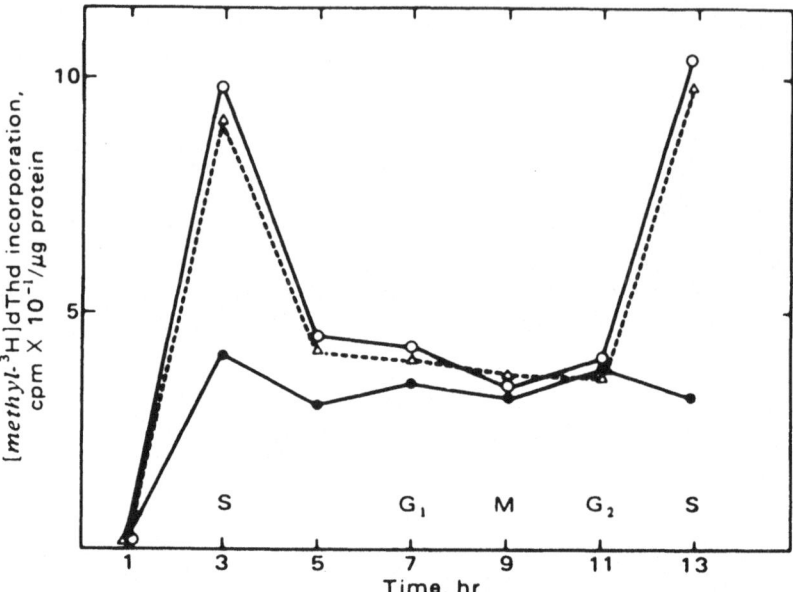

Fig. 2. Role of mevalonate in DNA replication. ∘ control; • 2.5 μM compactin added 1 h; △ 2.5 μM compactin added 1 h; 0.4 mM mevalonate added 15 min before harvest

hibit DNA synthesis. Moreover, when we attempted to reverse the compactin-induced inhibition of DNA synthesis by adding either cholesterol or mevalonate, there was once more a striking dissociation in the effect of these compounds. Cholesterol itself was totally ineffective in reversing the compactin-induced inhibition of DNA synthesis. By contrast, adding mevalonate not only reversed this inhibition, but in many cases, actually caused an "over-shoot" of DNA synthesis.

These findings led to the conclusion that the pathway of cholesterogenesis serves two functions in controlling cell growth and the cell cycle (26,22). The first is a requirement for cholesterol itself which, as shown by Chen et al. (19) and Kandutsch and Chen (20) and Brown and Goldstein (21), involves a function in cell growth, presumably for membrane synthesis. Independent of this cholesterol requirement, however, the data in Fig. 2 indicate that mevalonate plays an essential role in controlling DNA replication. Our finding that mevalonate, in addition to its role in cholesterol synthesis, is required for DNA replication has subsequently been widely confirmed, specifically by Habenicht et al. (27), Larson et al. (28), Perkins et al. (29), and most recently by Fairbanks et al. (30). Larson has shown a similar effect of mevalonate in malignant cells (31).

The mechanism of this striking effect of mevalonate on DNA synthesis remains to be established. We have, therefore, recently directed our attention to the enzyme that is most closely associated with DNA replication, namely DNA polymerase. In these studies we made use of a rat hepatoma HTC cell line grown in tissue culture. As shown in Table 3, one again sees in these tumor cells that compactin treatment causes an almost complete blockage of DNA synthesis. DNA synthesis is controlled in large part by the activity of DNA polymerase, which is known to exist as at least three distinct enzymes. Alpha-DNA polymerase is the enzyme that is believed to be primarily responsible for DNA replication. Beta-polymerase plays a key role in DNA repair. One can readily determine the relative amount of these two DNA polymerases in cell nuclei by use of specific inhibitors of alpha- and beta-polymerase, respectively, aphidicolin and dideoxythmidine. With this technique it was found that compactin inhibited alpha-DNA polymerase by between 71 and 92% (average 81%) (Table 3). In contrast, beta polymerase is not significantly affected by the presence of compactin. This effect of mevalonate depletion on the alpha-polymerase is relatively specific in that RNA synthesis is inhibited by compactin by only 21%, whereas protein synthesis, as shown in Table 3, is not significantly affected by compactin.

Table 3. Effect of mevalonate depletion on DNA synthesis, DNA polymerase, RNA, and protein synthesis in HTC cells

	Percent inhibition by compactin
DNA synthesis	86%
DNA polymerase alpha	81%
DNA polymerase beta	6%
RNA	21% (av)
Protein	2%

To summarize, these data have shown that blocking mevalonate synthesis results in an average 86% inhibition of DNA synthesis which is paralleled by an 81% inhibition of alpha-polymerase. This effect of mevalonate depletion is specific for DNA polymerase alpha in that compactin causes no detectable inhibition of polymerase beta.

We have repeated these observations in a nontumor cell (Table 4), the 3T3 fibroblast, where compactin again caused an almost complete inhibition of DNA synthesis, which also is reversed by the addition of mevalonate. In this cell, too, compactin causes a comparable (86%) inhibition of DNA polymerase alpha with, again, a complete reversal of the inhibition produced by the addition of mevalonate. The marked inhibition of polymerase alpha which follows mevalonate depletion is therefore demonstrated in both tumor and nonmalignant cells.

Table 4. Effect of mevalonate depletion on DNA synthesis and DNA polymerase alpha in 3T3 mouse fibroblasts

	Percent inhibition
DNA synthesis	86%
DNA polymerase alpha	86%

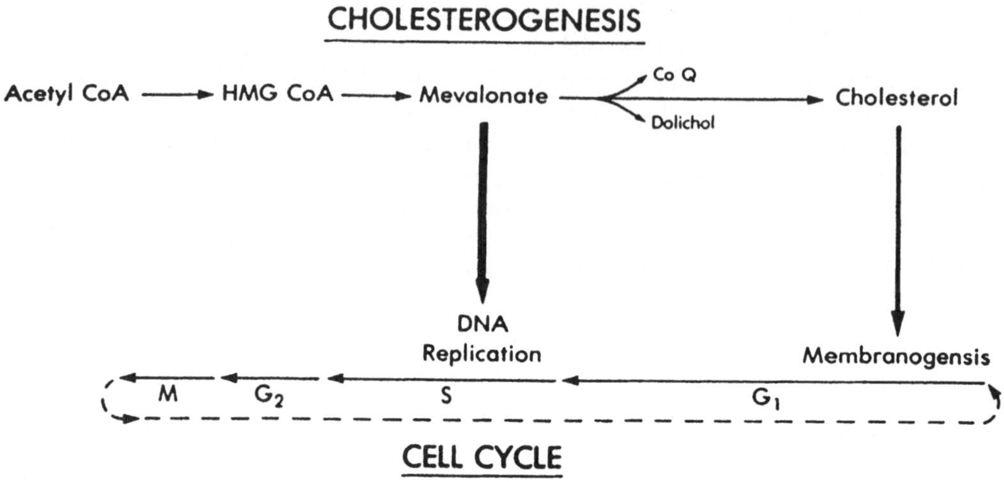

CHOLESTEROGENESIS

Acetyl CoA $\longrightarrow$ HMG CoA $\longrightarrow$ Mevalonate $\longrightarrow$ $\begin{array}{c}\text{Co Q}\\\text{Dolichol}\end{array}$ $\longrightarrow$ Cholesterol

DNA
Replication

Membranogensis

M G_2 S G_1

CELL CYCLE

Fig. 3. Role of cholesterol and mevalonate in the cell cycle and DNA synthesis

In summary (Fig. 3), previous studies have demonstrated that cholesterol and, apparently, mevalonate are required for cell growth. Our laboratory has shown that mevalonate, completely independent of its role in cholesterol synthesis, is also required for DNA replication. In the present studies we have demonstrated that both in normal and malignant cells inhibition of mevalonate synthesis causes a marked and comparable inhibition of DNA polymerase alpha. These studies therefore lead to the conclusion that loss of mevalonate-dependent alpha-polymerase can account completely for our previous observation that mevalonate depression results in over an 80% inhibition of de novo DNA synthesis. While the detailed mechanism by which mevalonate deprivation causes loss of DNA polymerase alpha activity remains to be elucidated, these findings suggest that mevalonate or a nonsterol isoprene product of mevalonate may play a previously unsuspected role in regulating DNA polymerase alpha, and hence, DNA replication in both normal and malignant cells.

References

1. Siperstein MD, Fagan VM (1964) Deletion of the cholesterol negative feedback system in liver tumors. Cancer Res 24:1108–1115
2. Siperstein MD, Goest MJ (1960) Studies on the site of the feedback control of cholesterol synthesis. J Clin Invest 39:642–652
3. Siperstein MD, Fagan VM (1966) Feedback control of mevalonate synthesis by dietary cholesterol. J Biol Chem 241:602–609
4. Siperstein MD, Fagan VM, Morris HP (1966) Further studies on the deletion of the cholesterol feedback system in hepatomas. Cancer Res 26:7–11
5. Siperstein MD (1970) Regulation of cholesterol biosynthesis in normal and malignant tissues. Curr Top Cell Regul 2:65–100

6. Siperstein MD, Gyde AM, Morris HP (1971) Loos of feedback control of hydroxymethyl-glutaryl coenzyme A reductase in hepatomas. Proc Natl Acad Sci USA 68:315–317
7. Goldfarb S, Pitot HC (1971) The regulation of β-hydroxy-β-methylglutaryl coenzyme A reductase in Morris hepatomas 5123C, 7800 and 9618A². Cancer Res 31:1879–1882
8. Polsky FI, Brown MS, Siperstein MD (1973) Feedback control of cholesterol synthesis in circulating granulocytes and deletion of feedback control in a granulocytic leukemia. J Clin Invest 52:65a
9. Sabine JR, Abraham S, Chaikoff IL (1967) Control of lipid metabolism in hepatomas: insensitivity of rate of fatty acid and cholesterol synthesis by mouse hepatoma BW7756 to fasting and to feedback control. Cancer Res 27:793–799
10. Sabine JR (1975) Defective control of lipid biosynthesis in cancerous and precancerous liver. Proc Biochem Pharmacol 10:269–307
11. Elwood JC, Morris HP (1968) Lack of adaptation in lipogenesis by hepatoma 9121. J Lipid Res 9:337–341
12. Kandutsch AA, Hancock RL (1971) Regulation of the rate of sterol synthesis and the level of β-hydroxy-β-methylglutaryl coenzyme A reductase activity in mouse liver and hepatomas. Cancer Res 31:1396–1401
13. Siperstein MD (1966) Deletion of the cholesterol negative feedback system in precancerous liver. J Clin Invest 45:1073
14. Wiley MH, Siperstein MD (1976) The control of cholesterol synthesis in normal and malignant cells. In: Criss WE, Ono T, Sabine JR (eds) Control mechanisms in cancer. Raven, New York, pp 343–350
15. Siperstein MD (1970) Regulation of cholesterol biosynthesis in normal and malignant tissues. Curr Top Cell Regul 2:65–100
16. Horton BJ, Horton JD, Sabine JR (1972) Metabolic controls in precancerous liver. II. Loss' of feedback control of cholesterol synthesis measured repeatedly in vivo during treatment with the carcinogens N-2-fluorenylacetamide and aflatoxin. Eur J Cancer 8:437–443
17. Horton BJ, Horton JD, Sabine JR (1973) Metabolic controls in precancerous liver. V. Loss of control of cholesterol synthesis during feeding of the hepatocarcinogen 3'-methyl-4-dimethylaminoazobenzene. Eur J Cancer 9:573–576
18. Horton BJ, Sabine JR (1971) Metabolic controls in precancerous liver: defective control of cholesterol synthesis in rats fed N-2-fluorenylacetamide feeding. Eur J Cancer 9:459–465
19. Chen HW, Kandutsch AA, Waymouth C (1974) Inhibition of cell growth and oxygenated derivatives of cholesterol. Nature (London) 251:419–421
20. Kandutsch AA, Chen HW (1977) Consequences of blocked sterol synthesis in cultured cells. J Biol Chem 252:409–415
21. Brown MS, Goldstein JL (1974) Suppression of 3-hydroxy-3-methylglutaryl coenzyme A reductase activity and inhibition of growth of human fibroblasts by 7-ketocholesterol. J Biol Chem 249:7306–7314
22. Huneeus VQ, Wiley MH, Siperstein MD (1979) Essential role for mevalonate synthesis in DNA replication. Proc Natl Acad Sci USA 76:5056–5060
23. Trentalance A, Leoni S, Mangiantini MT, Spagnuolo S, Feingold K, Hughes-Fulford M, Siperstein MD, Cooper AD, Erickson SK (1984) Regulation of 3-hydroxy-3-methylglutaryl coenzyme A reductase and cholesterol synthesis and esterification during the first cell cycle of liver regeneration. Biochem Biophys Acta 794:142–151
24. Kaneko I, Hazama-Shimada Y, Endo A (1978) Inhibitory effects on lipid metabolism in cultured cells of ML-236B, a potent inhibitor of 3-hydroxy-3-methylglutaryl coenzyme A reductase. Eur J Biochem 87:313–321
25. Quesney-Huneeus V, Galick HA, Siperstein MD, Erickson SK, Spencer TA, Nelson JA (1983) The dual role of mevalonate in the cell cycle. J Biol Chem 258:378–385
26. Siperstein MD (1984) Role of cholesterogenesis and isoprenoid synthesis in DNA-replication and cell growth. J Lipid Res 25:1462–1468
27. Habenicht AJR, Glomset JA, Ross R (1980) Relation of cholesterol and mevalonic acid to the cell cycle in smooth muscle and Swiss 3T3 cells stimulated to divide by platelet-derived growth factor. J Biol Chem 255:5134–5140

28. Larson RA, Chung J, Scanu AM, Aychnin S (1982) Neutrophils are required for the DNA synthetic response of human lymphocytes to mevalonic acid: evidence suggesting that a nonsterol product of mevalonate is involved. Proc Natl Acad Sci USA 79:3028–3032
29. Perkins SL, Ledin SF, Stubbs JD (1982) Linkage of the isoprenoid biosynthetic pathway with induction of DNA synthesis in mouse lymphocytes: effects of compactin on mitogen-induced lymphocytes in serum-free medium. Biochem Biophys Acta 711:83–89
30. Fairbanks KP, Witte LD, Goodman DS (1984) Relationship between mevalonate and mitogenesis in human fibroblasts stimulated with platelet-derived growth factor. J Biol Chem 259:1546–1551
31. Larson RA, Yachnin S (1984) Mevalonic acid induces DNA synthesis in chronic lymphocytic leukemia cells. Blood 64:257–262

Turnover of HMG CoA Reductase is Influenced by Phosphorylation

R. A. PARKER [1,2], S. J. MILLER [1], and D. M. GIBSON [1]

3-Hydroxy-3-methylglutaryl coenzyme A (HMG CoA) reductase (E.C. 1.1.1.34) in most metabolic contexts is the limiting step in cholesterol formation in mammalian liver and other tissues. The properties and significance of this enzyme have been extensively reviewed (Gibson 1985; Kennelly and Rodwell 1985).

HMG CoA reductase is a glycoprotein of 97,000 daltons embedded in the endoplasmic reticulum with a long cytoplasmic extension that is the site of catalytic conversion of HMG CoA to mevalonate (Liscum et al. 1985) (Fig. 1). The enzyme is subject to both long-term (induction/repression; degradation) and short-term control (lipids in the endoplasmic reticulum and reversible phosphorylation) (review: Gibson 1985). The present abbreviated review focuses on the regulation of HMG CoA reductase by phosphorylation and the influence this may have on the degradation of the enzyme.

The catalytic capacity of microsomal reductase falls rapidly on incubation with $ATP(Mg^{2+})$ and a cAMP-, Ca^{2+}-independent protein kinase (cytosolic or microsomal reductase kinase). Activity is. restored with protein phosphatases. Phosphatases inactivate reductase kinase and the latter activity is brought back with $ATP(Mg^{2+})$ and a second protein kinase (reductase kinase kinase) (Ingebritsen et al. 1981) (Fig. 2). The modulating enzymes in the bicyclic system have been purified and further characterized by several laboratories (Kennelly and Rodwell 1985). In addition to reductase kinase, protein kinase-C and calmodulin-dependent protein kinase phosphorylate and inactivate HMG CoA reductase (Beg et al. 1985).

The bicyclic system is operational in intact hepatocytes (Ingebritsen et al. 1979) and fibroblasts (Beg et al. 1986). Both HMG CoA reductase and reductase kinase are sensitive to endocrine control in hepatocytes. Glucagon added to hepatocytes suspended in simple buffered media engenders phosphorylation of both HMG CoA reductase and kinase, as the expressed activity of the former is decreased and the latter increased. Insulin promotes dephosphorylation. The total activity of HMG CoA reductase (as-assayed after incubating isolated microsomes with protein phosphatase) also varies in these hepatocytes: insulin slows apparent degradation, while glucagon hastens the loss of total enzyme.

---

1 Department of Biochemistry, Indiana University School of Medicine, Indianapolis, IN 46223, USA
2 Bristol-Myers Pharmaceutical Research and Development Division, Evansville, IN 47721, USA

Drugs Affecting Lipid Metabolism
Ed. by R. Paoletti et al.
© Springer-Verlag Berlin Heidelberg 1987

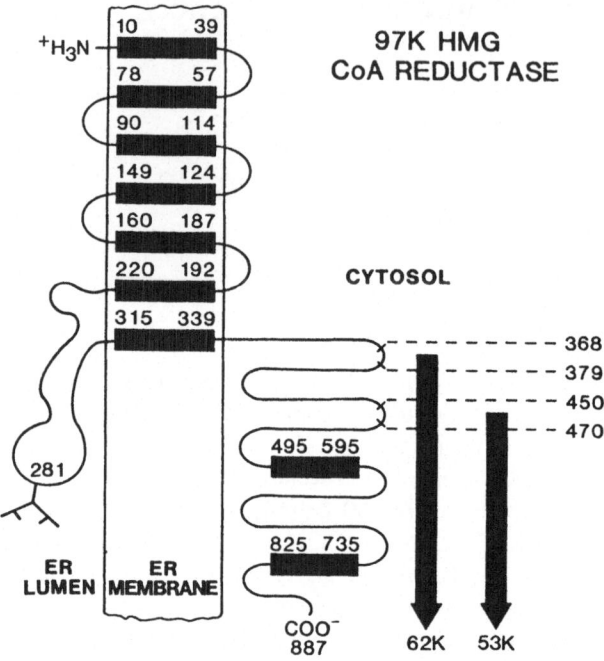

97K HMG
CoA REDUCTASE

Fig. 1. Model of native 97 kD HMG CoA reductase in endoplasmic reticulum membrane. (Redrawn from Liscum et al. 1985)

Mevalonate gives rise to nonsterol and sterol products (including cholesterol and oxysterols) that act as negative feedback signals for cholesterol synthesis principally by impairing reductase synthesis and enhancing its degradation (review: Gibson 1985).

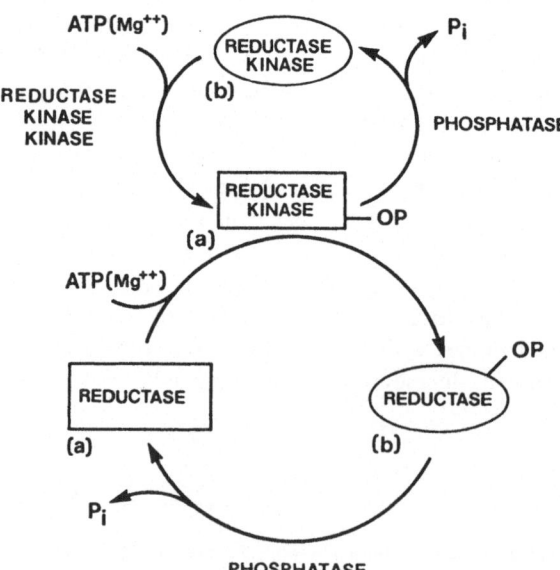

Fig. 2. Regulation of HMG CoA reductase activity through a bicyclic phosphorylation system

Mevalonate given to animals or added to hepatocyte incubations also actuely diminishes the expressed activity of HMG CoA reductase by increasing the degree of phosphorylation. In these studies the acute loss of expressed activity precedes a fall in total activity (more recently analyzed as enzyme mass by immunoblotting). Beg et al. (1986) have reported a similar sequence of events in fibroblast monolayer cultures in response to added low-density lipoprotein, 25-hydroxycholesterol, and 7-ketocholesterol.

In the present research we show that the specific degradative rate of reductase in hepatocyte suspension is diminished 40–50% with insulin, propylamine, methylamine, amino acid mixtures, and monensin (K_D is normally 0.31 h^{-1}; t-1/2, 2.2 h). These effects are considered diagnostic of degradation in the autophagic-lysosomal system of liver. Degradation of HMG CoA reductase is enhanced in the presence of mevalonate and 25-OH cholesterol, both of which cause an acute drop in expressed activity before total enzyme activity and mass (specific immunoblots) begin to fall. Addition of propylamine or monensin lowers the rate of degradation when added with mevalonate and 25-hydroxycholesterol.

Variation of total and expressed HMG CoA reductase activities in intact animals supports the view that phosphorylation of reductase precedes its degradation (review: Gibson 1985). Most notable of these are the studies of Easom and Zammit (1984) who followed the diurnal variation of liver reductase in rats. As expected the total activity rose during the dark (feeding) period (net synthesis of enzyme) and fell during progressive starvation. These investigators found that there was also a rise in expressed reductase activity concomitant with total activity. Of interest was the fall in expressed activity which preceded the decline in total activity. These changes were probably linked to variation in the insulin-glucagon ratio attending the feeding schedule.

We have recently tried to construct an in vitro model for degradation of HMG CoA reductase (Parker et al. 1984). Freezing and thawing of microsomes releases a soluble, enzymatically-active species of HMG CoA reductase of 50–60,000 daltons that has been extensively employed in many reductase studies (reviews: Gibson 1985; Kennelly and Rodwell 1985). Liscum et al. (1983) observed that the liver cytosolic, calcium-dependent, leupeptin-sensitive protease, calpain, releases similar soluble forms of the enzyme as diagrammed in Fig. 1. In our hands treatment of liver microsomes with calpain releases both soluble and membrane-bound, enzymatically-active species of reductase that appear on SDS-PAGE and subsequent immunoblots as a doublet in the range of 50–60,000 daltons. Cleavage by calpain of the microsomal, native 97,000 dalton phosphorylated HMG CoA reductase released more of the soluble species in comparison with the dephosphorylated form of the enzyme, as determined by activity measurements and immunoblots (Parker et al. 1984). With low levels of calpain the 97,000 dalton substrate diminished correspondingly. In recent studies it has become apparent that the upper (heavier?) band of the doublet product of calpain action is phosphorylated and is more readily released from microsomal membranes during centrifugation of the reaction mixture. In prolonged incubations of microsomal reductase with calpain the lower, unlabeled band intensifies, while the higher band diminishes (immunoblot evidence). This suggests that calpain cleaves successively at two points on the cytosolic domain of native reductase (Fig. 1), i.e., on either side of the phosphorylation site(s). Results obtained with calpain remain at best a model demonstration and do not prove that an extralysosomal or calcium-dependent protease is involved in the intracellular degradative path of HMG CoA reductase.

In intact cells the acute fall in HMG CoA reductase expressed activity (increased phosphorylation) is brought about by metabolic products of mevalonate, including sterols. There is evidence that the modulating enzymes themselves may be influenced by the addition of mevalonate to cells, the net result being that HMG CoA reductase is increasingly phosphorylated (review: Gibson 1985). On the other hand, membrane-bound reductase may be looked upon as a receptor for one or more products of mevalonate metabolism. In the binding of these putative negative feedback signals to HMG CoA reductase the enzyme could become a better substrate for phosphorylation. This viewpoint gains some credence in recent reports that the secondary structure of the beta-adrenergic receptor and the visual pigment rhodopson are strikingly similar to each other (Benovic et al. 1986) and to HMG CoA reductase (Liscum et al. 1985). Further, it has been observed that when the beta-adrenergic receptor is occupied by a beta-against and rhodopsin is illuminated, both become more readily phosphorylated on their cytosolic domains by a specific protein kinase and become refractory (Benovic et al. 1986). Whatever mechanisms prove to be operational, HMG CoA reductase is acutely responsive to mevalonate-product feedback and to the insulin-glucagon ratio by a change in the phosphorylation state of the enzyme. The mass of HMG CoA reductase in turn depends on the rate of enzyme formation (which is also sensitive to mevalonate-product feedback) and the rate of degradation which, at least in part, appears to be tuned to the degree of enzyme phosphorylation. Both acute and long-term parameters affect the flux of HMG CoA through the reductase enzyme step, which in most situations determines the rate of cholesterol formation.

References

Beg ZH, Stonik JA, Brewer HB (1985) Phosphorylation of hepatic HMG CoA reductase and modulation of its enzymic activity by calcium-activated and phospholipid-dependent protein kinase. J Biol Chem 260:1682–1687

Beg ZH, Reznikow DC, Avigan J (1986) Regulation of HMG CoA reductase activity in human fibroblasts by reversible phosphorylation: modulation of enzymatic activity by low density lipoprotein, sterols and mevalonolactone. Arch Biochem Biophys 244:310–322

Benovic JL, Mayor F Jr, Somers RL, Caron MG, Lefkowitz Rj (1986) Light-dependent phosphorylation of rhodopsin by beta-adrenergic receptor kinase. Nature (London) 321:869–872

Easom RA, Zammit VA (1984) Diurnal changes in the fraction of HMG CoA reductase in the active form. Biochem J 220:739–745

Gibson DM (1985) Reversible phosphorylation of hepatic HMG CoA reductase in endocrine and feedback control of cholesterol biosynthesis. In: Preiss B (ed) Regulation of HMG CoA reductase. Academic Press, London, New York, p 79–132

Ingebritsen TS, Geelen MJH, Parker RA, Evenson KJ, Gibson DM (1979) Modulation of HMG CoA reductase activity, reductase kinase activity and cholesterol synthesis in rat hepatocytes in response to insulin and glucagon. J Biol Chem 254:9986–9989

Ingebritsen TS, Parker RA, Gibson DM (1981) Regulation of liver HMG CoA reductase by a bicyclic phosphorylation system. J Biol Chem 256:1138–1144

Kennelly PJ, Rodwell VW (1985) Regulation of HMG CoA reductase by reversible phosphorylation-dephosphorylation. J Lipid Res 903–914

Liscum L, Cummings RD, Anderson RGW, DeMartino GN, Goldstein JL, Brown MS (1983) HMG CoA reductase: a transmembrane glycoprotein of the endoplasmic reticulum with N-linked high-mannose oligossacharides. Proc Acad Sci USA 80:7165–7169

Liscum L, Finer-Moore J, Stroud RM, Luskey KL, Brown MS, Goldstein JL (1985) Domain structure of HMG CoA reductase, a glycoprotein of the endoplasmic reticulum. J Biol Chem 260:522–530

Parker RA, Miller SJ, Gibson DM (1984) Phosphorylation of microsomal HMG CoA reductase increases susceptibility to proteolytic degradation in vitro. Biochem Biophys Res Commun 125:629–635

Regulation of ACAT

K. A. MITROPOULOS [1]

1 Introduction

Cellular cholesteryl esters are the products of acyl-CoA: cholesterol acyltransferase (ACAT). The enzyme can be isolated from tissue or cellular homogenates with the microsomal fraction and its activity determined. The liver microsomal fraction, isolated by standard techniques, consists mainly of two vesicle populations: those derived from plasma membrane and those derived from endoplasmic reticular membranes (rough and smooth). The major part of non-esterified cholesterol present in the microsomal fraction is associated with plasma membrane vesicles, whereas ACAT is associated with vesicles derived from rough endoplasmic reticular membranes (Balasubramaniam et al. 1978). "Activity" in the microsomal fraction is assayed under optimal conditions with respect to acyl-CoA but at the concentration of non-esterified cholesterol in the environment of the enzyme in endoplasmic reticular membrane vesicles. The rate at which this activity changes can, hence, be useful as an index of the rate at which cholesterol gets transferred to the immediate environment of the enzyme.

2 Cholesterol Availability

In liver or other tissues, as well as in various cells in culture, the influx of lipoprotein cholesterol is associated with an increase of several fold in the concentration of cholesteryl esters, in ACAT activity but only with moderate increase in the concentration of non-esterified cholesterol. The rat liver microsomal ACAT also operates below substrate saturation and activity responds linearly to cholesterol availability over a large range (Synouri-Vrettakou and Mitropoulos 1983a). We have, therefore, used the rat liver microsomal fraction to investigate transport of cholesterol to ACAT, the mechanism of intervesicular transfer of cholesterol and its modulation by various cellular factors.

---

1 MRC Epidemiology and Medical Care Unit, Northwick Park Hospital, Harrow, Middlesex, Great Britain

Drugs Affecting Lipid Metabolism
Ed. by R. Paoletti et al.
© Springer-Verlag Berlin Heidelberg 1987

2.1 The Role of Plasma Membranes in the Supply of Cholesterol to ACAT

Figure 1 shows the relation between activity and preincubation time in mixtures that contained the microsomal fraction and a plasma membrane supplement or the microsomal fraction without supplement. This plasma membrane preparation was obtained from the liver of a rat that was injected 15 h earlier with (^{3}H)cholesterol. The rate of increase in ACAT activity is higher in the presence of the plasma membrane supplement. The amount of (^{3}H)cholesteryl oleate has increased linearly with preincubation time. The data in Fig. 1 demonstrate that most of the increase in the rate observed in the presence of the additional plasma membrane can be attributed to the (^{3}H)cholesteryl oleate formed. This would mean that there is net transfer of (^{3}H)cholesterol from the plasma membrane supplement to the ACAT substrate pool that results in the increase of the preincubation-dependent rate. Moreover, the

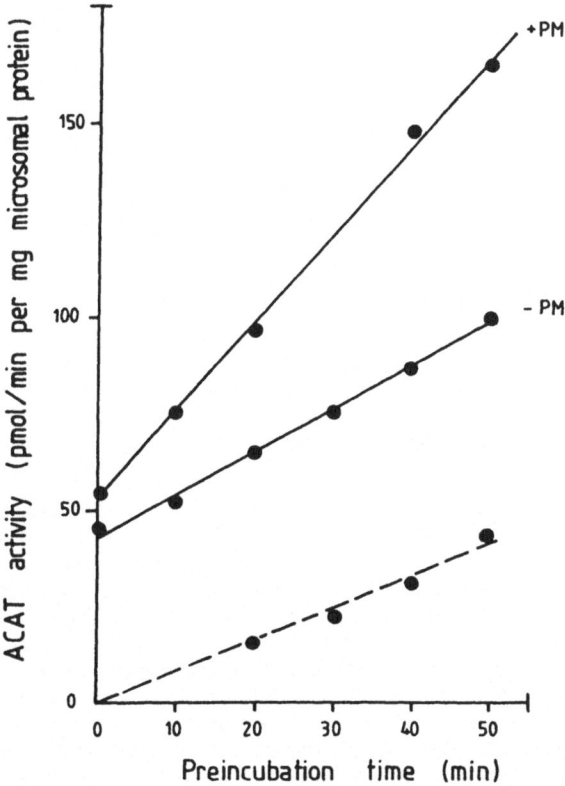

Fig. 1. The preincubation time-dependent increase in microsomal ACAT in the absence and in the presence of a plasma membrane supplement. Both preincubations contained 1.8 mg ml^{-1} microsomal protein and in the mixture that contained the plasma membrane supplement (+PM) the ratio of microsomal to plasma membrane protein was 1.1. The activity of 5'-nucleotidase (a plasma membrane enzyme) was 56 and 350 nmol mg^{-1} protein min^{-1} in the microsomal and in the plasma membrane preparation respectively. The (^{3}H)cholesteryl oleate (– – – –) synthesized during the ACAT assay was determined from the radioactivity incorporated into cholesteryl ester and the specific radioactivity of (^{3}H)cholesterol in the plasma membrane supplement (2.3 dmp/pmol)

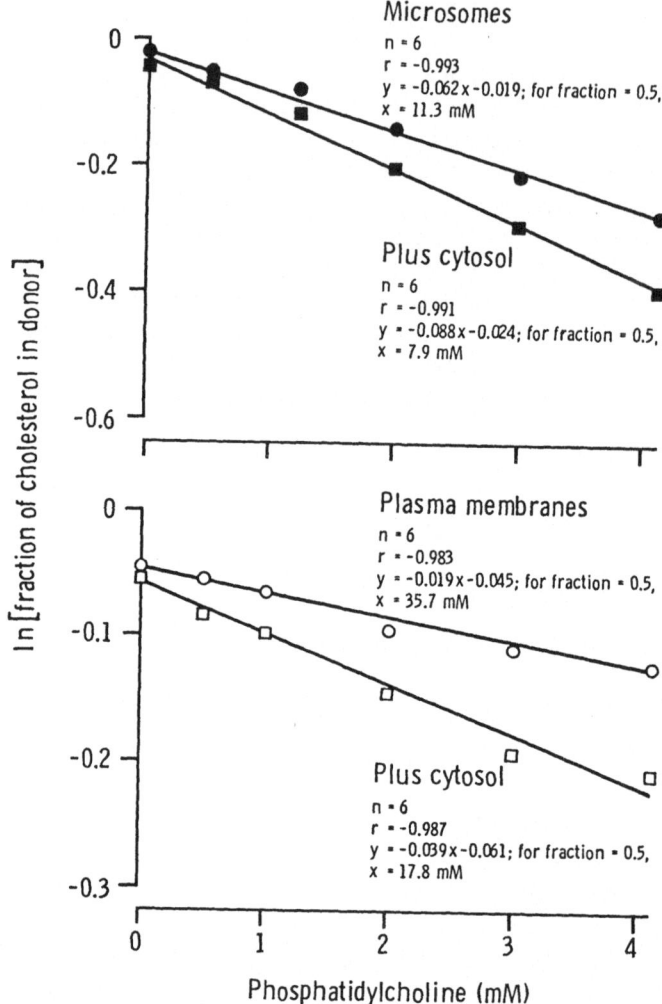

Fig. 2. Effect of cytosol on the efflux of cholesterol from the microsomal or a plasma membrane preparation incubated in the presence of liposomal acceptor. The microsomal fraction (2.18 mg protein ml^{-1} mixture) or the plasma membrane preparation (1.24 mg ml^{-1} mixture) were incubated for 50 min in the absence or presence of cytosol (1.36 mg ml^{-1} mixture) and the presence of the indicated concentration of phosphatidylcholine liposomes. Portions of the mixture were removed at the end of the incubation period to separate the subcellular membranes from the liposomal vesicles and to determine the non-esterified cholesterol associated with the re-isolated acceptor. The original microsomal fraction contained 52.5 nmol mg^{-1} protein of non-esterified cholesterol and had activity of 5'-nucleotidase, 39 nmol min^{-1} mg^{-1}; of NADPH cytochrome c oxidoreductase, 78 nmol min^{-1} mg^{-1}; and of ACAT, 52 pmol min^{-1} mg^{-1} protein. The plasma membrane preparation contained 193 nmol mg^{-1} protein of non-esterified cholesterol and had activity of 5'-nucleotidase, 244 nmol min^{-1} mg^{-1}; and of NADPH cytochrome c oxidoreductase 1.1 nmol min^{-1} mg^{-1} protein

rate in the absence of the supplement can be attributed to the transfer of cholesterol from plasma membrane vesicles present in the microsomal fraction to the ACAT substrate pool. The present and previous results (Mitropoulos et al. 1984) suggest that there is a chemical potential difference for cholesterol between plasma membrane and endoplasmic reticular membrane vesicles and this difference is responsible for transfer of cholesterol to the ACAT substrate pool. It is likely that this transfer of cholesterol also operates in vivo and therefore plasma membrane is important in determining the size of the cholesterol pool available to ACAT and the rate of cholesterol esterification.

2.1.1 Reverse Transfer of Cholesterol

The preincubation of the microsomal fraction in the presence of phosphatidylcholine liposomes results in transfer of microsomal cholesterol to the artificial cholesterol acceptor (Fig. 2). The first-order rate constant describing the transfer of microsomal cholesterol to the liposomal vesicles increased with the concentration of acceptor, whereas the rate of ACAT change due to preincubation has progressively decreased with increasing ratio of liposomal to microsomal phospholipid (Mitropoulos et al. 1984). The latter observation indicates a decreased availability of cholesterol to the ACAT enzyme which can be attributed to a competition by the artificial acceptor with the ACAT-containing membranes for the cholesterol desorbed from the plasma membranes.

2.1.2 Forward Transfer of Cholesterol and Increase in ACAT Activity

The preincubation of the microsomal fraction in the presence of phosphatidylcholine: cholesterol liposomes results in the transfer of liposomal cholesterol to the microsomal vesicles and to the membranes that contain ACAT (Synouri-Vrettakou and Mitropoulos 1983a). The kinetics of cholesterol transfer to the microsomal vesicles and those of modulation of ACAT activity are consistent with the possibility that during the preincubation liposomal cholesterol is transferred to plasma membrane vesicles and that the rate of transfer of plasma membrane cholesterol to the ACAT substrate pool increases with increasing concentration of cholesterol in plasma membrane vesicles.

2.1.3 The Role of Cytosol in the Forward and Reverse Transfer of Cholesterol

Cytosol present in the preincubation mixture increased the preincubation-dependent rate of change of ACAT activity both in the absence and in the presence of cholesterol/ phospholipid liposomes (Mitropoulos and Venkatesan 1984). In the latter case the presence of cytosol also increased the rate of transfer of liposomal cholesterol to cellular membrane vesicles (see also Table 1). Moreover, in preincubations of the microsomal fraction with phosphatidylcholine liposomes the presence of cytosol has increased the efflux of microsomal cholesterol (Fig. 2) and consequently has decreased the rate of change in ACAT due to preincubation. It is unlikely that the effects of

Table 1. The effect of cytosol on the rate of transfer of liposomal cholesterol to the rat liver microsomal or plama membrane preparations

Preincubation mixture[a]	10^{-3} x rate constant (min^{-1})	
	No cytosol	+ cytosol
Microsomas[b] + liposomes	4.1	7.3
Plasma membranes[c] + liposomes	0.9	3.7

[a] The microsomal fraction (2.1 mg protein ml^{-1}) or the plasma membrane preparation (0.6 mg protein ml^{-1}) were incubated for various periods in the presence of (^{3}H)cholesterol/phosphatidylcholine (1:1,mol/mol; specific radioactivity 10 dpm/pmol) liposomes at a ratio of liposomal to hepatic membrane cholesterol of 1 and in the absence or the presence of cytosol (1.2 mg protein ml^{-1}). At the end of the incubation period a portion of the mixture was removed to determine the liposomal cholesterol associated with the re-isolated subcellular membranes. These values were used to determine the first-order rate constant (Synouri-Vrettakou and Mitropoulos 1983a).

[b] The microsomal fraction contained 72 nmol mg^{-1} protein of non-esterified cholesterol and had activity of 5'-nucleotidase, 42 nmol min^{-1} mg^{-1}; of NADPH cytochrome c oxidoreductase, 82 nmol min^{-1} mg^{-1} and of ACAT, 56 pmol min^{-1} mg^{-1} protein.

[c] The plasma membrane preparation contained 140 nmol mg^{-1} protein of non-esterified cholesterol and had activity of 5'-nucleotidase, 316 nmol min^{-1} mg^{-1}; and of NADPH cytochrome c oxidoreductase, 2.5 nmol min^{-1} mg^{-1} protein.

cytosol on intervesicular cholesterol transfer, in the present system, are due to a specific interaction with a donor or acceptor membrane. The data rather suggest that cytosolic factors increase the concentration of monomeric cholesterol and this results that in a higher rate of transfer to the appropriate acceptor.

2.1.4 Interaction of Endoplasmic Reticular Membranes with Plasma Membranes and the Regulation of Cholesterol Transfer

Kinetics of intervesicular cholesterol transfer are overall consistent with a mechanism that involves desorption of cholesterol molecules from the donor membrane, diffusion of monomeric cholesterol through the aqueous phase and absorption by the acceptor membrane. This mechanism requires that at high ratio of acceptor to donor concentrations, desorption of cholesterol determines the rate of transfer. However, Table 1 shows that the rate of transfer of liposomal cholesterol to a plasma membrane preparation is three- to four-fold lower than that to the microsomal fraction. Moreover, in the presence of an artificial cholesterol acceptor efflux from a plasma membrane preparation is lower than that from the microsomal fraction (Fig. 2). These results demonstrate that the ratio of endoplasmic to plasma membranes is important in determining rates of forward and reverse cholesterol transfer.

3 Other Regulatory Mechanisms

Several reports have shown that rat liver or intestinal ACAT in the microsomal fraction from these tissues can be reversibly inactivated or activated by preincubation under conditions that favour dephosphorylation or phosphorylation respectively (Suckling and Stange 1985). However, it has recently been reported that preincubation conditions that are expected to alter the ratio of phosphorylated to dephosphorylated enzyme result also in modulation of cholesterol intervesicular transfer and that such modulation can be responsible for the observed activation/inactivation (Mitropoulos and Venkatesan 1984). The hypothesis that the enzyme can be modulated in vitro by covalent modification has to be considered tentatively until the purified enzyme in a reconstituted system is shown to be modulated by this modification.

Hydroxysterols, including some steroid hormones, have been shown to modulate ACAT activity in membrane fractions from various tissues or in various cells in culture. For instance, progesterone inhibits ACAT activity in membrane preparations (Synouri-Vrettakou and Mitropoulos 1983b) from a number of tissues, although its concentration, even in steroid hormone-producing tissues, is not enough to be able to inhibit ACAT (Schuler et al. 1981).

4 Synthetic Inhibitors of ACAT

Investigations directed towards finding compounds that inhibit cholesteryl ester accumulation in cells of the arterial wall yielded some interesting competitive inhibitors of ACAT (Heider et al. 1983). Apart from pharmacological potential these compounds are proving useful to investigate the specificity of ACAT (Ross et al. 1984) and for modifying fluxes of cholesterol that take place within the cell (Tabas et al. 1986).

5 Conclusions

The data from the study of intervesicular transfer of cholesterol in the rat liver microsomal fraction can be relevant to the economy of cellular cholesterol in vivo. It is likely that ACAT also operates in vivo below substrate saturation with respect to cholesterol and that activity responds readily to expansion of the substrate pool. The steady-state concentration of cholesterol in plasma membrane is determined by the interaction of this membrane with the endoplasmic reticular membranes. The concentration of cholesterol in plasma membrane will, in turn, determine the rate of transfer of non-esterified cholesterol to the ACAT substrate pool and therefore will determine the rate of synthesis of cholesteryl esters. A small increase in the plasma membrane concentration of cholesterol is expected to result in a large increase in the rate of cholesterol transfer and of cholesteryl ester synthesis. Consistent with the above, the considerable changes in ACAT activity that result from various dietary or environmental conditions are associated with only a small increase in cellular non-esterified cholesterol.

References

Balasubramaniam S, Venkatesan S, Mitropoulos KA, Peters TJ (1978) The submicrosomal localization of acyl-coenzyme A:cholesterol acyltransferase and its substrate, and of cholesteryl esters in rat liver. Biochem J 174:863–872

Heider JG, Pickens CE, Kelly LA (1983) Role of acyl CoA:cholesterol acyltransferase in cholesterol absorption and its inhibition by 57–118 in the rabbit. J Lipid Res 24:1127–1134

Mitropoulos KA, Venkatesan S (1984) Conditions that may result in (de-)phosphorylation of hepatic acyl-CoA: cholesterol acyl transferase result also in modulation of substrate supply *in vitro*. Biochem J 221:685–695

Mitropoulos KA, Venkatesan S, Synouri-Vrettakou S, Reeves BEA, Gallagher JJ (1984) The role of plasma membranes in the transfer of non-esterified cholesterol to the acyl-CoA:cholesterol acyltransferase substrate pool in the liver microsomal fraction. Biochim Biophys Acta 792: 227–237

Ross AC, Go KJ, Heider JG, Rothblat GH (1984) Selective inhibition of acyl coenzyme A:cholesterol acyltransferase by compound 58-035. J Biol Chem 259:815–819

Schuler LA, Toaff ME, Strauss JF (1981) Regulation of ovarian cholesterol metabolism: control of 3-hydroxy-3-methylglutaryl coenzyme A reductase and acyl coenzyme A:cholesterol acyl transferase. Endocrinology 108:1476–1486

Synouri-Vrettakou S, Mitropoulos KA (1983a) Acyl-coenzyme A:cholesterol acyltransferase. Transfer of cholesterol to its substrate pool and modulation of activity. Eur J Biochem 133: 299–307

Synouri-Vrettakou S, Mitropoulos KA (1983b) On the mechanism of the modulation *in vitro* of acyl-CoA:cholesterol acyltransferase by progesterone. Biochem J 215:191–199

Suckling KE, Stange EF (1985) Role of acyl-CoA:cholesterol acyltransferase in cellular cholesterol metabolism. J Lipid Res 26:647–671

Tabas I, Weiland DA, Tall AR (1986) Inhibition of acyl coenzyme A:cholesterol acyl transferase in J774 macrophages enhances down-regulation of the low density lipoprotein receptor and 3-hydroxy-3-methylglutaryl-coenzyme A reductase and prevents low density lipoprotein-induced cholesterol accumulation. J Biol Chem 261:3147–3155

Dietary and Pharmacological Control of Cholesterol 7α-Hydroxylase

E. Bosisio [1]

1 Introduction

The 7α-hydroxylation of cholesterol is the initial and major rate-limiting step in the conversion of cholesterol into bile acids. The enzyme system cholesterol 7α-hydroxylase, which catalyzes the introduction of the hydroxyl group in the 7α-position of the cholesterol molecule, is a mixed-function oxidase, cytochrome P 450-dependent and is located in the smooth endoplasmic reticulum of the liver cell (Myant and Mitropoulos 1977).

Due to the importance of this enzyme in the control of bile acid synthesis, a variety of studies has been performed on the possible factors which modulate the enzyme activity.

This review will take into consideration some aspects, both physiological and pharmacological, of the regulation of cholesterol 7α-hydroxylase and relevance will be given to the correlation between data obtained in vitro by measuring cholesterol 7α-hydroxylase activity and result derived from studies in vivo on bile acid synthesis, under analogous treatments.

2 Physiological Control of Cholesterol 7α-Hydroxylase

It has been shown by several investigators (Myant and Mitropoulos 1977) that in rats the activity of cholesterol 7α-hydroxylase follows a variation due to the light cycle, with a peak at the middle of the dark period. Studies from our laboratory evidenced that a similar fluctuation of the enzyme activity occurs also in the hamster (Fig. 1). Among the signals that seem to cause and maintain the diurnal variation of the activity of cholesterol 7α-hydroxylase, hormones of the pituitary-adrenal axis appear to play a role (Gibbons et al. 1982).

Hormones from the thyroid and pancreas glands also have some influence on the rate of synthesis of bile acids and affect the activity of cholesterol 7α-hydroxylase. The injection of thyroxine stimulated the enzyme, while thyroidectomy resulted in a

---

1 Institute of Pharmacological Sciences, University of Milan, 20129 Milan, Italy

Drugs Affecting Lipid Metabolism
Ed. by R. Paoletti et al.
© Springer-Verlag Berlin Heidelberg 1987

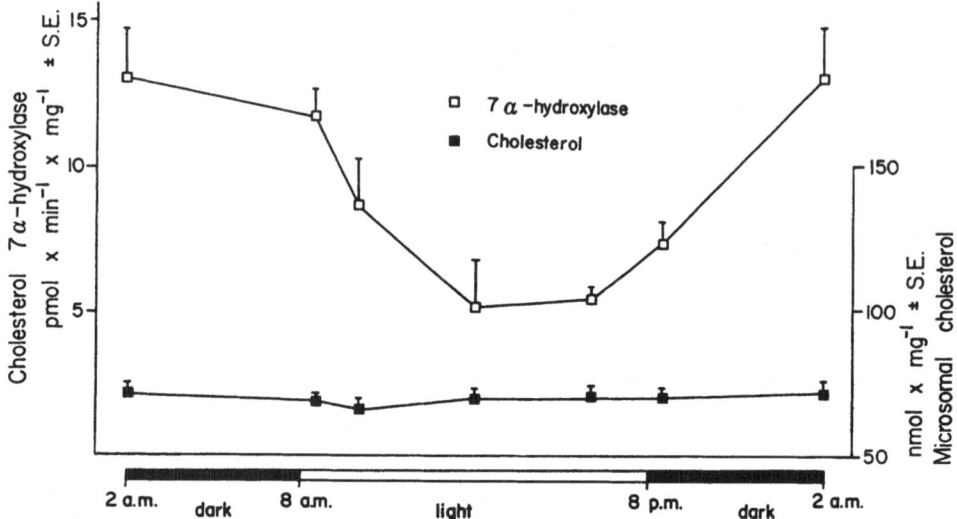

Fig. 1. Diurnal rhythm of cholesterol 7α-hydroxylase in the hamster

decrease (Gibbons et al. 1982). In diabetic rats cholesterol 7α-hydroxylase activity was higher than in normal rats and the administration of insulin caused a down-regulation of the enzyme towards control values (Subbiah and Yunker 1984).

Concerning species and sex, most of the investigations on the mechanism regulating cholesterol 7α-hydroxylase have been carried out using the male rat as experimental model. There is, however, evidence pointing out the possibility of species differences in the regulation of bile acid formation and metabolism.

The levels of enzyme activity can change in relation to the animal species since, as compared to rats, rabbits and hamsters synthesized less 7α-hydroxycholesterol, while guinea-pig liver was much more active than the rat liver, although the microsomal cholesterol content was comparable among the species (Cighetti et al. 1983). In human liver cholesterol 7α-hydroxylase was almost undetectable under normal

Table 1. Microsomal cholesterol and cholesterol 7α-hydroxylase activity in male and female animals[a]

	Cholesterol 7α-hydroxylase pmol × min^{-1} × mg protein^{-1}	Microsomal cholesterol nmol × mg protein^{-1}
Male rat (5)	11.1 ± 0.33	94.1 ± 6.5
Female rat (6)	13.2 ± 0.5	87.0 ± 3.6
Male hamster (6)	4.71 ± 1.1	62.8 ± 8.3
Female hamster (7)	2.94 ± 0.35	64.5 ± 8.5

[a] Results are the mean ± SE of the number of animals shown in parentheses.

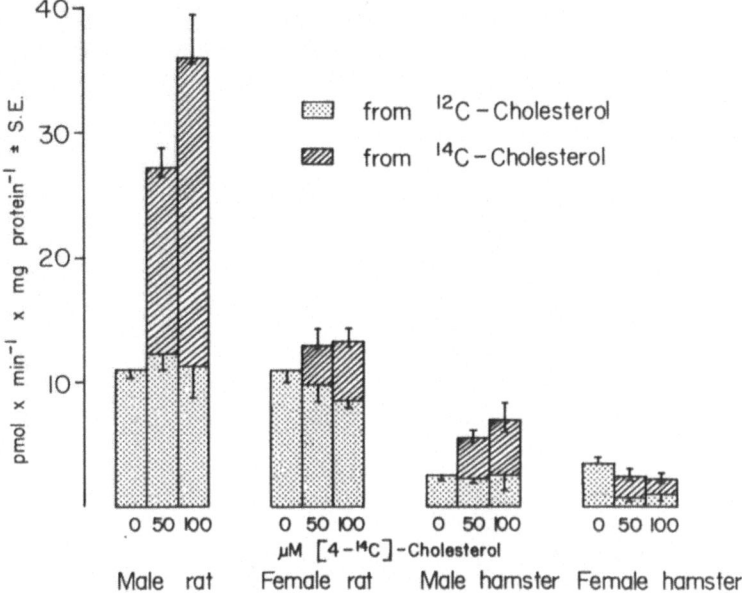

Fig. 2. Influence of ^{14}C-cholesterol on cholesterol 7α-hydroxylase in male and female animals

conditions, but it became measurable when stimulated by cholestyramine treatment (Bosisio et al. 1984).

There is no evidence for a difference in enzyme levels due to sex, at least in rat and hamsters, when the activity is expressed as the absolute mass of 7α-hydroxycholesterol formed from endogenous substrate (Table 1). Conversely, the enzyme activity is differently modulated in male and female animals by addition of ^{14}C-cholesterol in the incubation media (Fig. 2). In the male rat and hamster, the total activity increased with the addition of increasing amounts of exogenous cholesterol, but the amount of 7α-hydroxycholesterol formed from the endogenous substrate remained constant. This indicates that in the male rat and hamster, the enzyme is not saturated by the microsomal pool of the substrate.

The trend appears different in females of both species. With the increase of exogenous cholesterol, the conversion of the endogenous substrate decreased, whereas the total activity stayed constant, indicating that the enzyme is saturated by the endogenous cholesterol. This result seems to correlate well with the fact that females develop gallstones more frequently than males and with the observed dietary induction of gallstones in female hamsters.

In cholestyramine-treated female hamsters (Fig. 3) the hydroxylation of exogenous and endogenous cholesterol followed the same pattern as in male animals, indicating that the enzyme is no longer saturated under these conditions. Therefore, the effect of cholestyramine can be interpreted through the induction of enzyme synthesis, since thise drug is not known to affect the size of the substrate pool of the enzyme (Mitropoulos et al. 1973).

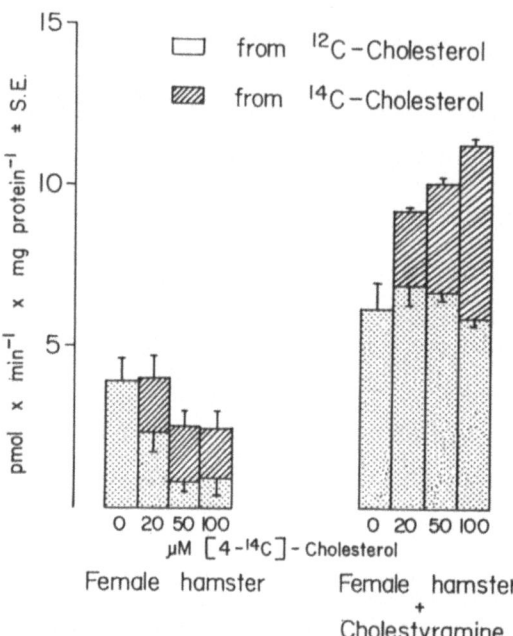

Fig. 3. Influence of ^{14}C-cholesterol on cholesterol 7α-hydroxylase in female hamsters treated with cholestyramine

3 Pharmacological Control of Cholesterol 7α-Hydroxylase

Hypolipidemic drugs act either on cholesterol synthesis and/or transport and therefore they may also interact with cholesterol 7α-hydroxylase.

Irrespective of the animal species, cholestyramine induced a marked increase of the enzyme activity (Cighetti et al. 1983) and of bile acid synthesis. Another circumstance which results in a partial interruption of the enterohepatic circulation is the ileal by-pass. This is also an example where the measurement of the enzyme activity in vitro and the evaluation of the bilde acid synthesis in vivo, are well correlated. Partial ileal by-pass was associated to an enhancement of bile acid synthesis in humans (Gibbons et al. 1982) and to a stimulation of cholesterol 7α-hydroxylase in rabbits, both under control and hypercholesterolemic regimens (Lovati et al. 1982).

The question whether clofibrate and related compounds affect bile acid synthesis is still debated. Angelin et al. (1976) found no effect of clofibrate on cholesterol 7α-hydroxylase and, along this line, it has been shown that bezafibrate does not change the enzyme activity (Bosisio et al. 1980). Other authors, however, under different experimental conditions, have shown a decrease of cholesterol 7α-hydroxylase in rats treated with clofibrate (Li et al. 1980b) and bezafibrate (Kritchevsky et al. 1980). Moreover, data obtained in vivo, concerning the effect of clofibrate on bile acid synthesis, indicate that clofibrate frequently but not consistently reduces the elimination of cholesterol as bile acids (Miettinen 1981).

Probucol was shown to decrease the enzyme activity, but the effect was evident only when the treatment was prolonged for 4 weeks (Balasubramaniam et al. 1981). The reduction of the 7α-hydroxylating activity is consistent with data in vivo which show that fecal bile acid excretion and concentration in bile are lower in probucol-treated rats than in controls (Li et al. 1980a).

The modifications of bile acid metabolism following treatments with the competitive inhibitors of HMG-CoA reductase are variable according to the animal model employed in the different studies. Actually, it is known that dogs and humans are good responders to these drugs in terms of lowering plasma cholesterol levels, while rats and hamsters are not. Data from Endo et al. (1979) and Björkhem (1986) indicate that compactin and mevinolin reduced cholesterol 7α-hydroxylase activity in rats, and decreased fecal bile acid excretion. In dogs, however, the increase of fecal bile acid output was not associated with a modification of cholesterol 7α-hydroxylase (Tsujita et al. 1979).

In hamsters, the effect of compactin treatment depends on the nutritional status of the animal (Bosisio et al. 1982). In control animals no alteration of the enzyme activity was evidenced between control and compactin-treated hamsters. On the contrary, when animals were fed a lithogenic diet, the enzyme activity was depressed by the drug. After acute treatment, compactin did not change the enzyme activity. In this study, results on the enzyme activity were well correlated with data on the amount of bile acids excreted in bile. When cholesterol 7α-hydroxylase was not affected, similar amounts of bile acids were excreted in bile of control and treated hamsters; when cholesterol 7α-hydroxylase was reduced by compactin, the biliary bile acid output was lower than in controls.

With regards to the effect of bile acid feeding on cholesterol 7α-hydroxylation, the discussion will be restricted to chenodeoxycholic (CDCA) and ursodeoxycholic acids (UDCA) which are clinically used for gallstone dissolution therapy.

Data available on CDCA feeding are consistent with the inhibition of bile acid synthesis and with the decrease of cholesterol 7α-hydroxylase activity either in humans and in experimental animals. There is only one report from Carulli et al. (1980), who did not find a significant change of the enzyme activity in gallstone patients treated for 7 days with CDCA. This discrepancy is perhaps due to the short-term administration in the experimental protocol.

Reports related to the influence of UDCA on bile acid synthesis are somewhat more controversial.

From the in vivo studies of Bergman et al. (1984) and Hardison and Grundy (1984) in controls and gallstone patients, there was no evidence that UDCA or its tauro-conjugated form might influence bile acid synthesis. Some data obtained in humans and hamsters, where cholesterol 7α-hydroxylase was evaluated, were consistent with an unchanged bile acid synthesis under UDCA treatment (Carulli et al. 1980 and Singhal et al. 1984). On the other hand, data from other laboratories showed a decrease of the enzyme activity with UDCA (Danielsson 1973 and Pearlman et al. 1979).

Certainly more data are needed to determine whether bile acid structure has any direct influence on the regulation of bile acid synthesis, but the trend seems to indicate a major effect of CDCA, as compared to UDCA, on the bile acid synthetic rate.

4   Control of Cholesterol 7α-Hydroxylase by Diet

Both the amount and the composition of the diet have a significant effect on bile acid metabolism and the activity of cholesterol 7α-hydroxylase. When animals are starved, there is a marked fall in the activity of cholesterol 7α-hydroxylase and a reduction of the amplitude of the diurnal rhythm (Gibbons et al. 1982).

Although fasting is associated with a depression of cholesterol 7α-hydroxylase activity, it is advisable to perform the enzyme assay in the fasting state if the activity observed in laboratory animals is to be compared to that in surgical liver biopsies.

The response to cholesterol feeding differs with respect to species (Table 2). Rats compensate for the excess cholesterol by increasing the rate of synthesis of bile acids, as demonstrated in vivo or as can be assessed by measuring cholesterol 7α-hydroxylase. Rabbits and hamsters do not possess such a compensatory mechanism which is confirmed by the fact that the hypercholesterolemic diet does not raise cholesterol 7α-hydroxylase activity in these two species. The question whether a cholesterol-bile acid feedback exists in man still awaits a definitive answer. Studies from Quintao et al. (1971) gave negative results on the possibility that humans might have the ability to enhance the conversion of cholesterol to bile acids after cholesterol feeding, but recent observations by Lin and Connors (1980) seem to indicate that, in long-term feeding, dietary cholesterol promotes an increase of bile acid excretion:

Table 2. Cholesterol 7α-hydroxylase in control and cholesterol-fed animals[a]

Animal	Cholesterol 7α-hydroxylase pmol $\times$ min^{-1} $\times$ mg protein $^{-1}$	
	Control	Cholesterol-fed
Rat (6)	7.4 ± 2.0	28.1 ± 1.4 [c]
Rabbit (4)	n.d.	n.d.
Hamster (7)	2.9 ± 0.4	1.9 ± 0.5

[a]   Results are the mean ± SE of the number of animals shown in parentheses.
[b]   n.d. = Lower than 0.3 pmol $\times$ min^{-1} $\times$ mg protein^{-1}.
[c] $p < 0.01$ vs controls.

5   Conclusion

In summary, from the analysis of the data reviewed in these pages, it appears that the circumstances where cholesterol 7α-hydroxylating activity coincides with the rate of synthesis of bile acids are predominant. Therefore, it is possible to conclude that the assay of cholesterol 7α-hydroxylase is an adequate approach for an estimation of bile acid biosynthesis.

References

Angelin B, Bjorkhem I, Einsarsson K (1976) Effects of clofibrate on some microsomal hydroxylations involved in the formation and metabolism of bile acids in rat liver. Biochem J 156: 445–448

Balasubramaniam S, Beins DM, Simons LA (1981) On the mechanisms of plasma cholesterol reduction in the rat given probucol. Clin Sci 615–619

Bergman K, Epple-Gutsfeld M, Leiss O (1984) Differences in the effects of chenodeoxycholic and ursodeoxycholic acid on biliary lipid secretion and bile acid synthesis in patients with gallstones. Gastroenterology 87:136–143

Björkhem I (1986) Effects of mevinolin in rat liver: evidence for a lack of coupling between synthesis of hydroxymethylglutaryl CoA reductase and cholesterol 7α-hydroxylase activity. Biochim Biophys Acta 877:43–49

Bosisio E, Catapano AL, Cighetti G, Paoletti R (1980) In: Greten H, Lang PD, Schettler G (eds) Lipoprotein and coronary heart disease. Witzstrock, New York, Baden Baden Cologne, pp 86–91

Bosisio E, Cighetti G, Di Padova C, Rovagnati P, Galli Kienle M, Galli G, Paoletti R (1982) Effects of compactin (ML-236B) on biliary lipid composition and cholesterol catabolism in the hamster. Pharmacol Res Commun 14:577–592

Bosisio E, Cighetti G, Galli Kienle M, Tritapepe R, Galli G (1984) HMGCoA reductase and cholesterol 7α-hydroxylase in human liver. Life Sci 34:2075–2081

Carulli N, Ponz de Leon M, Zironi F, Pinetti A, Smerieri A, Iori R, Loria P (1980) Hepatic cholesterol and bile acid metabolism in subjects with gallstones: comparative efffects of short-term feeding of chenodeoxycholic and ursodeoxycholic acid. J Lipid Res 21:35–43

Cighetti G, Bosisio E, Galli G, Galli Kienle M (1983) The effect of cholestyramine on liver HMGCoA reductase and cholesterol 7α-hydroxylase in various laboratory animals. Life Sci 33:2483–2488

Danielsson H (1973) Influence of dietary bile acids on formation of bile acids in rat. Steroids 22:667–676

Endo A, Tsujita Y, Kuroda M, Tanzawa K (1979) Effects of ML-236B on cholesterol metabolism in mice and rats: lack of hypocholesterolemic activity in normal animals. Biochim Biophys Acta 575:266–276

Gibbons GF, Mitropoulos KA, Myant NB (1982) Biochemistry of cholesterol. Elsevier/North Holland. Biomedical Press Amsterdam New York, Oxford, pp 280–290

Hardison WGM, Grundy SM (1984) Effect of ursodeoxycholate and its taurine conjugate on bile acid synthesis and cholesterol absorption. Gastroenterology 87: 130–135

Kritchevsky D, Tepper SA, Story JA, Mueller M, Klurfeld DM (1980) Effect of bezafibrate on cholesterol metabolism in the rat. In: Greten H, Lang PD, Schettler G (eds) Lipoproteins and coronary heart diease. Witzstrock, New York, Baden Baden, Cologne, pp 92–95

Li JR, Holets RJ, Kottke BA (1980a) Effect of probucol on cholesterol metabolism in the rat. Atherosclerosis 36:559–565

Li JR, Kim DN, Lee KT, Reiner JM, Thomas WA (1980b) Effect of clofibrate cholestyramine, cholesterol and feeding pattern on the diurnal variation of cholesterol 7α-hydroxylation in swine. Exp Mol Pathol 32:52–60

Lin DS, Connors WE (1980) The long-term effects of dietary cholesterol upon the plasma lipids, lipoproteins, cholesterol absorption and sterol balance in man: the demonstration of feedback inhibition of cholesterol biosynthesis and increased bile acid excretion. J Lipid Res 21:1042–1052

Lovati MR, Mora M, Bosisio E, Majone G, Galli Kienle M, Galli G, Sirtori CR (1982) Cholesterol and bile acid metabolism in normal and cholesterol fed rabbits after partial ileal by-pass. Exp Mol Pathol 37:156–165

Miettinen TA (1981) Effects of hypolipidemic drugs on bile acid metabolism in man. In: Paoletti R, Kritchewsky D (eds) Academic Press London, New York; Adv Lipid Res 18:65–97

Mitropoulos KA, Balasubramaniam S, Myant NB (1973) The effect of interruption of the entero-heptaic circulation of bile acids and of cholesterol feeding on cholesterol 7α-hydroxylase in relation to the diurnal rhythm in its activity. Biochim Biophys Acta 326:428–438

Myant NB, Mitropoulos KA (1977) Cholesterol 7α-hydroxylase. J Lipid Res 18:135–153

Pearlman BJ, Bonorris GG, Phillips MJ, Chung A, Vimadadal S, Marks JW, Schoenfield LJ (1979) Cholesterol gallstone formation and prevention by chenodeoxycholic and ursodeoxycholic acids. A new hamster model. Gastroenterology 77:634–641

Quintao E, Grundy SM, Ahrens EM (1971) Effects of dietary cholesterol on the regulation of total body cholesterol in man. J Lipid Res 12:233–247

Singhal AK, Cohen BI, Finver-Sadowsky J, McSherry CK, Mosbach EH (1984) Role of hydrophylic bile acids and of sterols on cholelithiasis in the hamster. J Lipid Res 25:564–570

Subbiah MTR, Yunker RL (1984) Cholesterol 7α-hydroxylase of rat liver: an insulin sensitive enzyme. Biochem Biophys Res Commun 124:896–902

Tsujita Y, Kuroda M, Tanzawa K, Kitano N, Endo A (1979) Hypolipidemic effects in dogs of ML-236B, a competitive inhibitor of 3-hydroxy, 3-methylglutarylcoenzyme A reductase. Atherosclerosis 32:307–313

Recent Advances in Metabolic Pathways of Bile Acid Synthesis

N. B. JAVITT [1]

After a number of years of being considered a well-established branch of biochemistry, we now have found that there is considerable less certainty about the pathways of bile acid synthesis from cholesterol, particularly in regard to human metabolism.

Perhaps the major event that has occurred as part of this renaissance is the recognition that the initiation of bile acid synthesis from cholesterol can occur either by 7α-hydroxylation or by 26-hydroxylation (Anderson et al. 1972). Because there is some indication that the oxidation of the side chain of cholesterol can also occur in nonhepatic tissues, I will refer to it as the extrinsic pathway. It is the pathway for side-chain oxidation of C-27 sterols that is providing new perceptions with regard to the specific role of subcellular organelles and appears to generate intermediates that can have a regulatory role in the synthesis of cholesterol.

The classical or traditional pathway of bile acid synthesis begins in the liver with cholesterol 7α-hydroxylase (Björkhem 1985), a microsomal P-450 enzyme, and is referred to as the intrinsic pathway. There is considerable evidence in animal species that it is the rate-limiting enzyme, although it has neither been sequenced nor cloned and evidence derived from studies of liver cells in culture indicates that it is not regulated directly by bile acids (Davis et al. 1983). The major evidence supporting its role as the rate-limiting enzyme is derived from the knowledge that cholestyramine treatment which induces bile acids synthesis certainly increases the activity of cholesterol 7α-hydroxylase and if one couples this information to the more limited evidence that other enzymes in the intrinsic pathway have a higher rate of activity in the basal state which are not induced by biliary drainage, then the concept that 7α-hydroxylase is rate-limiting is reasonably sound.

The enzyme is thought to have a relatively short half-life of 2-3 h and there is some evidence that the activity can be increased, at least under certain circumstances, by an increase in the cell cholesterol concentration. In addition to the relatively short half-life that provides rapid modulation in activity, experimental evidence exists for a phosphorylation-dephosphorylation mechanism and modulation by sulfhydryl compounds in the cytosol such as glutathione.

The next step in the intrinsic pathway is the conversion of 7α-hydroxycholesterol to 7α-hydroxy-4-ene-3-one. An enzyme of approximately 46 000 daltons has been

---

1 Division of Hepatic Diseases, New York University Medical Center, 550 First Avenue, New York, NY 10016, USA

Drugs Affecting Lipid Metabolism
Ed. by R. Paoletti et al.
© Springer-Verlag Berlin Heidelberg 1987

solubilized and partially purified from rabbit liver which acts rather specifically on 7α-hydroxycholesterol and does not appear to have activity towards hormones such as androsten-3β-ol that frequently undergo the same type of transformation.

Following this transformation, one of two events can occur. One event is a 12α-hydroxylation which leads to cholic acid synthesis. In humans, hamsters, and rabbits this is a predominant pathway, but in guinea pigs it is totally absent. In the rat, two types of trihydroxy bile acids occur owing to hydroxylation either at the 12α or 6 position. Although the enzyme is thought to be one of the microsomal P-450 enzymes, it has recently been reported to be present in the mitochondria of human fetal liver (Gustaffson 1985).

In the absence of an additional hydroxylation of the steroid ring the next step is the stereospecific transformation of the planar allylic ring structure to 5β-cholestane by two enzymes working in concert, a delta 4-3-oxidosteroid reductase and a 3α-hydroxysteroid dehydrogenase. Both of these enzymes are found in the cytosol and require NADPH. Their activity results in the formation of, respectively, 5β-cholestane-diol and triol, precursors of chenodeoxycholic and cholic acids.

However, in order to convert these C-27 sterols into C-24 bile acids we need the help of enzymes located in the mitochondria and in the peroxisomes.

The first step in this pathway in humans appears to be a stereospecific hydroxylation of the terminal methyl group which by custom is referred to as 26-hydroxylation, but by rules of nomenclature could also be referred to as a C-27 hydroxylation (Atsuta and Okuda 1982). There is much to learn about this mitochondrial enzyme which appears to subserve the oxidation of a variety of C-27 substrates and therefore has been referred to as a C-27 sterol-26-hydroxylase (Pedersen et al. 1979). The enzyme has been solubilized and purified and is clearly one of the mitochondrial P-450 enzymes requiring ferrodoxin, ferrodoxin reductase, and NADPH.

Because of the apparently broad substrate specificity, it initiates the oxidation of the two major intermediates in bile acid synthesis to form the 26-triol and tetrol and, in addition, generates 26-hydroxycholesterol. This latter compound is associated with the circulating lipoproteins (Javitt et al. 1981) and is a potent inhibitor of HMG CoA reductase (Esterman et al. 1983). We (Esterman et al. 1984) and others (Skrede et al. 1986) have found evidence for its occurrence in nonhepatic cells in culture, which leads us to think that the enzyme may also occur in extrahepatic tissues. In the next several years we hope to have data on the quantitative significance of the extra-hepatic pathway.

The role of this enzyme in human bile acid synthesis has been delineated because of the finding that it is the molecular basis of a genetically determined disease referred to as cerebrotendinous xanthomatosis (Ofteboro et al. 1980). The biochemical abnormalities that have been found are a decrease in plasma 26-hydroxycholesterol, an increase in 5β-cholestane-triol within the liver cell (Björkhem et al. 1981), and a marked increase in the fecal and urinary excretion of tetrols and pentols that are referred to as bile alcohols (Shimazu et al. 1986). It is thought that the bile alcohol production per day, which is markedly increased, exceeds the production rate of bile acids in normal individuals and accounts, therefore, in part for the increased rate of cholesterol synthesis (Salen and Grundy 1973). However, owing to faulty regulation, the total cholesterol synthesis rate probably exceeds the production rate of both bile acids and bile alcohols. This imbalance between cholesterol synthesis rate and production of metabolites probably accounts for the accelerated deposition of cholesterol in tissues and accelerated atherosclerosis.

The feeding of either chenodeoxycholic or cholic acids to these individuals causes a disappearance of bile alcohol production presumably attributable to a correction in the faulty down-regulation of cholesterol synthesis (Koopman et al. 1985).

The finding of a variety of intermediates that are hydroxylated at the C-25 position is consistent with the presence of a microsomal C-25 hydroxylase. A similar enzyme has also been reported to be present in fibroblasts in cell culture (Saucier et al. 1985). Since some of these intermediates, when administered to normal individuals, are excreted in part as bile acids, it is reasonable to classify the 25-hydroxylase as an alternate pathway of synthesis which inefficiently generates some bile acid.

We return now to the probable role of the mitochondria in normal bile acid synthesis. Following the stereospecific oxidation at the terminal methyl group, the metabolism to the C-27 bile acid requires both an alcohol and aldehyde dehydrogenase. From a mechanistic point of view the reactions appear to parallel the steps in the oxidation of ethanol. Although alcohol dehydrogenase can be found in the cytosol, the predominant aldehyde dehydrogenase appears to be a mitochondrial enzyme.

Because the mitochondria are a site for the oxidation of fatty acids, the further degradation of the C-27 bile acids to C-24 bile acids was also thought to occur in mitochondria and to parallel the same enzymatic reactions. However, it has been shown more recently that the peroxisomes, which also have a major role in the oxidation of fatty acids, probably also account for the bulk of the metabolism of C-27 bile acids to C-24 bile acids. Evidence for a major role of peroxisomes in bile acid synthesis is derived from the knowledge that bile acid metabolism is altered in diseases affecting peroxisomal function. The most dramatic of these diseases is Zellweger's syndrome in which electron microphotographs of the liver from infants born with this genetically-determined defect fail to show these subcellular organelles (Goldfischer et al. 1985). The molecular basis of this disease, however, is related to the formation of plasmologens (Schram et al. 1986), constituents of membranes, and not a specific defect in the synthesis of an enzyme that catalyzes the oxidation of fatty acids or bile acids. For this reason, I do not classify it as a primary defect in bile acid synthesis. There is evidence that at least some of the enzymes that normally would be confined to peroxisomes are distributed in the cytoplasm of the cell and although functional, their activity is diminished.

However, the existence of the syndrome gives a unique insight into the normal pathway of bile acid synthesis. Analysis of biological fluids from these infants indicates an increased proportion of all C-27 bile acids (Parmentier et al. 1979) as well as 26-hydroxycholesterol. As the disease progresses and cirrhosis develops, there is a further increase in the proportion of C-27 bile acids. I interpret the increased excretion of these C-27 intermediates in bile acid synthesis as the consequence of the loss of their efficient transfer to peroxisomal enzymes which results in their increased loss from the hepatocyte and their appearance in increased amounts in plasma, bile, and urine.

The mechanisms leading to cirrhosis in Zellweger's syndrome are not explained, but are unlikely to be directly related to the altered bile acid metabolism.

As a result of normal peroxisomal activity, two types of C-24 bile acids have been generated. Chenodeoxycholic and cholic acids have been generated from the intrinsic pathway and 3β-hydroxy-5-cholenoic acid has been generated from the extrinsic pathway.

In normal individuals, the monohydroxy bile acid can be metabolized further to chenodeoxycholic acid (Javitt et al. 1986). The pathway for this transformation is not certain. As a working hypothesis, we think there is a microsomal C-24 steroid 7α-hydroxylase that catalyzes the formation of 3β, 7α-dihydroxy-5-cholenoic acid and that following this hydroxylation the same oxidoreductase system transforms the allylic structure to chenodeoxycholic acid.

We have shown this metabolic pathway to be present in hamsters, rabbits, and humans and because of the relatively large amounts of 3β-hydroxy-5-cholenoic acid that are normally found in fetal and neonatal life, it is possible that it has some biological significance. One possibility is that intermediates generated by the extrinsic pathway modulate the activity of HMG CoA reductase in contrast to those generated by the intrinsic pathway that do not down-regulate the activity of the enzyme. However, our major interest in 3β-hydroxy-5-cholenoic acid is its biological property of inducing a cholestatic syndrome. Infusion of this bile acid or its conjugates into animals causes an immediate reduction in bile flow with elevations of both conjugated bilirubin and bile acids in plasma, hallmarks of a cholestatic syndrome.

A few years ago we identified a child with a familial cholestatic syndrome that did not metabolize 3β-hydroxy-5-cholenic acid to chenodeoxycholic acid. At the time of study she was 5-years-old and it is possible that the defect was acquired as a result of the liver disease. Currently, with the help of a number of physicians who are caring for these patients, we are evaluating a much larger population to determine whether we have identified another inborn error of bile acid metabolism that is the molecular basis of some types of familial cholestasis.

In summary, a renaissance has occurred with regard to the metabolic pathways of bile acid synthesis in humans. There is evidence that both an intrinsic pathway beginning with 7α-hydroxylation of cholesterol in the liver and an extrinsic pathway beginning with 26-hydroxylation of cholesterol exist. Although the bulk of the bile acids generated from cholesterol are probably generated via the intrinsic pathway, it is possible that the intermediates generated via the extrinsic pathway modulate the activity of HMG CoA reductase.

To guide us in discerning both the pathways of bile acid synthesis and the biological role of the bile acids, the molecular basis of genetically determined defects have been identified and their phenotypic expression characterized. I believe this approach will continue to expand and provide unique insights on the interrelationship of bile acid and cholesterol metabolism.

References

Anderson KE, Kok E, Javitt NB (1972) Bile acid synthesis in man: metabolism of 7α-hydroxy-cholesterol-^{14}C and 26-hydroxycholesterol-[3]H. J Clin Invest 51:112–117

Atsuta Y, Okuda K (1982) Partial purification and characterization of 5β-cholestane-3α,7α-12α-triol and 5β-cholestane-3α,7α-diol 27 monooxygenase. J Lipid Res 23:345–351

Björkhem I (1985) Mechanisms of bile acid biosynthesis in mammalian liver. In Danielsson H, Sjovall J (eds) Sterols and bile acids. Elsevier, Amsterdam New York Oxford, pp 231–278

Björkhem I, Ofteboro H, Skrede S, Pedersen J (1981) Assay of intermediates in bile acid bio-synthesis using isotope dilution-mass spectrometry: hepatic levels in the normal state and in cerebrotendinous xanthomatosis. J Lipid Res 22:191–200

Davis RA, Highsmith WE, MacNeal MM, Schexnayder J-C, Kuan J-CW (1983) Bile acid synthesis by cultured hepatic cells: inhibition by mevinolin but not by bile acids. J Biol Chem 258: 4079–4082

Esterman AL, Baum H, Javitt NB, Darlington GJ (1983) 26-hydroxycholesterol: regulation of hydroxymethylglutaryl CoA reductase in Chinese hamster ovary cell culture. J Lipid Res 24:1304–1309

Esterman AL, Kok E, Javitt NB (1984) 26-hydroxycholesterol: extrahepatic synthesis in aortic smoothmuscle cell culture. Clin Res 32:280A

Goldfischer S, Collins J, Rapin I, Coltoff-Schiller B, Chang C-H, Nigro M, Black VH, Javitt NB (1985) Peroxisomal defects in neonatal-onset and x-linked adrenoleukodystrophies. Science 227:67–70

Gustaffson J (1985) Bile acid synthesis during development: mitochrondial 12α-hydroxylation in human fetal liver. J Clin Invest 75:604–607

Javitt NB, Kok E, Burstein S, Cohen B, Kutscher J (1981) 26-hydroxycholesterol. Identification and quantitation in human serum. J Biol Chem 256:12644–12646

Javitt NB, Kok E, Carubbi F, Blizzard T, Gut M, Byon C-Y (1986) Bile acid synthesis: metabolism of 3β-hydroxy-5-cholenoic acid to chenodeoxycholic acid. J Biol Chem 261:12486–12489

Koopman BJ, Wolthers BG, van der Molen JC, Waterreus RJ (1985) Bile acid therapies applied to patients suffering from cerebrotendinous xanthomatosis. Clin Chem Acta 152:115–122

Ofteboro H, Björkhem I, Skrede S, Schreiner A, Pedersen JI (1980) Cerebrotendinous xanthoma-tosis: a defect in mitochondrial 26-hydroxylation required for normal biosynthesis of cholic acid. J Clin Invest 65:1418–1430

Parmentier GG, Janssen GA, Eggermont EA, Eyssen HJ (1979) C_{27} bile acids in infants with coprostanic acidemia and occurrence of a 3α,7α12α-trihydroxy-5 -C_{29} dicarboxylic bile acid as a major component in their serum. Eur J Biochem 102:173–183

Pedersen JI, Björkhem I, Gustaffson J (1979) 26-hydroxylation of C_{27}-steroids by soluble liver mitochondrial cytochrome P-450. J Biol Chem 254:6464–6469

Salen G, Grundy SM (1973) The metabolism of cholestanol, cholesterol and bile acids in cerebro-tendinous xanthomatosis. J Clin Invest 52:2822–2835

Saucier SE, Kandutsch AA, Taylor FR, Spencer AA, Phirwa S, Gayen AK (1985) Idenfication of regulatory oxysterols,24(s),25-epoxycholesterol and 25-hydroxycholesterol, in cultured fibro-blasts. J Biol Chem 260:14571–14579

Schram AW, Struland A, Takashi H, Wanders RJA, Schutgens RBH, van den Bosch H, Tager JM (1986) Biosynthesis and maturation of peroxisomal β-oxidation enzymes in fibroblasts in rela-tion to the Zellweger syndrome and infantile Refsum disease. Proc Natl Acad Sci USA 83: 6156–6158

Shimazu K, Kuwabara M, Yoshii M, Kihira K, Takeuchi H, Nakano I, Ozawa S, Onuki M, Hatta Y, Hoshita T (1986) Bile alcohol profiles in bile, urine, and feces of a patient with cerebro-tendinous xanthomatosis. J Biochem 99:477–483

Skrede S, Björkhem I, Kvittingen EA, Buchmann MS, Lie SO, East C, Grundy S (1986) Demonstra-tion of 26-hydroxylation of C_{27}-steroids in human skin fibroblasts, and a deficiency of this activity in cerebrotendinous xanthomatosis. J Clin Invest 78:727–735

Bile Acid Sequestrants: Do They Have a Future?

S. M. GRUNDY [1]

1 Introduction

It has been known for many years that interruption of the enterohepatic circulation (EHC) of bile acids will lower the plasma cholesterol. Interruption of the EHC can be achieved in several ways, but the usual means is by use of resins (sequestrants) that bind bile acids in the intestine and prevent their reabsorption. The bile acid sequestrants most commonly used are cholestyramine and colestipol. Cholestyramine was the drug used in the Lipid Research Clinics (LRC) Coronary Primary Prevention Trial (CPPT) (1,2). This trial tested cholestyramine vs placebo in about 4000 men with hypercholesterolemia who were treated for 7 years. Cholestyramine produced a significant lowering of the plasma cholesterol, and rates of coronary heart disease (CHD) were reduced significantly at $p < 0.05$. This trial has generally been accepted as strong support for the "lipid hypothesis", namely, that a lowering of the plasma cholesterol will decrease the risk for CHD. Furthermore, analysis of the trial results provided no consistent evidence that lowering of plasma cholesterol by bile acid sequestrants is accompanied by significant side effects. On the other hand, despite the success of the CPPT, doubts remain whether bile acid sequestrants have a significant future for the treatment of hypercholesterolemia. They are bulky and inconvenient to take. They can cause gastrointestinal distress, and they almost uniformly cause constipation. Also, they are only moderately effective for lowering of plasma cholesterol; and finally they are expensive. The potential usefulness of bile acid sequestrants may be further limited by the discovery of a new class of drugs for cholesterol lowering, namely, the inhibitors of 3-hydroxy-3-methylglutaryl coenzyme A (HMG CoA) reductase. These latter drugs appear to be devoid of many of the drawbacks of bile acid resins. Therefore, we might consider the future for resins as agents for treatment of hypercholesterolemia.

1 Center for Human Nutrition, Department of Clinical Nutrition, University of Texas Health Science Center at Dallas, 5323 Harry Hines Boulevard, Dallas, TX 75235, USA

Drugs Affecting Lipid Metabolism
Ed. by R. Paoletti et al.
© Springer Verlag Berlin Heidelberg 1987

2 Mechanisms of Bile Acid Sequestrants

Although it is generally assumed that the lowering of plasma cholesterol will decrease the rate of formation of atherosclerosis, the question always remains whether the mechanism of cholesterol lowering is compatible with this therapeutic goal. For example, a drug might alter the tissue metabolism of cholesterol or the structure of lipoproteins in such a way as to promote rather than to retard the development of atherosclerosis. The results of the CPPT provide strong support for the concept that bile acid sequestrants lower the plasma cholesterol in an efficacious way. However, there is increasing evidence that resins produce multiple effects on lipoproteins, and each of these actions might be examined for their potential efficacy.

2.1 LDL Receptor Activity

One action of bile acid resins appears to be to increase the activity of LDL receptors. In experimental animals, bile acid sequestration increases the binding of LDL to receptors on isolated liver cell membranes (3). Isotope kinetic studies in humans suggest that receptor-mediated clearance of LDL likewise is enhanced (4). This action apparently is secondary to depletion of the liver cell of cholesterol and stimulation of the activity of HMG CoA reductase. The latter changes have been shown to be associated with an increased synthesis of LDL receptors (5). Despite the fact that large quantities of hepatic cholesterol are converted to bile acids, the action of resins to lower the plasma level of LDL is limited because of a compensatory increase in the synthesis of cholesterol (6). However, the degree of fall in plasma cholesterol is a function of the extent of interruption of the EHC, and high doses of sequestrants are more efficacious than low doses.

The action of sequestrants to stimulate the synthesis of receptors might be expected to be limited to the liver. In this case, most of the cholesterol removed from the blood-stream should enter the liver and be excreted as bile acids. This would seem to be a mechanism that should *not* increase the danger of enhanced entry of LDL into the arterial wall. Of interest, a few studies suggest that receptor activity in extra-hepatic cells also can be enhanced by sequestrant therapy (7), most of the excess cholesterol removed from the plasma probably enters the liver. This mechanism for cholesterol lowering would appear to be effective for retardation of atherogenesis.

2.2 Depletion of LDL of Cholesterol

Another mechanism for cholesterol lowering by sequestrants has been reported by Witztum et al. (8,9). These workers have shown that the LDL particles circulating during resin therapy are smaller and denser than normal. On the average, the LDL are partially depleted of cholesterol, presumably LDL cholesterol ester. The mechanism whereby this occurs has not been determined with certainty. Witztum et al. (9) have postulated that smaller LDL have less affinity for LDL receptors than larger LDL, and when receptor activity is enhanced, the larger LDL are selectively removed from the circulation. Another possibility is that the massive conversion of cholesterol to bile acids in the liver during resin therapy creates a negative cholesterol balance that makes less cholesterol available for entry into LDL particles. Finally, the smaller LDL may be partly the result of an increased secretion of triglycerides into plasma during sequestrant therapy, as will be discussed below; patients with elevated triglycerides are known to have abnormally small LDL (10).

Is it beneficial to have abnormally small LDL? Some investigators have speculated that the unusually small LDL in patients with hypertriglyceridemia may be more atherogenic than normal LDL. Vega and Grundy (10) have shown that smaller, denser LDL filter more readily into extravascular spaces than larger LDL. On the other hand, small LDL carry less cholesterol per particle than larger LDL, and thus for every particle filtered into the arterial wall, less cholesterol (which might accumulate in the arterial wall) is carried. It thus is difficult to say in the balance whether small LDL are more or less atherogenic than normal-sized LDL. It is interesting to note that therapy with mevinolin in contrast to bile acid sequestrants does not produce a significant change in the composition and, presumably, particle size of LDL (11).

2.3 Stimulation of Triglyceride Production

Another curious effect of bile acid sequestrants is a stimulation of secretion of very low density lipoprotein (VLDL)-triglycerides (TG) (12). The mechanisms whereby resins promote the synthesis of VLDL-TG have not been elucidated. Whether the increased synthesis of cholesterol accompanying resin therapy induces a corresponding increase in the synthesis of triglycerides by the liver is unknown. Also unknown is whether the synthesis of VLDL-TG by the liver is associated with a heightened production of VLDL-apolipoprotein B (apo B). The production of VLDL-apo B during treatment with resins has not been measured, but if the production is high, we might ask whether this effect would be entirely benign, for an overproduction of VLDL-apo B has been claimed by some to be an atherogenic process.

Certainly not all patients treated with sequestrants develop hypertriglyceridemia; only a portion of them do, particularly those ingesting alcohol. But this does not necessarily mean that even those with normal plasma triglycerides do not have overproduction of VLDL-apo B. Patients can have excessive input of VLDL-apo B without having elevated plasma triglycerides, and in these patients, premature CHD is common (13). Is it possible that bile acid sequestrants induce an analogous situation? The occurrence of overproduction of VLDL-apo B, variable or intermittent hypertriglyceridemia, increased fractional clearance of LDL, and abnormally small and dense LDL are characteristics of (1) a subgroup of patients with premature CHD (13,14) and (2) patients treated with resins. Again, the results of the CPPT offer reassurance that bile acid sequestrants do not heighten the risk for CHD, but rather reduce the risk. On the other hand, one must ask whether the effects of resins on lipoprotein metabolism other than that to increase the activity of LDL receptors to some extent mitigate the risk-reducing action of these drugs.

3 Side Effects of Bile Acid Sequestrants

The difficulties of adherence to bile acid resins are well known. Because of their bulk and consistency, they are a nuisance to take for extended periods. In some people they cause significant gastrointestinal distress. In most they cause some degree of constipation, and not uncommonly, the constipation is so severe that the medication must be discontinued. Resins bind some drugs in the intestine and thereby reduce their efficacy. The difficulty of adhering to sequestrant therapy is illustrated by the fact that only a small fraction of men in the CPPT continued to take cholestyramine after termination of the trial, even though the drug was demonstrated to effectively reduce the risk for CHD. But beyond these difficulties, it is important to ask whether sequestrants may have serious side effects if they are ingested for years.

Perhaps the most important question is whether prolonged ingestion of resins will increase the likelihood of cancer. The most likely tumor would seem to be cancer of the colon, although other cancers of the gastrointestinal tract are a possibility. An increased prevalence of gastrointestinal cancers was not proven to occur in resin-treated patients of the CPPT, but numerically, more tumors of the gastrointestinal tract were noted in patients receiving cholestyramine than in placebo-treated patients. There are several mechanisms by which resins might be carcinogenic. First, the resins themselves might induce or promote the formation of tumors. Second, the constipation resulting from resins therapy may reduce the colonic transit time and allow more exposure of the colonic mucosa to potentially cancer-forming substances. And third, the passage of increased quantities of bile acids through the colon could be carcinogenic, because bile acids have a putative tumorigenic action. We can be reassured in part because the prevalence of colon cancer was not increased in the CPPT during the 7 years of treatment. Nevertheless, this study did not rule out the possibility that sequestrants enhanced the formation of benign polyps that someday might undergo a malignant transformation. Similar changes may have occurred in the stomach or other portions of the gastrointestinal tract. Bile acid sequestrants, like any drug taken on a chronic basis, always have the potential of being carcinogenic, but on the basis of current data, they appear to have a low level of carcinogenicity, if any.

Another possible side effect of resins is the formation of cholesterol gallstones. Fortunately, this adverse response is rare, although it does occur occasionally. By draining bile acids from the EHC, the solubility of cholesterol in bile may be reduced leading to formation of cholesterol crystals and gallstones. In the CPPT, there was no indication of excessive formation of gallstones, and hence this side effect appears to be relatively rare.

4 Sequestrants with Improved Efficacy

One of the major drawbacks of bile acid sequestrants is their bulkiness. For this reason, they are inconvenient to take for long periods of time, especially for many years. If resins could be made more efficient, i.e., if they could be made to bind more bile acids for a given mass of resin, they might have a greater utility. Thus far, however, despite repeated attempts, little progress seemingly has been made toward improvement of the efficiency of sequestrants. One problem is that resins are non-specific and bind other acidic anions besides bile acids. To date, sequestrants that are specific for bile acids have not been developed. To do so would be a worthy project, but whether a major effort will be made in this direction is uncertain. Presumably, any new resin of this type would have to be tested thoroughly, which for a new drug is extremely expensive under present circumstances. Therefore, it is doubtful that the pharmaceutical industry will commit significant funds to such a project until the potential for HMG CoA reductase inhibitors have been resolved.

Another area for possible development of resins is to prepare new formulations of the available agents. It may be possible to improve the consistency of resins or to incorporate them into foods. The latter might make the sequestrants more palatable, although it would have the disadvantage of adding calories to the preparation. Regard-less of any improvements that can be made in the dosage or formulation of resins, the problem of constipation cannot be solved, unless of course a bulk laxative is added to the preparation. The latter, however, would increase the total bulk of the pre-paration and thus would further reduce its palatability.

5 Potential for Low-Dose Resins

Even if resins cannot be made more efficient in their binding of bile acids, considera-tion can be given to the possibility that less than maximal doses of sequestrants may be satisfactory under certain circumstances. Considerable data indicate that the response to sequestrants is proportional to dose. However, the incremental response may not be linear. For example, increasing the dose of cholestyramine from 8 to 16 g day^{-1} may not double the response, nor will raising the intake to 32 day^{-1} induce another doubling of response. Indeed, Angelin and Einarsson (15) reported that a sizable decrease in plasma cholesterol will occur in hypercholesterolemic patients at a dose of 10 g day^{-1} of cholestyramine. In their study, the plasma cholesterol concentration fell by an average of 45 mg dl^{-1} during resin therapy. Thus, the potential for low doses of sequestrants has not been fully evaluated. If lower doses can be shown to be moderately effective in lowering of plasma LDL, they could prove useful in patients with less severe forms of hypercholesterolemia.

6 Sequestrants vs HMG CoA Reductase Inhibitors

Several investigations have revealed that the bile acid-binding resins do not lower the plasma LDL levels as much as the new HMG CoA reductase inhibitors. At standard doses the reductase inhibitors appear to be approximately twice as potent for LDL lowering as the bile acid resins (16-18). The enhanced potency of reductase inhibitors, furthermore, is not accompanied by the inconvenience of sequestrants or constipation. Furthermore, it is likely that the reductase inhibitors will be less expensive than sequestrants.

For these reasons, the reductase inhibitors have a clear edge over sequestrants in the current market for cholesterol-lowering drugs. This situation, however, could change if the reductase inhibitors should prove to have significant side effects. The reductase inhibitors, in contrast to the sequestrants, are absorbed internally and theoretically could be toxic in several organs. They are known to produce abnormalities of liver function tests in some patients. They may reduce the synthesis of cholesterol in many tissues, the consequences of which are not fully understood. Whether they have a direct toxic effect that might cause hepatic damage, neurological disorder, or cataracts remain to be determined. If significant untoward side effects were to develop with reductase inhibitors, then the door to the increased use of bile acid sequestrants might be reopened.

One type of patient who may be a candidate for bile acid sequestrants is one with only moderate hypercholesterolemia in whom drugs that act internally are deemed too risky. In other words, until the safety of HMG CoA reductase inhibitors is well established, it may be wise to employ sequestrants – possibly in low doses – for those in whom hypercholesterolemia is not severe. Since there are many more patients with hypercholesterolemia of the moderate rather than severe variety, then the number of individuals who might be candidates for sequestrants could be greater than generally realized. Consequently, from the present until that time when the safety of reductase inhibitors is firmly established, there may be a very large number of candidate patients for sequestrants, i.e., individuals with only moderate hypercholesterolemia.

7 Sequestrants + Reductase Inhibitors

For patients with severe hypercholesterolemia, such as those with familial hypercholesterolemia, therapy with reductase inhibitors alone may not be sufficient to lower levels of LDL to the desirable range. However, studies in our laboratory (17,19) and others (20) have demonstrated that the combination of reductase inhibitors and bile acid sequestrants can cause a 50% reduction in LDL-cholesterol, which for most patients will lower the LDL level to the desirable range. Thus, the combination of these two types of drugs may have a usefulness in patients with severe hypercholesterolemia. This combination has the theoretical advantage of promoting the synthesis of LDL receptors in two ways simultaneously: reductase inhibitors by inhibiting the formation of cholesterol and sequestrants by stimulating the catabolism of cholesterol. The addition of sequestrants to therapy with reductase inhibitors provides a distinct increment in response for LDL lowering that apparently cannot be achieved by increasing the dosage of reductase inhibitors.

Recent studies from our laboratory (21) also have shown that this same combination produces a marked decrease in LDL levels in patients with primary moderate hypercholesterolemia. The reduction in LDL-cholesterol again approximates about 50%. Although this drug combination almost certainly will not be required for patients in whom the only risk factor is moderate hypercholesterolemia, combined drug therapy is worthy of consideration for high-risk patients, i.e., for those with established CHD or the combination of hypercholesterolemia with other risk factors such as smoking, hypertension, or diabetes mellitus.

8 Summary

The above considerations suggest that it may be premature to discount the potential usefulness of bile acid sequestrants. These agents have been proven to be efficacious for prevention of CHD by the LRC-CPPT. Their major mechanism of action is to increase the activity of LDL receptors which appears to lower the LDL level in an advantageous way. They have been shown to be relatively safe in the CPPT, and they have the great advantage of being confined to the gastrointestinal tract. There is no convincing evidence that they increase the risk of gastrointestinal malignancy. While the bile acid sequestrants are not potent enough to be used alone in the treatment of familial hypercholesterolemia, they may be adequate for the much more common form of elevated LDL, i.e., primary moderate hypercholesterolemia.

The use of sequestrants in combination with HMG CoA reductase inhibitors appears to be a particularly powerful combination for treatment of severe hypercholesterolemia or even for high-risk patients with moderate hypercholesterolemia.

The sequestrants as used today have certain definite disadvantages that limit their usefulness. They are expensive, and because of their bulkiness, they are inconvenient to take. They often cause gastrointestinal discomfort, particularly constipation. Therefore, modification of current dosage or dose forms may be required. The potential for lower doses in patients with only moderate hypercholesterolemia needs to be explored. New forms of the agents, particularly those with greater efficacy need to be developed. If this could be accomplished, the sequestrants could play a significant role in primary and/or secondary prevention of CHD.

References

1. Lipid Research Clinics Program (1984) The lipid research clinics coronary primary prevention trial results. I. Reduction in incidence of coronary heart disease. JAMA 251:351−364
2. Lipid Research Clinics Program (1984) The lipid research clinic primary prevention trial results. II. The relationship of reduction in incidence for coronary heart disease to cholesterol lowering. JAMA 251:365−374
3. Kovanen PT, Bilheimer DW, Goldstein JL, Haramillo JJ, Brown MS (1981) Regulatory role for hepatic low density lipoprotein receptors in vivo in the dog. Proc Natl Acad Sci USA 78:1194−1198
4. Packard CJ, Shepherd J (1982) The hepatobiliary axis and lipoprotein metabolism: effects of bile acid sequestrants and ileal bypass surgery. J Lipid Res 23:1081−1098
5. Goldstein JL, Brown MS (1977) The low-density lipoprotein pathway and its relation to atherosclerosis. Annu Rev Biochem 46:897−930

6. Grundy SM, Ahrens EH, Jr, Salen G (1971) Interruption of the enterohepatic circulation of bile acids in man: comparative effects of cholestyramine and ileal exclusion on cholesterol metabolism. J Lab Clin Med 78:94–121

7. McNamara PJ, Davidson NO, Fernandez S (1980) In vitro cholesterol synthesis in freshly isolated mononuclear cells of human blood: effect of clofibrate and/or cholestyramine. J Lipid Res 21:65–71

8. Witztum JL, Schonfeld G, Wiedman SW, Giese WE, Dillingham MA (1979) Bile sequestrant therapy alters the composition of low density and high denstiy lipoproteins. Metabolism 28:221–229

9. Witztum JL, Young SG, Elam RL, Carew TE, Fisher M (1985) Cholestyramine-induced changes in low density lipoprotein composition and metabolism. I. Studies in the guinea pig. J Lipid Res 26:92–103

10. Vega GL, Grundy SM (1986) Kinetic heterogeneity of low density lipoproteins in primary hypertriglyceridemia. Arteriosclerosis 6:395–406

11. Grundy SM, Vega GL (1985) Influence of mevinolin on metabolism of low density lipoproteins in primary moderate hypercholesterolemia. J Lipid Res 26:1464–1475

12. Beil U, Grundy SM, Crouse JR, Zech L (1982) Triglyceride and cholesterol metabolism in primary hypertriglyceridemia. Arteriosclerosis 2:44–57

13. Kesaniemi YA, Beltz WF, Grundy SM (1985) Comparisons of metabolism of apolipoprotein B in normal subjects, obese patients, and patients with coronary heart diesease. J Clin Invest 76:586–595

14. Vega GL, Grundy SM (1985) Low density lipoprotein metabolism in hypertriglyceridemic and normolipidemic patients with coronary heart disease. J Lipid Res 26:115–126

15. Angelin B, Einarsson K (1981) Cholestyramine in type IIa hyperlipoproteinemia. Atherosclerosis 38:33–38

16. Yamamoto A, Sudo H, Endo A (1980) Therapeutic effects of ML-236B in primary hypercholesterolemia. Atherosclerosis 35:259–266

17. Bilheimer DW, Grundy SM, Brown MS, Goldstein JL (1983) Mevinolin and colestipol stimulate receptor-mediated clearance of low density lipoprotein from plasma in familial hypercholesterolemia heterozygotes. Proc Natl Acad Sci 80:4124–4128

18. Illingworth DR, Sexton GJ (1984) Hypocholesterolemia effects of mevinolin in patients with heterozygous familial hypercholesterolemia. J Clin Invest 74:1972–1978

19. Grundy SM, Vega GL, Bilheimer DW (1985) Influence of combined therapy with mevinolin and interruption of bile acid reabsorption on low density lipoproteins in heterozygous familial hypercholesterolemia. Ann Int Med 103:339–343

20. Mabuchi H, Sakai T, Sakai Y, Yoshimura A, Watanabe T, Wakasugi T, Koizumi J, Takeda R (1983) Reduction of serum cholesterol in heterozygous patients with familial hypercholesterolemia. Additive effects of compactin and cholestyramine. N Engl J Med 308:609–613

21. Vega GL, Grundy SM (1987) Treatment of primary moderate hypercholesterolemia with lovastatin (mevinolin) and colestipol. JAMA 257:33–38

Transport of Cholesterol and Cholesterol Esters by HDL

D. STEINBERG [1]

1 Introduction

The epidemiologic evidence establishing a low HDL cholesterol level as an independent risk factor for atherosclerosis and coronary heart disease is convincing beyond all possible doubt. However, the *mechanisms* by which a low HDL level predisposes to and a high HDL level protects against atherosclerosis are not firmly established. By all odds the most widely accepted hypothesis is that put forward by Glomset almost 20 years ago, namely, the hypothesis that HDL is involved in "reverse cholesterol transport" (1). Based on his studies of lecithin-cholesterol acyltransferase, Glomset saw that HDL could accept free cholesterol from tissues and that this free cholesterol could be converted to the ester form by the action of LCAT. This would then allow the surface of the HDL molecule to accept another molecule of *free* cholesterol from the cells, and so on. This hypothesis is supported by many cell culture studies showing that HDL can act as an acceptor of cholesterol from cells overloaded with cholesterol [reviewed by Reichl and Miller (2)]. On the other hand, lipoprotein-deficient serum or even serum albumin alone can also facilitate the removal of cholesterol from cholesterol-loaded cells. Studies by Oram and colleagues, demonstrating a set of saturable binding sites on cells in culture that are up-regulated when the cells are loaded with cholesterol, also support the hypothesis (3,4) and several other groups have either observed specific, saturable binding of HDL or have actually partially purified a putative HDL receptor (see ref. 2). However, the results of these in vitro studies cannot be fully evaluated until there is a translation to the in vivo situation.

The difficulty, of course, in obtaining the in vivo evidence we so sorely need is that every component of the HDL macromolecule is labile in the sense that it exchanges with one or more other pools in the plasma or in the tissues. For example, free cholesterol in HDL exchanges rapidly with free cholesterol in other lipoproteins and in cells. The same is true for the phospholipids in HDL. The cholesterol ester of HDL, once believed to be a rock-stable marker, in most species exchanges very, very rapidly with cholesterol esters and/or triglycerides in other lipoprotein fractions. Every one of the apolipoproteins associated with HDL can exchange with the apoprotein on VLDL, chylomicrons, or both. It is because of this chimeric quality of

---

1 Department of Medicine, M-013D, University of California, San Diego, La Jolla, CA 92093, USA

Drugs Affecting Lipid Metabolism
Ed. by R. Paoletti et al.
© Springer-Verlag Berlin Heidelberg 1987

HDL that we have not yet been able to obtain clear, convincing data on the quantitative aspects of HDL metabolism that might establish its role in "reverse cholesterol transport".

Despite the incomplete circle of evidence, it has become almost conventional wisdom that anything that raises the HDL level is good for you and anything that lowers it is bad for you. That may well be true. On the other hand, in the absence of hard data on the kinetics of HDL transport of cholesterol such a conclusion is premature. One could even conceive of a situation in which HDL might be actively engaged in reverse cholesterol transport to the extent that its steady state concentration in the plasma was *lower* than it might be in an individual in whom the process of reverse cholesterol transport was proceeding at a more leisurely pace. I do not think we can take epidemiologic correlations, parlay them with an hypothesis about reverse cholesterol transport that however plausible remains an hypothesis, and come out with a dogma that reads "Raise HDL!". And yet that may be perfectly correct. In view of the technical difficulties of obtaining the kinds of in vivo kinetic data that we need, I would like to suggest that a more likely route to the resolution we so badly need would be experiments in vivo in experimental animals showing that raising HDL levels *does* slow the progess of atherogenesis. Just as we now have incontrovertible evidence that *raising* LDL levels induces atherosclerosis in animals, we ought to be able to obtain evidence that *lowering* HDL levels does something similar. Studies of that kind are feasible within the context of today's technology and I hope that something like that will be done soon.

One reason for insisting on the definitive in vivo experiments, as discussed above, is that there are some situations in which one ought to expect serious consequences as a result of low HDL levels but one doesn't find that. In the case of Tangier disease, for example, we do *not* see the severe rapidly accelerated lesion development that we see in patients with homozygous familial hypercholesterolemia — or even that we see in heterozygous familial hypercholesterolemia. Tangier patients do show some lipid accumulation, but it is notable that it occurs in tissues that are *not* favored sites of lipid accumulation due to elevation of the LDL level (tonsils and other lymphoidal tissue). We are evidently not dealing with a symmetrical relationship between high LDL (or beta-VLDL), on the one hand, and low HDL, on the other hand. Before leaving our discussion of patients with Tangier disease, we should stress that the absolute steady state concentration of HDL in these patients is very low, *but the turnover is extremely high* (5). In fact, the net flux of HDL through the plasma compartment may be as great as it is in a normal individual due to a much, much higher fractional catabolic rate. As a result, it is conceivable that the reverse cholesterol transport system is working just fine even though the steady state level in the plasma is so vanishingly low. If that is the case, it would at one and the same time salvage the reverse cholesterol transport hypothesis and inform us that *steady state levels* of HDL cannot be taken to reflect faithfully the rate at which reverse cholesterol transport is taking place.

If LCAT is a key component of the proposed reverse cholesterol transport system, then one would expect accelerated atherosclerosis in patients with LCAT deficiency. However, again we do not see as severe a syndrome as we see in patients with familial hypercholesterolemia. There is some lipid accumulation in these patients, particularly in the kidney, and that can have serious clinical consequences, offering some support for the Glomset hypothesis. However, the pattern of lipid deposition is quite different from that which we see when there is an excess rate of delivery of cholesterol from LDL. As far as I know, there is no evidence for accelerated athero-sclerosis in patients with LCAT deficiency. However, the number of patients that have been autopsied is small and it may be too early to draw firm conclusions.

Finally, it should be pointed out that there are a couple of families with extremely rare genetic disorders characterized by an almost complete absence of apo A-I in whom obviously premature atherosclerosis has been documented (6). The severe atherosclerosis in these patients, in the absence of any elevation of LDL or other evident risk factors, constitutes perhaps the strongest evidence that low HDL levels can *in some way* evoke premature atherosclerosis.

2 Possible Pathways for Reverse Cholesterol Transport

Even if we accept that HDL is involved in reverse cholesterol transport, what can we say about the precise mechanisms by which the cholesterol picked up from peripheral tissues gets back to the liver? To understand that, we need to have studies on, first, exactly how HDL itself gets taken up by the liver and to what extent, and second how cholesterol initially accepted by HDL (perhaps nascent HDL) may be transferred to other lipoprotein fractions for secondary delivery to the liver. There are perhaps four different ways in which this might be accomplished:

1. By Hepatic Uptake of the Entire HDL Macromolecule in Analogy with the Uptake of Holo-LDL via Its Receptor. Most thinking tends along these lines. The very strong evidence for the mechanisms by which holo-LDL is taken up has tended to make us all think in terms of analogous mechanisms when we begin to deal with other lipoproteins. Yet we already know that uptake of holo-lipoprotein particles is not the norm. Certainly VLDL is not taken up primarily as an intact particle. Instead, it undergoes progressive modification through the action of lipoprotein lipase (and, later, by hepatic lipase) constantly altering its composition before it is ultimately either taken up by the liver as a remnant particle or converted to LDL. In any case, one of the potential mechanisms for transporting tissue cholesterol back to the liver is to have HDL pick it up and get engulfed by some mechanism (still not too well defined) at the hepatocyte surface. Studies by Glass and co-workers (7) have shown that the liver is the dominant site of uptake of HDL apoprotein A-I, at least in the rat. That is certainly compatible with HDL playing the role postulated for it in reverse cholesterol transport. On the other hand, without precise quantitative data on how much cholesterol leaves the liver in HDL, how much is picked up from the tissues, how much is transferred to other lipoprotein fractions, and so on, we can't really be sure that a dominant uptake of apoprotein A-I from HDL into the liver establishes anything at all about the amount of peripheral tissue cholesterol deposited in the liver. The studies of Glass and co-workers were done in the rat because the rat has virtually no cholesterol ester transfer protein in the plasma and so at least the cholesterol ester of HDL appears to stay put. Whether the findings in the rat apply also to animals that *do* have the cholesterol ester transfer protein remains to be established.

2. Transfer of Cholesterol Ester from HDL to VLDL (or IDL) with Subsequent Hepatic Uptake of the Acceptor Lipoprotein. Most animals, including man, have high plasma levels of cholesterol ester transferase protein. Thus, it is quite possible that the cholesterol picked up from tissues and converted to cholesterol ester in HDL is subsequently transferred to VLDL and then taken up as a part of the uptake of whole lipoprotein particles into the liver. Again, in the absence of some way of assessing rates of transfer and rates of uptake in absolute terms, we can't be sure how important this postulated mechanisms may be. It might be pointed out, however, that there *are* animals that have very low levels of cholesterol ester transferase activity, including the rat and the pig. They do *not* have spontaneous atherosclerosis and the rat, at least, is very resistant to the development of experimental atherosclerosis. While by no means conclusive, these observations cast some doubt on the primacy of reverse cholesterol transport mediated by the cholesterol ester transfer protein as an essential mechanism for protection against atherosclerosis.

3. Selective Uptake of Cholesterol Ester from HDL Without Uptake of the Entire Particle. Studies in this laboratory have revealed a novel mechanisms by which cholesterol may be delivered to the liver by way of HDL (8). What we have shown is that there is a greater hepatic uptake of cholesterol esters from HDL than of apoprotein A-I. This was shown in the rat in vivo (7) and in a number of cell types, including human hepatoma cells in vitro (9). For many years it was thought that the cholesterol esters in lipoproteins were stable and nonexchangeable. That conclusion was based largely on studies in the rat, which we only learned much later is almost unique in having very little cholesterol ester transferase activity. Thus, the dogma arose that cholesterol esters do *not* transfer and therefore the idea that cholesterol esters might be delivered to tissues by lipoproteins without uptake of the lipoprotein in its entirety seemed quite implausible. But now we know that such a process does occur in vivo, at least in the rat, and studies with cultured cells suggest that it may occur in other species. However, to establish whether or not it occurs in vivo and to what extent will require highly sophisticated multicompartmental analyses of kinetic studies, taking into account all of the many transfer and exchange reactions that are going on simultaneously.

4. "Channeling" of Cholesterol to Liver by Way of HDL$_c$. Mahley and co-workers (10) have pointed out the unique development of an HDL species in cholesterol-fed animals that could play a role in reverse cholesterol transport. The so-called HDL$_c$ appears in the plasma of cholesterol-fed animals and even, to some extent, in the plasma of cholesterol-fed man (11). This species of HDL is particularly rich in apoprotein E and is taken up avidly by the liver. As shown by Pitas et al. (12), molecules that contain 4 mol or more of apo E per particle bind to the apo B/E receptor even more tightly than does LDL itself. Presumably this is because increasing the number of binding sites per particle geometrically increases the affinity constant. If cholesterol were accepted from cells by nascent HDL, and if that HDL progressively increased its apo E content at the expense of other apoproteins, a species would be generated that would then be very rapidly and selectively taken up by the liver along with the cholesterol it had acquired from peripheral cells.

3 Summary

The complexity of this problem is well illustrated by the several quite different mechanisms for cholesterol delivery via HDL that have been postulated. Until these mechanisms can be evaluated in vivo we are still guessing which ones are operative and to what extent. What we badly need are new experimental attacks that will resolve the ambiguities surrounding current approaches. Until we can definitively establish how HDL is involved in processes that may influence atherogenesis, it is difficult to advocate intervention to raise HDL without regard to the underlying mechanism. Is it possible that a high HDL level is protective, not because of anything the HDL does, but because it simply reflects effective handling of chylomicrons and VLDL by lipoprotein lipase? Is it possible that a low HDL level predicts risk because those individuals are depositing cholesterol into peripheral tissues (including the artery) at a high rate and their HDL "system" is overloaded? Until the definitive animal model study, mentioned above, or a human intervention study is done, we should be cautious about doing any more than using HDL levels as an extremely valuable predictor of risk.

References

1. Glomset JA (1968) The plasma lecithin: cholesterol acyltransferase reaction. J Lipid Res 9:155—67
2. Reichl D, Miller NE (1986) The anatomy and physiology of reverse cholesterol transport. Clin Sci 70:221—231
3. Oram JF, Albers JJ, Cheung MC, Bierman EL (1981) The effect of subfractions of high density lipoprotein on cholesterol efflux from cultured fibroblasts. J Biol Chem 256:8348—8356
4. Biesbroeck R, Oram JF, Albers JJ, Bierman EL (1983) Specific high affinity binding of high density lipoprotein to cultured human skin fibroblasts and arterial smooth muscle cells. J Clin Invest 71:525—539
5. Schaefer EJ, Blum CB, Levy RI et al. (1978) Metabolism of high-density lipoprotein apolipoprotein in Tangier disease. N Engl J Med 299:905—910
6. Schaefer EJ, Heaton WH, Wetzel MG, Brewer HB, Jr (1982) Plasma apolipoprotein A-I absence associated with a marked reduction of high density lipoproteins and premature coronary artery disease. Arteriosclerosis 2:16—26
7. Glass C, Pittman RC, Weinstein DB, Steinberg D (1983) Dissociation of tissue uptake of cholesterol ester from that of apoprotein A-I of rat plasma high density lipoprotein: selective delivery of cholesterol ester to liver, adrenal, and gonad. Proc Natl Acad Sci USA 80:5435—5439
8. Pittman RC, Steinberg D (1984) Sites and mechanisms of uptake and degradation of high density and low density lipoproteins. L Lipid Res 25:1577—1585

9. Pittman RC, Steinberg D (1985) A novel mechanism by which high density lipoprotein selectively delivers cholesterol esters to the liver. In: Greten H, Windler E, Beisiegel U (eds) Receptor-mediated uptake in the liver. Springer, Berlin Heidelberg New York Tokyo, pp 108–119

10. Mahley RW (1978) Alterations in plasma lipoproteins induced by cholesterol feeding in animals including man. In: Dietschy JM, Gotto AM, Ontko JA (eds) Disturbances of lipid and lipoprotein metabolism. Am Physiol Soc, Bethesda, MD, pp 181–197

11. Mahley RW, Innerarity TL, Bersot TP, Lipson A, Margolis S (1978) Alterations in human high-density lipoproteins, with or without increased plasma-cholesterol, induced by diets high cholesterol. Lancet Oct 14:807–809

12. Pitas RE, Innerarity TL, Arnold KS, Mahley RW (1979) Rate and equilibrium constants for binding of apo-E HDL_C (a cholesterol-induced lipoprotein) and low density lipoproteins to human fibroblasts: evidence for multiple receptor binding of apo-E HDL_C. Proc Natl Acad Aci USA 76:3211–2315

Role of HDL in the Metabolism of the Plasma Lipoproteins

S. EISENBERG [1]

High density lipoproteins (HDL) accounts for about one-half of total plasma lipo-protein mass in middle-aged humans, and more than two-thirds of the lipoprotein mass in children. It is undoubtedly the major lipoprotein family in many animal species. In normolipidemic adults, the plasma HDL mass is about 300-400 mg, similar to the combind mass of VLDL + LDL. Since more HDL than VLDL or LDL is dis-tributed to extravascular spaces, it is reasonable to assume that even in adult humans, HDL is the major plasma lipoprotein. If lipoprotein mass is expressed in particle number, then the plasma contains 10-20 times more HDL particles than VLDL or LDL. In view of this predominance of HDL, it is surprising that investigations of the biological behavior of the HDL system began only during the 1970s, and many features of the system are still obscure. Accumulated knowledge of the physiology und patho-physiology of the HDL has been summarized in 1984 (1). The purpose of the present text is to update the earlier text, with special emphasis on the role that HDL plays in plasma fat transport processes.

HDL consist of several populations of spherical particles — HDL_1, HDL_2, HDL_3 and HDL_4 — that differ in density, size, and lipid and protein composition (1). All populations contain a relatively small-sized cholesteryl ester (CE) core, surrounded by a 20 A wide outer layer of phospholipids (PL), free cholesterol (FC), and apo-proteins A-I, A-II, C and E. Various amounts of triglycerides (TG) are also found in HDL. A hallmark of HDL particles is the large proportion of surface to core com-ponents: more than 80% of the mass and volume of HDL is accounted for by surface material. Hence, in terms of fat transport, HDL appears to be a very inefficient vehicle for transfer of lipids through the bloodstream. These puzzling considerations contrast with the predominance of HDL in the lipoprotein system and its universal presence in all animal species. It thus appears that the function of HDL must be viewed in terms different from those employed for other lipoproteins.

Studies published in the early 1970s established that HDL participates in the plasma fat transport system as an acceptor of constituents released from the surface of lipolyzed TG-rich lipoproteins, mainly apolipoproteins (2). It has, moreover, been shown that the functionally important C apoproteins return to VLDL when newly synthesized particles enter the circulation (2). A few years later it became apparent

---

1   Lipid Research Laboratory, Department of Medicine B, Hadassah University Hospital, Jerusalem, Israel

Drugs Affecting Lipid Metabolism
Ed. by R. Paoletti et al.
© Springer Verlag Berlin Heidelberg 1987

that HDL is not simply a "reservoir" of surface constitutents generated from the lipolysis process, but in fact is formed "surface remnants" that are obligatory during the shrinkage of lipolyzed large TG-rich lipoproteins (3). In the same year we showed another link between plasma fat transport systems and HDL, namely, incorporation of proteins and lipids (mainly PL and FC) released from lipolyzed VLDL by HDL_3 (4). This reaction caused conversion of HDL_3 to HDL_2-like lipoproteins, and led us to suggest that when LCAT is active, true HDL_2 particles are formed.

The investigations described above provide a basis for the hypothesis that HDL is the final product of the metabolism of the surface domain of TG-rich lipoproteins, as much as LDL is the final product of the core domain (5). Once formed, however, HDL is the center of intense metabolic activity and none of the HDL constituents (in humans) is associated with the particle throughout its circulating time. An example is the lecithin (phosphatidylcholine, PC) moiety of HDL. PC molecules are the "backbone" of HDL precursors, either synthesized by cells or generated from the surface of lipolyzed TG-rich lipoproteins. Some PC molecules are released from chylomicrons upon their entrance to the bloodstream and some may originate from cell membranes. Once in HDL, PC molecules exchange with the same molecules in other lipoproteins (VLDL, LDL) and cell membranes, a reaction that, at least in part, depends on the presence of plasma lipid exchange proteins (6). The major route of HDL-PC metabolism is, however, transferase and hydrolase activities: LCAT, lipoprotein lipase and the hepatic lipase use PC as a substrate. Thus, there is a continuous flux of PC molecules in HDL: entrance and consumption, and the lifetime of PC in HDL is less than 1 day compared with 5 days for apoproteins (1). A similar situation exists for FC molecules. FC enters the HDL system with nascent particles, is generated by the lipolysis process and is transferred from other lipoproteins and cell membranes. FC in HDL exchanges with FC in other lipoproteins and cell membranes, and is consumed by the LCAT reaction (1). The lifetime of FC in HDL is no longer than a few hours. A not dissimilar situation exists for apoproteins when transfer and exchange reactions cause profound alterations of the apoprotein profile of the lipoprotein (1). Dissociation of the metabolic fate of HDL lipids and HDL apoproteins clearly exists (7).

Of particular interest is the reaction that causes neutral (CE and TG) lipid exchange and is mediated by the specific lipid transfer protein (LTP) present in human plasma (8). LTP catalyzes both exchange and transfer reactions. In the first reaction, there is no change of the CE content of lipoprotein. The second, however, causes loss of CE molecules that in part or on the whole are replaced by TG. Acquired TG molecules then can be hydrolyzed by plasma lipases (lipoprotein lipase and hepatic lipase) resulting in net loss of core-lipid molecules from HDL. Through the combined action of lipid transfer and TG hydroylsis, HDL_2 can lose 70-80% of its core-lipid molecules and a "reverse conversion" of $HDL_2 \rightarrow HDL_3$ occurs (9). This cycle provides another link between the HDL system and the VLDL $\rightarrow$ IDL $\rightarrow$ LDL delipidation cascade: supply of CE molecules to lower density lipoproteins.

The dynamic nature of HDL is best described as equilibrium between conversion and reverse-conversion processes. Supply of free cholesterol and phospholipids (from lipolysis and cell membranes) followed by cholesterol esterification is responsible for conversion of small to larger particles, while CE-TG exchange, together with hydrolysis of transferred TG, causes reverse-conversion of HDL populations. Recently, we took advantage of the absence of lipid transfer activity in the rat to study such processes in vivo (10). Rapid conversion of human HDL$_3$ to HDL$_2$ and then to apo E-rich HDL$_1$ is evident in intact rats, while delay of the conversion process is found when the animals are injected with human lipid transfer proteins. An important feature of the experiment is the association of the injected human HDL$_3$-CE with apo E (in HDL$_1$) of rat origin. This finding indicates that the HDL conversion and reverse-conversion cycle must involve alteration of the surface coat of the lipoprotein, including major apoprotein exchange. The source of apo E is, undoubtedly, lipolyzed TG-rich lipoproteins. It thus appears that when HDL acquires CE molecules and becomes larger, apo E molecules displace apo A-I from the outer coat of the lipoprotein. Whether reverse conversion is associated with back-transfer of apo A-I to HDL and release of the apo E molecules, is unknown.

Figure 1 demonstrates the central position of HDL in plasma fat transport processes. The figure emphasizes the dynamic behaviour of individual HDL lipid and protein components. Each component forms its own "mini-cycle" that is coordinated with the other cycles, and together determine HDL levels and HDL subpopulation distribution. As is evident, almost all plasma events that regulate fat transport affect HDL. Triglyceride transport in chylomicrons supplies HDL with A apoproteins, phospholipids and free cholesterol, while VLDL provides apo C, apo E and very considerable amounts of surface lipids. Reciprocity is achieved by cholesterol esterification in HDL and CE transfer to lower density lipoproteins. When all systems operate at high efficiency,

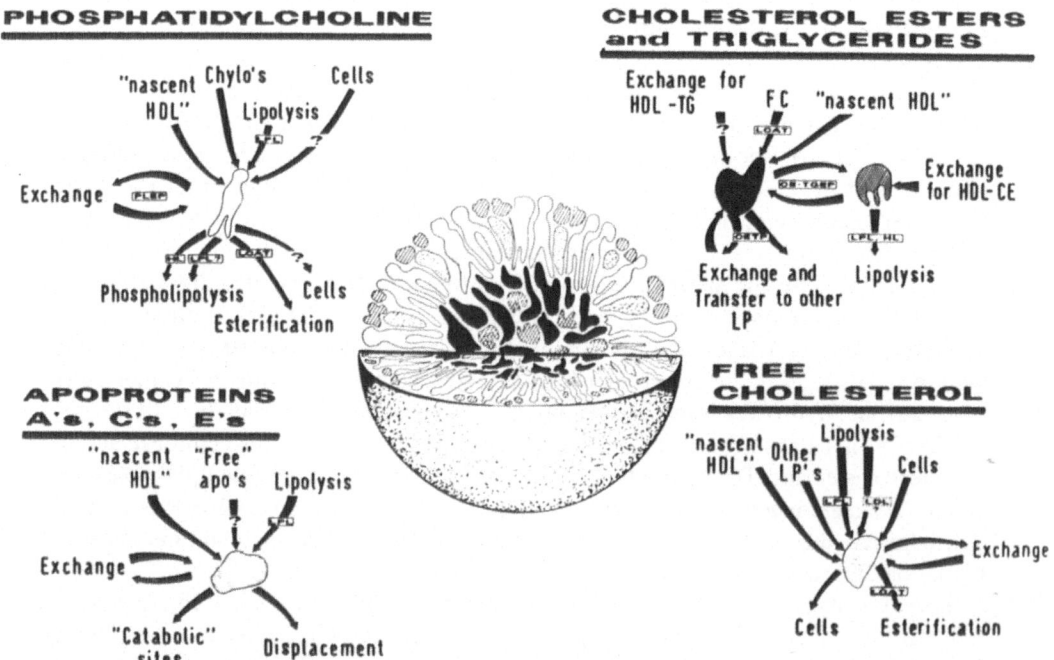

Fig. 1. The central position of HDL in plasma fat transport systems. (Reproduced with permission from The Journal of Lipid Research 1984. J Lipid Res 25:1017)

HDL levels are expected to be high, HDL_2 (and HDL_1) will be formed and their steady state levels also will be high. It is therefore suggested that high HDL levels reflect a basically "healthy" plasma fat transport system that does not cause atheroma formation.

References

1. Eisenberg S (1984) High density lipoprotein metabolism (JLR review). J Lipid Res 25: 1017–1058
2. Eisenberg S, Bilheimer DW, Lindgren FT, Levy RI (1973) On the metabolic conversion of human plasma very low density lipoprotein to low density lipoprotein. Biochim Biophys Acta 326:361–377
3. Chajek T, Eisenberg S (1978) Very low density lipoprotein. Metabolism of phospholipids, cholesterol, and apolipoprotein C in the isolated perfused rat heart. J Clin Invest 61:1654–1665
4. Patsch JR, Gotto AM, Olivecrona T, Eisenberg S (1978) Formation of high density lipoprotein$_2$-like particles during lipolysis of very low density lipoprotein in vitro. Proc Natl Acad Sci USA 75:4519–4523
5. Eisenberg S (1980) Plasma lipoproteins interconversion. Ann NY Acad Sci 248:30–47
6. Eisenberg S (1978) Effect of temperature and plasma on the exchange of apolipoproteins and phospholipids between rat plasma very low and high density lipoproteins. J Lipid Res 19:229–236
7. Glass C, Pittman RC, Weinstein DB, Steinberg D (1983) Dissociation of tissue uptake of cholesterol ester from that of apoprotein A-I of rat plasma high density lipoprotein: selective delivery of cholesterol ester to liver, adrenal and gonad. Proc Natl Scad Si USA 80:5435–5439
8. Eisenberg S, Deckelbaum R (1986) Intravascular lipoprotein remodelling: neutral lipid transfer proteins. Clin Biochem (in press)
9. Deckelbaum JR, Eisenberg S, Oschry Y, Granot E, Sharon I, Bengtsson-Olivecrona G (1986) Modeling of human plasma high density lipoproteins: roles of neutral lipid exchange and triglyceride lipases. J Biol Chem 261:5201–5208
10. Gavish D, Oschry Y, Eisenberg S (1987) In vivo conversion of human HDL_3 to HDL_3 and apo E-rich HDL_1 in the rat: effects of lipid transfer proteins. J Lipid Res 28:257–267

The Structure of ApoB-100: Structure-Function Studies

A. M. Gotto, W. A. Bradley, L. Chan, S. H. Gianturco, H. J. Pownall, J. T. Sparrow, and C. Y. Yang[1]

The plasma apolipoproteins constitute a unique group of molecules which are now known to have multifunction relationships with plasma lipid transport. It is the purpose of this review to cover more recent aspects of research in which the structure of apolipoprotein B is related to its function.

The functions of apolipoproteins include the binding, transport, and solubilization of lipids in stable macromolecular emulsions called the plasma lipoproteins; the transport of lipids from cells into the circulation and lymph; the recognition of receptors on the surface of cells to facilitate the cellular uptake and removal of plasma lipoproteins and possibly lipid as well; and the activation of specific enzymes involved in regulating lipoprotein metabolism.

All of the apoproteins are involved in lipid binding and transport. ApoB-100 and apoE are ligands for recognition by lipoprotein receptors. The LDL receptor recognizes apoB-100 or apoE, while apoE appears to be recognized by the hepatic receptor for chylomicron remnants. ApoC-II is an obligatory activator of the enzyme lipoprotein lipase, while apoA-II is believed to be the activator of hepatic lipase. Lecithin cholesterol acyl transferase (LCAT) can be activated by apoA-I, apoC-I, and probably apoA-IV. ApoD has been implicated in some studies in cholesteryl ester exchange activity.

The basic structural determinant for lipid binding by most of the apolipoproteins is an amphipathic helix, first described by Segrest et al. (1973). The amphipathic helix appears to have been preserved in evolution through a common block of 33 amino acids which consist of 3 repeats of 11 residues, each repeat starting with a hydrophobic residue. ApoA-I, an activator of LCAT, may be viewed as a prototype of this structure, according to Breslow (1986). ApoA-I has six such repeat units.

The amphipathic content or amphipilicity of a segment of apolipoprotein may be expressed on the basis of a parameter known as the hydrophobic moment, as described by Eisenberg et al. (1982). Pownall et al. (1983) have applied the helical amphiphatic moment to make predictions about the secondary structures of the apolipoproteins and to classify proteins. Integral membrane proteins fall into one group, for example, those spanning a membrane; they have a high degree of hydrophobicity, but a relatively low helical amphipathic moment. The surface-seeking peptides, of which the plasma

---

1 Baylor College of Medicine, The Methodist Hospital, Houston, TX 77030, USA

Drugs Affecting Lipid Metabolism
Ed. by R. Paoletti et al.
© Springer-Verlag Berlin Heidelberg 1987

apolipoproteins are examples, have higher values for the helical amphipathic moment, but are less hydrophibic.

The amphipathic helix is located at the surface between an aqueous exterior and a hydrophobic interior. It is likely that polar residues on the outer, hydrophilic surface cause the plasma apolipoproteins to be more closely related to surface-associating proteins rather than to integral membrane proteins.

Ponsin et al. (1984) have separated the amphipathic content from the hydrophobic content of peptides through the use of the acylation of the amino terminus. As hydrophobicity is increased, the potency as an LCAT activator increases as well. Furthermore, Ponsin et al. (1986) established in studies with the rat, a correlation between the degree of hydrophobicity of the peptide and its metabolic fate. Thus, a 15-amino acid peptide with no acyl group attached or with only a 4-carbon acyl group attached, was metabolized primarily as a peptide and cleared by the kidney. The residence time of the peptide increased progressively as did the length of the fatty acyl group. The most hydrophobic acyl peptide studied had residence times very similar to those of HDL; the liver and the adrenals where the primary sites of metabolism.

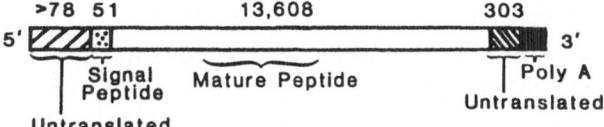

Fig. 1. The structure of human apoB-100 mRNA

ApoB-48 is made in the intestine; apoB-74 and apoB-26 are thought to be fragments of apoB-100; ApoB-100 is the largest of the apoB proteins, and is made in the liver. The complete amino acid sequence of apoB-100 cDNA has now been determined by Chan et al. (1986). The structure of human apoB-100 mRNA is shown in Fig. 1. The length of the cDNA is 14.1 kb and codes for 4563 amino acids. This includes a signal peptide of 27 amino acids. A schematic representation of the regions of the cDNA is shown in Fig. 2. The mature protein contains 4536 residues of which 2336 residues were determined directly by sequencing of apoB tryptic peptides. The release of

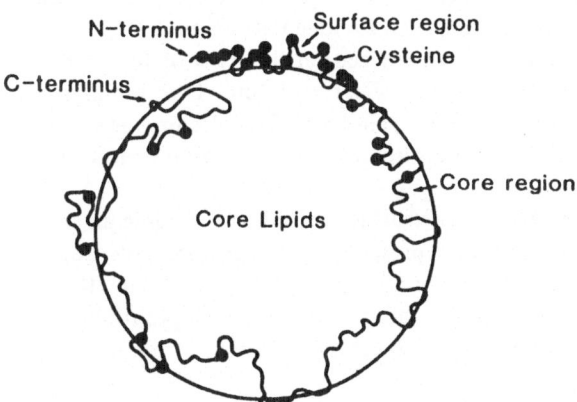

Fig. 2. A model of apoB configuration on LDL

specific peptides of LDL by treatment with trypsin make it possible to map domains of apoB designated as trypsin-accessible and trypsin-inaccessible. The mature peptide contained a high degree of hydrophobicity, 0.916 kcal per residue and 20% beta-pleated sheet structure; this had previously been predicted for apoB on the basis of analysis with infrared spectrocopy and circular dichroism. A dot-matrix analysis of the structure of apoB indicated the presence of long internal repeat units. ApoB, therefore, like other apolipoproteins, contains internal repeating units. However, in apoB, these units do not show a typical pattern.

Homology was examined between residues 11 and 43 of human apoA-I with apoB-100. Homology was found with six segments of apoB having degrees of homology from 24 to 30%. It is not possible at this time to determine whether these segments are actually homologous with apoA-I. As with all the other apolipoproteins, apoB-100 had segments which could be divided into common blocks of 3 repeats of 11 residues each starting with a hydrophobic residue. These were segments 581-613 and 1318-1350.

The cystine residues were not evenly distributed. For example, 12 of the 25 cystine residues are in the amino terminal-500 residues. Seven of the cystines were in very close proximity and were ten residues or less apart from the next cystine. These residues are most likely involved in maintaining the structure of apoB.

Knott et al. (1985) identified a partial cDNA sequence of apoB-100 with homology to residues 140-150 of apoE, the putative ligand that is recognized by the apoB/E receptor. Negatively charged groups bind to one or more postively charged domains within apoB-100 or apoE. We prepared a chemically synthetic peptide which corresponds to residues 3345-3381 of apoB-100. Gianturco et al. (1983) have previously shown that hypertriglyceridemic VLDL binds to the LDL receptor via apoE on the surface, and that this ability to bind to the receptors is destroyed by treating with trypsin. Thus, trypsin-treated hypertriglyceridemic particles are unable to bind to the LDL receptor. Adding exogenous apoE restores the ability to bind to the receptor. The peptide 3345-3381 bound to the hypertriglyceridemic VLDL and restored the ability to bind to the LDL receptor. Another peptide, 4154-4189, which bound lipids, but which does not have homology to residues 140-150 of apoE, was unable to bind to the LDL receptor. Thus, synthetic peptide apoB 3345-3381 contains a domain with the ability to bind to the LDL receptor on cultured human fibroblasts. This peptide contains a cluster of positively charged amino acids. The existence of other receptor-binding domains of apoB-100 has not been excluded at this time.

Various other regions of apoB-100 were synthesized on the basis of the predicted amphipathic helix or hydrophobic beta-pleated sheet structure. Our findings suggested that some peptides bound to phospholipid, cholesterol, cholesteryl ester with a beta-pleated sheet configuration, while the others bound to lipid via amphipathic alpha-helical structure.

The model for LDL which we propose shows surface peptides containing an amphipathic helix structure which we postulate to represent the trypsin-accessible domains of the protein. Other parts of the protein-containing beta-pleated sheet structure are shown to be buried within the lipid, unlike all of the other apolipoproteins, and these we postulate would represent the trypsin-inaccessible regions.

The knowledge of the structure of the apoB-100 can be used for further studies to determine the structure/function relationship between lipid binding, transport, interaction with receptors, and epitope expression. Finally, it is anticipated that these structure/function analyses will make it possible to elucidate the role of apoB-100 in atherosclerosis.

References

Breslow JL (1986) Lipoprotein genetics and molecular biology. In: Gotto AM, Jr (ed) New comprehensive biochemistry. Elsevier/North Holland, Biomedical Press, Amsterdam (in press)

Chen SH, Yang CY, Chen PF, Setzer D, Tanimura M, Li WH, Gotto AM, Jr, Chan L (1986) The complete cDNA and amino acid sequence of human apolipoprotein B-100. J Biol Chem 261: 12918–12921

Eisenberg D, Weiss R, Terwilliger TC (1982) The helical hydrophobic moment: a measure of the amphiphilicity of a helix. Nature (London) 299:371–374

Gianturco SM, Gotto AM, Jr, Hwang SC, Karlin JB, Lin AH, Prased SC, Bradley WA (1983) Apolipoprotein E mediates uptake of S_f 100-400 hypertriglyceridemic very low density lipoprotein receptor pathway in normal human fibroblasts. J Biol Chem 258:4526–4533

Knott TJ, Rall SC, Jr, Inneraritiy TL (1985) Human apolipoprotein B: structure of carboxyl-terminal domains, sites of gene expression, and chromosomal localization. Science 230:37–43

Ponsin G, Strong K, Gotto AM, Jr, Sparrow JT, Pownall HJ (1984) In vitro binding of synthetic acylated lipid-associating peptides to high-density lipoproteins: effect of hydrophobicity. Biochemistry 23:5337–5342

Ponsin G, Sparrow JT, Gotto AM, Jr, Pownall HJ (1986) In vivo interaction of synthetic acylated apopeptides with high density lipoproteins in rat. J Clin Invest 77:559–567

Pownall HJ, Knapp RD, Gotto AM, Jr, Massay JB (1983) Helical amphipathic moment: application to plasma lipoproteins. FEBS Lett 159: 17–23

Segrest JP, Jackson RL, Morrisett JD, Gotto AM, Jr (1973) A molecular theory of lipid-protein interactions in the plasma lipoproteins. FEBS Lett 38:247–253

A Genetic Marker in the Apolipoprotein AI/CIII Gene Complex Associated with Hypercholesterolaemia

C. C. Shoulders[1], M. J. Ball[2], J. I. Mann[2], and F. E. Baralle[3]

1 Introduction

In 1983 we reported a strong association between hypertriglyceridaemia and a restriction fragment length polymorphism (RFLP) associated with the apolipoprotein AI gene which was not present in normolipidaemic individuals (Rees et al. 1983). This RFLP arises because of the existence of a polymorphic nucleotide in the 3' non-coding region of the linked apo-CIII gene that creates an additional cleavage site for the restriction enzyme SacI (Shoulders and Baralle 1986). Since our original observation, two groups have reported a significant increase in the incidence of the variant apo-AI/CIII allele (S2) in patients with severe coronary heart disease (CHD), and in survivors of myocardial infarction compard with healthy controls (Ferns et al. 1985; Rees et al. 1985). However, this increased frequency could not entirely be explained by a greater number of hypertriglyceridaemic individuals in these groups. In view of these observations and the fact that Kessling and Humphries (1985) could not confirm our initial findings (Rees et al. 1983), we have tried to clarify the relationship between the S2 allele and hyperlipidaemia by studying newly recruited patients with well-definied forms of hyperlipidaemia.

2 Patients, Methods and Results

The frequency of the S2 allele was determined in 58 unrelated adults, who had primary hypercholesterolaemia (plasma cholesterol >7.6 mmol l^{-1}, LDL cholesterol >5 mmol l^{-1} and fasting triglycerides <2.0 mmol l^{-1}) and 26 patients with primary hypertriglyceridaemia (LDL cholesterol <4.7 mmol l^{-1}, fasting triglycerides >3.3 mmol l^{-1}). Thirty-two healthy normolipidaemic volunteers (plasma cholesterol <6 mmol l^{-1}, triglycerides <1.6 mmol l^{-1}) with no family history of premature CHD were used as controls. All patients were assessed in the Lipid Clinic at the John Radcliffe

---

1   Sir William Dunn School of Pathology, University of Oxford, OX1 3 RE, Great Britain
2   Lipid Clinic, John Radcliffe Hospital, Oxford, Great Britain
3   Instituto Sieroterapico, Milan, Italy

Drugs Affecting Lipid Metabolism
Ed. by R. Paoletti et al.
© Springer-Verlag Berlin Heidelberg 1987

Table 1. Frequency of the S2 allele in patients with hyperlipidaemia[a]

Patients[b]	No.	Plasma [cholesterol] (mmol l^{-1})	Plasma [triglycerides] (mmol l^{-1})	No. with S2 allele	Frequency of S2 allele (%)
Normolipidaemic[b]	31	3.7- 6.0	<1.6	1	3.2
Hypercholesterolaemic, with xanthomas	29	7.6-10.5	<2.0	2	6.9
Hypercholesterolaemic, no xanthomas	29	7.6- 9.0	<2.0	9	31.0[c]
Hypertriglyceridaemic	26	4.0- 7.0	3.3-8.0	12	46.2[c]

[a] Genotyping carried out as previously described.
[b] All participating individuals were of English descent.
[c] Difference from normolipidaemic controls statistically significant, $p < 0.005$. chi square test.

Hospital, and those who were hypercholesterolaemic were divided into two groups according to the presence or absence of tendon xanthomas. Twenty-nine of the 58 hypercholesterolaemic patients had tendon xanthomas and were regarded as' having a definite diagnosis of familial hypercholesterolaemia. The genotype analysis were performed on DNA prepared from venous blood. The DNA was digested with the restriction enzyme SacI, fractionated by gel electrophoresis and blotted onto nitro-cellulose filters. These filters were subsequently hybridised to radiolabelled apo-AI/CIII gene-specific probes (Shoulders and Baralle 1986).

The results show a strikingly increased frequency of the apo-AI/CIII S2 allele in patients with primary hypertriglyceridaemia. The frequency of the allele is also high in patients with hypercholesterolaemia who do not have tendon xanthomas. On the other hand, the difference between controls and patients with tendon xanthomas and definite familial hypercholesterolaemia is not statistically significant.

The results of our study are shown in Table 1. They confirm our previous findings of a significant association between the S2 allele and hypertriglyceridaemia in an English population. More importantly, it shows for the first time that there is a highly statistically significant association between this genetic marker and primary hyper-cholesterolaemia in patients without tendeon xanthomas. This association may be of greater clinical relevance and may help to explain the reported increased incidence of the apo-AI/CIII S2 allele in patients with CHD. The mechanism by which the as yet uncharacterised genetic defect linked to this allele prediposes to the different types of hyperlipidaemia remains to be explained.

Acknowledgement. This work was supported by the British Heart Foundation.

References

Fern GAA, Stocks J, Ritchie C, Galton DJ (1985) Genetic polymorphisms of apolipoprotein CIII and insulin in survivors of myocardial infarction. Lancet 1985 (ii): 300–303

Kessling AM, Humphries SE (1985) Interpretation of presence of S2 allele. Lancet (letter) 1985; (ii): 510–511

Rees A, Shoulders CC, Stocks J, Galton DJ, Baralle FE (1983) DNA polymorphism adjacent to the human apo AI gene: relation to hypertriglyceridaemia. Lancet 1983; (i): 444–446

Rees A, Stocks J, Williams LG, Caplin JL, Jowett NI, Camm AJ, Galton DJ (1985) DNA polymorphisms in the apolipoprotein CIII and insulin genes. Altherosclerosis 58:269–275

Shoulders CC, Baralle FE (1986) Genetic polymorphisms in the apo AI/CIII complex. Methods Enzymol 128, pt A: 727–745

Cellular and Molecular Biology of Apolipoproteins: Receptor-Mediated Regulation of Cholesterol Metabolism

R. W. MAHLEY [1]

1 Introduction

Lipoprotein receptors, especially those within the liver, play a central role in the regulation of cholesterol metabolism. Specific apolipoproteins (apo-) serve as the ligands for these receptors (for review, see Brown and Goldstein 1983, 1986; Mahley and Innerarity 1983; Mahley et al. 1984).

2 Lipoprotein Receptors

The liver possesses at least two distinct lipoprotein receptors: the apo-B,E(LDL) receptor, which is similar, if not identical, to the low-density lipoprotein (LDL) receptor of extrahepatic cells, and the apo-E receptor (Hui et al. 1981, 1986a; Mahley et al. 1981). The hepatic apo-B,E(LDL) receptor interacts with both apo-B- and apo-E-containing lipoproteins. The apo-E receptor is distinct from the apo-B,E(LDL) receptor in that it interacts with apo-E-containing lipoproteins, but not with apo-B-containing LDL. We have postulated that the apo-E receptor may function specifically in the uptake of chylomicron remnants and may represent the chylomicron remnant receptor.

Recently, the canine and human liver apo-E receptor has been isolated and purified (Hui et al. 1986a) and, using recombinant DNA techniques, partially sequenced (Hui et al. 1986b). The purified apo-E receptor is identified as one distinct protein with an M_r = 56,000 on sodium dodecyl sulfate-polyacrylamide gels. This purified M_r = 56,000 apo-E receptor displayed specific binding activity for apo-E-containing lipoproteins and did not bind LDL (Hui et al. 1986a). It is a protein distinct from the apo-B,E(LDL) receptor.

---

1 The Gladstone Foundation Laboratories for Cardiovascular Disease, Cardiovascular Research Institute, Departments of Pathology and Medicine, University of California, San Francisco, CA 94140, USA

Drugs Affecting Lipid Metabolism
Ed. by Paoletti et al.
© Springer-Verlag Berlin Heidelberg 1987

3 Characterization of the Receptor-Binding Domains of Apolipoproteins E and B

Over the past several years we have studied in detail the structure of the ligands for the lipoprotein receptors in an attempt to define the receptor-binding domains of apo-E and apo-B (for review, see Mahley and Innerarity 1983; Mahley et al. 1984, 1985). The receptor-binding domain of apo-E has been elucidated using three different approaches. The first was a genetic approach that took advantage of identifying naturally occurring mutants of apo-E that disrupt receptor-binding activity. Single amino acid substitutions near the middle of the apo-E molecule prevent normal binding and are associated with the genetic disorders type III hyperlipoproteinemia (Mahley and Innerarity 1983; Mahley et al. 1984, 1985). The second was a biochemical approach in which the binding activity of apo-E fragments was examined (Innerarity et al. 1983). The third, an immunological approach, determined the epitope of apo-E antibodies that inhibit receptor binding (Weisgraber et al. 1983).

A fourth approach, site-directed mutagenesis, is now being used. Apolipoprotein E produced in transformed *Escherichia coli* has the full biological activity of authentic apo-E in plasma (Vogel et al. 1985). With this technology, it is now possible to systematically change specific amino acids throughout the protein to define more precisely the receptor-binding domain of apo-E.

These data have established that the region of apo-E reponsible for mediating its binding to the receptor is in the vicinity of residues 140 to 150. Key arginine and lysine residues appear to interact directly with the receptor (Innerarity et al. 1983, 1984; Mahley and Innerarity 1983; Weisgraber et al. 1983) (Fig. 1). Also shown in Fig. 1 is the repeating sequence within the apo-B,E(LDL) receptor that has been postulated to represent its ligand-binding domain (Brown and Goldstein 1986). An

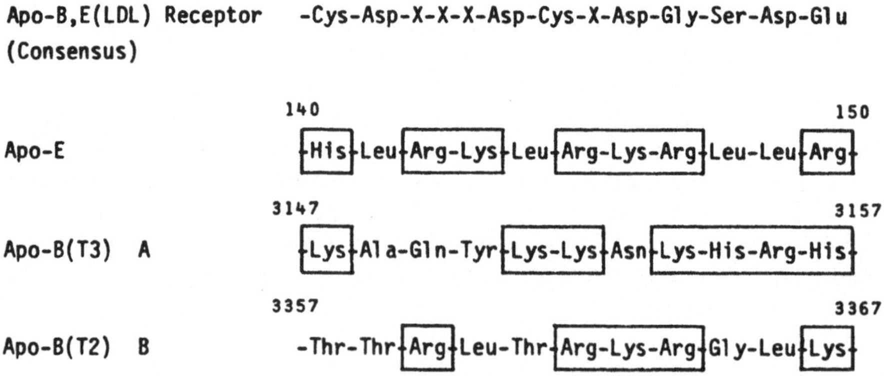

Apo-B,E(LDL) Receptor   -Cys-Asp-X-X-X-Asp-Cys-X-Asp-Gly-Ser-Asp-Glu
(Consensus)

Apo-E   140 ... 150
-His-Leu-Arg-Lys-Leu-Arg-Lys-Arg-Leu-Leu-Arg-

Apo-B(T3)   A   3147 ... 3157
-Lys-Ala-Gln-Tyr-Lys-Lys-Asn-Lys-His-Arg-His-

Apo-B(T2)   B   3357 ... 3367
-Thr-Thr-Arg-Leu-Thr-Arg-Lys-Arg-Gly-Leu-Lys-

Fig. 1. Receptor-binding sequences. Basic amino acid residues of the ligands that may be involved in mediating binding to the receptor are within the *boxes*. An ionic interaction between the basic residues of apo-E and apo-B and the acidic residues (aspartate and glutamate) of the receptor has been postulated. [Reproduced with permission from Cardiovascular Disease: Molecular and Cellular Mechanisms, Prevention, Treatment (L. Gallo, ed.), Plenum Publishing, New York, 1987]

ionic interaction between the basic residues of apo-E and the acidic residues (aspartate and glutamate) of the receptor has been hypothesized.

A natural extension of our studies defining the receptor-binding domain of apo-E was to attempt to identify the receptor-binding domain of apo-B (Knott et al. 1985, 1986; Blackhart et al. 1986). Several lines of evidence indicate that two regions enriched in the basic amino acids arginine and lysine may form the receptor-binding domain of apo-E. These two basic regions are located on either side of a thrombin cleavage site that separates the T2 (carboxyl-terminal peptide of apo-B100) from the T3 peptide (the midportion of apo-B100). These regions may lie in close proximity within the primary structure of apo-B100 and contribute to the formation of a single receptor-binding domain. The amino acid sequences of these two regions are shown in Fig. 1.

4 Conclusion

The postulated receptor-binding domain of apo-B bears a striking homology with the receptor-binding domain of apo-E (Fig. 1). It is postulated that the basic amino acid residues (arginine and lysine) of apo-E and apo-B interact ionically with the acidic residues (glutamate and aspartate) of the apo-B,E(LDL) and apo-E receptors. As more is learned about the structure of apo-E and apo-B, we will be better able to understand receptor-ligand interactions. Such studies will continue to help elucidate the mechanisms whereby the different apolipoproteins participate in cholesterol homeostasis and the role of the various receptors in the overall regulation of cholesterol metabolism.

References

Blackhart BD, Ludwig EH, Pierotti VR, Caiati L, Onasch MA, Wallis SC, Powell L, Pease R, Knott TJ, Chu ML, Mahley RW, Scott J, McCarthy BJ, Levy-Wilson B (1986) Structure of the human apolipoprotein B gene. J Biol Chem 261:15364–15367

Brown MS, Goldstein JL (1983) Lipoprotein receptors in the liver. Control signals for plasma cholesterol traffic. J Clin Invest 72:743–747

Brown MS, Goldstein JL (1986) A receptor-mediated pathway for cholesterol homeostasis. Science 232:34–47

Hui DY, Innerarity TL, Mahley RW (1981) Lipoprotein binding to canine hepatic membranes. Metabolically distinct apo-E and apo-B,E receptors. J Biol Chem 256:5646–5655

Hui DY, Brecht WJ, Hall EA, Friedman G, Innerarity TL, Mahley RW (1986a) Isolation and characterization of the apolipoprotein E receptor from canine and human liver. J Biol Chem 261:4256–4267

Hui DY, Hall EA, Brecht WJ, Innerarity TL, Mahley RW (1986b) Molecular cloning and identification of the ligand binding domain of the human apolipoprotein E receptor. Arteriosclerosis 6:551a

Innerarity TL, Friedlander EJ, Rall SC, Jr, Weisgraber KH, Mahley RW (1983) The receptor binding domain of human apolipoprotein E: binding of apolipoprotein E fragments. J Biol Chem 258:12341–12347

Innerarity TL, Weisgraber KH, Arnold KS, Rall SC, Jr, Mahley RW (1984) Normalization of receptor binding of apolipoprotein E2. Evidence for modulation of the binding site conformation. J Biol Chem 259:7261–7267

Knott TJ, Rall SC, Jr, Innerarity TL, Jacobson SF, Urdea MS, Levy-Wilson B, Powell LM, Pease RJ, Eddy R, Nakai H, Byers M, Priestley LM, Robertson E, Rall LB, Betsholtz C, Shows TB, Mahley RW, Scott J (1985) Human apolipoprotein B: structure of carboxyl-terminal domains, sites of gene expression, and chromosomal localization. Science 230:37–43

Knott TJ, Pease RJ, Powell LM, Wallis SC, Rall SC, Jr, Innerarity TL, Blackhart B, Taylor WR, Lusis AJ, McCarthy BJ, Mahley RW, Levy-Wilson B, Scott J (1986) Human apolipoprotein B: complete cDNA sequence and identification of structural domains of the protein. Nature (London) 323:734–738

Mahley RW, Innerarity TL (1983) Lipoprotein receptors and cholesterol homeostasis. Biochim Biophys Acta 737:197–222

Mahley RW, Hui DY, Innerarity TL, Weisgraber KH (1981) Two independent lipoprotein receptors on hepatic membranes of the dog, swine, and man. Apo-B,E and apo-E receptors. J Clin Invest 68:1197–1206

Mahley RW, Innerarity TL, Rall SC, Jr, Weisgraber KH (1984) Plasma lipoproteins: apolipoprotein structure and function. J Lipid Res 25:1277–1294

Mahley RW, Innerarity TL, Rall SC, Jr, Weisgraber KH (1985) Lipoproteins of special significance in atherosclerosis: insights provided by studies of type III hyperlipoproteinemia. Ann NY Acad Sci 454:209–221

Vogel T, Weisgraber KH, Zeevi MI, Ben-Artzi H, Levanon AZ, Rall SC, Jr, Innerarity TL, Hui DY, Taylor JM, Kanner D, Yavin Z, Amit B, Aviv H, Gorecki M, Mahley RW (1985) Human apolipoprotein E expression in *Escherichia coli:* structural and functional identity of the bacterially produced protein with plasma apolipoprotein E. Proc Natl Acad Sci USA 82:8696–8700

Weisgraber KH, Innerarity TL, Harder KJ, Mahley RW, Milne RW, Marcel YL, Sparrow JT (1983) The receptor binding domain of human apolipoprotein E: monoclonal antibody inhibition of binding. J Biol Chem 258:12348–12354

APOA-I and APOA-II Metabolism and Coronary Artery Disease

G. Ghiselli, J. Heibig, F. Turturro, A. M. Gotto, Jr., and E. Wittels[1]

1 Introduction

A number of epidemiological studies (1—4) indicate that there is a negative association between the level of high density lipoprotein in plasma and the prevalence of coronary artery disease (CAD) in populations. In addition, the level of apoA-I, the major HDL apolipoprotein, has been found to be significantly decreased in different group populations including survivors of myocardial infarction, subjects with angina pectoris, and others with coronary artery disease as documented by angiography (1—4). In one study, apoA-I was found to be a superior predictor of CAD in individual patients compared to concentrations of HDL cholesterol (4,5)

The turnover of two apoA-I variants, apoA-I Tangier and apoA-I Milano, has been recently investigated (6—8). In both cases, the variants have faster catabolism in plasma than normal apoA-I. The Tangier and Milano apoA-I diseases are both characterized by dramatically low levels of HDL in plasma. These results support the idea that the catabolism of HDL is dependent upon the metabolic behavior of apoA-I. The relationship between the metabolism of apoA-I and the level of serum HDL has also been shown during conditions of physiological variation (9,10), or drug and dietary treatment (11—15). Such data as are available indicate that alterations in the metabolism of apoA-I frequently are associated with alterations in that of apoA-II (6,7).

Although there has been wide speculation on the mechanisms by which apoA-I and HDL protect against premature atherosclerosis, no data are available on the metabolic basis of lowered apoA-I levels in subjects with CAD. HDL and HDL apolipoprotein concentrations in plasma are decreased in hyperlipoproteinemic states, yet the majority of subjects with CAD is normolipidemic. In this study, we have then addressed the question of whether the hypoalfalipoproteinemia in normolipidemic CAD patients may be explained on the basis of a metabolic defect of apoA-I and apoA-II in plasma.

1 Baylor College of Medicine and the Methodist Hospital Department of Medicine, Houston, TX 77030, USA

Drugs Affecting Lipid Metabolism
Ed. by. R. Paoletti et al.
© Springer-Verlag Berlin Heidelberg 1987

2 Experimental Design

Male Caucasian subjects in their 40's and 60's were recruited among those undergoing coronary angiography for chest pain or suspected coronary artery disease. The angiography was performed at the Veterans Administration Hospital in Houston. Complete clinical history was obtained for each subject. The angiographic results were read by two independent observers without prior knowledge of lipid or lipoprotein plasma levels. Coronary angiography was performed by the Judkins technique in which multiple views of the right and left coronary arteries are recorded (16). Coronary artery disease was regarded as clinically important when a stenosis greater than 50% was observed in the left anterior descending coronary artery, the left circumflex artery, or the right coronary artery. Results were tabulated as described by Gensini on the basis of the percentage of stenosis in each of these vessel segments (17). Patients were classified according to the angiographic results and the severity of the coronary obstruction into four different groups with clinically important coronary obstruction observed at none, one, two, or all three of the major coronary arteries. Lipid, lipoprotein lipid, and apolipoprotein levels were determined on samples of plasma collected in the fasting state at the same time the angiography was performed. The normolipidemic subjects (with triglycerides less than 200 mg dl^{-1} and cholesterol less than 250 mg dl^{-1}) represented 92% of the total population recruited.

In all the groups lipid, lipoprotein, and apolipoprotein values were normally distributed. The three groups with increasing severity of coronary obstruction had lower HDL cholesterol, apoA-I, and apoA-II levels in plasma, as compared to subjects without significant CAD (see Table 1). When variables such as age, diabetes, smoking and drinking habits, and body mass were considered, no simple correlation could be found between any one of these variables taken individually and the presence (or absence) and severity of the coronary obstruction. Drug consumption taken as a variable did not appear to be a determinant for lipid, lipoprotein, and apolipoprotein

Table 1. Lipid and Apolipoprotein concentrations (mg dl^{-1})

Subjects	No.	Age	Chol	TG	HDL	A-I	A-II
Normals	44	52 ± 2[a]	188 ± 7	107 ± 10	38 ± 2	107 ± 3	31 ± 1
Disease							
1 vs	35	55 ± 2	191 ± 6	130 ± 9	31 ± 1[b]	96 ± 4	25 ± 1[b]
2 vs	47	56 ± 1	190 ± 5	133 ± 10	33 ± 1[b]	85 ± 3[b]	26 ± 1[b]
3 vs	66	59 ± 1[b]	197 ± 6	133 ± 9	31 ± 1[b]	87 ± 3[b]	27 ± 1[b]
Mean	148	57 ± 1	192 ± 2	132 ± 4	31 ± 1[b]	89 ± 1[b]	26 ± 1[b]

[a] Mean ± SEM.
[b] $p < 0.01$ vs normals.
HDL cholesterol concentration was determined following the LRC guidelines. ApoA-I and apoA-II concentrations were determined by the means of specific radioimmunoassays.

Table 2. Lipids and apolipoprotein concentrations (mg dl^{-1}) and apolipoprotein fractional catabolic rates (day^{-1}) in plasma

Subjects	Chol	TG	HDL	A-I	A-II	A-I FCR	A-II FCR
Normals	195 ± 15[a]	106 ± 11	52 ± 5	105 ± 5	28 ± 1	0.21 ± 0.01	0.19 ± 0.01
CAD	193 ± 25	151 ± 14	33 ± 5[b]	86 ± 2[b]	23 ± 2[b]	0.27 ± 0.02	0.24 ± 0.02[b]
Hyper-Tg	313 ± 46	848 ± 358	27 ± 3[b]	75 ± 2[b]	20 ± 2[b]	0.29 ± 0.02[b]	0.27 ± 0.02[b]

[a] Mean ± SEM (n = 6 for each group).
[b] $p < 0.01$ vs normals.

levels in the different groups. For the purpose of the turnover study, subjects with significant obstruction observed at one, two, or three of the coronary arteries were regarded as a single group since no differences in the lipid, lipoprotein, and apolipoprotein levels were observed among them.

ApoA-I and apoA-II turnover studies (18) were carried out in normolipidemic healthy subjects recruited from the local medical school which served as the control group, in six normolipidemic subjects with CAD randomly recruited from the subject population previously screened, and in a group of patients with low HDL cholesterol plasma concentration and hypertriglyceridemia. Two of the hypertriglyceridemic patients had CAD. The concentration of lipids and apolipoproteins of the subjects participating in the turnover studies are reported in Table 2. The residence time of radioiodinated-apoA-I was significantly decreased from 4.7 days in the controls to 3.7 days in the CAD group and 3.5 days in the hypertriglyceridemic group. The residence time for radioiodinated-apoA-II was also significantly reduced from 5.3 days in the control, to 4.3 days in the CAD group and to 3.8 days in the hypertriglyceridemic group. The fractional catabolic rate for apoA-I and apoA-II was then significantly higher both the CAD and in the hypertriglyceridemic groups as compared to the control; the values are also presented in Table 2. The degradation rate of apoA-I and apoA-II was similar in different groups of subjects ans averaged 7.7 (mg kg.^{1} day^{-1}), 8.0 and 6.7 for apoA-I, respectively, in the control, CAD and hypertriglyceridemic subjects groups. ApoA-II degration rate averaged 1.7 mg kg^{-1} day^{-1} in all three of the groups.

3 Discussion

A number of experimental results support the view that decreased levels of apoA-I in plasma are deleterious and accelerate the development of the atherosclerotic process (19,20). It is believed that the clearance of cholesterol from the arterial wall is promoted by HDL and apoA-I is a determinant of this process (20). Thus, the removal of cholesterol from the peripheral tissues including the arteries could be inefficient when the level of apoA-I in plasma is low. The aim of the proposed study has been to shed light on the metabolic processes leading to a decreased apoA-I plasma level in patients with CAD. In vivo turnover studies have been carried out.

Our results are consistent with the idea that low levels of HDL in the plasma of normolipidemic subjects with CAD are achieved through faster catabolism of apoA-I and apoA-II. The amount of apolipoprotein synthesized (or degraded) is not affected. The mechanisms of the accelerated catabolism can be only postulated at present. Perhaps, the threshold concentrations for HDL clearance in plasma have been readjusted to a lower level. The liver and the kidney appear to be the major sites for catabolism of apoA-I (21). ApoA-I catabolism in the liver may have some specificity and there are claims that it is receptor-mediated (22,23). On the other hand, in the kidney, clearance of apoA-I may involve a relatively nonspecific process of glumerular filtration and tubular secretion of that fraction of the apolipoprotein not bound to lipoproteins (21). In these organs, the catabolism of apoA-I and apoA-II may be dependent upon HDL composition, as this will affect the interaction of HDL with the putative hepatic receptor or the rate of apoA-I dissociation, and thus kidney clearance. There are reports that the distribution and chemical composition of the subfraction of HDL is abnormal in subjects with CAD (24,25).

It has been previously reported that akin to our CAD subjects the catabolism of apoA-I and apoA-II is accelerated in subjects with low HDL levels and hypertriglyceridemia (28–30). The data we have obtained in a similar group of subjects are in agreement with this conclusion. We speculate that normolipidemic CAD subjects and hypertriglyceridemic subjects may then share some common metabolic defect. It is well documented that the rate of conversion of VLDL to LDL is impaired in hypertriglyceridemia (30) due to a reduced lipoprotein-lipase activity. Lipoprotein-lipase is a major determinant of the HDL turnover rate and level in plasma (29–32). It has been postulated (30) that by acting on VLDL and making available surface lipid components to HDL, lipoprotein-lipase modifies the HDL subfraction distribution and composition. This alters the conformation of apoA-I and apoA-II at the surface of HDL particles, ultimately affecting their rate of catabolism. In fact, HDL composition is abnormal in hypertriglyceridemia. Similarly to the subjects with hypertriglyceridemia, VLDL catabolism may be impaired in CAD subjects, whether or not hypertriglyceridemia is expressed. This hypothesis is under current active investigaton.

References

1. Heiss G, Tyroler HA (1983) In: Proc Workshof on Apolipoprotein quantification 1983. NIH Publ No 83–1266:7–24
2. Thompson G (1984) Apoproteins: determinants of lipoprotein metabolism. Br Heart J 51:585–588
3. Blackburn H (1983) The meaning of a new marker for coronary artery disease. N Engl J Med 309:426–428
4. Maciejko JJ, Holmes DR, Kottke BA, Zinsmeister AR, Dinh DM, Mao SJT (1983) Apolipoprotein A-I as a marker of angiographically assessed coronary artery disease. N Engl J Med 309:385–389
5. Kukita H, Hiwada K, Kokubu T (1984) Serum apolipoprotein A-I, A-II and B levels and their discriminative values in relatives of patients with coronary artery disease. Atherosclerosis 51:261–267

6. Schaefer EJ (1984) Clinical, biochemical and genetic features in familial disorders of high density lipoprotein deficency. Arteriosclerosis 4:303–322
7. Schaefer EJ, Blum CB, Levy RI, Jenkins LL, Alaupovic P, Foster DM, Brewer HB (1978) Metabolism of high density lipoprotein apolipoproteins in Tangier disease. N Engl J Med 299:905–910
8. Ghiselli G, Summerfield JA, Schaefer EJ, Sirtori C, Jones EA, Brewer HB (1982) Abnormal catabolism of apolipoprotein A-I-Milano, a cause of high density lipoprotein deficiency. Clin Res 30:291A
9. Schaefer EJ, Zech LA, Jenkins LL, Bronzert TJ, Rubalcaba EA, Lindgren FT, Aamodt RL, Brewer HB (1982) Human apolipoprotein A-I and A-II metabolism. J Lipid Res 23:850–862
10. Schaefer EJ, Foster DM, Zech LA, Lindgren FT, Brewer HB, Levy RI (1983) The effects of estrogen administration on plasma lipoprotein metabolism in premenopausal females. J Clin Endocrinal Metabol 57:262–267
11. Shepherd J, Packard CJ, Patsch JR, Gotto AM, Jr, Taunton OD (1979) Effects of nicotinic acid therapy on plasma high density lipoprotein subfraction distribution and composition and on apolipoprotein A metabolism. J Clin Invest 63:858–867
12. Sheperd J, Packard CJ, Morgan HG, Third JLHC, Stewart JM, Lawrie TDV (1979) The effect of cholestyramine on high density lipoprotein metabolism. Atherosclerosis 33:433–444
13. Saku K, Gartside PS, Hynd BA, Kashyap ML (1985) Mechanism of action of gemfibrozil on lipoprotein metabolism. J Clin Invest 75:1702–1712
14. Nestel PJ, Tada N, Fidge NH (1980) Increased catabolism of high density lipoprotein in alcoholic hepatitis. Metabolism 29:101–104
15. Shepherd J, Packard CJ, Patsch JR, Gotto AM, Jr, Taunton OD (1978) Effects of dietary polyunsaturated and saturated fat on the properties of high density lipoproteins and the metabolism of apolipoprotein A-I. J Clin Invest 61:1582–1592
16. Ad Hoc Committee for Grading Coronary Artery Disease, Council on Cardiovascular Surgery. Am Heart Assoc, Comm Rep Circul 1975, 51 Suppl:5–40
17. Gensini GG (1975) Coronary arteriography. Futura, Mount Kisco, NY, pp 269–274
18. Ghiselli G, Rohde MF, Ranenbaum S, Krishnan S, Gotto AM, Jr (1985) Origin of apolipoprotein A-I polymorphism in plasma. J Biol Chem 260:15662–15668
19. Miller GJ, Miller NE (1975) Plasma high density lipoprotein concentration and development of ischaemic heart disease. Lancet I:16–19
20. Steinberg D (1978) The rediscovery of high density lipoprotein: a negative risk factor in atherosclerosis. Eur J Clin Invest 8:107–109
21. Glass CK, Pittman RC, Keller GA, Steinberg D (1983) Tissue sites of degradation of apoprotein A-I in the rat. J Biol Chem 258:7161–7167
22. Nakai T, Otto PS, Kennedy DL, Whayne TF (1976) Rat high density lipoprotein subfraction (HDL₃) uptake and catabolism by isolated rat liver parenchymal cells. J Biol Chem 251: 4914–4921
23. Rifici VA, Eder HA (1984) A hepatocyte receptor for high density lipoproteins specific for apolipoprotein A-I. J Biol Chem 259:13814–13818
24. Miller NE, Hammett F, Saltissi S, Rao S, Van Zeller H, Coltart J, Lewis B (1981) Relation of angiographically defined coronary artery disease to plasma lipoprotein subfractions and apolipoproteins. Br Med J 282:1741–1744
25. Puchois P, Bertrand M, Lablance JM, Fruchart JC (1985) Decrease of plasma apoA-I in coronary artery disease is related to lipoprotein particles that contain apoA-I but not apoA-II. Atheroslersosis 513a
26. Barth JD, Jensen H, Hugenholtz PG, Birkenhager JC (1983) Post-heparin lipases, lipids and related hormones in men undergoing coronary arteriography to assess atherosclerosis. Atherosclerosis 48:235–241
27. Breier C, Muhlberger V, Drexel H, Herold M, Lisch HJ, Knapp E, Braunsteiner H (1985) Essential role of post-heparin lipoprotein lipase activity and of plasma testosterone in coronary artery disease. Lancet I:1242–1244
28. Fidge N, Nestel P, Ishikawa T, Reardon M, Billington T (1980) Turnover of apoproteins A-I and A-II of high density lipoprotein and the relationship to other lipoproteins in normal and hyperlipidemic individuals. Metabolism 29:643–653

29. Rao SN, Magill P, Miller NE, Lewis B (1980) Plasma high density lipoprotein metabolism in subjects with primary hypertriglyeridemia: altered metabolism of apoproteins A-I and A-II. Clin Sci 59:359–362
30. Magill P, Rao SN, Miler NE, Nicoll A, Brunzell J, St. Hilaire J, Lewis B (1982) Relationships between the metabolism of high density and very low density lipoproteins in man: studies of apolipoprotein kinetics and adipose tissue lipoprotein lipase activity. Eur J Clin Invest 12:113–120
31. Nikkila EA, Taskinen MR, Kekki M (1978) Relation of plasma high density lipoprotein cholesterol to lipoprotein-lipase activity in adipose tissue and skeletal muscle of man. Atherosclerosis 29:497–501
32. Eisenberg S (1984) High density lipoprotein metabolism. J Lipid Res 25:1017–1058

Hypocholesterolemic Drugs and Lipoprotein Metabolism

D. STEINBERG [1]

1 Introduction

The primary reason for studying drugs that affect lipid metabolism is, of course, the hope of generating useful pharmacologic agents. In addition, however, studies of new agents have taught us important lessons regarding normal lipid and lipoprotein metabolism. In this presentation I wish to take up three "case histories" in the latter category: (1) bile acid sequestrants; (2) inhibitors of cholesterol biosynthesis, particularly inhibitors of HMG CoA reductase; (3) probucol.

2 Bile Acid Sequestrants

Cholestyramine, the prototype bile acid sequestrant, was developed in the hope that draining bile acids from the system would decrease the available stores of precursor cholesterol in the liver and thus reduce the output of cholesterol in plasma lipoproteins. Cholestyramine (and colestipol) proved to be very effective hypocholesterolemic drugs, widely used in clinical medicine. Unexpectedly, the rate of *removal* of LDL was found to be increased in cholestyramine-treated patients (1). Subsequent studies in several laboratories established clearly that both in man and in experimental animals the increased rate of removal was due in major part to an increase in the number of LDL receptors, particularly in the liver (2—4). There may be some decrease in lipoprotein production as well, but the capacity of the liver to compensate with an increase in cholesterol biosynthesis probably limits that effect. In any case, the finding of enhanced receptor expression focused attention on a general area in which drugs may be usefully developed for treating hypercholesterolemia (5).

What I wish to concentrate on here, however, is the lesson learned from the studies of Witztum and co-workers (6,7) that taught us something about, first, the heterogeneity of LDL and, second, the importance of LDL receptor-mediated removal of IDL and possibly small VLDL molecules. The first clue that treatment with bile acid

1 Department of Medicine, M-013D, University of California, San Diego, La Jolla, CA 92093 USA

Drugs Affecting Lipid Metabolism
Ed. by R. Paoletti et al.
© Springer-Verlag Berlin Heidelberg 1987

sequestrants might be more complex than appreciated came from studies showing that LDL in patients treated with colestipol was distinctly abnormal in size and composition (6). The cholesterol content fell to a significantly greater extent than did the apoprotein-B content. Later studies by Witztum anc co-workers in the choles-tyramine-treated guinea pig showed that there were similar changes in composition of LDL and that the molecules were distinctly smaller (7). Kinetic studies confirmed that the fractional catabolic rate (FCR) of LDL in treated guinea pigs was greater than that in untreated animals. In addition, these investigators compared the catabolic rate of LDL from untreated animals with that of LDL isolated from animals under treatment with cholestyramine. Surprisingly, it was found that when ^{125}I-LDL from untreated guinea pigs was injected simultaneously with ^{131}I-LDL from *treated* guinea pigs, the *former* consistently showed a greater FCR, i.e., the smaller LDL in the treated animals was removed *more slowly* than the larger, native LDL from the control animals. As shown in Table 1, this was true whether the two forms of LDL were injected into a normal, untreated recipient or were injected into a cholestyramine-treated recipient. It can seen from the data in Table 1 that the FCR in the treated animals was always greater than that of the same LDL species injected into untreated animals. This is consistent with the findings in several laboratories that treatment with bile acid sequestrants increases the rate of removal of LDL and that that increase is largely receptor-mediated. When liver plasma membrane fractions were prepared from the livers of cholestyramine-treated and untreated guinea pigs, it was found that the number of LDL receptors by this membrane-binding assay was indeed enhanced by cholestyramine treatment.

If we look at the data of Table 1 in another way, there is a lesson. If we accept the premise on which all turnover studies are based, namely, that the fraction labeled and then studied is kinetically homogeneous, then the valid comparison of LDL turnover in untreated and treated animals would be a comparison of the number in the upper left-hand corner with the number in the lower right-hand corner. In other words, one should measure the turnover of the LDL species *that is actually found in the test animal.* In the normal, untreated guinea pig, the relevant FCR would be that of LDL from untreated guinea pigs, namely, 0.110. Similarly, the relevant FCR for the treated guinea pigs would be that obtained by injecting its *own* LDL (i.e., LDL prepared from treated animals) and that value was 0.120. There is a slight increase but it is modest and probably not significant compared to the increase obtained

Table 1. Fractional catabolic rate of LDL from untreated guinea pigs or cholestyramine-treated guinea pigs injected into untreated or treated recipients. [Data of Witztum et al. (7)]

	I*-LDL from untreated donor	I*-LDL from treated donor	
Untreated recipient (n=10)	0.110 (± 0.006)	0.095 (± 0.006)	$p < 0.01$
Treated recipient (n=9)	0.155 (± 0.010)	0.120 (± 0.009)	$p < 0.05$
	$p < 0.001$	$p < 0.05$	

when turnover of the same LDL fraction is compared in untreated and treated animals. Those values would be 0.110 versus 0.155 (using normal LDL) or 0.095 versus 0.120 (using "cholestyramine-LDL").

These studies established that treatment with a bile acid sequestrant enhances receptor-mediated removal of LDL, *but at the same time causes a very striking change in the LDL, generating a fraction that is removed from the plasma more slowly than normal LDL.* The net result *is* the desired lowering of the LDL but, interestingly, the effect would have been greater had the LDL *not* changed its properties. Most studies of LDL turnover in man are, of course, carried out by sampling the patient's plasma, isolating and labeling the LDL fraction, and then reinjecting it as quickly as possible to avoid changes of LDL on storage. Therefore, during the control period the patient receives his own normal LDL; during the treatment period he again receives his own LDL but now this LDL may have been modified by virtue of the treatment. This is the pattern followed in virtually all clinical studies. If the treatment never altered the behavior of the LDL, it would be a perfectly acceptable protocol. When, however, the LDL (or whatever lipoprotein fraction one is investigating) is altered in composition and in biological properties, then it ceases to be a suitable protocol. (Below we shall discuss the way in which probucol treatment also can, at least in rabbits, generate a form of LDL that differs metabolically from normal.) The lesson, of course, is that we must check with each new pharmacologic agent or intervention to make sure that the labeled ligand we use is appropriate. Kesaniemi, Witztum, and Grundy have carried out similar studies in patients and with similar results (unpublished results).

One way to explain the findings just discussed is to postulate that LDL is normally homogeneous, at least kinetically, but that cholestyramine in some way generates a species of LDL that differs considerably in size and composition. This new LDL species may itself be homogeneous and behave kinetically as a single entity. If that were the case, the FCR values using autologous LDL in both the untreated and treated states would give correct values for turnover.

Another interpretation, however, is that both species of LDL are actually present together in the untreated state and that the larger, less dense molecules are preferentially removed during cholestyramine treatment. If that were the case, then the necessary assumption of homogeneity made in all kinetic analyses is not met and the calculated flux of LDL is not correct, at least in the untreated state. And, of course, we now have evidence that this is the case. The kinetic inhomogeneity of LDL was pointed out as early as 1975 by Phair et al. (8). Actually, we should have known from the fact that the urine/plasma ratios, which should be constant when a homogeneous plasma component is undergoing degradation, do *not* remain constant. This was observed in a number of laboratories, but never commented on. It is interesting how we scientists willfully blink at things that "don't fit". There is an important lesson there, too.

Witztum and co-workers (7) propose the following thesis to rationalize all of the available data on the action of bile acid sequestrants. First, LDL in the normal animal is heterogeneous, consisting of large particles that are more rapidly metabolized, and smaller particles, that are more slowly metabolized. Second, treatment with colestipol or cholestyramine induces an increase in the number of LDL receptors on the hepatic cells, as first proposed by Slater et al. (2) and by Kovanen et al. (3) and this increases the rate of removal of LDL from the plasma. Third, the larger LDL particles are preferentially removed and therefore reduced out of proportion to the reduction in the small LDL, thus accounting for the observed altered composition of the LDL fraction. Finally, the increase in LDL receptors increases the removal not only of LDL but also of IDL and perhaps very small VLDL particles. This results in a decrease in the input of apo-B into the LDL fraction and an apparent decrease in "production" of LDL (but not necessarily of total apo-B).

This accounts for all of the observed findings but leaves one question unanswered: What is the *true* rate of turnover of LDL in the normal state? It is now postulated that LDL is heterogeneous, consisting of at least two different fractions (lighter, larger) and (denser, smaller). Until we know the kinetic relationships of these two subfractions we do not really necessarily have a valid figure for LDL turnover in the normal animal. Consequently, we cannot with assurance partition the effects of bile acid sequestrants with regard to the relative importance of decreased input from larger, triglyceride-rich molecules, on the one hand, and increased removal of LDL molecules, on the other hand.

3 Inhibitors of Cholesterol Biosynthesis

The proposition that plasma cholesterol levels might be controlled by introducing inhibitors of cholesterol biosynthesis was put forward by Cottet and co-workers (9) and by Steinberg and Fredrickson (10) in the 1950s. Many investigators were highly skeptical about the feasibility of inhibiting so vital a pathway as that of cholesterol biosynthesis for the lifetime of the patient, and their skepticism was understandable. The compounds initially proposed were either quite ineffective or had unacceptable side effects (reviewed in 11). It was at the second of these symposia in 1960 that we reported on the mechanism of action of triparanol (12), which was the first clinically applied inhibitor of cholesterol synthesis. We pointed out that because of the accumulation of desmosterol, which was itself taken up into atherosclerotic lesions, the drug was most unlikely to be useful. Indeed, it turned out to be even worse than that! We and others suggested that the ideal point of attack, if this approach were to be successful, would be at the rate-limiting reaction, i.e., at HMG CoA reductase. A large number of analogs of hydroxymethylglutaric acid, of mevalonic acid and of intermediates on up to squalene were synthesized and tested (11), but none of them really worked in vivo. An important breakthrough came in 1977 when Endo and co-workers discovered a naturally occurring analog of hydroxymethylglutaric acid which was astonishingly potent, inhibiting reductase at concentrations as low as 2 μM (13). Of most importance, this inhibitor entered cells very readily and was extremely potent in vivo as well as in vitro. Soon thereafter Alberts and co-workers (14) developed a structurally similar inhibitor, mevinolin (now redesignated lovastatin), and there are now several analogs in this series under active investigation. Since mevinolin and other inhibitors have been amply discussed in other sessions at this Symposium I will limit myself to a brief discussion of its mechanism of action and a few words about its quite remarkable efficacy.

Lovastatin and its congeners are structural analogs of hydroxymethylglutaric acid and act as competitive inhibitors of the reductase. Cells treated with the inhibitor respond by enhancing their rates of synthesis of the enzyme, but not sufficiently to maintain normal rates of cholesterol biosynthesis. The cells respond to the resulting partial cholesterol depletion by increasing their LDL receptor number. One would anticipate that inhibition of cholesterol synthesis would decrease production as well and there is some evidence that this occurs, but it may not be the most important effect. Certainly the combination of cholestyramine and mevinolin markedly increases LDL receptor number in the liver (5).

The clinical efficacy of the reductase inhibitors is remarkable. Yamamoto and co-workers showed that treatment with ML236B (compactin) could reduce LDL levels by up to 40% even in patients with heterozygous familial hypercholesterolemia (15). Illingworth reported similar results with mevinolin (lovastatin) (16). The combination of the two causes a perfectly astonishing response (16,17). Mean lowering of cholesterol levels of 50% or more have been reported (16) and this has been the experience in our own clinic (Witztum, Steinberg, and co-workers, unpublished results). One patient under our care, a patient with heterozygous familial hypercholesterolemia, decreased her plasma total cholesterol level from about 450 to less than 200 and decreased her LDL cholesterol concentration by approximately 70% with total resolution of xanthelasma and extensor tendon xanthomas in the hand.

A series of reductase inhibitors is under intensive investigation in the laboratory and in the clinic. Release for clinical use could come as early as 1988. It is probably not too much to say that the development of this class of compounds will probably revolutionize drug management of hypercholesterolemia.

4 Studies on the Mechanism of Action of Probucol

Probucol, widely used in the clinic for a number of years, has its primary effect on LDL but also consistently lowers HDL levels. The mechanisms underlying these effects have yet to be elucidated in detail. Nestel and Billington (18) and Kesaniemi and Grundy (19) found that the fractional catabolic rate of LDL was increased in probucol-treated patients. Although not tested directly, the implication was that there had been an increase in receptor number, in analogy with the mechanism of action of cholestyramine and of mevinolin discussed above. However, it was then observed that probucol was effective (perhaps even more effective) in patients who *lacked* the LDL receptor, i.e., patients with homozygous familial hypercholesterolemia (20,21). This caught our attention and was the basis for our own investigations. We reasoned that either probucol was working by mechanisms not involving the Brown and Goldstein LDL receptor or that, at least, it must have an additional way of effecting increases in fractional catabolic rate.

During his tenure as a Fogarty fellow in our laboratory in La Jolla, Dr. Marek Naruszewicz undertook studies of the mechanism of action of probucol in the receptor-deficient rabbit (WHHL rabbit) (22). He showed that treatment with probucol decreased total plasma cholesterol levels by an average of 23% and LDL cholesterol levels by 36%. He also showed that the composition of LDL in the treated animals was different from that in the untreated animals. Under probucol treatment apo-B levels in the LDL fraction decreased by only 10%, whereas the cholesterol decrased by 36%. Thus, there was a very significant increase in the cholesterol:apo-B ratio, a change reminiscent of those observed in animals and patients treated with bile acid sequestrants.

Kinetic studies were then carried out. Initially these were done by preparing a large batch of LDL from untreated WHHL rabbits, iodinating it, and then injecting it into both untreated and treated recipient WHHL rabbits. Using this protocol, no significant difference in FCR was observed between the treated and untreated groups. Since the plasma level of LDL had decreased, however, the result implied a decrease in production. At this point we discussed our results with Dr. Grundy. From those discussions we realized that the protocol used in his studies was different from that used in those first WHHL rabbit studies and that the difference was quite analogous to what was discussed above with respect to cholestyramine and other drug-treatment protocols. They had used LDL from the patient when he was *untreated* and then a different sample of his LDL drawn from him when he *was* treated; in our first rabbit studies we had used a single batch of labeled LDL from untreated animals to asses turnover in both the treated and untreated recipients. Informed by the experience of Witztum et al. with cholestyramine (7) we decided to repeat our studies using a protocol analogous to that used in clinical studies. As shown in Table 2, when this was done we found that the FCR in the treated animals was always greater than that in the untreated animals, regardless of whether the donor LDL came from a treated or an untreated donor.

Table 2. Comparison of plasma FCR for labeled LDL prepared either from untreated or from probucol-treated WHHl rabbits. [Data of Naruszewicz et al. (22)]

Expt. No.	Recipient rabbits	Source of labeled LDL		
		Untreated WHHL rabbit	Probucol-treated WHHL rabbit	Increase in FCR due to probucol
		Fractional catabolic rate (hr^{-1})		
1	Probucol-treated WHHL	0.019	0.028	+47%
	Untreated WHHL	0.018	0.028	+56%
	Untreated normal NZW	0.062	0.102	+64%
2	Probucol-treated WHHL	0.019	0.032	+68%
	Untreated WHHL	0.027	0.031	+29%
	Untreated normal NZW	0.065	0.093	+43%
3	Untreated normal NZW	0.067	0.098	+46%

One possible explanation for these findings, of course, might be that the WHHL rabbit is *not* completely deficient in LDL receptor activity and that the increase in FCR is attributable to induction of new receptors. Against this interpretation is the fact that in WHHL rabbits no significant difference has been found between the FCR of native LDL and that of blocked LDL (i.e., LDL-treated so as not to be recognized by the LDL receptor) (23). Still, the possibility that LDL receptors might be undetectable and yet be enhanced by treatment cannot be absolutely ruled out. If, on the other hand, we assume that the WHHL rabbit expresses no significant receptor activity, then we have to look for an explanation that is independent of the LDL receptor. Let us return to this question in a moment.

Examining the data in Table 2 from another point of view, we see that the FCR for LDL taken from treated donors was always greater than that of LDL taken from untreated donors. This was independent of whether the recipient was treated or not treated. In other words, it appeared that the LDL from treated animals had been altered in some fashion such that its kinetic behavior was different from that of LDL from untreated animals. This was also true when the recipient was a wild-type New Zealand white rabbit. Finally, it was shown that the uptake and degradation of LDL from probucol-treated animals in cultured fibroblasts was greater than that of LDL from untreated animals.

To summarize, the data are compatible with the hypothesis that treatment with probucol somehow alters the chemical and physical properties of LDL such that its removal from the plasma compartment is accelerated and that this increased removal does not require the B/E receptor of Brown and Goldstein to be operative. The removal could be occurring by way of alternative receptors if the LDL undergoes modification in vivo or it could be occurring via nonspecific, receptor-independent mechanisms. It should be emphasized that these findings are so far limited to studies in receptor-deficient rabbits and have not been as yet extended to other animal species or to man. Furthermore, there may for some reason be heterogeneity in the LDL prepared from probucol-treated animals. A possibility that deserves consideration is that the incorporation of probucol into a lipoprotein itself changes its kinetic properties. Probucol is carried in the plasma compartment primarily in lipoproteins, is distributed among them more or less in proportion to their lipid content, and exchanges readily among them. If the probucol contained in the labeled LDL injected were to transfer out of the LDL after entry into the recipient plasma, that might change its catabolic rate and account for the observed slowing in the rate of removal after an hour or two. In more recent studies, again in WHHL rabbits, the responses to probucol treatment have been variable and much less than previously observed (M. Naruszewicz et al., unpubl. results). The reasons for the variability are under investigation.

5 Probucol as an Antioxidant

We have discussed elsewhere the possibility that oxidative modification of LDL may play a role in the generation of foam cells in early atherosclerotic lesions (24). Oxidative modification can convert native LDL to a form recognized by the acetyl LDL or "scavenger" receptor (25) and then uptake becomes much more rapid than that of native LDL and foam cell formation can occur (26). Space does not permit a

Table 3. LDL from probucol-treated patients is protected against oxidative modification[a]. [Data of Parthasarathy et al. (27)]

LDL sample	Rate of degradation by mouse peritoneal macrophages	
	LDL from untreated normal subjects	LDL from probucol-treated subjects
	(μg 5 h^{-1} mg^{-1})	
Unincubated	1.31 ± 0.27	1.27 ± 0.05
Incubated 24 h without cells	1.30 ± 0.22	1.10 ± 0.24
Incubated 24 h with endothelial cells	9.34 ± 2.40	1.14 ± 0.28

[a] ^{125}I-labeled LDL samples were incubated for 24 h in the absence or presence of endothelial cells. Aliquots were then transferred to dishes containing freshly-harvested mouse peritoneal macrophages. After 5 h, trichloroacetic acid-soluble ^{125}I was measured as in index of degradation.

detailed discussion of the mechanisms involved, but the modification involves several changes beginning with the generation of free radicals, generated either by cells in culture or by copper-containing media even in the absence of cells. Cells that have been shown to catalyze the modification include endothelial cells, smooth muscle cells, and macrophages themselves. Addition of alpha-tocopherol or butylated trihydroxytoluene inhibits not only the peroxidation of the lipids but also the conversion to a form recognized by the acetyl LDL receptor. Without regard to mechanisms, however, the essential point in the present context is that antioxidants appear to arrest the process totally.

Recently, Parthasarathy et al. have shown that probucol is itself a potent antioxidant, even more potent than alpha-tocopherol and butylated hydroxytoluene, and prevents oxidative modification of LDL even at concentrations as low as 2 μM (27). At 20 μM it inhibits it completely. Even more important, the LDL taken from probucol-treated patients, treated with conventional dosages of probucol, is extremely resistant to oxidative modification in vitro. As shown in Table 3, LDL from control subjects previously incubated with endothelial cells for 24 h was rapidly degraded in a subsequent incubation with macrophages, about seven times faster than unincubated LDL. In striking contrast, the LDL from probucol-treated patients was not degraded any faster than unincubated LDL. Furthermore, the generation of lipid peroxides was almost completely inhibited in the case of LDL from probucol-treated patients (data not shown).

Now it is only an hypothesis that oxidative modification of LDL in vivo is important in the generation of foam cells. There is no doubt that peroxidation of lipids occurs in vivo and that peroxidized lipids are present in atheromatous lesions, but this obviously falls short of proof that it is pathogenetically relevant. To the extent that it is a viable hypothesis, one can speculate that compounds that have antioxidant activity, such as probucol, could conceivably play a protective role over and above any effects they have on LDL levels. Further studies are needed.

References

1. Levy R, Langer T (1972) Hypolipodemic drugs and lipoprotein metabolism. Adv Exp Med Biol 26:155–163
2. Slater HR, Packard CJ, Bicker S, Shepherd J (1980) Effects of cholestyramine on receptor-mediated plasma clearance and tissue uptake of human low density lipoproteins in the rabbit. J Biol Chem 265:10210–10213
3. Kovanen P, Bilheimer DW, Goldstein JL, Jaramillo JJ, Brown M (1981) Regulatory role for hepatic low density lipoprotein receptors in vivo in the dog. Proc Natl Acad Sci USA 78: 1194–1198
4. Hui D, Innerarity TL, Mahley W (1981) Lipoprotein binding to canine hepatic membranes. J Biol Chem 256:5646–5655
5. Brown MS, Kovanen PT, Goldstein JL (1981) Regulation of plasma cholesterol by lipoprotein receptors. Science 212:626–635
6. Witztum JL, Schonfeld G, Weidman SW, Giese WE, Dillingham MA (1979) Bile sequestrant therapy alters the compositions of low-density and high-density lipoproteins. Metabolism 28:221–229
7. Witztum JL, Young SG, Elam RL, Carew TE, Fisher M (1985) Cholestyramine-induced changes in low density lipoprotein composition and metabolism. I. Studies in the guinea pig. J Lipid Res 26:92–103
8. Phair RD, Hammond MG, Bowden JA, Fried M, Fisher WR, Berman M (1975) Federation Proc 34:2263–2270
9. Cottet J, Mathivat A, Redel J (1954) Etude thérapeutique d'un hypocholestérolémiant de synthèse: l'acide phényl-éthyl-acétique. Presse Med 62:939–941
10. Steinberg D, Fredrickson DA (1956) Inhibitors of cholesterol biosynthesis and the problem of hypercholesterolemia. Ann NY Acad Sci 64:579–589
11. Steinberg D (1962) Chemotherapeutic approaches to hyperlipidemia. Adv Pharmacol 1: 59–159
12. Steinberg D, Avigan J (1961) Inhibitors of cholesterol biosynthesis with special reference to the mechanism of action of MER-29. In: Garattini S, Saoletto R (eds) Drugs affecting lipid metabolism. Academic Press, London New York, pp 132–143
13. Endo A, Tsujita A, Kuroda M, Tanzawa K (1977) Inhibition of cholesterol synthesis in vitro and in vivo by ML-236A and ML-236B, competitive inhibitors of 3-hydroxy-3-methylglutaryl-coenzyme A reductase. Eur J Biochem 77:31–36
14. Alberts AW, Chen J, Kuron G (1980) Mevinolin: a highly potent competitive inhibitor of hydroxymethylglutaryl-coenzyme. A reductase and a cholesterol-lowering agent. Proc Natl Acad Sci USA 77:3957–3961
15. Yamamoto A, Sudo H, Endo A (1980) Therapeutic effects of ML-236B in primary hypercholesterolemia. Atherosclerosis 35:259–266
16. Illingworth DR (1984) Mevinolin plus colestipol in therapy for severe heterozygous familial hypercholesterolemia. Ann Int Med 101:598–604
17. Mabuchi H, Sakai T, Sakai Y (1983) Reduction of serum cholesterol in heterozygous patients with familial hypercholesterolemia: additive effects of compactin and cholestyramine. N Engl J Med 308:609–613
18. Nestel PJ, Billington T (1981) Effects of probucol on low density lipoproteins removal and high density lipoprotein synthesis. Atherosclerosis 38:203–209
19. Kesaniemi YA, Grundy SM (1984) Influence of probucol on cholesterol and lipoprotein metabolism in man. J Lipid Res 25:780–790
20. Baker SG, Joffe BI, Mendelssohn D, Seftel HG (1982) Treatment of homozygous familial hypercholesterolemia with probucol. S Sfr Med J 62:7–11
21. Yamomoto A, Matsuzawa Y, Kishino B, Hayashi R, Hirobe K, Kikkawa T (1983) Effects of probucol on homozygous cases of familial hypercholesterolemia. Atherosclerosis 48:157–166
22. Naruszewicz M, Carew TE, Pittman RC, Witztum JL, Steinberg D (1984) A novel mechanism by which probucol lowers low density lipoprotein levels demonstrated in the LDL receptor-deficient rabbit. J Lipid Res 25:1206–1213

23. Steinbrecher UP, Witztum JL, Kesaniemi YA, Elam RL (1983) Comparison of glucosylated low density lipoprotein with methylated or cyclohexane-treated low density lipoprotein in the measurement of receptor-independent low density lipoprotein catabolism. J Clin Invest 71:960–964
24. Steinberg D (1987) Current theories of the pathogenesis of atherosclerosis. In: Steinberg D, Olefsky J (eds) Hypercholesterolemia and atherosclerosis. Churchill Livingstone, New York, pp 5–23
25. Goldstein JL, Ho YK, Basu JK, Brown MS (1979) Binding site on macrophages that mediates uptake and degradation of acetylated low density lipoprotein, producing massive cholesterol deposition. Proc Natl Acad Sci USA 76:333–337
26. Steinbrecher UP, Parthasarathy S, Leake DS, Witztum JL, Steinberg D (1984) Modification of low density lipoprotein by endothelial cells involves lipid peroxidation and degradation of low density lipoprotein phospholipids. Proc Natl Acad Sci USA 81:3883–3387
27. Parthasarathy S, Young SG, Witztum JL, Pittman RC, Steinberg D (1986) Probucol inhibits oxidative modification of low density lipoprotein. J Clin Invest 77:641–644

Hypotriglyceridemic Drugs and Lipoprotein Catabolism

S. EISENBERG [1]

In the last DALM Symposium, we described in detail our findings of abnormal lipoprotein systems in human subjects with dyslipoproteinemia, predominantly hypertriglyceridemia (HTG) (1). These studies, recently published (2,3), can be summarized
as follows: HTG states are associated with excessive lipid transfer reactions (4). The
lipids that are transferred are the hydrophobic cholesteryl ester (CE) and triglyceride
(TG) molecules. The reaction thus causes excessive enrichment of VLDL with CE
(5,6), while both LDL and HDL lose CE and acquire TG (1,2,7). Since acquired TG
in LDL and HDL are hydrolyzed by plasma lipases (8,9), the particles become smaller
and denser as compared to the normal lipoproteins. On the basis of these observations,
we suggested abnormal metabolism of HTG-lipoproteins (1). For VLDL, we showed
that the large and less dense populations, VLDL-I and VLDL-II, contain 50–150%
more CE molecules than present in LDL, and pointed out that such particles cannot
complete the VLDL → LDL conversion cascade. Hence, in HTG, "remnants" of these
VLDL populations must be cleared from the plasma independently of the LDL pathway
(1,5). HTG-LDL is CE-poor, TG-rich, small and dense lipoprotein, and the cholesterol
content of the LDL decreases with a curvilinear relation to plasma TG levels. Such
LDLs, therefore, are expected to be inefficient regulators of cellular metabolic activities that depend on cholesterol influx, e.g., cholesterol synthesis and B,E receptor
protein activity. These postulates have been further investigated in our laboratory
during the last 3 years, using LDL and cultured human skin fibroblasts.

LDL has been isolated from the plasma of normo- and HTG-human subjects and the
same HTG-subjects during triglyceride lowering therapy with Bezafibrate by zonal
centrifugation procedures (2,10). Six abnormalities are found in HTG-LDL. The
HTG lipoprotein is denser and smaller than N-LDL, is relatively enriched with protein
(predominantly apo B) and triglycerides, and contains significantly less free and
esterified cholesterol. These six abnormalities are strongly and significantly related
to the degree of triglyceridemia, and tend to revert towards normal when plasma
TG levels are reduced. When tested in the fibroblast system, abnormal metabolism
of HTG-LDL is clearly evidend (10). The first abnormality is defective binding of
HTG-LDL to the fibroblast B,E receptor. In up-regulated cells, specific binding of
HTG-LDL is reduced by about 30%, and the degree of binding of the HTG-LDL to the

---

1   Lipid Research Laboratory, Department of Medicine B, Hadassah University Hospital, Jerusalem,
    Israel

Drugs Affecting Lipid Metabolism
Ed. by Paoletti et al.
© Springer-Verlag Berlin Heidelberg 1987

receptor appears to reflect the compositional and/or structural abnormalities of the lipoprotein. For example, the lower the CE to the protein ratio in LDL (a measure of compositional and structural abnormalities), less LDL binds to the receptor (10). Consequent to the decreased binding, we observed decreased entry, and decreased degradation of the LDL by the cells. A second abnormality is the lower capacity of HTG-LDL to down-regulate cellular cholesterol synthesis and cellular B,E receptor activity. The two metabolic abnormalities depend on the trigylceridemic state and revert toward normal when plasma TG levels are lowered.

The reduced affinity of HTG-LDL to the B,E receptor is of particular interest. To the best of our knowledge, this is the first instance when abnormal metabolism of LDL is due to reversible compositional and/or structural abnormalities of the lipoprotein. Undoubtedly, this finding must reflect reversible conformational changes of the apo-B moiety that, in the HTG state, affect the receptor protein recognition site of the B-apoprotein. Two studies were performed in order to elucidate the mechanisms responsible for defective binding of HTG-LDL to the B,E receptor. In the first investigation, we compared the immunoreactivity of HTG-LDL tested by monoclonal antibodies to apo-B with the ability of the lipoprotein to bind to the fibroblast receptor (11). The study was repeated with LDLs isolated from the same patients during treatment with Bezafibrate (BZ-LDL). When the different LDLs are tested with an antibody directed toward epitopes unrelated to the receptor recognition site of the apo-B, all preparations (HTG, BZ, and N) react similarly. HTG-, but not BZ-LDL, however, exhibits reduced reactivity when tested with monoclonals specific for epitopes at, or near the receptor recognition site of the apo-B molecule. A strong and significant relationship, moreover, is found between the decreased immunoreactivity of LDLs and their capacity to interact with the fibroblast B,E receptor. Thus, we suggest that indeed abnormalities of HTG-LDL are responsible for its decreased affinity towards the B,E receptor, and that triglyceride lowering therapy (with Bezafibrate) restores the full ability of the lipoprotein to interact with the receptor. In the second investigation, we studied the biological reactivity (in fibroblasts) of LDLs isolated from the plasma of patients with type I hyperlipoproteinemia (12). The plasma of these patients contains two LDL populations. The first (type I-LDL-1) is of normal density and size, while the second (type I-LDL-2) consists of small-sized and dense LDL particles. Both LDLs are of abnormal composition, i.e., enriched with TG and depleted of CE molecules. When tested in the fibroblast system, type I-LDL-1 binding and degradation is about one-half that of N-LDL; type I-LDL-2 metabolism is even further reduced. These findings strongly indicate that TG enrichment alone can cause conformational changes at the receptor-binding region of apo-B, and that structural abnormalities, mainly a greater radius of curvature, further depress the binding process. As stated above, hypotriglyceridemic therapy corrects both the structural/compositional and the metabolic abnormalities of HTG-LDL.

The reduced ability of HTG-LDL to regulate cellular metabolic activity is clearly due to the low number of cholesterol molecules that enter the cells with the lipoprotein (10). In that initial study, we were able to show that when the number of cholesterol molecules that enter the cells is increased (e.g., by incubating the cell with different concentrations of HTG- or N-LDL), the degree of regulation of cellular

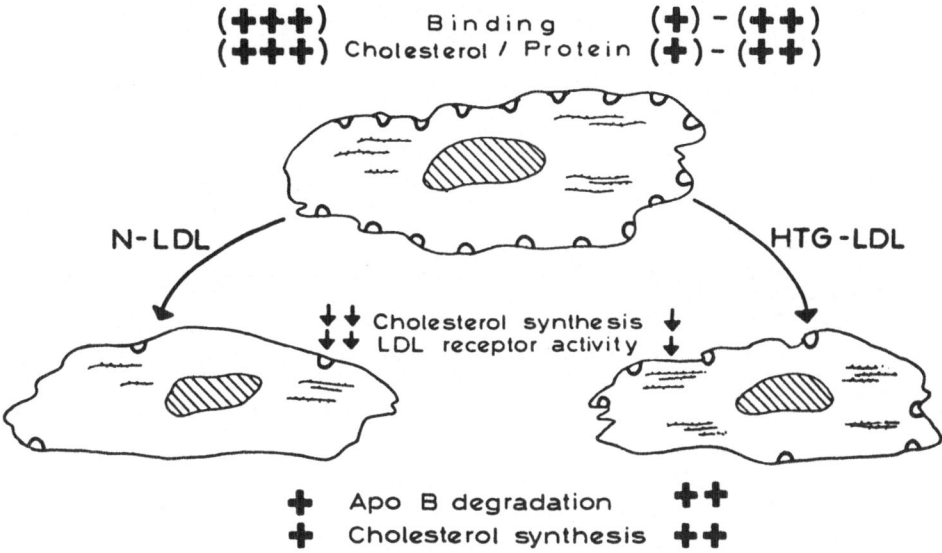

Fig. 1. The regulation of LDL catabolism in N and HTG states

metabolic activities is solely dependent on the cholesterol influx into the cells and is not affected by type of LDL used. When using a constant concentration of LDL protein, furthermore, cellular cholesterol synthesis is strongly and linearly related to the cholesterol content of the LDL (10). It thus follows that cellular cholesterol synthesis in humans various in inverse proportion to the cholesterol content of the LDL. More recently, we demonstrated a similar situation with regard to the regulation of LDL receptor activity (unpublished). Again, a very strong correlation is found between the LDL-CE content and the degree of regulation of the LDL receptor activity. The observation that Bezafibrate therapy corrects the abnormally low regulation capacity of HTG-LDL, strongly supports the hypothesis that the abnormal composition of the LDL is responsible for these abnormalities.

The studies described above point toward *two* defects of HTG-LDL metabolism that operate in opposite directions. On the one hand, HTG-LDL does not bind well to the receptor. On the other hand, however, cells exposed to HTG-LDL contain more receptors than cells exposed to N-LDL. How then, is HTG-LDL metabolized by cells? To answer this question, we have compared the actual metabolism of HTG-LDL in cells regulated by HTG-LDL to that of BZ- or N-LDL in cells regulated by the respective lipoproteins (13). The study demonstrated that under these conditions, actually more HTG-LDL is degraded in spite of its decreased affinity to the receptor. Figure 1 presents graphically the cellular behavior of HTG-LDL as elucidated by our investigations. Because HTG-LDL does not bind well to the B,E receptor and contains small amounts of cholesterol, the lipoprotein does not down-regulate cellular metabolic activities effectively. Cells regulated with HTG-LDL, therefore, contain more receptors and synthesize more cholesterol than cells regulated with N- or BZ-LDL. This causes increased degradation of HTG-LDL by the cells. In spite of the increased degradation, however, cholesterol inflow with HTG-LDL is still small, and the higher rates of cellular cholesterol synthesis and cellular receptor activity persist. Hypotriglyceridemic therapy appears to normalize these processes.

References

1. Eisenberg S (1985) Hypertriglyceridemia and atherosclerosis: analysis of an abnormal lipoprotein system and potential beneficial effects of triglyceride lowering therapy. Adv Exp Med Biol 183:73–84
2. Eisenberg S, Gavish D, Oschry Y, Fainaru M, Deckelbaum RJ (1984) Abnormalities in very low, low and high density lipoproteins in hypertriglyceridemia. Reversal towards normal with Bezafibrate treatment. J Clin Invest 74:470–482
3. Gavish D, Oschry Y, Fainaru M, Eisenberg S (1986) Change in very low-, low-, and high-density lipoproteins during lipid lowering (Bezafibrate) therapy: studies in type IIA and type IIB hyperlipoproteinemia. Eur J Clin Invest 16:61–68
4. Eisenberg S, Deckelbaum R (1986) Intravascular lipoprotein remodelling: neutral lipid transfer proteins. Clin Biochem (in press)
5. Oschry Y, Olivecrona T, Deckelbaum RJ, Eisenberg S (1985) Is hypertriglyceridemic very low density lipoprotein a precursor of normal low density lipoproteins? J Lipid Res 26:158–167
6. Eisenberg S (1985) Preferential enrichment of large-sized very low density lipoprotein populations with transferred cholesteryl esters. J Lipid Res 26:487–494
7. Deckelbaum RJ, Granot E, Oschry Y, Rose L, Eisenberg S (1984) Plasma triglyceride determines structure-composition in low and high density lipoproteins. Arteriosclerosis 4:226–231
8. Deckelbaum R, Eisenberg S, Oschry Y, Olivecrona T (1982) Reverse modification of human plasma low density lipoprotein toward triglyceride rich precursors: a mechanism for losing excess cholesterol ester. J Biol Chem 257:6509–6517
9. Deckelbaum RJ, Eisenberg S, Oschry Y, Granot E, Sharon I, Bengtsson-Olivecrona G (1986) Modelling of human plasma high density lipoproteins: roles of neutral lipid exchange and triglyceride lipases. J Biol Chem 261:5201–5208
10. Kleinman Y, Eisenberg S, Orschry Y, Gavish D, Stein O, Stein Y (1985) Defective metabolism of hypertriglyceridemic low density lipoprotein in cultured human skin fibroblasts. J Clin Invest 75:1796–1803
11. Kleinman Y, Schonfeld F, Gavish D, Oschry Y, Eisenberg S (1987) Hypolipidemic therapy modulates expression of apolipoprotein B epitopes on low density lipoproteins. J Lipid Res 28:540–548
12. Kleinman Y, Oschry Y, Berger GMB, Eisenberg S (1986) Familial lipoprotein lipase deficiency: abnormal lipoproteins and defective metabolism of low density lipoproteins in cultured human skin fibroblasts (submitted)
13. Kleinman Y, Oschry Y, Eisenberg S (1987) Abnormal regulation of LDL receptor activity and abnormal cellular metabolism of hypertriglyceridemic low density lipoprotein. Normalization with Bezafibrate therapy. Eur J Clin Invest (in press)

Lipolysis and Lipoprotein Metabolism

P. NILSSON-EHLE [1]

1 Introduction

The intravascular metabolism of lipoproteins represents a complex interplay between enzymatic reactions, nonenzymatic transfer of lipoprotein components between lipoprotein particles, and receptor-mediated uptake of lipoprotein particles in various organs. Three lipases, hormone-sensitive lipase (HSL), lipoprotein lipase (LPL), and hepatic lipase (HL) represent important points of regulation in lipoprotein metabolism and are consequently major determinants for the concentrations of plasma · lipoproteins.

2 Physiological Roles of Lipolytic Enzymes

HSL and LPL have been known as discrete enzyme activities for almost 40 years. Both have been purified and characterized, and their metabolic regulation is known in some detail (Nilsson-Ehle 1982; Belfrage et al. 1984; Smith and Pownall 1984). The main function of HSL is to regulate the rate of release of free fatty acid from adipose tissue depots. LPL is active mainly in the degration of triglyceride-rich lipoproteins, thus regulating the rate of removal of triglyceride from the bloodstream into adipose, muscle, heart, and other tissues.

HL was isolated only 12 years ago, has been purified, and characterized, but is function and regulation are not known in detail. It has been proposed that HL participates in the later steps of the delipidation of VLDL particles (Goldberg et al. 1982), specifically in the degradation of IDL particles to LDL. Other investigators, however, promote a role of HL in the transformation of HDL_2 to HDL_3 particles (Nikkilä et al.1982).

1   Department of Clinical Chemistry, University of Lund, 22185 Lund, Sweden

Drugs Affecting Lipid Metabolism
Ed. by. R. Paoletti et al.
© Springer-Verlag Berlin Heidelberg 1987

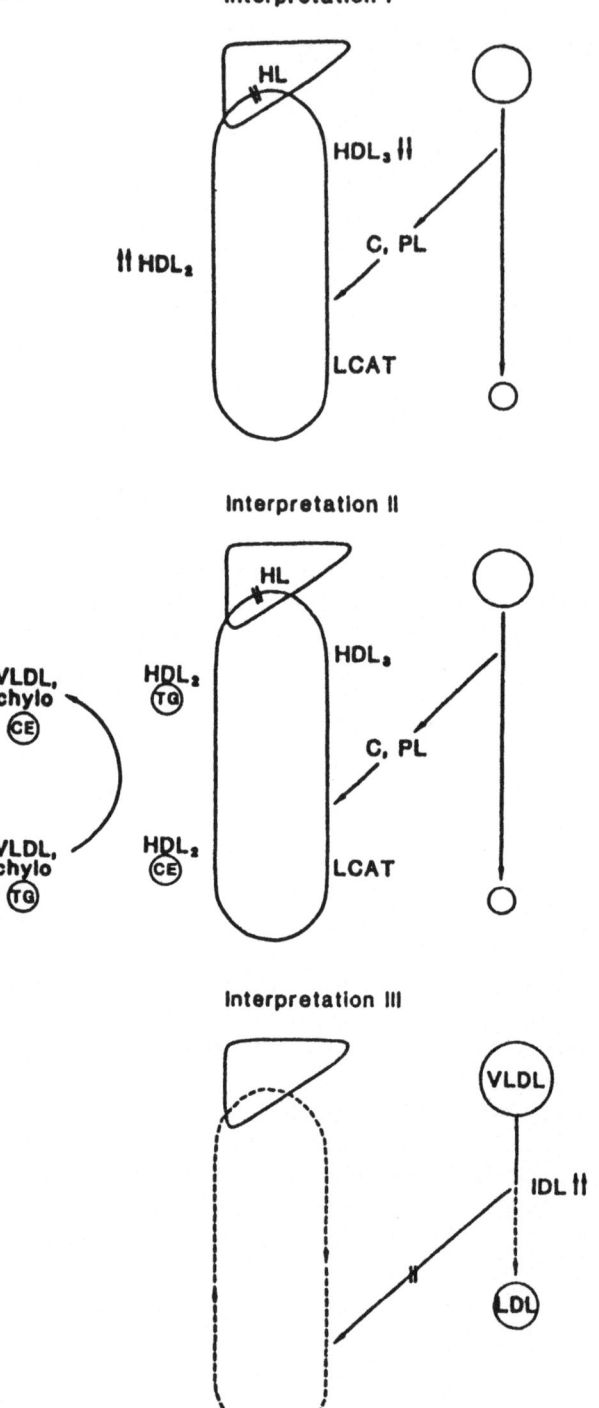

Fig. 1. Outline of lipoprotein metabolism and hypthetical model for disturbances in hepatic lipase deficiency

3 Hepatic Lipase Deficiency

The function of HL in man can be further discussed in the light of the biochemical findings in two brothers with total lack of hepatic lipase (Carlson et al. 1986). They are now in their sixties and asymptomatic. The index case was localized in 1975 by a health survey with moderate hypertriglyceridemia, which was largely accounted for by an enormous accumulation of triglycerides in HDL. The biochemical basis for this aberration was demonstrated only recently, when assays of postheparin lipase activities demonstrated total lack of HL with normal or only slightly reduced LPL activities. The only brother of the index case turned out to have essentially identical findings.

The most striking alteration in the lipoprotein pattern occurs in the HDL fraction, which is almost exclusively present as large triglyceride-rich particles. VLDL had the characteristics of IDL, i.e., an increased molar ratio triglyceride/cholesterol, and β-mobility. LCAT activity and neutral lipid transfer was normal. All apolipoproteins were present at essentially normal concentrations.

These findings are most readily compatible with the view that the physiological function of HL in man is to catalyze the degradation of HDL_2 triglyceride (Fig. 1). The accumulation of HDL_2 triglyceride would then result from a continuous exchange of cholesteryl ester and triglyceride between HDL_2 and triglyceride-rich· lipoproteins. The accumulation of IDL would represent a back-up phenomenon of surface components in partly degraded VLDL particles. The patients' HDL system, which contains almost exclusively large HDL_2 particles, would be poor acceptors for surface components which are normally transferred to HDL_3 from triglyceride-rich particles during degradation.

4 Lipoprotein Metabolism and Smoking

The functions of LPL and HL, as outlined above, predict that LPL would have a VLDL-lowering, HDL-raising function, whereas HL would have an HDL-lowering function. Epidemiological data, as well as observations from numerous longitudinal experimental and clinical studies support this view (Table 1).

The pathogenetic mechanisms behind one of the major risk factors for coronary heart disease, smoking, have not been elucidated. It has been pointed out that lower HDL levels among smokers than among nonsmokers might be of importance. We have recently performed a series of experiments to delineate the relationship between smoking, HDL levels, and lipase activities in more detail.

We studied male heavy smokers (more than 20 cigarettes daily) for 6 weeks after cessation of smoking (Stubbe, Eskilsson, and Nilsson-Ehle 1984). Within a week after cessation HDL levels started to rise to reach an average level of 20–30% above the initial level after 2 weeks. In subjects who resumed smoking HDL rapidly dropped again. The rise was most pronounced for the cholesteryl ester fraction, whereas the increase in other HDL components (triglyceride, phospholipid, apolipoprotein A1) was less prominent. Subsequent studies using zonal ultracentrifugation demonstrated

Table 1. HL, LPL, and HDL levels

	HL	LPL	HDL
Hypothyroidism	↓↓	↓	↑
treatment	↑↑	↑	↓
Hyperthyroidism	↑↑	–	↓
treatment	↓↓	–	↑
Diabetes	–	↓	↓
treatment	–	↑	↑
Long-term alcohol intake	–	↑	↑
Exercise	(↓)	(↑)	(↑)
Oxandrolone	↑	–	↓

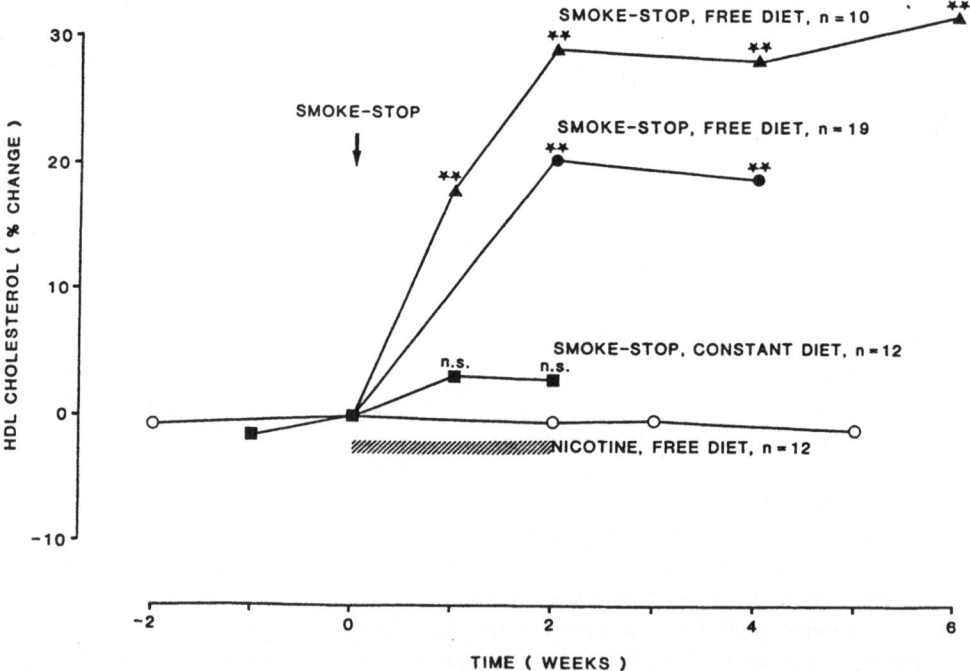

Fig. 2. Effects of smoking cessation and nicotine on plasma HDL cholesterol concentrations. *Solid symbols* indicate three groups of heavy smokers who stopped smoking at time 0. The mean (±SD) initial HDL cholesterol levels were 0.82 ± 0.12, 0.97 ± 0.26, and 1.00 ± 0.11 mmol^{-1}. *Open symbols* indicate nonsmokers who consumed nicotine

that the rise was confined mainly to the HDL_2 subfraction. The concentrations of other lipoprotein fractions were essentially unchanged.

The rise in HDL cholesterol was paralleled by a rise in LPL activity (measured in postherapin plasma) and a positive correlation between these two variables indicated that the rise in HDL may be accounted for by the rise in enzyme activity. Cessation of smoking was accompanied by spontaneous increase in caloric intake by about 10%. A positive correlation between the increase in dietary fat intake and the rise in HDL cholesterol levels indicated that the change in diet might contribute to the lipoprotein alterations.

When these experiments were repeated (Fig. 2), but with strict control of diet to keep caloric intake and caloric distribution constant, the rise in HDL was only marginal and we could register no change in LPL or HL activities. Obviously, these data indicate that dietary habits strongly influence the relationship between smoking and HDL levels. It is difficult to visualize that the moderate change in caloric intake as such would lead to this rather dramatic changes in HDL levels. However, analyses of the dietary records show that the extra calories were mainly supplied in snacks between meals. This change in feeding pattern, rather than the excess calories as such, might conceivably be more powerful in modulating LPL activity and thus HDL concentrations.

Another experiment (Fig. 2) addressed the possible role of nicotin in this context. Young healthy volunteers consumed nicotin (2 mg eight times daily for 2 weeks as nicotin chewing gum) and were closely followed 2 weeks before, during nicotin intake, and 3 weeks after. No changes in lipoprotein levels or lipase activities were observed. Similarly, we could register no effects on these variables in short-term experiments in which we monitored the effects of two cigarettes in smokers and nonsmokers.

Acknowledgments. Studies from the author's laboratory were supported by the Swedish Tobacco Company, the Swedish Medical Council (04966), and the Medical Faculty, University of Lund.

References

Belfrage P, Fredricksson G, Stralfors P, Tornqvist H (1984) Adipose tissue lipases. In: Borgström B, Brockmann HL (eds) Lipases. Elsevier, Amsterdam, pp 365–416

Carlson LA, Homquist L, Nilsson-Ehle P (1986) Deficiency of hepatic lipase activity in postheparin plasma in familial hyper-a-triglyceridemia. Acta Med Scand 219:435–447

Goldberg IJ, Lee NA, Paterniti JR, Ginsberg HN, Lindgren FT, Brown WV (1982) Lipoprotein metabolism during acute inhibition of hepatic triglyceride lipase in the cynomolgus monkey. J Clin Invest 70:1184–1194

Nikkilä EA, Kuusi T, Taskinen MR (1982) Role of lipoprotein lipase and hepatic endothelial lipase in the metabolism of high density lipoproteins: a novel concept on cholesterol transport in HDL cycle. In: Carlsson LA, Pernow B (eds) Metabolic risk factors in cardiovascular disease. Raven, New York, pp 205–215

Nilsson-Ehle P (1982) Regulation of lipoprotein lipase. In: Carlsson LA, Pernow B (eds) Metabolic risk factors in cardiovascular disease. Raven, New York, pp 49–57

Smith LC, Pownall HJ (1984) Lipoprotein lipase. In: Borgström B, Brockman HL (eds) Lipases. Elsevier, Amsterdam, pp 263–305

Stubbe I, Eskilsson I, Nilsson-Ehle P (1984) High density lipoprotein concentrations increase after stopping smoking. Br Med J 284:1511–1513

Regulation of Lipoprotein Lipase Activity: Its Role in Lipid-Lowering Therapies

TH. OLIVECRONA [1], S. R. PRICE [2], P. H. PEKALA [2], R. O. SCOW [3], S. S. CHERNICK [3], H. SEMB [1], S. VILARÓ [4], and G. BENGTSSON-OLIVECRONA [1]

1 Introduction

Lipoprotein lipase (LPL) is the first, and probably the rate-limiting, enzyme in metabolism of triglyceride-rich lipoproteins (Cryer 1981). There are strong correlations between LPL activity and the plasma concentrations of HDL and LDL, suggesting that LPL has an important role also for the cholesterol-rich lipoproteins (Deckelbaum et al. 1984). Hence, LPL should be a prime target for lipid-lowering therapies. It is well documented that LPL activity is regulated by feeding/fasting, by exercise, and by a variety of hormones (Cryer 1981). For rational design of drugs we need to know the molecular mechanisms involved in its regulation. The aim of this chapter is to outline present information in the area. Space is limited; we have therefore chosen to present a few extreme examples to illustrate the principles.

Many of the enzymes in energy metabolism are regulated at several levels; synthesis/degradation, phosphorylation/dephosphorylation, allosteric control. LPL is also regulated at several levels, but the situation is unusual in that the enzyme is released from the hormone-sensitive cells and then transferred to the capillary endothelium. Here, the enzyme is activated by apolipoprotein CII, and can be inhibited by lipolytic products (Olivecrona and Bengtsson 1983). Much of the regulation of LPL must, however, be exerted in the lipase-producing cells. Two main mechanisms have been proposed.

2 Regulation of LPL Synthesis

A dramatic example of regulation of LPL synthesis came from studies on the effects of a monokine on LPL in 3T3-L1 adipocytes (Price et al. 1986a). Exposure of the cell cultures to medium from endotoxin-stimulated macrophages decreased incorpora-

---

1 Department of Physiological Chemistry, University of Umeå, S-901 87 Umeå, Sweden
2 Department of Biochemistry, East Carolina University School of Medicine, Greenville, NC 27834, USA
3 Laboratory of Cellular and Developmental Biology, National Institute of Diabetes, and Digestive and Kidney Diseases, National Institutes of Health, Bethesda, MD 20892, USA
4 Departament de Bioquimica i Fisiologia, Universitat de Barcelona, Spain

Drugs Affecting Lipid Metabolism
Ed. by R. Paoletti et al.
© Springer-Verlag Berlin Heidelberg 1987

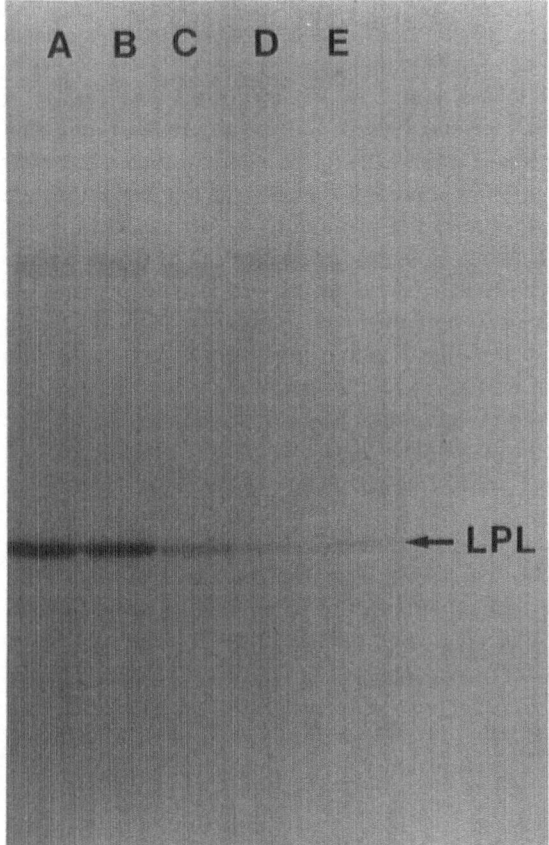

Fig. 1. Suppresion of LPL synthesis in 3T3-L1 adipocytes by crude monokine and purified cachectin. 3T3-L1 cells were cultured and induced to differentiate into adipocyte-like cells, and were exposed to 12.5 ng pure cachectin ml^{-1} or an equivalent amount of the crude monokine preparation. The cells were pulse-labeled with 75 μCi ^{35}S-methionine for 30 min. LPL was immunoprecipitated and analyzed by SDS-PAGE and fluorography. The fluorogram is shown. Lane *A,* control cells not exposed to monokine; lanes *B, C,* and *D,* cells exposed to the crude monokine preparation for 1, 3, and 17 h, respectively; lane *E,* cells exposed to purified cachectin for 17 h. Purified LPL from bovine milk was coelectrophoresed and stained for identification of immunoprecipitated protein (Price et al. 1986a)

tion of ^{35}S-methionine into immunoprecipitable LPL 80% in 3 h and 95% in 17 h (Fig. 1). LPL activity decreased in parallel to the synthesis, 75% in 3 h and more than 90% in 17 h. In contrast, there was only a minor decrease in total protein synthesis. These effects were also obtained with the purified monokine, named cachectin (Fig. 1). This monokine is closely related to tumor necrosis factor (TNF, Beutler et al. 1985), and recent studies have shown that also recombinant TNF suppresses LPL synthesis in 3T3-L1 adipocytes (Price et al. 1986b).

3 Release of LPL from the Lipase-Producing Cells

Several investigators have proposed that a main site of regulation is the release of LPL from lipase-producing cells. For instance, Nilsson-Ehle et al. (1976) found that adipocytes from rats given insulin released more LPL to the medium than adipocytes from controls rats. Spooner et al. (1979) reported that insulin increased the rate at which 3T3-L1 adipocytes released LPL to the medium from low values to 3–8% of cell total LPL per hour. This rate would, however, not be sufficient to cause rapid redistribution of the lipase to extracellular sites as happens in vivo in response to feeding. We speculated that the primary event might be transfer of LPL to the cell surface, and that spontaneous release into the medium might be an artifact of the in vitro system. In vivo the enzyme travels on to the endothelium by some as yet undefined mechanism. We have therefore reevaluated the system. Plates with 3T3-L1 adipocytes were cooled on ice, heparin was added, and the amount of LPL released was measured. It seems unlikely that heparin could recruit lipase from intracellular compartments to the medium in the cold. Therefore, the release by heparin should give an estimate of how much LPL was located at the cell surface. Figure 2 shows that 3T3-L1 adipocytes deprived of insulin for several days displayed relatively low LPL activity, and only 7 ± 4% of this could be released by heparin in the cold. Insulin treatment of these cells for 4 h (at 37° C) increased the release to 36 ± 5%, but caused

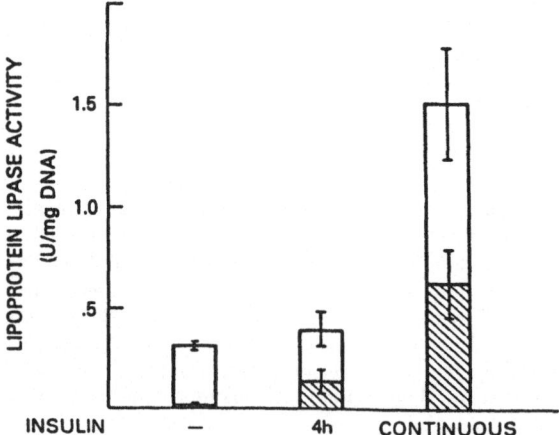

Fig. 2. Effect of insulin on the distribution of LPL in 3T3-L1 adipocytes. Ten plates with mature 3T3-L1 adipocytes were incubated for 5 days in medium with no insulin and only 5% fetal calf serum, as described by Spooner et al. (1979). Then, insulin was added to five of the plates to give a final concentration of 10^{-7} M. Four hours later the medium was changed in all plates and the plates were placed on a slurry of crushed ice in water. Two min later, a sample (2 ml) of the medium was removed, passed through a 25-μm filter and used for immediate assay of LPL activity, which was found to be negligible. Enough heparin was then added to give a concentration of 20 μg ml^{-1} in the remaining medium (5 ml). Ten min later, a new sample of medium was taken, filtered, and used for assay of LPL. The plates were drained, rinsed, and the cells were scraped off and processed into acetone-ether powders which were later assayed for LPL activity. Mean ± SD of five plates in each group. The *white portion of the bar* indicates the amount of LPL activity in cells, the *hatched portion* indicates the amount in the postheparin medium

only a small increase in total LPL activity. In cells grown continuously with insulin, the fraction that could be released by heparin was the same (39 ± 5%) as that in cells treated with insulin for 4 h, but total LPL activity was about three times higher, presumably because insulin had stimulated synthesis of LPL. These experiments demonstrate that there are mechanisms that rapidly change the location of LPL. Friedman et al. (1986) have recently reported that LPL is translocated to the cell surface in rat heart cells exposed to β-adrenergic stimuli.

Thus, there are at least two mechanisms which regulate LPL activity in cultured 3T3-L1 adipocytes. Do these mechanisms also operate in vivo? We have explored this in animal models. Injection of a single dose of recombinant TNF resulted in suppression of LPL activity in adipose tissue of three species: rats, mice, and guinea pigs (Semb et al. 1987). We studied this further in the guinea pig, and found that TNF suppressed incorporation of ^{35}S-methionine into immunoprecipitable LPL, and that this decrease in synthesis correlated closely to the decrease in LPL activity. There was no decrease in general protein synthesis or redistribution of the enzyme within

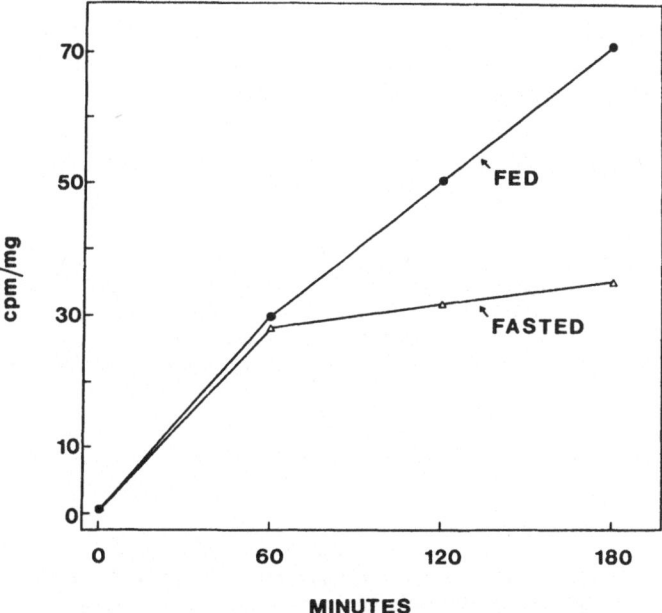

Fig. 3. Incorporation of ^{35}S-methionine into LPL in epididymal fat pads from fed and fasted rats. Fat pads from fed or fasted rats were incubated with 10^{-7} M insulin and ^{35}S-methionine (45 μCi ml^{-1}). Labeling was stopped by chilling the tubes in crushed ice. The fat pads were homogenized in buffer with detergents and protease inhibitors, and radioactivity in total protein was determined. An aliquot of the homogenate was immunoprecipitated using a chicken antiserum to bovine LPL. A goat anti-chicken IgG coupled to Sepharose was used as the second antibody. The immunoprecipitates were washed and then dissolved in sample buffer and separated by SDS-PAGE. The gels were dried and exposed on Kodak X-omat films at -70°C. The ^{35}S-labeled, immunoprecipitated rat LPL was found to have the same mobility as ^{125}I-labeled bovine LPL. The bands corresponding to LPL were cut out, eluted in a scintillation cocktail and counted. For details on metholody, see Semb and Olivecrona (1986)

the tissue. We have compared the effects of TNF on adipose tissue LPL to those of fasting (Semb et al. 1987). Total protein synthesis in adipocytes decreased on fasting, but relative synthesis of LPL did not change. There was redistribution of the enzyme in the tissue, however, with a smaller fraction located outside the adipocytes. This indicates a change in turnover of the enzyme. To explore this possibility fat pads from fed or fasted rats were incubated with ^{35}S-methionine, and LPL was isolated at a series of times by immunoprecipitation (Fig. 3). Radioactivity in total proteins increased continuously in both groups. In the fat pads from fasted rats LPL radioactivity leveled off after 1 h. This is similar to the results of Speake et al. (1985) and indicates that LPL is being degraded within the tissue. In contrast, in fat pads from fed rats radioactivity in LPL increased continuously for 3 h. Thus, LPL protein was turned over less rapidly in the fed state. When the pads were incubated in heparin-containing media, more labeled LPL was released from the fed than from the fasted pads. Taken together, these results suggest that more LPL is released from the adipocytes in the fed state, whereas more of the enzyme is degraded in the fasted state. Thus, the translocation of LPL demonstrated in 3T3-L1 adipocytes seems to operate also in vivo, and may explain why a much larger fraction of LPL activity is found outside adipocytes in the fed than in the fasted state.

I summary, these studies demonstrate two mechanisms which regulate LPL in cell culture systems and in vivo. TNF presumably acts on transcription of the LPL gene, whereas feeding/fasting modulates intracellular transport and turnover of the lipase protein. Using these techniques it should now be possible to determine how other agents, e.g. corticosteroids and catecholamines, regulate LPL activity, and to begin to design drugs that can modulate this process.

Acknowledgments. Parts of this work were carried out when Thomas Olivecrona was a visiting scientist and Gunilla Bengtsson-Olivecrona a guest worker at the National Institutes of Health. Dr. Vilaró is presently a visiting scientist at the University of Umeå. The work was supported in part by grants to T.O. and G.B.-O. from the Swedish Medical Research Council (B13–727), and to T.O. and H.S. from the Lions in Umeå Research Fund (399/85) and by grants to P.P. from NIH (GM 32892).

References

Beutler B, Greenwald D, Hulmes JD, Chang M, Pan YC-E, Mathison J, Ulevitch R, Cerami A (1985) Identity of tumor necrosis factor and the macrophage-secreted factor cachectin. Nature (London) 316:552–554

Cryer A (1981) Tissue lipoprotein-lipase activity and its action in lipoprotein metabolism. Int J Biochem 13:525–541

Deckelbaum RJ, Olivecrona T, Eisenberg S (1984) Plasma lipoproteins in hyperlipidemia: roles of neutral lipid exchange and lipase. In: Carlson LA, Olsson AG (eds) Treatment of hyperlipoproteinemia. Raven, New York, pp 85–93

Friedman G, Chajek-Shaul T, Stein O, Noe L, Etienne J, Stein Y (1986) β-Adrenergic stimulation enhances translocation, processing and synthesis of lipoprotein lipase in rat heart cells. Biochim Biophys Acta 877:112–120

Nilsson-Ehle P, Garfinkel AS, Schotz MC (1976) Intra- and extracellular forms of lipoprotein lipase in adipose tissue. Biochim Biophys Acta 431:147–156

Olivecrona T, Bengtsson G (1983) Lipases in milk. In: Borgström B, Brockman H (eds) Lipases. Elsevier, Amsterdam, pp 205–261

Price SR, Olivecrona T, Pekala PH (1986a) Regulation of lipoprotein lipase synthesis in 3T3-L1 adipocytes by cachectin. Further proof for identity with tumor necrosis factor. Biochem J 240:601–604

Price SR, Olivecrona T, Pekala PH (1986b) Regulation of lipoprotein lipase synthesis by recombinant tumor necrosis factor. The primary regulatory role of the hormone in 3T3-L1 adipocytes. Arch Biochem Biophys 251:738–746

Semb H, Olivecrona T (1986) Lipoprotein lipase in guinea pig tissues. Molecular size and rates of synthesis. Biochim Biophys Acta 878:330–337

Semb H, Peterson J, Tavernier J, Olivecrona T (1987) Multiple effects of tumor necrosis factor on lipoprotein lipase in vivo. J Biol Chem 262:8390–8394

Speake BK, Parkin SM, Robinson DS (1985) Lipoprotein-lipase in rat adipose tissue. Biochim Biophys Acta 840:419–422

Spooner PM, Chernick SS, Garrison MM, Scow RO (1979) Insulin regulation of lipoprotein lipase activity and release in 3T3-L1 adipocytes. J Biol Chem 254:10021–10029

Lipolysis and Antilipolytic Drugs at Different Ages

R. M. GAION and P. DORIGO [1]

1 Introduction

Adipose tissue plays a major role in caloric homeostasis by shifting between lipid mobilization and storage. Its primary functions can be summarized as follows:

1. Hydrolysis of circulating triglycerides combined with plasma proteins in the form of lipoproteins and uptake of fatty acids thus released.
2. Glucose uptake and utilization for the synthesis of lipids.
3. Lipid storage mainly in the form of triglycerides.
4. Lipolysis and fat mobilization.

All these processes are tightly controlled by hormones and neurotransmitters as well as by a number of endogenous modulators, such as adenosine, prostaglandins, and fatty acids.

A typical feature of aging is the enlargement of fat depots due to fat cell hyperplasia and hypertrophy (Hartman et al. 1971; Masoro et al. 1979). The metabolic alterations associated with these morphologic changes have been widely investigated in rodents and mainly in rats. In this animal model:

1. Extracellular lipoprotein lipase activity and fatty acid uptake from chylomicron triglycerides decrease with aging (Hartman 1977).
2. Glucose uptake and utilization are also reduced (Holm et al. 1975).
3. A linear correlation exists between age and triglyceride cell content (Hartman et al. 1971).
4. The basal rate of lipolysis decreases with age (Holm et al. 1975) even though a cell size-related increase has been demonstrated (Hartman et al. 1971).

Besides these changes in the basal state of the tissue, age-related alterations of the metabolic responses to external stimuli can be even more important from a physiological and pharmacological point of view.

In fat cells from old rats the sensitivity of glucose oxidation to insulin is reduced (Holm et al. 1975), while its stimulation by PGE_1 increases (Chang and Roth 1981).

---

1 Department of Pharmacology, University of Padua, Largo E. Meneghetti 2, 35100 Padua, Italy

Drugs Affecting Lipid Metabolism
Ed. by R. Paoletti et al.
© Springer-Verlag Berlin Heidelberg 1987

Also changes in the lipolytic response to stimulatory and inhibitory agents are well documented and they will be illustrated later.

A link among these alterations, the changes in circulating hormones that occur with aging and the development of age-associated diseases, has been often suggested and the pharmacological implications should be carefully evaluated.

A limited number of relevant findings will be now considered for illustrating the influence of age on lipolysis as well as the changes in the antilipolytic effect of drugs.

2 Age and Stimulation of Lipolysis in Fat Cells

The ability of glucagon, ACTH, and catecholamines to stimulate lipolysis in fat cells markedly decreases with age (Hartman et al. 1971; Holm et al. 1975; Masoro et al. 1979; Hoffman et al. 1984) due to a reduced activation of fat cell adenylate cyclase by these hormones (Cooper and Gregerman 1976). However, while the age-associated decrease in the response to glucagon and to ACTH appears to be related to a decreased number of receptors on the cell membrane (Cooper and Gregerman 1976), the number of beta-receptors is not affected by aging or is even increased (Hoffman et al. 1984). Thus, a partial loss of some component of the adenylate cyclase complex that allows coupling of beta-receptors to adenylate cyclase is likely to account for the diminished response of aging cells to beta-agonists (Cooper and Gregerman 1976; Hoffman et al. 1984). This view is supported by the decline in the stimulation of the enzyme by fluoride and by GMP-P(NH)P (Forn et al. 1970; Cooper and Gregerman 1976) and by the lack of age-related changes in the lipolytic response to a cyclic AMP analog, 8-(4-chlorophenylthio)adenosine $3',5'$ monophosphate cyclic (Hoffman et al. 1984). Alternatively or additionally, an increased activity of inhibitory pathways, e.g., those mediated by adenosine, has been proposed (Hoffman et al. 1984), but the stimulation of lipolysis by the adenosine antagonist, theophylline is not affected by aging (Carpene et al. 1983).

Among the other enzymes involved in the lipolytic process, protein kinase activity was found to increase in fat cells from old rats as compared to young animals (DiGirolamo et al. 1977). Also the activity of cyclic AMP phosphodiesterase increases with age (Forn et al. 1970), possibly contributing to the reduced ability of hormones to stimulate cyclic AMP accumulation and lipolysis.

3 Age and Antilipolytic Drugs

The best-known example of age-dependent changes in the antilipolytic action of drugs is the decreased effect of insulin on stimulated lipolysis (Olefsky 1977) that is associated with a reduced effect of the hormone on glucose transport and oxidation (Holm et al. 1975; Olefsky 1976) and correlate well with the decreased number of insulin receptors in adipocytes from old rats (Olefsky 1976).

In hamsters, aging was shown to reduce the antilipolytic potency of phenyl-isopropyl adenosine (Carpene et al. 1983), a P_1-receptor agonist that inhibits cyclic AMP accumulation and lipolysis by activating the N_i regulatory protein coupled to the catalytic subunit of adenylate cyclase. Activation of N_i protein is also the mechanism by which alpha-adrenergic agents exert their antilipolytic effect. In hamster and rabbit fat cells, alpha-adrenergic agonists show an increased effectiveness with aging (Lafontan 1979; Carpene et al. 1983) and also the number of $alpha_2$-adreno-ceptors on hamster fat cell membrane increases (Carpene et al. 1983), possibly con-tributing to the reduced lipolytic effect of catecholamines.

Nicotinic acid inhibits lipolysis by interacting with specific putative receptors linked to the N_i regulatory protein of adenylate cyclase. In rat adipose tissue aging increases the sensitivity of norepinephrine- and theophylline-induced lipolysis to nicotinic acid (Caparrotta et al. 1983). The possibility that parallel changes in receptor number and/or affinity occur has not been examined so far.

4 Considerations on Some Factors Possibly Involved in the Altered Responses of Lipolysis

The examples reported above indicate that the ability of hormones and neurotrans-mitters to activate the lipolytic process generally decreases with age, due either to a loss of membrane receptors or to functional alterations of the receptor-adenylate cyclase complex. On the other hand, the response to antilipolytic agents is reduced or increased, depending on the receptors involved. Considering the ability of cir-culating hormones to regulate their own, as well as other receptors in many tissues, age-related modifications of hormone levels in plasma are likely to play some role in the altered metabolic responses of adipose tissue. In this respect it is interesting to note that some striking similarities exist between the changes induced by aging and by hypothyroidism. In fat cells from hypothyroid rats adenylate cyclase activity, cyclic AMP accumulation, and lipolysis, both basal and stimulated by catecholamines, are lower than normal, while the lipolytic response to theophylline and to dibutyryl-cyclic AMP is reduced only slightly (Krishna et al. 1968; Armstrong and Stouffer 1974). Like in old animals, the activity of cyclic AMP phosphodiesterase is increased in fat cells from hypothyroid rats (Armstrong and Stouffer 1974). Hypothyroidism does not alter the number of beta-adrenergic receptors, but reduces the ability of GPP(NH)P to activate adenylate cyclase (Malbon 1980). Moreover, in adipose tissue from hypothyroid human subjects the balance between the responsiveness of stimulatory beta-receptors and inhibitory $alpha_2$-receptors to catecholamines is shifted toward a predominance of the latter ones (Rosenqvist and Efendic 1971). These similarities between the metabolic features of aging and hypothyroidism suggest that the decrease in T_3 serum levels that occurs in humans with aging (Westgren et al. 1976), together with other hormonal changes, might play some role in the control of adipose tissue function.

Another important point is the equilibrium among local modulators of lipolysis, such as adenosine, fatty acids and prostaglandins. Reduced blood flow, hypoalbuminemia, and alterations of fatty acid substrates for prostaglandin biosynthesis are only some of the age-dependent factors that may influence the formation and removal of these modulators in adipose tissue. This point has been poorly investigated and its understanding will contribute to prevent or correct in a selective manner the alterations of lipid metabolism due to aging.

References

Armstrong KJ, Stouffer JE (1974) Effects of thyroid hormone deficiency on cyclic adenosine 3':5'-monophosphate and control of lipolysis in fat cells. J Biol Chem 249:4226–4231

Caparrotta L, Fassina G, Gaion RM, Tessari F (1983) Age related changes in the antilipolytic effects of nicotinic acid in rat adipose tissue. Br J Pharmacol 80:663–670

Carpene C, Berlan M, Lafontan M (1983) Influence of development and reduction of fat stores on the antilipolytic α_2-adrenoceptor in hamster adipocytes: comparison with adenosine and β-adrenergic receptors. J Lipid Res 24:766–774

Chang WC, Roth GS (1981) Changes in prostaglandin E_1 stimulation of glucose oxidation in rat adipocytes during maturation and aging. Life Sci 28:623–627

Cooper B, Gregerman RI (1976) Hormone-sensitive fat cell adenylate cyclase in the rat: influence of growth, cell size, and aging. J Clin Invest 57:161–168

DiGirolamo M, Owens JL, Patrik JG, Kuo JF (1977) Protein kinase activity of isolated rat adipocytes as related to cell size. Proc Soc Exp Biol Med 154:513–516

Forn JN, Schonhofer S, Skidmore IF, Krishna G (1970) Effect of aging on the adenylate cyclase and phosphodiesterase activity in isolated fat cells of rat. Biochim Biophys Acta 208:304–309

Hartman AD (1977) Lipoprotein lipase distribution in rat adipose tissue: effect on chylomicron uptake. Am J Physiol 232:E316–E323

Hartman AD, Cohen AI, Richane CJ, Hsu T (1971) Lipolytic response and adenyl cyclase activity in rat adipocytes as related to cell size. J Lipid Res 12:498–505

Hoffman BB, Chang H, Farahbakhsh ZT, Reaven GM (1984) Age-related decrement in hormone-stimulated lipolysis. Am J Physiol 247:E772–E777

Holm G, Jacobson B, Björntorp P, Smith U (1975) Effects of age and cell size on rat adipose tissue metabolism. J Lipid Res 16:461–464

Krishna G, Hynie S, Brodie BB (1968) Effects of thyroid hormones on adenyl cyclase in adipose tissue and on free fatty acid metabolism. Proc Natl Acad Sci 59:884–889

Lafontan M (1979) Inhibition of epinephrine-induced lipolysis in isolated while adipocytes of aging rabbits by increased alpha-adrenergic responsiveness. J Lipid Res 20:208–216

Malbon CC (1980) The effects of thyroid status on the modulation of fat cell β-adrenergic receptor agonist affinity gy guanine nucleotides. Mol Pharmacol 18:193–198

Masoro EJ, Bertrand H, Liepa G, Yu BP (1979) Analysis and exploration of age-related changes in mammalian structure and function. Fed Proc 38:1956–1961

Olefsky JM (1976) The effects of spontaneous obesity in insulin binding, glucose transport and glucose oxidation in isolated rat adipocytes. J Clin Invest 57:842–851

Olefsky JM (1977) Insensitivity of large rat adipocytes to the antilipolytic effects of insulin. J Lipid Res 18:459–464

Rosenqvist U, Efendic S (1971) Stimulatory effect in vitro of prostaglandin E_1 on noradrenaline-induced lipolysis in subcutaneous adipose tissue from hypothyroid subjects. Acta Med Scand 190:341–345

Westgren U, Burger A, Ingemansson S, Melander A, Tibblin S, Wahlin E (1976) Blood levels of 3,5,3'-triiodothyronine and thyroxine: differences between children, adults and elderly subjects. Acta Med Scand 200:493–495

Fatty Acid-Binding Proteins of Various Tissues

J. H. VEERKAMP and R. J. A. PAULUSSEN [1]

1 Introduction

The cytosol (105,000 xg supernatant) of nearly all tissues investigated appears to contain one or more proteins with a molecular weight of approximately 14,000 that bind long-chain fatty acids or their CoA and carnitine esters with a rather high affinity. The concentration of these fatty acid-binding proteins (FABPs) is high (30–50 μg mg^{-1} cytosolic protein) in tissues with a high capacity for fatty acid oxidation (liver, heart) or lipid metabolism (intestinal epithelium). Most investigations have been concentrated on the FABPs from these tissues (reviewed in Bass 1985; Glatz and Veerkamp 1985; Glatz et al. 1985a). FABPs occur in the rat in at least three tissue-specific forms (liver, intestinal and heart FABP), but two distinct forms may be present in one tissue (e.g. liver and intestinal forms in intestine). Liver FABP was also named Z-protein or protein A.

Intracellular transport of the poorly soluble long-chain fatty acids was originally suggested to be the main function of FABPs (Fig. 1). Evidence has been obtained that these proteins promote the release of fatty acids from the plasma membrane, and may modulate rates of fatty acid utilization and uptake in specific intracellular membranes. The FABPs may also regulate or protect enzyme activities or membrane transport by sequestration of modulating or noxious fatty acid (derivatives) in normal and pathological conditions.

Understanding of the role and function of FABPs in the various cell types in vivo requires more knowledge of their (tissue-) specific structure, their interaction with fatty acid derivatives and with membranes, and their metabolism. In this report we will present the results of an assay of fatty acid-binding activity and some data on the purification, physico-chemical properties and binding characteristics of FABPs from various rat, human and porcine tissues. Also the immunochemical reactivity of a series of tissue cytosols and FABP preparations with specific antisera against purified FABPs will be given. Details on specific methods are given elsewhere (Glatz et al. 1985a, b; Paulussen et al. 1986).

---

1 Department of Biochemistry, University of Nijmegen, P.O. Box 9101, 6500 HB Nijmegen, The Netherlands

Drugs Affecting Lipid Metabolism
Ed. by R. Paoletti et al.
© Springer-Verlag Berlin Heidelberg 1987

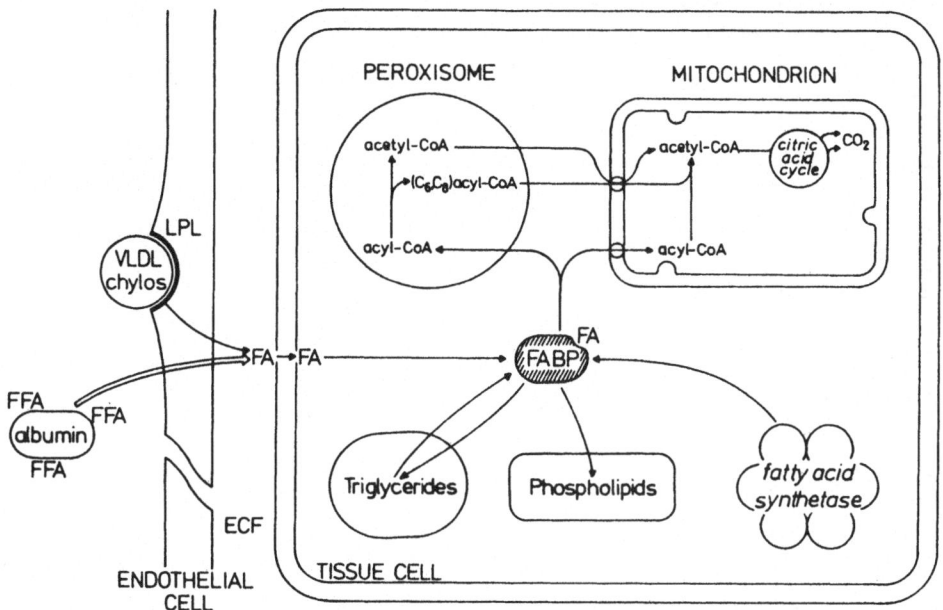

Fig. 1. Putative role of FABP in the cellular metabolism of long-chain fatty acids (Glatz and Veerkamp 1985)

2 Fatty Acid-Binding Capacity

We determined the total fatty acid-binding capacity of de-albuminized cytosolic proteins of various tissues with a radio-chemical assay, using Lipidex 1,000 for delipidation at 37°C and for separation of unbound and protein-bound fatty acids at 0°C. Maximal binding and apparent binding affinity were derived by Scatchard analysis of binding isotherms at 37°C, using 6–10 fatty acid concentrations. The oleic acid-binding capacity amounts to 1.6–4.4 pmol μg^{-1} cytosolic protein in seven rat tissues (Fig. 2). Conditions that appear to increase this binding capacity are starvation (liver), female sex (liver) and clofibrate feeding (liver and kidney) (Paulussen et al. 1986). We found no effect of post-natal age and only a minor effect of the diurnal cycle in rat heart and liver.

The high binding capacity in liver and heart is in accordance with data on FABP content with immunochemical analysis (Bass et al. 1985; Bass and Manning 1986). The rather high binding capacity in the other rat tissues does not fit with literature data on tissue expression of FABPs and on the relative levels of mRNA species (Gordon et al. 1985; Bass and Manning 1986). This indicates the presence of other FABP forms in these tissues rather than the ones investigated in the above mentioned studies.

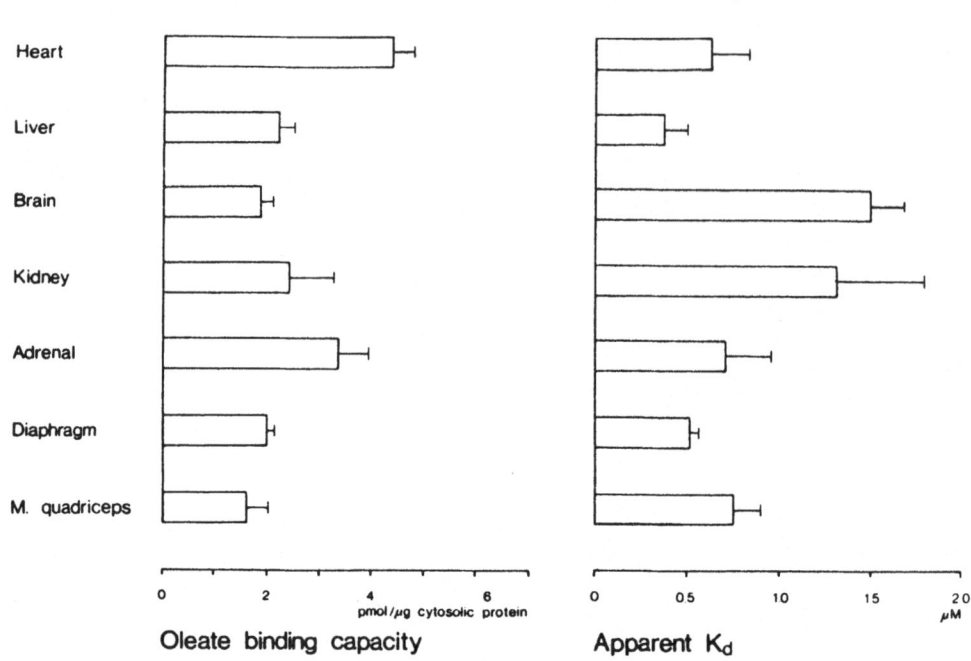

Fig. 2. Distribution of cytosolic fatty acid-binding activity in rat tissues. Maximal binding and apparent binding affinity are shown

3   Purification and Characterization of Fatty Acid-Binding Proteins

Isolation of FABPs from tissues other than heart and liver may provide information about the possible occurrence of more tissue-specific forms. These preparations can also be used to produce specific antisera for investigation of tissue expression and for immunochemical analyses. We purified FABP from four rat tissues (liver, heart, skeletal muscle and kidney) and from human and pig heart by gel filtration of 105,000 x g supernatants on Sephacryl S-200, followed by anion exchange chromatography on DEAE-Sephacel or DEAE-cellulose. The liver and kidney FABP preparations had to be re-fractionated on a Sephadex G-50 column to obtain pure FABP. The purification factors varied between 9- to 22-fold, starting with cytosol (after de-albuminization). The yield was 10–20%. FABP from rat brain could not be obtained in a pure form with this procedure, even after an additional gel-filtration step. The FABP preparations were homogeneous, as could be concluded from a single band on SDS-polyacrylamide gels. The apparent molecular weights of the FABPs varied between 14 and 17 kD. The isoelectric points for FABP from rat liver, heart, muscle and kidney were 5.5, 5.0, 5.2 and 4.9 respectively, in accordance with their chromatographic behaviour. pI values were 5.2 and 4.9 for human and pig heart FABP respectively.

The amino acid composition of the FABP preparations showed marked differences. Human and rat liver FABP contain cysteine and a rather high number of lysine residues in contrast to the heart preparations. Pig heart FABP differs markedly from FABP of human and rat heart, and also from bovine heart FABP (Jagschies et al. 1985).

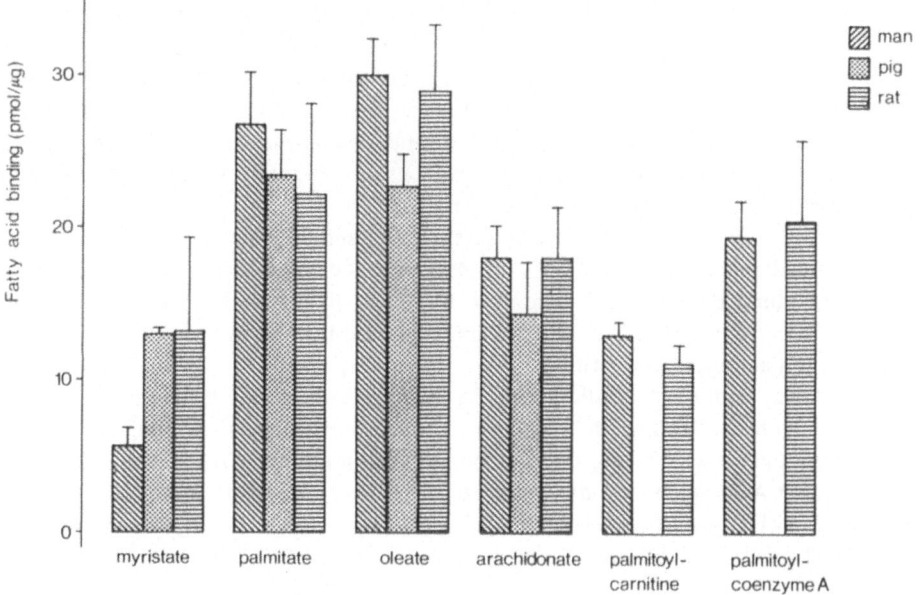

Fig. 3. Binding activity of heart FABP preparations for various long-chain fatty acids, palmitoyl-CoA and palmitoyl-carnitine

Table 1. (^{14}C) Palmitate binding to albumin and rat liver FABP in the presence of various fatty acids and agents

Fatty acid (agent) added	Agent concentration	1 μM	4 μM	
		FABP	FABP	Albumin
Palmitic acid		69	42	54
Oleic acid		67	36	38
Linoleic acid		65	39	44
Erucic acid		75	35	66
Arachidonic acid		67	33	44
Palmitoyl-CoA		94	61	90
α-Br-palmitic acid		53	25	44
Flavaspidic acid		56	39	95
Clofibric acid		101	99	95

Assay system: 5 μg protein, 1 μM (1-^{14}C) palmitate (130 dpm/pmol) in 0.5 ml 10 mM K-phosphate (pH 7.4). Separation of bound and unbound (^{14}C) palmitate with Lipidex 1000. Values (in % of control) are means of three experiments.

It contains relatively more glutamic acid (+ glutamine), proline and alanine and less threonine and lysine. Complete structural analyses of the FABP forms from human liver and rat liver, heart and intestine were published recently (Gordon and Lowe 1985; Sacchettini et al. 1986). The secundary structure of liver and heart FABPs show a prevalence of β-structure. A concentration-dependent self-association as observed for pig heart FABP (Fournier and Rahim 1983) was not found for FABP from rat or bovine heart (Jagschies et al. 1985; Offner et al. 1986).

We studied the binding of various fatty acids and of palmitoyl esters to heart FABPs by radiochemical assay (Fig. 3). Myristic and arachidonic acid bound to a lesser extent than palmitic and oleic acid, but their affinity did not differ markedly (range of Kd values, 0.2–1.0 μM). Palmitoyl-carnitine showed a lower binding maximum, but a comparable affinity.

Competitive inhibition studies of (^{14}C)palmitate binding to albumin and rat liver FABP were performed in the presence of various ligands. All long-chain fatty acids inhibit palmitate binding more strongly than palmitoyl-CoA (Table 1). These data for liver FABP agree with those reported by Bass (1985), but are in contrast with those obtained by co-elution on gel filtration and by equilibrium dialysis. Flavaspidic acid and α-Br-plamitic acid were also inhibitory in contrast to clofibric acid.

Table 2. Reactivity of cytosolic proteins with specific anti-FABP antisera

Antiserum / Tissue	Anti-rat liver	Anti-rat heart	Anti-rat muscle	Anti-human heart	Anti-pig heart
Rat					
liver	3+	–	–	+	–
heart	–	3+	–	+	+
muscle	tr	–	3+	–	–
kidney	–	–	–	–	–
adrenal	tr	–	–	–	–
spleen	–	–	–	–	–
intestine	3+	–	–	–	tr
adipose tissue	tr	tr	+	+	tr
lung	–	–	–	–	–
diaphragm	tr	2+	2+	tr	–
brain	–	–	–	–	–
Pig					
liver	+	–	–	tr	tr
heart	tr	–	–	+	3+
muscle	–	–	–	tr	+
brain	–	–	–	–	tr
Man					
liver	tr	–	–	–	+
heart	tr	–	tr	3+	3+
muscle	tr	tr	–	2+	3+
brain	–	–	–	–	tr

4 Immunochemical Studies

With the pure FABP preparations specific antibodies could be raised in rabbits. They were used in an enzyme-linked immunosorbent assay to look for the presence of different FABP forms in cytosols of various rat tissues (Table 2) and to compare the cross-reactivity of the isolated FABP preparations from human, rat and pig tissues. FABP preparations from human heart and skeletal muscle reacted with anti-human heart FABP and to a lesser extent with anti-pig heart FABP serum. Pig heart FABP shows cross-reactivity with human, but not with rat heart FABP antiserum. Antisera against rat liver and muscle FABP reacted only with their antigen. Anti-rat heart FABP reacted to an equal extent with rat heart and kidney FABP, but only slightly with human heart and muscle FABPs.

5 Conclusions

The presence of various FABP forms in different tissues suggests an adaptation to specific functions, related to the tissue in which they are expressed. Comparative studies in vitro and in vivo may be useful to solve this question. The availability of pure FABP preparations from various tissues and of their antisera allow more adequate functional and localization studies. The total fatty acid-binding capacity is not subject to acute changes, but more activity is inducible under conditions of increased fatty acid utilization. The induction of both mRNA and protein differ, however, for liver and intestinal FABP in intestine (Bass et al. 1985). This may also indicate different functions of the FABP forms in the cell. Although the physico-chemical properties of the isolated FABPs differ markedly, no large differences are present in their binding of long-chain fatty acids and acyl esters.

References

Bass NM (1985) Function and regulation of hepatic and intestinal fatty acid-binding proteins. Chem Phys Lipids 38:95−114

Bass NM, Manning JA (1986) Tissue expression of three structurally different fatty acid-binding proteins from rat heart muscle, liver and intestine. Biochem Biophys Res Commun 137:929−935

Bass NM, Manning JA, Ockner RK, Gordon JL, Seetharam S, Alpers DH (1985) Regulation of the biosynthesis of two distinct fatty acid-binding proteins in rat liver and intestine. J Biol Chem 260:1432−1436

Fournier NC, Rahim MH (1983) Self-aggregation, a new property of cardiac fatty acid-binding protein. J Biol Chem 258:2929−2933

Glatz JFC, Veerkamp JH (1985) Intracellular fatty acid-binding proteins. Int J Biochem 17: 13−22

Glatz JFC, Paulussen RJA, Veerkamp JH (1985a) Fatty acid-binding proteins from heart. Chem Phys Lipids 38:115−129

Glatz JFC, Janssen AM, Baerwaldt CCF, Veerkamp JH (1985b) Purification and characterization of fatty acid-binding proteins from rat heart and liver. Biochim Biophys Acta 837:57–66

Gordon JL, Lowe JB (1985) Analyzing the structures, functions and evolution of two abundant gastrointestinal fatty acid-binding proteins with recombinant DNA and computational techniques. Chem Phys Lipids 38:137–158

Gordon JL, Elshourbagy N, Lowe JB, Liao WS, Alpers DH, Taylor JM (1985) Tissue specific expression and developmental regulation of two genes coding for rat fatty acid-binding proteins. J Biol Chem 260:2995–2998

Jagschies G, Reers M, Unterberg C, Spener F (1985) Bovine fatty acid-binding proteins. Eur J Biochem 152:537–545

Offner GD, Troxler RF, Brecher P (1986) Characterization of a fatty acid-binding protein from rat heart. J Biol Chem 261:5584–5589

Paulussen RJA, Jansen GPM, Veerkamp JH (1986) Fatty acid-binding capacity of cytosolic proteins of various rat tissues. Biochim Biophys Acta 877:342–349

Sacchettini JC, Said B, Schulz H, Gordon JI (1986) Rat heart fatty acid-binding protein is highly homologous to the murine adipocyte 422 protein and the P2 protein of peripheral nerve myelin. J Biol Chem 261:8218–8223

Interaction of Lipid-Lowering Drugs with Fatty Acid-Binding Proteins

N. M. Bass and R. K. Ockner [1]

1 Molecular Forms and Function of the Fatty Acid-Binding Proteins

Concepts regarding the structure and function of the low molecular weight cytosolic fatty acid-binding proteins (FABPs) have expanded considerably over the past decade (for reviews see Bass 1985; Glatz et al. 1985; Ockner 1986). To date, three FABPs, the products of separate genes, have been characterized in some detail regarding their structure, tissue distribution, and regulation (Table 1). Although many details

Table 1. Comparative Properties of the FABPs

	Liver FAPB	Intestinal FABP	Heart FABP
Synonyms	L-FABP, hFABP, Sterol carrier protein (SCP), Z protein	I-FABP, gFABP	M-FABP
Size (kD)	14.2	15.1	15.0
Tissue expression	Liver, small intestine	Small intestine	Heart and skeletal muscle, testis, ovary, brain, kidney
Induction by: Hypolipidemic drugs	+++	+	0
Female sex steroid hormones	++	0	+
High fat diet	+++	++	?

1 Department of Medicine and Liver Center, HSW 1120, Box 0538, University of California, San Francisco, CA 94143, USA

Drugs Affecting Lipid Metabolism
Ed. by R. Paoletti et al.
© Springer-Verlag Berlin Heidelberg 1987

of the function of the FABPs await clarification, a considerable body of evidence supports a broad role for these abundant proteins in the transport, utilization, and cellular economy of long-chain fatty acids, and also in protecting several aspects of cellular function from the potentially injurious effects of fatty acids and their CoA esters. Factors which have been found to modulate FABP concentrations include sex steroid hormones, dietary fat, and peroxisome-proliferating hypolipidemic drugs and phthalic acid esters. This chapter will discuss aspects of our current knowledge regarding the mechanisms and significance of the interaction of the latter group of agents with the FABPs.

2 Quantitative Effects of Lipid-Lowering Drugs on the FABPs

Administration of several lipid-lowering drugs including clofibrate, nafenopin, tiadenol, and phthalic acid esters to rats has been shown by several groups to increase the concentration of L-FABP in liver (Fleischner et al. 1975; Renaud et al. 1978; Kawashima et al. 1982; Bass et al. 1985a; McTigue et al. 1986) and in small intestinal epithelium (Bass et al. 1985a; Kawashima et al. 1985). During clofibrate feeding,

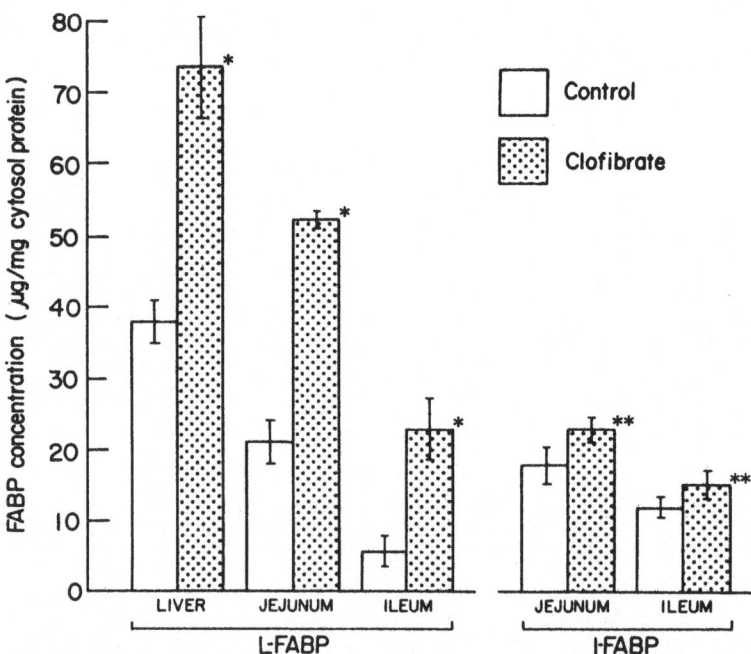

Fig. 1. Induction of L-FABP and I-FABP by clofibrate. Treateds rats received clofibrate 0.35% (w/w) in chow for 10 days; controls rats were pair-fed normal chow for the same period. L-FABP and I-FABP were each measured in the 105,000 x g supernatant (cytosol) of homogenates of liver and small intestinal epithelium by specific immunoassays. The *bars* indicate mean values ± SD; * $P < 0.001$; ** $P < 0.05$. (Data from Bass et al. 1985a)

L-FABP increases to a new steady state by 10 days, consistent with the measured half-life of this protein of 3.1 days, while L-FABP in intestine reaches a new steady state in the relatively short period of 2–3 days, reflecting the more rapid turnover of L-FABP in gut and the rapid renewal rate of intestinal epithelial cells (Bass 1985). L-FABP comprises 3–4% of the cytosol protein of male rat liver, 2% of proximal-, and 1% of distal-intestinal epithelial cytosol protein. Following clofibrate administration, L-FABP concentration in cytosol is increased two- to three fold (Fig. 1). I-FABP is present in almost identical amounts to L-FABP in intestine. However, in contrast to L-FABP it shows only a 25% increase in concentration with clofibrate treatment (Fig. 1), while the concentration of M-FABP in rat heart is entirely unaffected by clofibrate (Glatz et al. 1985).

The differing responses of the three FABPs to clofibrate as well as to other influences (Table 1) invites the speculation that they each perform specialized functions in fatty acid metabolism. A physiological significance for the increase in L-FABP in liver following clofibrate administration was proposed by Renaud et al. (1978) who found that fatty acid uptake by isolated perfused liver increased to a similar magnitude as L-FABP following clofibrate treatment. However, these findings do not resolve the question as to whether the increase in L-FABP is, at least in part, a determinant of or a secondary response to the increase in hepatocellular fatty acid flux.

3 Relationship Between L-FABP Induction and Peroxisome Proliferation

Although the various hypolipidemic agents which induce L-FABP may differ substantially in structure (e.g., clofibrate, tiadenol, and phthalate esters), they share in common the property of inducing a striking proliferation of peroxisomes (Reddy et al. 1982). Thus, hypolipidemic drugs such as probucol and niceritrol, which are not peroxisome proliferators, also fail to induce L-FABP (Kawashima et al. 1982), while the extent to which clofibrate and similarly acting agents induce hepatic peroxisomes shows a strong correlation with the magnitude of L-FABP increase (Kawashima et al. 1983). Also, the time course of L-FABP induction during clofibrate administration (Bass 1985) closely parallels the time course of increased peroxisomal palmitoyl-CoA oxidation in the livers of rats fed peroxisome proliferators (Hashimoto 1982). Furthermore, both peroxisomes (Ishii et al. 1980) and L-FABP (Bass 1985) are induced by feeding diets rich in long-chain tricylglycerols. The close apparent relationship between L-FABP and peroxisomal responses to these pharmacological and dietary influences raises the possibility of a functional dependence between L-FABP and peroxisomal β-oxidation of long-chain fatty acids. Indeed, experimental evidence has been provided by Appelkvist and Dallner (1980), suggesting that L-FABP may be involved in the transfer of acyl-CoA derivatives through the peroxisomal membrane to the enzymes of β-oxidation in the inner compartment. This important possibility clearly merits further investigation.

Of further interest is the fact that the livers of infants with the cerebrohepatorenal (Zellweger) syndrome which lack peroxisomes, contain normal levels of L-FABP

(Bass NM and Moser AE, unpubl. observations). Thus, L-FABP expression per se appears to be independent of the assembly of functional peroxisomes. However, the activities of several soluble peroxisomal matrix enzymes are similarly preserved in Zellweger patients even though membrane-bound peroxisomal enzymes are severely deficient (Wanders et al. 1984).

4   The Mechanism of L-FABP Induction by Lipid-Lowering Drugs

The increase in L-FABP in both liver and intestine that follows clofibrate feeding is the result of increased synthesis of the protein without alteration in its fractional degradation rate determined by double-label isotope incorporation (Bass et al. 1985a). Furthermore, increased L-FABP synthesis is attributable to a two fold increase in the mRNA specifying this protein determined by both cell-free translation of total cellular mRNA and Northern blot analysis using a cDNA probe for L-FABP (Bass et al. 1985a, McTigue et al. 1986). Increased gene transcription of peroxisomal fatty acid β-oxidation enzymes results from clofibrate administration (Reddy et al. 1986), and a similar mechanism could account for the accumulation of L-FABP mRNA, although this has not been directly examined.

Since both peroxisome proliferators and long-chain fatty acids induce peroxisomes and L-FABP, while binding of long-chain fatty acids to L-FABP at moderate affinity ($Kd \approx 1 \mu M$) is well established (Bass 1985), the possibility was raised as to whether L-FABP might act as a soluble "receptor" for both long-chain fatty acids and peroxisome proliferators, mediating their effects in a manner analogous to the steroid hormone receptors. However, we have been unable to show binding of ^{14}C-clofibrate or its metabolites to L-FABP in vitro or following in vivo administration (Bass NM and Ockner RK, unpublished observations). On the other hand, evidence has been presented recently for a specific, soluble receptor for nafenopin and structurally related drugs, present in hepatic cytosol at a concentration an order of magnitude lower than that of L-FABP (Lalwani et al. 1983).

The question remains of whether L-FABP induction by peroxisome proliferators represents the results of direct interaction of these agents with the genome in which L-FABP may exist as part of a "peroxisomal gene cluster", or whether this protein is induced in a secondary manner, possibly in response to an increase in fatty acid flux produced by increased peroxisomal β-oxidation. Recent immunohistochemical studies have provided information which may aid in resolving this issue. In uninduced male rats, L-FABP is predominantly expressed in the hepatocytes of the periportal (lobular zone I) and midlobular (zone II) areas. Following clofibrate treatment, L-FABP expression extends into the hepatocytes surrounding the central vein (zone III) as well (Bass et al. 1985b). Thus, the induction of L-FABP by clofibrate appears to involve, to an extent, the recruitment of synthesis in cells in which, under normal circumstances, the protein is expressed at relatively low levels. This is strikingly similar to the finding that I-FABP is only increased in ileal enterocytes by feeding a diet rich in fat (Ockner and Manning 1974).

When fluorescent derivatives of long-chain fatty acids were perfused through liver via the portal vein, predominant uptake by zones I and II of the hepatic lobule (i.e., corresponding to the normal distribution of L-FABP) was observed (Bass et al. 1985b, 1986). However, reversal of perfusion via the hepatic veins resulted in predominant uptake of fluorescent fatty acid by hepatocytes in lobular zone III where L-FABP is least expressed (Bass et al. 1986). These findings imply that fatty acid uptake governs FABP expression rather than the reverse. However, additional data are needed to conclusively resolve this question.

5 Conclusions

Lipid lowering drugs are important tools in the study of FABP regulation and function. Studies to date have indicated that the different molecular forms of FABP perform specialized functions in fatty acid utilization, and have provided important insights into the mechanism and significance of L-FABP regulation. A close relationship between L-FABP and peroxisomal β-oxidation has also been suggested by several studies, pointing to a possible role for this protein in mediating, at least in part, the mechanisms whereby peroxisome-proliferating drugs lower plasma lipids.

Acknowledgments. The authors express their thanks to Joan A. Manning for expert technical assistance and to Diana Fedorchak and Michael Karasik for preparing the manuscript. The authors are also grateful to Dr. Mitchell N, Cayen of Ayerst Research Laboratories for the gift of radio-labeled clofibrate.

This work was supported in part by research grants AM-32926, AM-13328, and Liver Core Center grant AM-26743 from the National Institutes of Health. NMB is the recipient of the American Gastroenterological Association/Janssen Pharmaceutica Research Scholar Award.

References

Appelkvist EL, Dallner G (1980) Possible involvement of fatty acid binding protein in peroxisomal β-oxidation of fatty acids. Biochim Biophys Acta 617:156−160

Bass NM (1985) Function and regulation of hepatic and intestinal fatty acid binding proteins. Chem Phys Lipids 38:95−114

Bass NM, Manning JA, Ockner RK, Gordon JE, Seetharam S, Alpers DH (1985a) Regulation of the biosynthesis of two distinct fatty acid-binding proteins in rat liver and intestine. J Biol Chem 260:1432−1436

Bass NM, Barker ME, Jones AL, Ockner RK (1985b) Zonal expression of fatty acid binding protein (FABP) corresponds with fatty acid uptake by zones 1 and 2 of the hepatic acinus. Hepatology 5:1101 (Abstr)

Bass NM, Manning JA, Ockner RK (1986) Hepatic zonal uptake of a fluorescent fatty acid derivative is determined by velocity and direction of flow. Gastroenterology 90:1710 (Abstr)

Fleischner G, Meijer DKF, Levine WG, Gatmaitan Z, Gluck R, Arias IM (1975) Effect of hypolipidemic drugs, nafenopin and clofibrate, on the concentration of ligandin and Z protein in rat liver. Biochem Biophys Res Commun 67:1401−1407

Glatz JFC, Paulussen RJA, Veerkamp JH (1985) Fatty acid binding proteins from heart. Chem Phys Lipids 38:115−129

Hashimoto T (1982) Individual peroxisomal β-oxidation enzymes. Ann NY Acad Sci 386:5–12

Ishii H, Fukumori N, Horie S, Suga T (1980) Effects of fat content in the diet on hepatic peroxisomes of the rat. Biochim Biophys Acta 617:1–11

Kawashima Y, Nakagawa S, Kozuka H (1982) Effect of some hypolipidemic drugs and phthalic acid esters on fatty acid binding protein in rat liver. J Pharm Dyn 5:771–779

Kawashima Y, Nakagawa S, Tachibana Y, Kozuka H (1983) Effects of peroxisome proliferators on fatty acid-binding protein in rat liver. Biochim Biophys Acta 754:21–27

Kawashima Y, Takegishi M, Watanuki H, Katoh H, Tachibana Y, Kozuka H (1985) Effect of clofibric acid and tiadenol on peroxisomal β-oxidation and fatty acid binding protein in intestinal mucosa of rats. Toxicol Appl Pharmacol 78:363–369

Lalwani ND, Fahl WE, Reddy HK (1983) Detection of a nafenopin-binding protein in rat liver cytosol associated with the induction of peroxisome proliferation by hypolipidemic compounds. Biochem Biophys Res Commun 116:388–393

McTigue J, Taylor JB, Craig RK, Christodoulides L, Ketterer B (1986) Effect of the peroxisome proliferator tiadenol on levels of mRNA for fatty acid-binding protein. Biochem Soc Trans 13:896–897

Ockner RK (1986) Cellular fatty acid binding proteins: evolving concepts of structure and function. Hepatol Rapid Lit Rev 16:VII–XV

Ockner RK, Manning JA (1974) Fatty acid-binding protein in small intestine. J Clin Invest 54: 326–335

Reddy JK, Warren JR, Reddy MK, Lalwani ND (1982) Hepatic and renal effects of peroxisome proliferators: biological implications. Ann NY Acad Sci 386:81–110

Reddy JK, Goel SK, Nemali MR, Carrino JJ, Laffler TG, Reddy MK, Sperbeck SJ, Osumi T, Hashimoto T, Lalwani ND, Rao MS (1986) Transcriptional regulation of peroxisomal fatty acyl-CoA oxidase and enoyl-CoA hydratase/3-hydroxyacyl-CoA dehydrogenease in rat liver by peroxisome proliferators. Proc Natl Acad Sci USA 83:1747–1751

Renaud G, Foliot A, Infante R (1978) Increased uptake of fatty acids by the isolated rat liver after raising the fatty acid binding protein concentration with clofibrate. Biochem Biophys Res Commun 80:327–334

Wanders RJA, Kos M, Roest B, Meijer AJ, Schrakamp G, Heymans HSA, Tegelaers WHH, van den Bosch H, Schutgens RBH, Tager JM (1984) Activity of peroxisomal enzymes and intracellular distribution of catalase in Zellweger syndrome. Biochem Biophys Res Commun 123:1054–1061

Animal Models for Hyperlipidemia-Induced Atherosclerosis

R. W. Wissler and D. Vesselinovitch [1]

1 Introduction

Immense strides have been made in the last 4 decades in defining, evaluating, and utilizing animal models for the study of various aspects of atherosclerosis. Several superb models of the advanced human atherosclerotic plaque have been developed (Wissler and Vesselinovitch 1986). Substantial contributions have been made in the study of the chemistry and the cellular and molecular pathobiology (Wissler 1984) as well as the molecular genetics of the atheromatous lesions.

There are many factors, some species-specific and others related to either abnormalities or disease states in the model, which can influence the type of response which occurs when hyperlipidemia or other risk factors are present. Variations which are likely to affect the response to nutritional hyperlipidemia are a deranged catabolic mechanism of cholesterol excretion and/or an unusual reticuloendothelial response to high levels of circulating lipoproteins. Frequently both of these metabolic disorders are present simultaneously. In spite of many attempts to find other approaches almost all of the animal models of progressive atherosclerosis which resemble human plaques are the result of chronic induced hypercholesterolemia.

2 Review of Progress from 1908 to 1986

Animal models have been used in the study of experimental atherosclerosis for at least 75 years. The rabbit provided the first successful model of atherosclerosis from which most of the others have developed. These include mouse, rat , guinea pig, hamster, fowl, dog, swine, and a variety of nonhuman primates (Fig. 1). However, these animals vary greatly in their sensitivity to nutritional hyperlipidemia and/or background of spontaneous atherosclerosis against which the dietary effects have to be evaluated (Wissler and Vesselinovitch 1978; Vesselinovitch 1979) (Fig. 2). In general, atherosclerosis in rat and many of the canine species cannot be produced by diet

---

1 The Specialized Center of Research in Atherosclerosis and Department of Pathology, The University of Chicago, Chicago, IL 60637, USA

Drugs Affecting Lipid Metabolism
Ed. by R. Paoletti et al.
© Springer-Verlag Berlin Heidelberg 1987

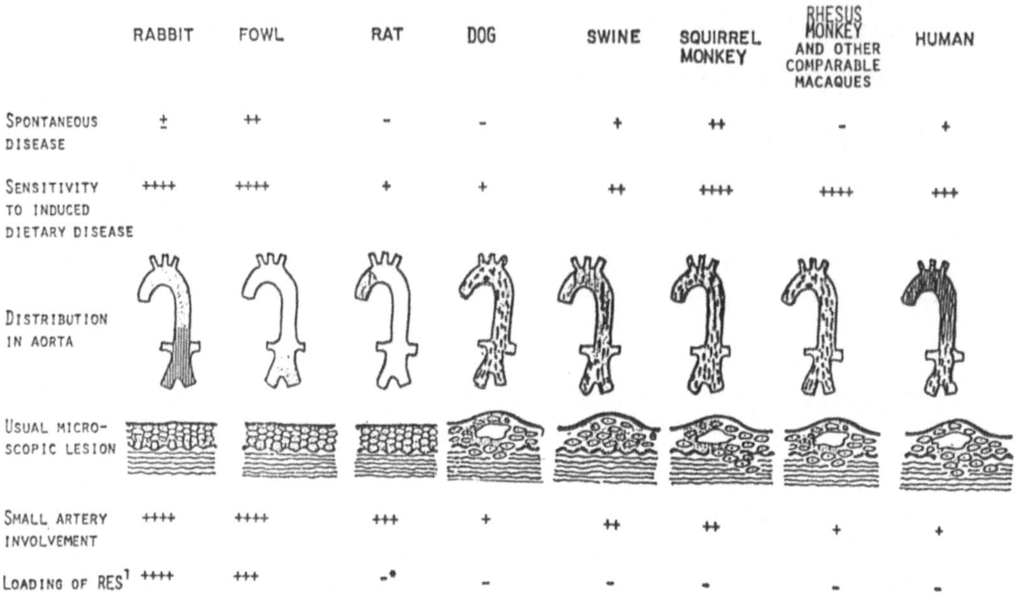

	RABBIT	FOWL	RAT	DOG	SWINE	SQUIRREL MONKEY	RHESUS MONKEY AND OTHER COMPARABLE MACAQUES	HUMAN
SPONTANEOUS DISEASE	±	++	-	-	+	++	-	+
SENSITIVITY TO INDUCED DIETARY DISEASE	++++	++++	+	+	++	++++	++++	+++
DISTRIBUTION IN AORTA								
USUAL MICRO-SCOPIC LESION								
SMALL ARTERY INVOLVEMENT	++++	++++	+++	+	++	++	+	+
LOADING OF RES[1]	++++	+++	-*	-	-	-	-	-

Fig. 1. A diagrammatic representation of some of the most notable features in the responses to hyperlipidemia of commonly used animal models of atherosclerosis which have been studied and reported.

[1] RES = reticuloendothelial system.

* Possible +++ with combined Na cholate, thiouracil and high fat + cholesterol feeding.

Raised lesions with either coconut oil and cholesterol rich ration or with thiouracil or other thyroid ablation plus high fat + cholesterol rich ration to produce sustained serum cholesterol over 700 mg%

rabbit > cynomolgus > rhesus > stumptail > green > baboon > dog > rat

|←——man——→|

Fig. 2. Sensitivity to dietary cholesterol in various mammalian species

alone; additional manipulations are needed if one is to be successful. On the contrary, rabbit, swine, and several of the nonhuman primates are quite responsive to dietary hyperlipidemia, which results almost always in moderate to severe atheromatous change.

In the past 25 years new models, most of these in nonhuman primates, have been demonstrated to be especially valuable in simulating human atherosclerotic disease (Strong 1976; Wissler and Vesselinovitch 1978; Vesselinovitch 1979). These inclue: baboon; squirrel, African green and patas monkeys; rhesus, cynomolgus, pigtail, and Celebes macaques. Their advantages and disadvantages have already been described in detail (Wissler and Vesselinovitch 1977). The studies in the 1950s in the baboon demonstrated that this species was rather resistant to atherosclerotic changes, while much more severe disease was demonstrated in the rhesus monkey, with simple additions to the diet of fat and cholesterol. The lesions, in addition to having gross and microscopic resemblance to human plaques, also produced the same complications which are associated with human atherosclerosis, namely thrombosis, ischemic peripheral vascular disease and coronary heart disease (Wissler and Vesselinovitch 1977, 1978; Vesselinovitch 1979).

In striking contrast to many other animal models, the topography of the disease in the rhesus monkey appears to be very similar to that in the human. It is quite clear that many of these species-related variable factors, when understood, can contribute to our understanding of the reactions of the arterial wall to different stimuli and therefore serve a useful purpose. On the contrary, if the peculiarities of the species are not appreciated, misconceptions may occur and results may be misinterpreted.

When several of these prototypes of animal models are compared and contrasted, to each other and with human disease, it appears that lesions in swine and in rhesus monkey are most similar to the human disease (Fig. 1) (Wissler and Vesselinovitch 1986). It now appears that there are very important smilarities in the components of advanced atheromatous lesions in young people suffering from homozygous familial hypercholesterolemia, older individuals with advanced disease, and advanced lesions in rhesus monkeys. It is apparent from this that time is not the main factor determining the architecture of the advanced atherosclerotic plaque. Experimental plaques which develop in a brief period in the relatively short life span of primates offer reliable models in which to study the effects of intervention on the major components of the lesions (Wissler et al. 1985). Other useful models for the study of the effect of nutritional hyperlipidemia are: the utilization of a special genetic strain of rabbits (Watanabe), a promising counterpart of homozygous familial hypercholesterolemia in humans; coconut oil-induced lesions in the dog which have been found to be closely related to the presence of broad beta-VLDL in the serum (Innerarity et al. 1982); and certain strains of swine lacking the von Willebrand's factor which develop a syndrome closely resembling von Willebrand's disease in humans (Griggs et al. 1981). These various models offer substantial advances in understanding mechanisms which are likely to influence the pathogenesis of atherosclerosis in humans. Perhaps the most exciting developments in this field relate to: (1) new insights into how hyper- and hyporesponders differ and (2) the development of genetic strains of baboons by McGill and co-workers which appear to have heritable characteristics affecting levels of serum LDL, HDL, and HDL_1 (Wissler and Vesselinovitch 1986).

It is quite clear that as one gains more experience with animals models, postulates emerge which, if fulfilled, will help greatly in establishing the validity and probable dependability of a model for studies of human disease. These primary postulates are: (1) progressive atherosclerosis is produced by a given stimulus; (2) the greater the stimulus, the greater the response; (3) lesion chemistry and plaque components as well as major functional effects should be similar to human; (4) lesions can be prevented and/or reversed by removing the stimulus or by intervention, i.e. drugs and/or diet; (5) no other conditions appear to produce the same results, i.e., no hyperlipidemia, no atherosclerosis (Wissler and Vesselinovitch 1986).

The value of some animal models can be increased or decreased by several factors. One important example of these which seems to influence the architecture of the atherosclerotic lesions, are circulating immune complexes which, if sustained, may alter the reaction of the artery wall to hyperlipidemia. As a consequence, the hemo-

Table 1. Seventy-five years of contributions to knowledge of atherosclerosis by use of animal models

1. Controlled evaluation of effects of dietary variables on disease at lesion level
   - Different types of food fats
   - Saturated and unsaturated
   - Penaut oil
   - Coconut and palm kernel oils
   - Fish oils
   - Calories
   - Cholesterol (quantity and dangers of oxidation)
   - Fats interact with proteins and fiber

2. Establishment of interrelationships and additive effects on lesions among major pathogenetic risk factors and demonstration of influence on lesions
   - Hypertension
   - Endocrine imbalance (incl. diabetes and hypothyroidism)
   - Cigarette products
   - Lipoprotein imbalances

3. Successful testing and documented evidence of interventions which help to prevent atherosclerotic plaque progression at the arterial lesion level
   - Dietary
   - Lipid lowering drugs
   - Exercise
   - Induced bradycardia
   - Induced hypotension
   - Ca^{++} channel block

4. Development of various methods for producing regression of lesions
   - Nutritional
   - Drug
   - Combined therapy

5. Studies of metabolic and cell pathobiological mechanisms by which risk factors and intervention influence the development of atherosclerosis at the artery wall and cellular level

dynamic factors which normally cause the plaque to form in an eccentric position, may be masked or overcome so that concentric and transmural lesions develop in muscular arteries (Wissler et al. 1985). These factors may alter the fundamental disease process by accelerating the lesion development, increasing the tendeny to thrombosis and focal dilatation of the arteries, and retarding the beneficial effects of lesion regression (squirrel, cynomolgus, pigtailed macaque).

Virus infections of the arterial lesion cells, especially endothelial or smooth muscle cells, accompanied by hyperlipidemia, may influence the rate of progression of the diesease, as demonstrated in some animal models (Fabricant et al. 1983).

Other factors which may alter and augment the atherogenic process are: chronic endotoxin injury, oxidized fatty acids, and oxidized cholesterol. Furthermore, increasing emphasis on the vascular effects of leukotrienes and free radicals may broaden this list of injurious agents.

A summary of the main overall contributions of animal models which have especially improved our understanding of atherosclerosis and its response to intervention is given in Table 1. As a result of these studies we now have a greater knowledge of some of the factors which can alter the main components of the atherosclerotic lesions. Many of these have proven to be as effective in the human disease as in experimental models of atherosclerosis. Data regarding regression and/or retardation as well as healing of some of the lesion components have been amply demonstrated in numerous models (Wissler and Vesselinovitch 1984).

3 Predictions of Future Progress

The future role of animal models in atherosclerosis research lies in the areas where momentum has already been generated. A few selected examples in which future work appears most promising are: (1) increasing emphasis on cellular and molecular mechanisms of lesion progression and regression; (2) extension of "combined therapy" and development of improved rationale for such treatment; (3) determining the significance of diffuse fibrous thickening and its contribution to the fate of lesions; (4) elucidating the mechanisms of hyper- and hyporesponders to lesion stimuli; and (5) better understanding of the pathobiological and molecular pathology effects of dietary, metabolic, and toxic factors at the arterial wall level.

Future studies of the mode of action of different currently available pharmaceutical agents will enable us to utilize two or three drugs simultaneously for therapy. It is now widely recognized that this approach may have several advantages such as: fewer side reactions, greater effectiveness, less cost to the patient, etc.

Moreover, the effects of various dietary fats on the fate of atherosclerotic lesions have been largely clarified through the use of animal models (Vesselinovitch et al. 1974, 1980). Future studies should increase our understanding of the mechanisms of the effects of various nutrients at the arterial wall level, and other variations in the metabolic handling of dietary factors due to phenotypic alterations in the host.

4 Summary

The development and use of animal models of human atherosclerosis, especially nonhuman primates, has increased markedly during the last 25 years. Comprehensive studies in dogs, swine, and nonhuman primates have led to an improvement in our knowledge of the pathogenesis (plaque evolution) from early to advanced lesions, as well as the clinical complications in human cardiovascular, cerebrovascular and other peripheral vessels.

The future holds great promise for extending knowledge using animal models of atherosclerosis so that it should be possible not only to take full advantage of existing models, but to apply ones which are being developed in inbred strains. The foremost general precept is that all models are valuable if their use is governed by thorough knowledge of their similarity or lack of similarity to human atherosclerosis. If the results of studies are to be extrapolated and/or applied to human situations, the physiological, biochemical, anatomical, and morphological features should be compared carefully to human disease in order to draw valid conclusions.

Acknowledgments. The authors are grateful for the contributions of many co-workers who have added to our understanding of animal models of atherosclerosis, especially those who have worked with us on many animal studies in many species in our laboratory. This manuscript was made possible by remarkable contributions from Gertrud Friedman, Joan King, and Delores Irvin. Grant support for partial support of many of our studies came from HL 15062.

References

Fabricant CG, Fabricant J, Minick CR, Litrenta NM (1983) Herpes virus-induced atherosclerosis in chickens. Fed Proc 42:2476

Griggs TR, Reddick RL, Seltzer D, Brinkhous KM (1981) Susceptibility of atherosclerosis in aorta and coronary arteries of swine with von Willebrand's disease. Am J Pathol 102:137–145

Innerarity TL, Pitas RE, Mahley RW (1982) Modulating effects of canine high density lipoproteins on cholesterol ester synthesis induced by β-very low density of lipoproteins in macrophages: possible in vitro correlates with atherosclerosis. Arteriosclerosis 2:114–124

Strong JP (ed) (1976) Atherosclerosis in primates. Karger, Basel

Vesselinovitch D (1979) Animal models of atherosclerosis, their contributions and pitfalls. Artery 5:193–206

Vesselinovitch D, Getz GS, Hughes RH, Wissler RW (1974) Atherosclerosis in the rhesus monkey fed three food fats. Atherosclerosis 20:303–321

Vesselinovitch D, Wissler RW, Schaffner TJ, Borensztain J (1980) The effects of various diets on atherogenesis in rhesus monkeys. Atherosclerosis 35:198–207

Wissler RW (1984) The pathobiology of the atherosclerotic plaque in the mid-1980s In: Malinow MR, Blaton VH (eds) Regression of atherosclerotic lesions. Plenum, New York, pp 5–20

Wissler RW, Vesselinovitch D (1977) Atherosclerosis in non-human primates In: Brandly CA, Cornelius CE, Simpson CF (eds) Advances in veterinary science and comparative medicine. Academic Press, London New York, pp 351–420

Wissler RW, Vesselinovitch D (1978) Evaluation of animal models for the study of the pathogenesis of atherosclerosis. In: Hauss WH, Wissler RW, Lehmann R (eds) Int Symp State of prevention and therapy in human arteriosclerosis and in animal models. Westdeutscher Verlag, Opladen, pp 13–29

Wissler RW, Vesselinovitch D (1984) Interaction of therapeutic diets and cholesterol-lowering drugs in regression studies in animals. In: Malinow MR, Blaton VH (eds) Regression of atherosclerotic lesions. Plenum, New York, pp 21–41

Wissler RW, Vesselinovitch D (1986) The development and use of animal models in atherosclerosis research. In: Gallo L (ed) Cardiovascular disease '86. Plenum, New York (in press)

Wissler RW, Vesselinovitch D, Davis HR, Lambert PH, Bekermeier M (1985) A new way to look at atherosclerotic involvement of the artery wall and the functional effects. Ann NY Acad Sci 454:9–22

Animal Species as Models of Spontaneous and Induced Hyperlipidemias: A Critical Appraisal

M. J. CHAPMAN [1]

1 Introduction

The search for spontaneous animal models of primary hyperlipidemias in man has become increasingly important given their relevance to the quest for new, pharmacologically-active, lipid-lowering compounds, and to studies of the onset, development, and regression of atherosclerosis. At a theoretical level, it may be argued that certain basic criteria be respected when selecting a mutant strain of an animal species as a model for a particular type of primary hyperlipidemia: suitable criteria might be ·
(1) that the genetically-transmitted mutation(s) characteristic of the human condition be present in the animal model, preferably with a similar transmission, and (2) that there be a clear resemblance in the quantitative and qualitative aspects of lipoprotein and apolipoprotein profiles in man and the animal model.

Animal models in which these two criteria are satisfied are unfortunately rare. This situation has arisen since, on the one hand, our lack of knowledge of the genetic basis of several of the most common forms of human hyperlipoproteinemia (Grundy 1986) makes it all but impossible to define appropriate animal counterparts; on the other, only a restricted number of genetically-inbred animal models of spontaneous hyperlipidemia is available, and in only a few of these has the basic biochemical lesion(s) been identified.

2 Diet-Induced Models of Hyperlipidemia

As a reflection of the lack of "spontaneous" animal models, there has been a plethora of studies in recent years of diet-induced hyperlipoproteinemia and experimental atherosclerosis in a diverse range of animal species, ranging from birds such as the pigeon (Barakat and St. Clair 1985) and Japanese quail (Chapman et al. 1975) to rodents (rat, Mahley and Holcome 1977; mouse, Morrisett et al. 1982; guinea pig, Chapman and Mills 1977), to the rabbit (Shore et al. 1974), the dog (Mahley 1978)

---

1 Groupe de Recherches INSERM sur les Lipoprotéines, Pavillon Benjman Delessert, Hôpital de la Pitié, 75651 Paris CEDEX 13, France

Drugs Affecting Lipid Metabolism
Ed. by R. Paoletti et al.
© Springer-Verlag Berlin Heidelberg 1987

and the pig (Mahley 1978) and to certain species of nonhuman primate, including the Patas monkey (Mahley 1978), the African green and cynomolgus monkeys (Rudel et al. 1986) and the chimpanzee (Blaton and Peeters 1975). Such investigations have, however, had a number of important consequences, which include the identification of atherosclerosis-susceptible and atherosclerosis-resistant strains (e.g., White Carneau and Show Racer pigeons; Barakat and St. Clair 1985), and the selective breeding of such strains (SEA- and REA-quail; Chapman et al. 1975).

Furthermore, these studies have led to the identification of lipoproteins which are apparently of high atherogenicity; these are notably β-migrating VLDL (Mahley 1982) and high molecular weight LDL (Rudel et al. 1986).

3 Spontaneous Models of Hyperlipidemia

Two basic groups of animal species which exhibit spontaneous, genetically-transmitted hyperlipidemias can be identified.

The first group is typified by the cyclical occurrence of elevated lipoprotein levels, and may be illustrated by the European badger (*Meles meles* L.). This animal displays a thyroid-dependent cycle, in which a marked hypercholesterolemia is evident during the winter months as a result of elevation of cholesteryl ester-rich lipoproteins spanning the density range of LDL and "light" LDL (Laplaud et al. 1980). This hypercholesterolemia may result from the down-regulation of hepatic LDL receptors in a hypothyroid state.

The second groups shows no pronounced cyclical phenomena and expresses a "continuous" hyperlipidemic condition. They may be divided into two subgroups, one of which includes the genetically-defined inbred strains, while the second is composed of species expressing a developmental and thus, transient hyperlipoproteinemia (e.g., suckling rat, Table 1).

The inbred strains presently available are essentially limited to two rodents, the mouse and rat, and to a herbivore, the rabbit. The potential of the mouse model has been well illustrated in the recent review of Lusis and LeBoeuf (1986) and offers several advantages which include the very large number of inbred strains available, the availability of genetic markers on mouse chromosomes, the ability to conduct genetic analyses, and thence to construct genetic maps. At present, more than ten genes have been identified in inbred mouse strains which are intimately associated with plasma transport and metabolism (Lusis and LeBoeuf 1986). By contrast, information on murine plasma lipoproteins is limited to structural parameters and mainly concerns the Swiss strain (Table 1), although data on genes controlling HDL levels and apolipoprotein structure (apo-AI and AII) is available in more than 40 strains (Lusis et al. 1983).

In the rat, few inbred strains with lipid transport abnormalities have been described to date (Table 1), and of these, none can be said to be thoroughly characterized at the biochemical and genetic levels at the present time. As has become evident in a recent review (Chapman 1986), the largest body of data on the chemistry, structure, and metabolism of lipoproteins in any animal species is that in the rat, and therefore

Table 1. Lipoprotein profiles in normal and inbred animal strains

Species	Strain/age[a]	Plasma concentration: mg dl^{-1}						
		VLDL	IDL	LDL	HDL$_1$	HDL	HDL$_2$	HDL$_3$
Mouse	Swiss	40–85	0–45	40–70	Present	250–570	300–380	150–265
Rat	W/S.D/H/Z	30–110	0–30	15–60	<20	70–250		(Low)
	Suckling	50	n.d.	200	125	220	220	–
	Zucker fatty	130 (S$_f$>400:110)	2	115	n.d.	442	n.d.	n.d.
	O-SHR "corpulent"	>95% TG >90% CHOL	↑	(tr)	↑	200–300	n.d.	n.d.
	Rico	90	73		(110)	285	n.d.	n.d.
	SW	97	36		(71)	212	n.d.	n.d.
	SDBB Wistar	220	115		n.d.	240	n.d.	n.d.
Rabbit	NZW control	206	46	59	n.d.	248	n.d.	n.d.
	WHHL	366	360	719	n.d.	48	n.d.	n.d.

n.d. = not determined; tr = trace; ↑ = elevated amounts present.

[a] Data in normolipidemic strains are taken from a recent review (Chapman 1986).
W = Wistar; S.D. = Sprague Dawley; H = Holtzman; Z = Zucker lean; O-SHR = obese spontaneously hypertensive; SDBB Wistar = spontaneously diabetic BB Wistar; NWZ = New Zealand White; WHHL = Watanabe heritable hyperlipidemic strains; TG =triglyceride; CHOL = cholesterol.
Data in Zucker fatty rats from Schonfeld et al. (1974), in corpulent rats from Weisgraber and Mahley (1983), in Rico and SW rats from Cardona (1986), and in SDBB Wistar rats from Patel et al. (1984), respectively.

rapid progress could be made in this rodent should new genetically-defined inbred strains suitable as models of human dyslipoproteinemias become available.

The lipoprotein profiles typical of normal and of variant, inbred strains of mice and rats are summarized in Table 1. Data in obese mice are lacking. In the rat, the mutant strains defined to date primarily express hypertriglyceridemia with associated obesity (Zucker fatty and obese-SHR); the spontaneously diabetic BB Wistar rat is also hypertriglyceridemic, but differs from the Zucker in that VLDL are the principal triglyceride-rich particles, whereas both chylomicrons ($S_f > 400$) and VLDL are elevated in the Zucker fatty strain. The Rico rat displays LDL (d $1.006-1.040$ g ml^{-1}) levels about twice as high as its normal (SW) counterpart, with elevated levels of both HDL$_1$ and total HDL of d $1.063-1.21$ g ml^{-1}. These increases apparently reflect an elevated rate of endogenous hepatic cholesterol synthesis (Cardona 1986).

The most extensively studied animal model of spontaneous hyperlipidemia, and more specifically, of familial hypercholesterolemia, is the WHHL rabbit, an inbred strains which lacks the cellular LDL receptor, and which exhibits markedly elevated plasma concentrations of IDL and LDL and diminished HDL (Table 1). These alterations are associated with premature atherosclerosis (Havel et al. 1982). For further details of the metabolic basis of these anomalies, the reader is referred to the elegant series of reports of Brown, Goldstein, and colleagues (Buja et al. 1983).

4 Conclusions

In conclusion, there would appear to be an urgent need for the development and characterization of genetically-defined animal models of human dyslipoproteinemias, for which only familial hypercholesterolemia can be said to possess an animal counterpart. Furthermore, substantial efforts should now be made to identify atherogenic lipoproteins in animal models and to determine their sites and mechanisms of action; only then can new horizons be opened for development of new therapeutic approaches to arrest and reverse atherosclerotic disease.

References

Barakat HA, St.Clair RW (1985) Characterisation of plasma lipoproteins of grain- and cholesterol-fed White Carneau and Show Racer pigeons. J Lipid Res 26:1252–1268

Blaton V, Peeters H (1975) The nonhuman primates as models for studying human atherosclerosis: studies on the chimpanzee, the baboon and the rhesus macacus. Adv Exp Biol Med 67:33–64

Buja LM, Kita T, Goldstein JL, Watanabe Y, Brown MS (1983) Cellular pathology of progressive atherosclerosis in the WHHL rabbit. An animal model of familial hypercholesterolemia. Arteriosclerosis 3:87–101

Cardona LE (1986) Biodynamics of cholesterol and genetic hypercholesterolemia. Thesis Paris-Sud, Orsay, Fr

Chapman MJ (1986) Comparative analysis of mammalian plasma lipoproteins. Methods Enzymol 128A:70–143

Chapman KP, Stafford WW, Day CE (1975) Animal model for experimental atherosclerosis produced by selective breeding of Japanese quail. Adv Exp Biol Med 67:347–356

Chapman MJ, Mills GL (1977) Characterisation of the serum lipoproteins and their apoproteins in hypercholesterolaemic guinea pigs. Biochem J 167:9–21

Grundy SM (1986) Hyperlipoproteinemias: metabolic derangements. In: Fidge NA, Nestel PJ (eds) Atherosclerosis VII. Excerpta Medica, Amsterdam, New York, Oxford, pp 133–136

Havel RJ, Kita T, Kotite L, Kane JP, Hamilton RL, Goldstein JL, Brown MS (1982) Concentration and composition of lipoproteins in blood plasma of the WHHL rabbit. Arteriosclerosis 2: 467–474

Laplaud PM, Beaubatie L, Maurel D (1980) A spontaneously seasonal hypercholesterolemic animal: plasma lipids and lipoproteins in the European badger (*Meles meles* L.). J Lipid Res 21:724–738

Lusis AJ, LeBoeuf RC (1986) Genetic control of plasma lipid transport: mouse model. Methods Enzymol 128A:877–894

Lusis AJ, Taylor BA, Wangenstein RW, LeBoeuf RC (1983) Genetic control of lipid transport in mice. J Biol Chem 258:5071–5078

Mahley RW (1978) Alterations in plasma lipoproteins induced by cholesterol feeding in animals including man. In: Dietschy JM, Gotto AM, Ontko JA (eds) Disturbances in lipid and lipoprotein metabolism. Am Physiol Soc, Bethesda, pp 181–198

Mahley RW (1982) Atherogenic hyperlipoproteinemia. The cellular and molecular biology of plasma lipoproteins altered by dietary fat and cholesterol. In: Havel RJ (ed) Medical clinics of North America: Lipid disorders, vol 66. Saunders, Philadelphia, pp 375–402

Mahley RW, Holcome KS (1977) Alterations of the plasma lipoproteins and apoproteins following cholesterol feeding in the rat. J Lipid Res 18:314–324

Morrisett JD, Kim MS, Patsch JR, Datta SK, Trentin JJ (1982) Genetic susceptibility and resistance to diet-induced atherosclerosis and hyperlipoproteinemia. Arteriosclerosis 2:312–324

Patel ST, Newman HAI, Yakes AJ, Thibert P, Falko JM (1984) Serum lipids and lipoprotein composition in spontaneously diabetic BB Wistar rats. J Lipid Res 28:1072–1083

Rudel LL, Parks JS, Johnson FL, Babiak J (1986) Low density lipoproteins in atherosclerosis. J Lipid Res 27:465–474

Schonfeld G, Felski C, Howald MA (1974) Characterisation of the plasma lipoproteins of the genetically obese hyperlipoproteinemic Zucker fatty rat. J Lipid Res 15:457–464

Shore VG, Shore B, Hart RG (1974) Changes in apolipoproteins and properties of rabbit very low density lipoproteins on induction of cholesterolemia. Biochemistry 13:1579–1585

Weisgraber KH, Mahley RW (1983) Characterisation of rat plasma lipoproteins. In: Lewis LA, Naito HK (eds) Lipoprotein studies of nonhuman species. CRC, Boca Raton, Fla, pp 103–132 (CRC Handbook of electrophoresis, vol 4)

Rabbit Models Hypo- or Hyperresponsive to Changes in Diet

A. C. Beynen [1], A. G. Lemmens [1], J. F. C. Glatz [2], M. B. Katan [2], and L. F. M. Van Zutphen [1]

1 Introduction

The feeding of a high-cholesterol diet to random-bred rabbits generally elicits marked differences in the response of plasma cholesterol between individuals. Individuals showing only small changes in the concentrations of plasma cholesterol (hyporesponders) can be discriminated from rabbits showing high degrees of hypercholesterolemia (hyperresponders). The mechanisms underlying hypo- and hyperresponsiveness have not yet been unravelled, but the availability of inbred strains of rabbits with defined, but different cholesterolemic responses to changes in diet may be of great importance in this respect. Table 1 shows that certain strains of rabbits are relatively sensitive or insensitive to cholesterol loading. In this communication we present the results of further studies with the strains displaying the most extreme (AX/J) and lowest response (IIIVO/J).

Table 1. Effect of dietary cholesterol on plasma cholesterol levels in inbred strains of rabbits [a]

Strain	Plasma cholesterol (mmol l^{-1})	
	Day 0	Day 21
IIIVO/J	0.7 ± 0.1	8.0 ± 1.9
WH/J	0.3 ± 0.0	8.0 ± 0.8
X/J	0.9 ± 0.3	17.3 ± 4.2
ACEP/J	0.7 ± 0.0	18.9 ± 3.5
OS/J	0.8 ± 0.1	29.4 ± 4.8
AX/J	0.6 ± 0.1	33.6 ± 3.1

[a] Results are expressed as means ± SE for 5 male animals per strain. Up until day 0 of the experiment all rabbits received a cholesterol-free, commercial diet; then, 0.5% (w/w) cholesterol was added to the diet. (Data taken from Van Zutphen and Fox 1977).

1 Department of Laboratory Animal Science, State University, P.O. Box 80.166, 3508 TD Utrecht, The Netherlands
2 Department of Human Nutrition, Agricultural University, De Dreijen 12, 6703 BC Wageningen, The Netherlands

Drugs Affecting Lipid Metabolism
Ed. by R. Paoletti et al.
© Springer-Verlag Berlin Heidelberg 1987

2 Effects of Dietary Components Other than Cholesterol

It is relevant to know whether the rabbit strains hypo- and hyperresponsive to dietary cholesterol are also hypo- and hyperresponsive respectively, to other dietary components that affect plasma cholesterol levels. Such information may provide clues to the mechanisms underlying the between-strain variation in cholesterolemic response to diet.

In the hypo- and hyperresponsive rabbit strains we have measured the response of their plasma cholesterol to saturated fatty acids provided by coconut fat versus poly-unsaturated fatty acids from corn oil. Cholesterol-free, semi-purified diets were used, and the fat source (10%, w/w) was the only variable. Table 2 shows that in the inbred rabbit strains hypo- and hyperresponsiveness to dietary cholesterol and to the type of fats coincided.

It is well known that in young, growing rabbits cholesterol-free, semi-purified diets containing casein as a protein source produce hypercholesterolemia, whereas no such effect is observed with soy protein. Table 2 illustrates that the rabbit strain hyperresponsive to dietary cholesterol also showed a significantly higher response of plasma cholesterol to casein (21%, w/w) than the hyporesponsive strain.

Table 2. Plasma cholesterol responses to dietary variables in inbred rabbits[a]

	Change in plasma cholesterol (mmol l^{-1})	
	Hyporesponder	Hyperresponder
Casein vs soy protein	-0.36 ± 0.08	1.28 ± 0.96*
Coconut fat vs corn oil	0.64 ± 0.77	4.93 ± 3.70*
Cholesterol (0.3%, w/w) vs no cholesterol	4.39 ± 2.25	15.20 ± 9.73*

[a] Results are expressed as means ± SD for 4 animals per strain. Change in plasma cholesterol is the difference between values at the end of the dietary periods which lasted 4 weeks each.

[b] * Significantly different from hyporesponders: $P < 0.05$ (Student's t-test). (After Beynen et al. 1986).

3 Fecal Steroid Excretion

In an attempt to gain insight into the mechanisms underlying the strain difference in cholesterolemic response to dietary cholesterol, we have measured the fecal excretion of steroids. Table 3 shows that on a cholesterol-free, commercial diet the hyporesponders excrete more bile acids than the hyperresponders, while the rates of excretion of neutral steroids were similar in both strains. Thus, the turnover of cholesterol is higher in the hypo- than hyperresponders. The transfer to a cholesterol-enriched, commercial diet caused a more pronounced increase in neutral steroid excretion in the hypo- than hyperresponders. Bile acid excretion was not affected in both strains.

Table 3. Fecal excretion of steroids by inbred rabbits fed a cholesterol-free or high-cholesterol (0.3%, w/w) commercial diet [a]

	Steroid excretion (μmol day^{-1})	
	Hyporesponder	Hyperresponder
Low cholesterol diet		
Bile acids	167 ± 31	88 ± 19
Neutral steroids	105 ± 15	101 ± 9
High cholesterol diet		
Bile acids	185 ± 54	107 ± 44
Neutral steroids	632 ± 155	437 ± 121

[a] Results expressed as means ± SD for 5 or 6 animals per strain. On day 0 the animals were transferred from the cholesterol-free to the high cholesterol diet. Feces of each rabbit were collected daily for 4 days (days -7 to -4 and days 28 to 31). Fecal steroids were analyzed as described (Glatz et al. 1985).

4 Cholesterol Absorption

One explanation of the enhanced excretion of neutral steroids after cholesterol feeding in hyporesponders (Table 3) could be a lower efficiency of absorption in these animals. We therefore assessed cholesterol absorption in the two strains of rabbits. Both on a cholesterol-free and high-cholesterol (0.08%, w/w) semi-purified diet the hyperresponders absorbed a significantly higher percentage of dietary cholesterol than the hyporesponders (Table 4).

Table 4. Cholesterol absorption in inbred strains of rabbits fed cholesterol-free or high cholesterol, semi-purified diets [a]

	Cholesterol absorption (%)	
	Hyporesponder	Hyperresponder
Cholesterol-free diet	84.8 ± 2.2	91.4 ± 2.0
High cholesterol diet	89.6 ± 1.0	92.5 ± 1.4
Cholesterol-free diet	85.0 ± 1.8	90.0 ± 1.0

[a] Results expressed as means ± SD for 5 or 6 animals per strain. On day 0 the animals were transferred from a cholesterol-free, commercial diet to the cholesterol free, semi-purified diet. On days 41 and 111 cholesterol was added and removed from the diet respectively. Cholesterol absorption was assessed using the (^{14}C)β-sitosterol/(^{3}H) cholesterol ratio method (St. Clair et al. 1981) during days 35 to 38, 97 to 100 and 189 to 192.

5 Underlying Mechanisms

Hyperresponsiveness to dietary cholesterol is associated with responsiveness to dietary saturated fatty acids and dietary casein, pointing to a common metabolic pathway. However, it is not yet possible to illustrate this in molecular terms. A possible chain of events is that the high efficiency of cholesterol absorption in the hyperresponders produces an increased flux of cholesterol into the liver. The increased liver cholesterol pools may cause a higher hepatic efflux of cholesterol after cholesterol consumption. Cholesterol feeding of rabbits has been shown to increase cholesterol output with apoprotein B containing lipoproteins by their perfused livers (MacKinnon et al. 1985). The stimulation of very low density lipoprotein (VLDL) production accounts for the increase in serum cholesterol. In both strains the excess of cholesterol in plasma after cholesterol feeding is located in the VLDL fraction, and probably represent β-VLDL particles, the increase in VLDL cholesterol being significantly higher in the hyperresponders than in their hyporesponsive counterparts (Beynen et al. 1984). The number of hepatic LDL receptors, which have high affinity for β-VLDL, will decrease (Kovanen et al. 1981) through down-regulation. The receptor-mediated VLDL clearance decreases, but the absolute amount of VLDL cholesterol taken up by the cells via the receptor and by the receptor-independent pathway increases because of the increased level of VLDL cholesterol. In this way a new equilibrium is reached in which VLDL production equals VLDL catabolism.

Acknowledgement. Thanks are due to I. Zaalmink for accurate production of this manuscript.

References

Beynen AC, Katan MB, Van Zutphen LFM (1984) Plasma lipoprotein profiles and arylesterase activities in two inbred strains of rabbits with high or low response of plasma cholesterol to dietary cholesterol. Comp Biochem Physiol 79B:401–406

Beynen AC, West CE, Katan MB, Van Zutphen LFM (1986) Hyperresponsiveness of plasma cholesterol to dietary cholesterol, saturated fatty acids and casein in inbred rabbits. Nutr Rep Int 33:65–70

Glatz FJC, Schouten FJM, Den Engelsman G, Katan MB (1985) Quantitative determination of neutral steroids and bile acids in human feces by capillary gas-liquid chromatography. In: Beynen AC, Geelen MJH, Katan MB, Schouten JA (eds) Cholesterol metabolism in health and disease. Studies in the Netherlands. Ponsen and Looijen, Wageningen, pp 103–112

Kovanen PT, Brown MS, Basu SK, Bilheimer DW, Goldstein JL (1981) Saturation and suppression of hepatic lipoprotein receptors: mechanism for the hypercholesterolemia of cholesterol-fed rabbits. Proc Natl Acad Sci USA 78:1396–1400

MacKinnon AM, Savage J, Gibson RA, Barter PJ (1985) Secretion of cholesteryl ester-enriched very low density lipoproteins by the liver of cholesterol-fed rabbits. Atherosclerosis 54:145–153

St.Clair RW, Wood LL, Clarkson TB (1981) Effect of sucrose polyester on plasma lipids and cholesterol absorption in African green monkeys with variable hypercholesterolemic response to dietary cholesterol. Metabolism 30:176–183

Van Zutphen LFM, Fox RR (1977) Strain differences in response to dietary cholesterol by JAX rabbits: correlation with esterase patterns. Atherosclerosis 28:435–446

Plasma Lipoproteins and Cholesterol Metabolism in a Strain of Hyperlipemic Rats

S. Fantappié, E. Bosisio, M. Crestani, G. Galli, and A. L. Catapano [1]

1 Introduction

The avilability of genetic models for animal hyperlipemia is one of the major problems in the study of hypolipidemic drugs. While the genetic influence on response to "western diets" can be adequately studied in several animal species, few models of genetically determined hypercolesterolemia are available. The Watanabe rabbit is a model for familial hypercholesterolemia for both heterozygote and homozygote forms of type IIa hypercholesterolemia related to a low density lipoprotein receptor deficiency (1). This animal model, therefore, is relevant only to the familial hypercholesterolemia, which accounts for a minority of the hypercholesterolemias.

The rat is a widely used animal in the early testing of potentially valuable hypolipidemic drugs. So far only two strains of hyperlipidemic rats have been described (2,3).

For this reason selective breeding of spontaneous hyperlipidemic rats was undertaken at the animal breeding unit of Carlo Erba-Farmitalia, Milano (Italy). From the selection a strain of rat with high plasma levels of cholesterol and triglicerides (HL) and a strain of rats with low lipid plasma levels, which are referred as LC, were obtained. In this chapter we report on the distinctive features of the hyperlipemia observed in these animals and investigations on some of the mechanisms underlying the development of hyperlipemia in the HL rats.

2 Methods

Animals. Twenty-two males and 23 females with spontaneous hypercolesterolemia were selected from different albino rats (ICEMICER SPF Carlo Erba) by brother-sister mating. The experiments described here were performed with the tenth offspring generation on male animals (400—500 g b.wt.). The weight gain in control and hyperlipidemic rats was similar.

---

1 Institute of Pharmacological Sciences, School of Pharmacy, University of Milano, 20129 Milano, Italy

Drugs Affecting Lipid Metabolism
Ed. by R. Paoletti et al.
© Springer-Verlag Berlin Heidelberg 1987

Rats whose plasma was used for the determination of plasma glucose, lipids, T3, T4 and for lipoprotein fractionation were fasted overnight before sacrifice. Animals whose liver was used for enzyme assay and liver lipid determination had free access to food.

Analytical Methods. Plasma cholesterol, triglyceride levels as well as lipoprotein cholesterol and phospholipids were determined with commercially available kits (Boehringer Mannheim and Behring, West Germany). Protein determinations in lipoprotein fractions were made according to Lowry et al. (4) in the presence of SDS to avoid lipid turbidity. Liver-free and total cholesterol levels were determined by gas chromatography. Liver triglycerides were determined using an enzymatic kit (Sclavo, Siena, Italy).

Lipoprotein Separation. Blood was collected in the presence of EDTA (1 mg ml^{-1}) from the abdominal aorta. Lipoprotein classes were separated by sequential ultracentrifugation according to the method of Havel et al. (5) at the following densities: VLDL 1.019, LDL 1.019–1.050 and HDL 1.050–1.21 g ml^{-1}.

Triton Test. 300 mg kg^{-1} Triton WR 1339 were injected i.v. in fasted rats. The plasma triglyceride concentration was measured at time 0 (before Triton injection), 40, 80 and 120 min after. The triglyceride increase in plasma was linear with time. The rate of triglyceride secretion was calculated from the slope of the regression line.

Determination of Liver Enzyme Activities. Liver microsomes were prepared following homogenization in 0.25 M sucrose, 1 mM EDTA and 50 mM NaF by centrifugation at 105,000 g. Microsomes were washed and suspended in the appropriate buffer as described for each enzyme determination. Protein concentration was determined according to Bradford (6). HMG CoA reductase was assayed by using the microsomal pellet suspended in 300 mM KCl, 1 mM EDTA, 5 mM dithiothreitol, 50 mM phosphate buffer pH 7.2 (7), under the conditions already described for the evaluation of the total and active from of the enzyme (7,8).

For both acyl CoA cholesterol acyltransferase (ACATase) and cholesterol-7α hydroxylase activity determination, the microsomal fraction was suspended in 0.1 M phosphate buffer, pH 7.4.

ACATase was determined by the rate of incorporation of (1-^{14}C) oleyl–CoA (10 μCi mol^{-1}, Amersham, UK 50 μM) into cholesteryl esters under conditions described by Synouri Vrettakou et al. (9). Incubation time was 10 min at 37°C and the reaction was stopped by the addition of 20 vol of chloroform methanol (2:1 vol/vol). The organic phase was chromatographed on silica gel 60 plates (Merck, Darmstadt, West Germany) with petroleum ether, hexane and acetic acid (70:30:1, vol/vol/vol). The bands at the Rf corresponding to cholesteryl ester were scraped. The amount of cholesteryl esters formed was calculated from the specific activity of the incubated (-^{14}C) oleyl-CoA (^{3}H). Cholesteryl oleate was used as an internal standard to correct for losses during extraction and separation on the cholesteryl esters formed. The cholesterol-7α hydroxylase activity was determined by the method of Sanghvi et al. (10) which measures the amount of 7-α hydroxycholesterol formed from endogenous cholesterol during incubations of liver microsomes. In all experiments described, zero-time control samples were prepared with heat-inactivated microsomes.

3   Results and Discussion

Plasma lipid concentrations in control (LC) and hyperlipemic rats (HL) are shown in Table 1. Significantly higher concentrations of plasma cholesterol and triglycerides ($p < 0.01$) were detected in the HL group. The distribution of cholesterol among different lipoprotein classes, however, was not different (see also Table 1), thus suggesting that a quantitative rather than a qualitative increase of lipoproteins occurs in the hyperlipemic strain. To address this question further we studied the chemical composition of VLDL, LDL and HDL (Fig. 1); no difference in the chemical composition of the lipoproteins was found. This finding is consistent with an increased number of lipoprotein particles and not with a specific increase of the cholesterol content per lipoprotein. This is also consistent with the finding that the lipoprotein pattern is normal as determined by electrophoresis (data not shown).

Table 1. Plasma Lipids and Lipoproteins in the LC and HL rat (mg dl^{-1}, n =8, mean ± SE, male animals)

			Cholesterol (%)		
	Cholesterol	Triglycerides	VLDL	LDL	HDL
LC	75 ± 2	59 ± 3	3.5	23.8	72.7
HL	238 ± 14**	185 ± 22**	4.1	23.6	72.3

** $p < 0.01$ Student's t-test.

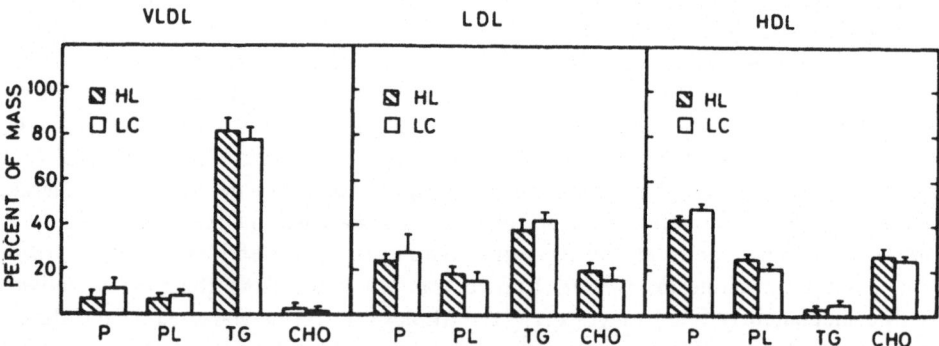

Fig. 1. Percent chemical composition of plasma lipoproteins in the LC and HL strain. Data are expressed as percent ± SD, n = 6. *P* protein; *PL* phospholipids; *TG* triglycerides; *CHO* cholesterol. Lipoproteins were isolated as described under methods

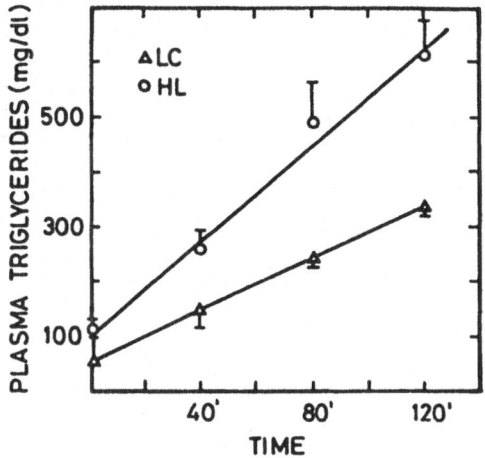

Fig. 2. Plasma triglyceride concentration after Triton WR 1339 injection. n = 5, X ± SD. Means 0, 40, 80 and 120 min were different at $p < 0.01$

To adress the question of whether the hypertriglyceridemia in these animals is due to an increased synthesis of triglyceride-rich lipoproteins, we determined the rate of triglyceride production in vivo using the Triton WR 1339 test. The data are reported in Fig. 2. A linear increase of plasma triglycerides was detected up to 3 h · after an injection 300 mg kg^{-1} of Triton WR 1339. The rate of triglyceride accumulation per 100 g b.wt. is about two-fold in the HL rat. This is consistent with an increased synthesis of triglycerides, most probably as VLDL by the liver. Whether the increase of plasma triglycerides in the HL rats is also due to a catabolic defect still remains to be determined.

To investigate the metabolic defects that may cause the hypercholesterolemia in these animals we determined the activity of the key enzymes of cholesterol metabolism in the liver. It is known that in cholesterol-fed, hypercholesterolemic animals the activity of the HMG CoA reductase is very low compared to controls, while the ACAT activity is largely increased, leading to cholesteryl ester storage within the

Table 2. Liver weight and liver lipid content in the LC and HL rat (n = 8. mean ± SE)

	Liver weight (% b.wt.)	Liver cholesterol (mg^{-1} tissue)	Liver TG (mg^{-1} tissue)
LC	3.08 ± 0.06	2.29 ± 0.11 (83.5)	7.08 ± 0.25
HL	3.83 ± 0.08**	2.38 ± 0.10 (85.5)	5.60 ± 0.22**

** $p < 0.01$ Student's t-test. Numbers in parentheses represent the percent of esterified cholesterol.

Table 3. Microsomal enzyme activities in HL and LC (n = 5, mean ± SD, p mol min^{-1} mg^{-1} protein)

	LC	HL
HMG CoA reductase		
Total	253 ± 42	266 ± 93
Active	59 ± 16	85 ± 18
Cholesterol 7-α Hydroxylase	30 ± 15	31 ± 11
ACAT	143 ± 32	232 ± 68

hepatocyte. In the HL strain liver cholesterol is not affected, and triglycerides are reduced (Table 2), thus showing that, at variance with the cholesterol-fed animals, no accumulation of lipids occurs in the liver. The activity of HMG CoA reductase, ACAT and 7-α hydroxylase was also not different from controls, thus suggesting that the metabolism of cholesterol in the liver is normal in the HL strain (Table 3).

Reasons other than an increased cholesterol synthesis must be therefore responsible for the hypercholesterolemia in this strain. Whether an LDL receptor deficiency (partial or total) occurs in these animals is unknown and is currently under investigation.

Acknowledgements. This work was supported in part by a grant to A.L.C. from the CNR, obiettivo N° 84.02241.56115.11485. Dr. Lovisolo and Marchi (Farmitalia-Carlo Erba) provided the animals, Miss Marta Colombo and Mrs. Silvana Magnani typed the manuscript.

References

1. Watanabe Y (1980) Serial inbreeding of rabbits with hereditary hyperlipidemia (WHHL-rabbit). Atherosclerosis 36:261–268
2. Müller KR, Li JR, Dinh DM, Subbiah MTR (1979) The characteristics and metabolism of a genetically hypercholesterolemic strain of rats (RICO). Biochim Biophys Acta 574:334–343
3. Imai Matsumura H, Miyajima, Oka K (1977) Serum and tissue lipids and glomerulonephritis in the spntaneously hypercholesterolemic (SHC) rat, with a note on the effects of gonadectomy. Atherosclerosis 27:165–178
4. Lowry OH, Rosebrough AL, Farr AL, Randall RJ (1951) Protein measurement with the Folin phenol reagent. J Biol Chem 193:265–275
5. Havel RJ, Eder HA, Bragdon JH (1955) The distribution and chemical composition of ultracentrifugally separated lipoproteins in human serum. J Clin Invest 34:1345–1353
6. Bradford M (1976) A rapid and sensitive method for the quantitation of microgram quantities of protein utilizing the principle of protein-dye binding. Anal Biochem 72:248
7. Cighetti G, Santaniello E, Galli G (1981) Evaluation of 3-hydroxy-3-methylglutaryl-CoA reductase activity by multiple-selected ion monitoring. Anal Biochem 110:153–158
8. Cighetti G, Galli G, Galli Kienle M (1983) A simple model for studies on the regulation of cholesterol synthesis using freshly isolated hepatocytes. Eur J Biochem 133:573–578
9. Synouri-Vrettakou S, Mitrapoulos KA (1983) Acal coenzyme A: cholesterol acyltransferase transfer of cholesterol to its substrate pool and modulation of activity. Eur J Biochem 133:299–307
10. Sanghvi A, Grassi E, Bartman C, Lester R, Galli-Kienle M, Galli G (1981) Measurement of cholesterol 7-α hydroxylase activity with selected ion monitoring. J Lipid Res 22:720–724

Spontaneous Hypertriglyceridemia in a Non-Obese Rat Model: The IVA-SIV Rat

M. R. Lovati, L. Allievi, C. Manzoni, C. Galli, and C. R. Sirtori[1]

1 Introduction

Among the many risk factors taken into consideration in the development of the atherosclerotic process, hypertriglyceridemia is the object of frequent debate (Böttiger and Carlson 1980; Hulley et al. 1980). The lack of adequate animal models of the disease has hampered the understanding of the mechanism(s) by which the metabolic disease may lead to atherosclerosis. Hypertriglyceridemia (HTG) is generally only studied by the aid of short-term dietary loads (fructose, ethanol) (Zavaroni et al. . 1982) as well as of animal models with spontaneous HTG. This latter is generally linked to obesity and/or diabetes (Koletsky et al. 1973; Reaven et al. 1974). Increased triglyceridemia has been reported in older animals (Tobey et al. 1981) and in the Nagase analbuminemic rat variant (Kikuchi et al. 1983).

Elevated fasting triglyceridemia can be documented in Sprague-Dawley rats from the Ivanovas Sieve colony (IVA-SIV) compared to standard male Sprague Dawleys provided by Charles River (Lovati et al. 1986; Zavaroni et al. 1986). This report provides a detailed analysis of the plasma lipid-lipoprotein composition, the sensitivity to dietary induction, as well as data on the mechanism(s) responsible for the HTG. Preliminary studies on platelet function in these animals are also the object of this report.

2 Plasma Lipids and Lipoproteins

All experiments were carried out on male rats (150–180 g b.wt.) of the Sprague-Dawley strain, provided either from Charles River (Calco, Italy) or from Sieve Ivanovas (Milano, Italy). Plasma lipid and lipoprotein levels in the IVA-SIV rats, compared to standard CR, are reported in Table 1. The lipoprotein pattern shows a marked increase of TG levels in the very low density lipoproteins (VLDL) of the IVA-SIV rats (approximately doubled) compared to the CR. A higher VLDL cholesterol can also be

---

1 Institute of Pharmacological Sciences, Via A. Del Sarto 21, 20129 Milan, Italy

Drugs Affecting Lipid Metabolism
Ed. by R. Paoletti et al.
© Springer-Verlag Berlin Heidelberg 1987

Table 1. Lipid levels in plasma and isolated lipoproteins from the IVA-SIV rats, compared to Charles River (CR) rats on a normal diet (in parenthesis, % of total)

		Chol (mg dl^{-1})	TG (mg dl^{-1})
Plasma	CR[a]	60.7 ± 4.7	69.9 ± 5.9
	IVA-SIV[a]	63.5 ± 5.9	*114.5 + 8.3***
VLDL	CR[b]	6.8 ± 1.5 (8.0)	49.6 ± 14.6 (58.5)
	IVA-SIV[b]	*13.7 + 1.8*** (9.4)	*97.1 + 11.3*** (67.2)
LDL	CR[b]	7.7 ± 2.2 (18.6)	8.8 ± 0.5 (21.2)
	IVA-SIV[b]	*4.3 ± 1.4** (12.5)	7.6 ± 1.4 (22.1)
HDL	CR[b]	39.1 ± 2.2 (22.6)	6.9 ± 0.7 (4.0)
	IVA-SIV[b]	43.1 ± 3.2 (22.8)	5.4 ± 1.1 (2.4)

* $p < 0.05$ vs CR; ** $p < 0.01$ vs CR.
[a]   x±SEM from 12 animals.
[b]   x±SEM from 4 pools of 3 animals each.

detected in this variant strain. Conversely, only small differences are noted in the lipid and protein content of the low density and high density (LDL and HDL) fractions.

On *agarose electrophoresis* (Fig. 1, lower panel), it appears that slow migrating lipoproteins (VLDL and LDL in the rat) have a higher mobility in the IVA-SIV. The electrophoresis of isolated lipoprotein fractions shows a larger amount of material, in particular in VLDL, but also in the LDL density range of the HDL fraction (most likely HDL$_1$), which is more intensely stained in the IVA-SIV. The hypothesis that the HDL$_1$ content may be increased in the IVA-SIV animals is supported by rate-zonal ultracentrifugal findings (Lovati et al. 1986). The apolipoprotein patterns in the two rat strains, determined by analytical IEF of the isolated VLDL and HDL fractions, as well as the apo-B percentage content of VLDL, do not differ.

Fructose administration induced, as indicated by Nikkilä and Ojala (1966), a significant HTG in both rat strains. The post-diet plasma TG levels were again 1.5-fold higher in the IVA-SIV animals (265 ± 20 vs 180 ± 15 mg dl^{-1}, $p < 0.01$); triglyceridemia was raised in the VLDL fraction. A significant elevation of cholesterol and TG levels in both strains was observed after a cholesterol-cholic acid diet (Nath diet) (Table 2). However, the CR rat was significantly more cholesterol inducible compared to the IVA-SIV, whereas the opposite was found for TG. The measurement of *post-*

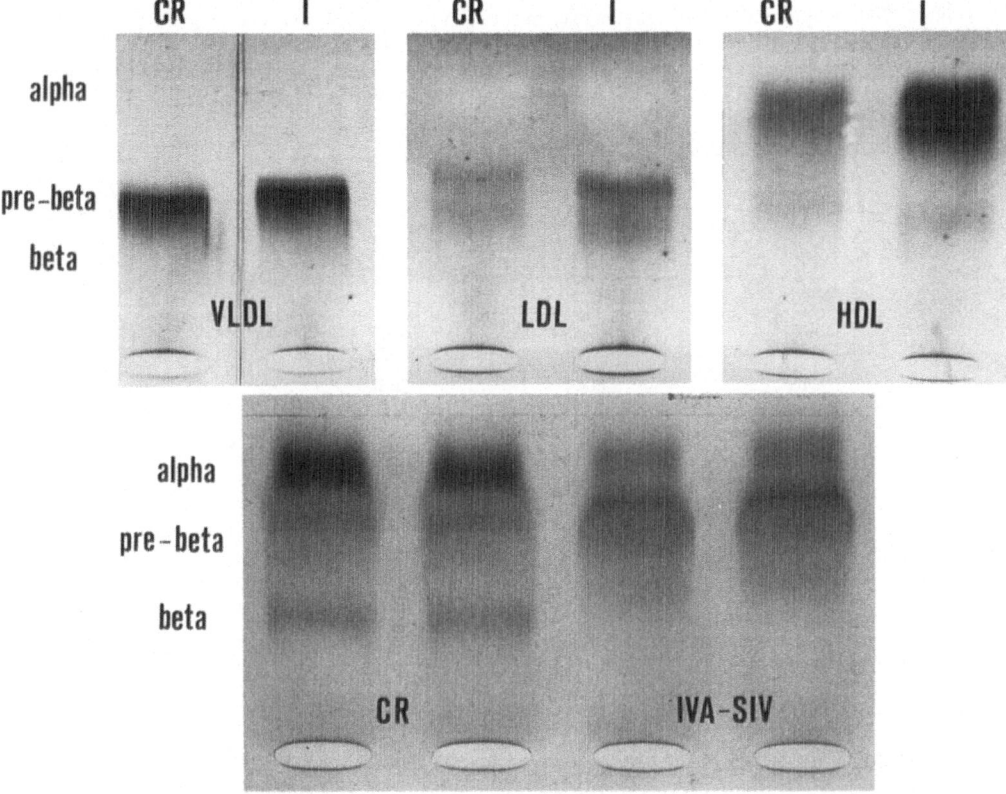

Fig. 1. Agarose electrophoretograms of sera (*lower panel*) and of isolated lipoproteins from Charles River (CR) and IVA-SIV (I) rats. The IVA-SIV show a larger amount of pre-beta migrating lipoproteins, as well as of HDL migrating in the beta position (most likely HDL$_1$)

absorptive plasma glucose, insulin and FFA concentrations resulted in higher fasting plasma glucose (115±3 vs 84±2 mg dl^{-1}), insulin (35±5 vs 19±2 μU ml^{-1}) and FFA (502.8±67.3 vs 337.2±18.2 mEq^{-1}) levels in the IVA-SIV. Furthermore, the IVA-SIV rats had an increased glycemic and insulinemic response to an oral glucose load (Zavaroni et al. 1986).

Turnover studies of VLDL-TG have documented a higher production rate when correlated with the plasma TG concentrations in the IVA-SIV animals. Since the lipoprotein lipase activity in both adipose and muscle tissues was not significantly different in the two strains, it appears that the HTG in IVA-SIV rats is due to an increased VLDL-TG secretion (Zavaroni et al. 1986).

Table 2. Plasma lipid and lipoprotein changes in the IVA-SIV and CR rats following the administration of fructose (A) and of a cholesterol-cholic acid (Nath) diet (B)

		A. Fructose (n=6 per group; x±SEM)	
		Chol	TG
Base	CR	47.6 ± 4.2	43.5 ± 12.6
	IVA-SIV	54.4 ± 3.8	*133.2+ 7.8**
Fructose	CR	40.1 ± 2.2	180.2 ± 14.6
	IVA-SIV	43.8+4.0	*264.8+19.8**

	B. Nath diet (n=12 per group; x+SEM)			
	Chol (mg dl⁻¹)	TG (mg dl⁻¹)	Chol (mg dl⁻¹)	TG (mg dl⁻¹)
	CR		IVA-SIV	
Plasma	310.0 ± 53.2	126.4 ± 15.8	*222.9+23.1*	*185.0+19.2*
VLDL	186.4 ± 23.8	94.7 ± 12.4	133.2 ± 18.5	*166.5+15.2**
LDL	120.1 ± 19.9	20.0 ± 3.0	86.8 ± 12.9	11.8 ± 4.6
HDL	3.2+ 0.7	10.2+ 3.7	*1.2+ 0.3**	9.0+ 1.8

* $p < 0.05$ vs CR; ** $p < 0.01$ vs CR.

3 Platelet Function

Preliminary studies on *platelet aggregability* after ADP stimulation show a significantly reduced response in the IVA-SIV animals compared to the CR when matched for age. The platelet number in the IVA-SIV rats is also markedly raised; a significantly higher prostacyclin (PGI_2) release from the arterial wall in the variant strain may account for the platelet hyposensitivity vs aggregating agents. Nevertheless, a higher sensitivity of platelets to Iloprost, a synthetic PGI_2 analogue was also documented in the IVA-SIV rats.

4 Conclusions

The IVA-SIV rat provides an animal model with striking analogies to human-type IV hyperlipoproteinemia. Characteristic features of these animals are, besides raised triglyceridemia, increased glucose, insulin (both basal and after-glucose challenge) and FFA levels. Interestingly, these dramatic biochemical abnormalities seem to be accompanied by a reduced aggregability of platelets, and by a significantly higher release of PGI_2 from the arterial walls.

Acknowledgements. This work was partially supported by grants from the Consiglio Nazionale delle Ricerche of Italy.

References

Böttiger LE, Carlson LA (1980) Risk factor for ischaemic vascular death for men in the Stockholm Prospective Study. Atherosclerosis 36:389–408

Hulley SB, Rosenman RH, Bawol RD (1980) Epidemiology as a guide to clinical decisions. The association between triglyceride and coronary heart disease. N Engl J Med 302:1383–1389

Kikuchi H, Tamura S, Nagase S (1983) Hypertriacylglycerolemia and adipose tissue lipoprotein lipase activity in the Nagase analbuminemic rat. Biochim Biophys Acta 744:165–170

Koletsky S (1973) Obese spontaneously hypertensive rats – a model for study of atherosclerosis. Exp Mol Pathol 19:53–60

Lovati MR, Franceschini G, Allievi L (1986) Endogenous hypertriglyceridemia in a non-obese rat model: plasma lipoproteins and dietary sensitivity. Metabolism 35:436–440

Nath N, Wiener R, Harper AE (1959) Diet and cholesterolemia. Part 1. Development of a diet for for the study of nutritional factors affecting cholesterolemia in rats. J Nutrit 67:289–307

Nikkilä EA, Ojala K (1966) Acute effects of fructose and glucose on the concentration and removal rate of plasma triglycerides. Life Sci 5:89–94

Reaven EP, Reaven GM (1974) Mechanism for development of diabetic hypertriglyceridemia in streptozotocin-treated rats. J Clin Invest 54:1167–1178

Tobey TA, Mondon CE, Reaven GM (1981) Age-related changes in very low density lipoprotein metabolism in normal rats. Metabolism 30:583–587

Zavaroni I, Chen YI, Reaven GH (1982) Studies of the mechanism of fructose-induced hypertriglyceridemia in the rat. Metabolism 31:1077–1083

Zavaroni I, Dall'Aglio E, Lovati MR (1986) Pathogenetic mechanisms of the endogenous hypertriglyceridemia in a non-obese rat model. Metabolism 35:383–385

Considerations for Controlled Clinical Testing of Safety and Efficacy of New Hypolipidemic Drugs

C. A. Dujovne and P. Krehbiel [1]

1 Introduction

Evaluating a new drug for the first time in patients requires knowledge of clinical trial methodology, the pharmacology of the drug, and the clinical condition being treated. When the new drug is designed to modify blood lipids this undertaking is complicated by many factors and circumstances affecting the accurate determination of safety and efficacy parameters in hyperlipidemic patients.

At our Clinic we have specialized in Phase 1 and early Phase II and III testing of brand new lipid-acting drugs in volunteers and patients (Dujovne et al. 1971, 1974, 1976, 1979, 1984a, 1986). We have developed a model operational structure using tools, methods, and personnel selected to maximize drug efficacy and safety and to yield the most accurate possible data.

2 Use of "Healthy" Dyslipidemics in Early Testing

The pathological consequences of dyslipidemia (abnormal serum lipid levels) start early in life, but morbidity may not be apparent until middle age or later. When detected early, the subjects are healthy individuals except for abnormal serum lipids. We proposed and implemented the use of such "healthy" human volunteers for Phase 1 studies. This approach offers early quantification of drug effects on lipid parameters and helps detect pharmacokinetic variables, which may be particular to hyperlipidemic individuals.

1 Lipid and Arteriosclerosis Prevention Clinic, Department of Medicine and Division of Clinical Pharmacology, The University of Kansas Medical Center, 1348 Bell Memorial Hospital, 39th and Rainbow Boulevard, Kansas City, KS 66103, USA

Drugs Affecting Lipid Metabolism
Ed. by. R. Paoletti et al.
© Springer-Verlag Berlin Heidelberg 1987

3 Identifying Variables to be Monitored During the Trial

A variety of interrelated factors can result in changes in serum lipids independently of the pharmacological intervention. Some are unavoidable seasonal changes, the various types of hyperlipidemias, and the large interindividual variability in baseline levels. Other factors may cause variations in serum lipids throughout a drug trial, and require strict monitoring. We have designed and implemented methods to deal with (1) diet instructions and constancy; (2) monitoring physical activity and smoking habits; and (3) documentation of concomitant use of nonlipid-acting drugs.

4 Uniform, Standardized Diet Instructions

We have produced ad 15 min videotape with diet recommendations. Patients are also given a specially prepared booklet containing: (1) instructions for Phase I through III American Heart Association low fat, low cholesterol diets; (2) graphic demonstration of portion sizes; (3) prototype menus; (4) instructions on how to fill out diet diaries. These are utilized to obtain information and documentation of diet constancy through-out the study.

5 Recording and Monitoring Diet Constancy

Patients fill out and return a set of four consecutive day records every 2 weeks. A nutritionist analyzes these, assisted by a computer program. The results reveal the quality and quantity of dietary components with a 90% confidence level (Jackson et al. 1986) (Fig. 1).

SOURCE	KCAL	PROT	CHO	FAT	CHOL	P/S RATIO
Day #1	1003	30	132	29	82	0.21
Day #2	1190	41	164	44	156	0.08
Day #3	1373	37	160	56	145	0.18
Day #4	1228	49	163	45	84	0.60
Total	4794	158	620	175	468	0.24
Average	1198	39	155	44	117	0.27

Fig. 1. Computer printout data generated from analysis of four consecutive days of diet records obtained every 2 weeks

6 Recording and Monitoring Physical Activity Constancy

A card listing the most common regular physical exercises on one side and a rating scale for usual daily activities on the other side is issued at each visit (Figs. 2 and 3).

PHYSICAL ACTIVITY AT WORK

Thinking about the things you died *at work* for the last two weeks, how would you rate yourself as to the *amount of physical activity* you did compared with the two weeks before your previous visit? Were you:

Much more active	1	
Somewhat more active	2	Circle number
About the same.	3	corresponding
Somewhat less active	4	to your change
Much less active	5	of activity
Not applicable	6	

PHYSICAL ACTIVITY OUTSIDE OF WORK

Now, thinking about the things you did *outside of work,* how would you rate yourself as to the *amount of physical activity* you did compared with the two weeks before your previous visit? Were you:

Much more active	1	
Somewhat more active	2	Circle number
About the same.	3	corresponding .
Somewhat less active	4	to your change
Much less active	5	of activity
Not applicable	6	

Fig. 2. Front side of physical activity-report card reporting amount of regular exercise routines per week

NAME_____

PERSONAL PHYSICAL ACTIVITY SHEET for the two weeks P

Please estimate the amount of activity each week (Circle way of measure)	Week 1	Week 2
Walking: yards, blocks, miles/time it took		
Jogging: total miles/ time it took		
Running: total miles/ time it took		
Stair Climbing: steps per day or week		
Trampoline: walking or jogging, minutes or hours		
Tennis: sets, minutes, hours		
Golf: holes, minutes, hours		
Cycling: miles, minutes, hours		
Rope Skipping: minutes, hours		
Skating: minutes, hours		
Dancing: minutes, hours		
Gardening: minutes, hours		
Lawn mowing: minutes, hours		
Swimming: minutes, hours		
Others:		

Fig. 3. Reverse side of physical activity report for a comparative rating of general physical activity

7 Documenting Frequency and Severity of Relevant Side Effects

Unbiased documentation of side effects is difficult. Patients, informed of potential side effects, may be inclined to change the frequency and/or severity of their complaints. We have obviated most of these problems by presenting a form to patients attending the clinic (Figs. 4 and 5). It renders itself to statistical tabulation of two

LIPID AND ARTERIOSCLEROSIS PREVENTION CLINIC   Date:_____

Name_____   Weight:_____

INDICATE IN THE BOXES THE APPROXIMATE NUMBER OF DAYS WITHIN THE LAST TWO WEEKS IN WHICH YOU
EXPERIENCED ANY OF THESE COMPLAINTS. (Write 1 if it was 1 day, 2 for 2 days and so on, up to
14, if it is everyday. (SEE BACK OF SHEET FOR RATING OF THE SEVERITY OF YOUR COMPLAINTS)

1.*	Less appetite	19†	Pain	37.*	Nausea
2.	More appetite	20.	Shortness of	38.	Belching
3.	Drowsiness		Breath	39.	Heartburn
4.	Dizziness	21.	Numbness	40.	Vomiting
5.	Nervousness	22.	Ringing in ears	41.	Abdominal or
6.	Weakness	23.	Blurred vision	42.	stomach pain
7.	Unable to sleep	24.	Skin rash	43.	Diarrhea or frequent
8.	Tiredness	25.	Hives		bowel movements
9.	Sleepiness	26.	Itching	44.	Constipation
10.	Lightheadedness	27.	Sweating	45.	Intestinal gas
11.	Forgetfulness	28.	Hot flashes	46.	Bloody or black stools
12.	Nightmares	29.	Dry mouth	47.	Dark urine
13.	Shakiness or	30.	Bad taste	48.	Painful urination
	tremors	31.	Easy bruising	49.	Get up at night to
14.	Confusion	32.	Muscle cramps		urinate
15.	Jitteriness	33.	Leg pains	50.	Chest pain
16.	Thirst	34.	Stuffy nose	51.	Fever or chills
17.	Headache	35.	Fast heartbeat	52.	Change in sexual
18.	Stiffness of	36.	Irregular heartbeat		ability and/or desire
	Joints			53.	Other--specify

REMEMBER: Have nothing but water for 12 hours before your clinic visits and also bring all
medicines you are taking in their original containers!
When was the last time you ate or drank ANYTHING except water?_____ Since your last visit
have you missed an appointment (answer yes or no)?_____had any illness?_____
changed your diet?_____taken any new medications or stopped old ones?_____

Fig. 4. Front side of symptoms checklist for reporting of number of days in the previous 2 weeks that symptom was present

EACH "NUMBER" BELOW THE ARROWS REPRESENTS THE "NUMBER" OF THE COMPLAINT OR
SYMPTOM ON THE FRONT PAGE. IF YOU HAVE GIVEN A NUMBER OF DAYS YOU EXPERIENCED
ANY OF THE COMPLAINTS, PLEASE MARK BELOW THE DEGREE OF SEVERITY BY PUTTING A
CIRCLE AROUND ONE OF THE NUMBERS ON THE LINE. 1, BEING THE MILDEST OR WEAKEST
AND 5 BEING THE WORST OR STRONGEST

↓						↓				
1.	1	2	3	4	5	27.	1	2	3	4	5
2.	1	2	3	4	5	28.	1	2	3	4	5
24.	1	2	3	4	5	52.	1	2	3	4	5
25.	1	2	3	4	5	53.	1	2	3	4	5

1. - VERY MILD   4. - SEVERE
2. - MILD   5. - VERY SEVERE
3. - MODERATE

Fig. 5. Reverse side of symptoms checklist for reporting the degree of severity

numbers for each of the 53 possible complaints checked at each visit. A number
(0–14) for the number of days the complaint was present in the 2 weeks prior to the
visit, and another number (1–5) for the severity of the respective complaint. This tool
detects irrelevant, regular complaints in some patients and reveals trends in appearance or disappearance of "real" side effects throughout the trial (Dujovne et al.
1984a, 1986).

8 Constancy in Intake of Concomitant and Study Drugs

We use a special card to monitor intake of concomitant medications and study drugs.
This facilitates compliance, and the effects of other drugs, such as antihypertensives
(Dujovne et al. 1984b) on serum lipids, can be monitored.

9 Quality Controls for Serum Lipid Parameter Measurements

Our laboratory is standardized with the Lipid Laboratory of the Center for Disease
Control for measurements of serum cholesterol levels in lipoprotein fractions, serum
triglycerides, and apoprotein levels. In addition, we utilize internal and commercial
standards, open and blind duplicate samples, and other stringent controls to assure
maximal accuracy and reproducibility of serum lipid parameter measurements.

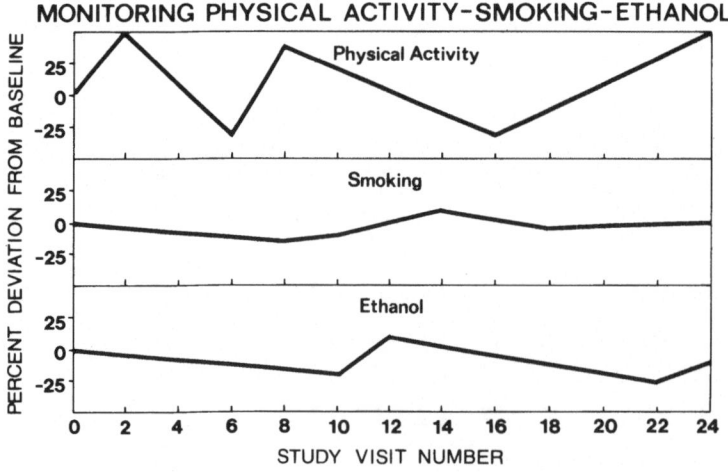

Fig. 6. End of trial report: chart showing variations in diet parameters as percent deviation from
baseline period at each study visit period

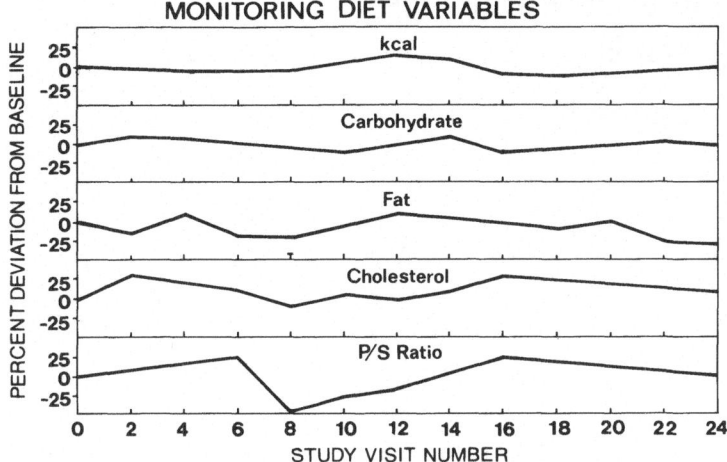

Fig. 7. End of trial report: chart showing variations in physical activity, smoking, and ethanol intake habits as percent deivation from baseline period at each study visit period

10 Maximizing Study Discipline and Compliance

All procedures in the protocol's flowsheet are reviewed and set up the day before the clinic. We use special checklists at every visit as a record of what was done to the patient throughout the trial. We provide patients with special wallets to keep all records which they must bring at every clinic appointment. At the end of a trial we can plot the variability shown in each of the parameters controlled on each patient throughout the trial (Figs. 6 and 7).

References

Dujovne CA, Weiss P, Bianchine JR (1971) Comparative clinical therapeutic trial with two hypolipidemic drugs, clofibrate and nafenopin (SU-13437). Clin Pharmacol Ther 12:117–125

Dujovne CA, Hurwitz A, Kauffman RE, Azarnoff DL (1974) Colestipol and clofibrate in hypercholesterolemia. Clin Pharmacol Ther 16:291–296

Dujovne CA, Azarnoff DL, Pentikainen P, Manion C, Hurwitz A, Hassanein K (1976) A two-year crossover therapeutic trial with halofenate and clofibrate. Am J Med Sci 272:277–284

Dujovne CA, Azarnoff DL, Pentikainen P, Manion C, Hurwitz A (1979) Clofibrate and nicotinyl alcohol tartrate in hyperlipoproteinemic patients. Am J Med Sci 277:255–261

Dujovne CA, Chernoff S, Krehbiel P, Jackson B. DeCoursey S, Taylor H (1984a) Low dose colestipol plus probucol for hypercholesterolemia. Am J Cardiol 53:1515–1518

Dujovne CA, DeCoursey S, Krehbiel P, Jackson B, Chernoff S (1984b) Serum lipids in normo- or hyperlipidemics after methyldopa or propranolol therapy. Clin Pharmacol Ther 36:157–162

Dujovne CA, Krehbiel P, DeCoursey S, Jackson B, Chernoff S, Pitterman A, Garty M (1984c) Probucol plus colestipol in treatment of hypercholesterolemia. Ann Int Med 100:477–482

Dujovne CA, Krehbiel P, Chernoff SB (1986) Controlled studies of the efficacy and safety of combined probucol-colestipol therapy. Am J Cardiol 57:36H–42H

Jackson B, Dujovne CA, DeCoursey S, Beyer P, Brown EF, Hassanein K (1986) Method to assess dietary compliance in outpatient clinical trials. J Am Diet Assoc (November 1986) 86:1531–1535

Desirable Lipoprotein Parameters in Early Drug Trials

C. R. SIRTORI and A. BONDIOLI [1]

1 Introduction

Lipid-lowering drugs may be tested very early in patients, in view of the overall "normality" of these subjects, aside from their specific biochemical disorder. Although current modalities for the early evaluation of new drug products require administration to "healthy" volunteers, it is becoming more and more common to carry out early testing in "healthy" subjects with elevated plasma lipid levels. In view of the high expectation for new drug products, active in primary hyperlipoproteinemias (The Lipid Research Clinics 1984), we personally consider as feasible and advisable to carry out the Phase I evaluation of new drugs for hyperlipidemia in patients. These should possibly never have received drug treatments or, in some cases, should be evaluated after stopping treatment with specific products (e.g. anion-binding resins).

The object of this brief review will be to analyze some of the laboratory problems in the early evaluation of lipid-lowering drugs. It is, in fact, mandatory to obtain as many pieces of information as possible from the initial testing. This is of particular significance in view of the time and cost of these trials, and particularly in order to obtain clear and non-ambiguous data.

2 Rise of HDL Levels: Significance and Proper Determination

Interest in compounds affecting the metabolism of high density lipoproteins (HDL), is certainly among the highest. This applies both to lipid-lowering drugs, as well as to any compound used for chronic treatments, e.g. contraceptive steroids, anti-hypertensive agents, gastrointestinal drugs, etc. (Sirtori and Franceschini 1984). The reasons for the rise of HDL may, indeed, be multiple (Table 1). By this token, whereas in some cases (e.g. activation of lipoprotein lipase, Nikkilä et al. 1977), the consequent rise of HDL may be part of a "beneficial" phenomenon, in others, e.g. the increase of HDL protein and cholesterol after pesticides (Carlson and Kolmodin-Hedman 1977), it may carry at least a doubtful connotation.

1 Center E. Grossi Paoletti, Institute of Pharmacological Sciences, University of Milano and Divisione Medica Vergani, Ospedale di Niguarda Ca' Granda, 20129 Milano, Italy

Drugs Affecting Lipid Metabolism
Ed. by R. Paoletti et al.
© Springer-Verlag Berlin Heidelberg 1987

Table 1. Mechanisms involved in the rise of HDL levels

1. Enzyme induction

2. Activation of lipoprotein lipase ($\uparrow$ HDL$_2$)

3. Reduction of hepatic lipase ($\uparrow$ HDL$_2$)

4. Excess dietary fat

5. β-adrenergic stimulation (possibly α_1 block)

Aside from the mechanism of rise of HDL (be it protein or cholesterol or both), it is crucial that the investigator is well aware of what is being determined. Several problems arise in an exact and meaningful determination of HDL levels (Table 2). In the first place, HDL-cholesterol (HDL-C) is not always a true representative of HDL levels. In the case of significant hypertriglyceridemia, cholesteryl esters in HDL are replaced to a large extent by triglycerides (Deckelbaum et al. 1984) and the HDL-C value represents an underestimation of the true HDL level. In this particular case, the determination of plasma (or HDL) apolipoprotein AI will provide a better estimate of HDL (see below).

In terms of the evaluation of the clinical significance of HDL-C levels, problems are more serious than generally appreciated (Table 2). Considerable variability has been reported when replicate samples have been examined in numerous clinical laboratories (Warnick et al. 1980); some improvement was noted when comparing similar trials carried out in 1980 and 1983 (Warnick et al. 1983). However, such findings are not reassuring when planning large multi-center trials.

A critical factor may be the selection of the proper *precipitating agent.* Although generally considered as superimposable, findings with the various precipitants may not provide similar information. It is well known that different concentrations of heparin may lead to considerable changes in the amount of apo B-containing lipoproteins in the precipitate (Warnick and Albers 1978). Since dextran-MgCl$_2$ and phosphotungstate (PTA) have become the most popular precipitating reagents, it

Table 2. Pitfalls in the determination of HDL levels

1. apo AI is a better index in hypertriglyceridemia

2. General errors (Warnick et al. 1983): SD 6.4 mg dl^{-1},
   range 3.4–13.6 mg dl^{-1}

3. *Precipitating agents:*
   PTA: precipitates AI, not HDL$_1$
   heparin: different concentrations
   dextran: precipitates HDL$_1$

Determination of HDL$_2$/HDL$_3$: polyanionic precipitation vs zonal (Simpson et al. 1982) HDL$_2$ $\uparrow$ 123%; HDL$_3$ $\downarrow$ 9.1%

should be underlined that the two interact differently with HDL-1, a receptorially active lipoprotein, potentially anti-atherogenic (Mahley 1983). Whereas, in fact, dextran-MgCl$_2$ exerts a significant interaction with apo-E and precipitates HDL-1 (thus leading to relatively lower HDL-cholesterol levels) (Gibson et al. 1984), PTA does not interact with HDL-1, detected in the supernatant after precipitation (Schmitz and Assmann 1982). On the other hand, PTA does precipitate, to some extent, apo AI, thus possibly affecting the relative distribution of apo AI-containing particles in HDL (Warnick et al. 1979).

A most delicate issue, potentially leading to grossly erroneous determinations, is that of *the separation of the HDL$_2$ and HDL$_3$ subfractions*. Aside from the still unsettled biological significance of changes in the HDL$_2$/HDL$_3$ ratio (Eisenberg 1985), knowledge of drug-induced changes may be important in the understanding of the mode of action of specific agents (Sirtori and Franceschini 1984). Although a simple and convenient precipitation technique has been described (Gidez et al. 1979), levels of HDL$_2$- and HDL$_3$-cholesterol, as determined by this method, carry no obvious relationship with values obtained by zonal ultracentrifugation (Table 2) (Simpson et al. 1982). This method should be discarded from further clinical use. Our personal experience indicates, moreover, that also the ultracentrifugal procedure for the separation of HDL$_2$ and HDL$_3$ at standard densities, is probably misleading (Franceschini et al. 1985). In fact, the sole preservation of serum at refrigerator temperature leads to gross changes in the relative proportion of the two subfractions and in their density profile. Such a caveat may explain the gross discrepancies in clinical studies related, e.g. to probucol and fibrates (Sirtori and Franceschini 1984). The case is particularly serious when examining turbid samples from type IV patients (Fig. 1). It is mandatory to preserve samples in an environment with a high salt con-

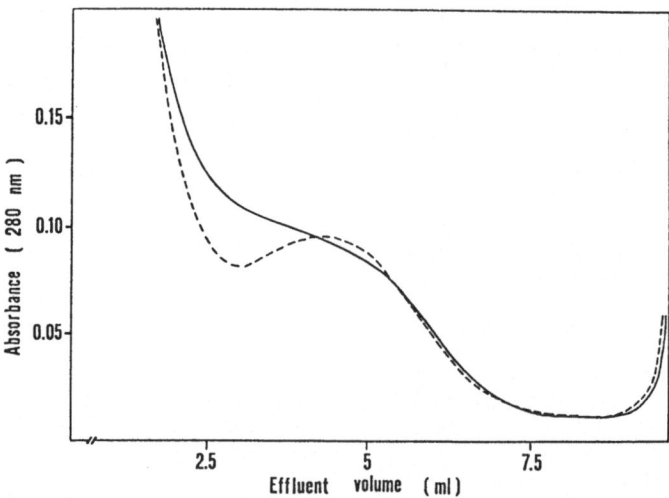

Fig. 1. Elution profiles of HDL subfractions from a severely hypertriglyceridemic patient (TG = 1,039 mg dl^{-1}). The sample was processed by zonal ultracentrifugation (Franceschini et al. 1985) after 7 days of storage of the serum at 4°C in the presence (– – – –) or absence (——–) of 5.1 m NaBr. No separation between HDL$_3$ and serum proteins is achieved in the sample stored in the absence of salt

centration (5.1 M NaBr) and to carry out the separation with a zonal method, e.g. with a swinging bucket rotor (Franceschini et al. 1985).

3 Determination of Apolipoproteins: When and Why

The growing knowledge on the protein components of plasma lipoproteins has raised interest in the possible drug effects. Although the role of apolipoproteins in atherosclerosis development is still controversial (Avogaro et al. 1979; Schmidt et al. 1985), the excess or deficiency of apolipoproteins may be linked to an enhanced atherosclerosis risk. Such may be the case of the hyperapo-β-lipoproteinemia in normo-lipidemic or hypertriglyceridemic patients (Sniderman et al. 1980), or the enrichment with apo-B in very low density lipoproteins (VLDL), described by us in patients with peripheral vascular disease (Franceschini et al. 1982). These types of pathologies may, however, enter to a very limited extent into early drug trials. The interest of apo-lipoprotein determinations at this stage should, therefore, be viewed with caution.

A good case for the determination of *apo AI*, as above indicated, is that of hyper-triglyceridemic samples, where erroneously low HDL-C levels may be obtained (Deckel-baum et al. 1984). Since low HDL-C appears to be stable in patients, even when total triglyceridemia is reduced (Witztum et al. 1980), the possibility of changes in apo AI levels after drug treatments is of special significance (Saku et al. 1986). The opposite case is that of *apo-B*, whose levels do not appear to be changed independently of total cholesterolemia, aside possibly after specific diet treatments (Mattson and Grundy 1985; Sirtori et al. 1986). The apo AI/B ratio has been suggested to be a good marker of atherosclerosis risk (Avogaro et al. 1979) and may be modified by some drug or dietary treatments (Sirtori et al. 1986, 1987).

Considerable emphasis has been placed on the determination of total or relative contents of *apo-C,* particularly in VLDL after treatments, e.g. with fibrates. Apo CII is the physiological activator of lipoprotein lipase (LPL) (Bengtsson and Olivecrona 1980), but it serves this role at very low concentrations; indeed, total apo CII levels are raised in hypertriglyceridemia (Schonfeld et al. 1979). The apo CII/CIII ratio is, however, definitely reduced in hypertriglyceridemic VLDL and it has been shown, also by us, that a rise of apo CII is specifically associated with fibrate treatment and it may represent a possible mode of action of these drugs (Franceschini et al. 1985). These findings, e.g. the increased apo CII/CIII$_1$ ratio in VLDL after treatment, should be today viewed with caution. Although they may provide further confirmation of the activating properties of some agents on LPL, the levels of apo CII in VLDL are inversely related to the triglyceride enrichment in these lipoproteins, and the relative rise of the CII/CIII ratio after drug treatment may only reflect the delipida-tion process (Kane et al. 1975).

Finally, an interesting case may be that of *apo-E,* an apolipoprotein specifically secreted by the liver, as well as a determinant of receptor activity (Mahley 1983). Apparently only in one study was a difference in apo-E transport detected after treatment. This was the case of *tiadenol,* an absorbable cholesterol-lowering drug (Baggio et al. 1979), shown by us to markedly reduce the apo-E content in VLDL

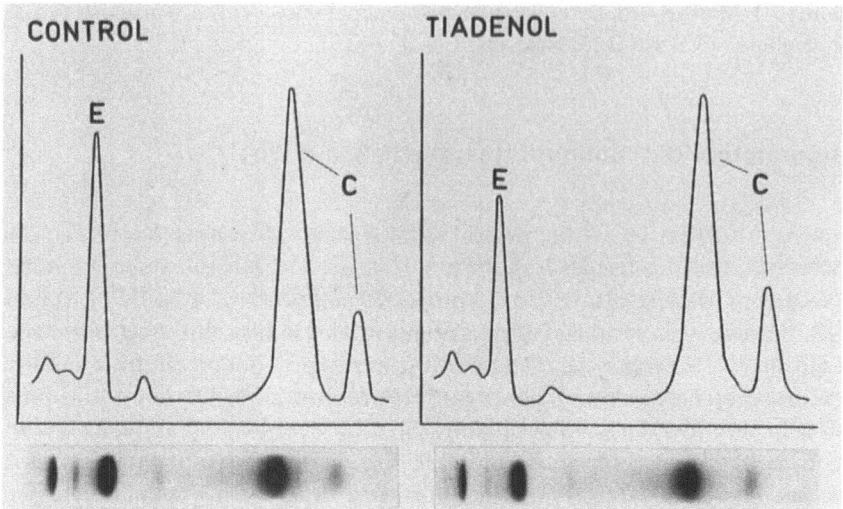

Fig. 2. Reduction of the apo-E content in VLDL after treatment with tiadenol in a patient with type IV hyperlipoproteinemia (Franceschini et al. 1981)

(Francheschini et al. 1981) (Fig. 2). Since tiadenol does not activate LPL and causes a significant accumulation of lipoprotein lipids in rodent liver (Kritchevsky et al. 1979), it seems likely that the reduction of VLDL-E may reflect a reduced liver lipoprotein secretion.

4   Early Drugs Trials: Protocols and Other Parameters

In our opinion, cross-over designs are preferable over the currently required randomized protocols for the early testing of drugs. They, in fact, allow a clear definition of even modest changes in plasma lipids and lipoproteins (Fig. 3) (Gaddi et al. 1984). Appropriate statistical analyses are available today for the elimination of carry-over effects or other background noises in the evaluation (Wallenstein and Fischer 1977).

Finally, the opportunity and rationale for the determination of *blood levels* of absorbable lipid-lowering drugs are still the object of debate (Gugler 1978). Recently, during a study on two formulations of *fenofibrate,* because of the poor bioavailability of one the two products, we were able to obtain widely dispersed blood levels, thus allowing correlations with the lipid/lipoprotein changes; analysis of the final data showed a significant correlation between the steady-state levels of fenofibrate and the reduction of both total and VLDL-triglyceridemia (Sirtori et al. 1985) (Fig. 4). This study, therefore, confirmed the primary activity of this drug on VLDL catabolism. This topic needs, however, further experimental and clinical work. Several of the new *HMG CoA(hydroxymethylglutaryl coenzyme A)-reductase inhibitors* have a very short plasma half-life, but a prolonged duration of activity (Pan et al. 1986). Moreover, there is evidence that *"retard"* formulations of specific lipid-lowering agents may not be as effective as the standard form with a rapid elimination (Knopp

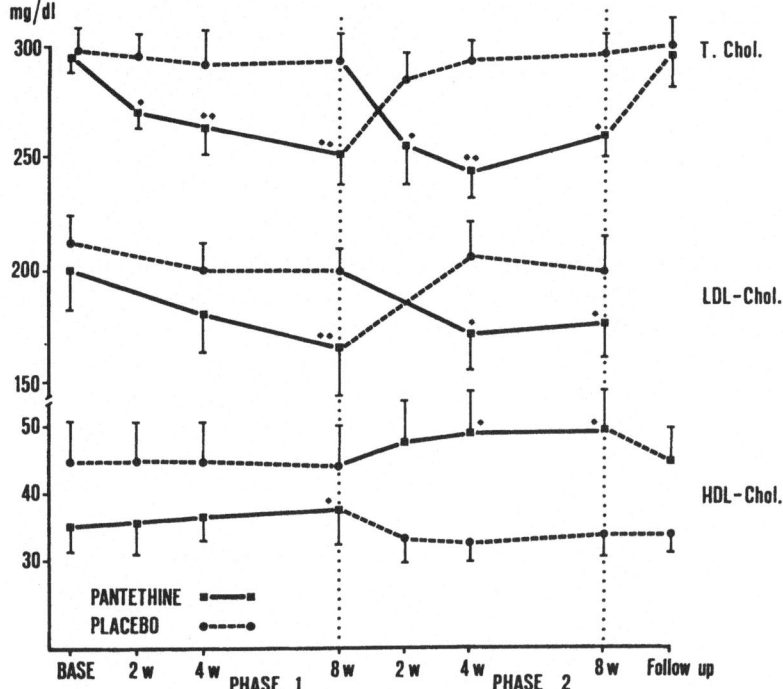

Fig. 3. Use of the cross-over design to evaluate the activity of pantethine in patients with type IIB hyperlipoproteinemia. In spite of the not pronounced effect on, e.g. HDL-cholesterol levels, the statistical analysis allows the definition of a significant pharmacodynamic activity (Gaddi et al. 1984)

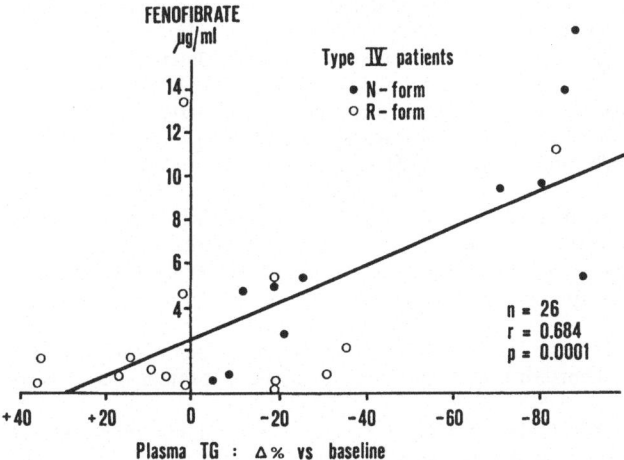

Fig. 4. Correlation between steady-state levels of fenofibrate in type IV patients treated with two different formulations of the drug and plasma triglyceride changes. The *N-form* achieved consistently higher levels vs the *R-form*. The wide dispersion of blood levels permits the detection of a significant correlation

et al. 1985). Lipid-lowering drugs may not adhere to the traditional concept of equilibration of the drug with a target, but rather act by a "hit and run" mechanism. Peak levels may, therefore, by more predictable for activity than steady-state levels.

5 Conclusions

At present, an accurate determination of cholesterol, triglyceride and HDL-cholesterol levels is probably what should be asked from a very early drug evaluation. Understanding of the meaning of the HDL cholesterol levels should also be made clear. Adding the apo AI determination may be necessary for the testing of drugs in severely hypertriglyceridemic patients. The HDL_2/HDL_3 ratio carries an important pharmacological significance, but is generally inaccurately determined. The assay of other apolipoprotein parameters does not generally add much information, but it should be performed in specific cases.

Acknowledgement. Supported in part by a grant from the Region Lombardia, Italy.

References

Avogaro P, Bittolo Bon G, Cazzolato G, Quinci GB (1979) Are apolipoproteins better discriminators than lipids for atherosclerosis? Lancet i:901–903

Baggio G, Briani G, Fellin R, et al. (1979) Effect of tiadenol on the concentration and composition of serum lipoproteins in familial hypercholesterolemia. Artery 5:486–498

Bengtsson G, Olivecrona T (1980) Lipoprotein lipase: some effects of activator proteins. Eur J Biochem 106:549–555

Carlson LA, Kolmodin-Hedman B (1977) Hyper-α-lipo-proteinemia in men exposed to chlorinated hydrocarbon pesticides. Acta Med Scand 201:375–376

Deckelbaum RJ, Granot E, Oschry Y, et al. (1984) Plasma triglyceride determines structure-composition in low and high density lipoproteins. Arteriosclerosis 4:225–231

Eisenberg S (1985) High density lipoprotein metabolism. J Lipid Res 25:1017–1058

Franceschini G, Poli A, Catapano A, et al. (1981) Pharmacological studies on tiadenol in type IV patients – evidence for a mechanism of action different from other lipid lowering drugs. Atherosclerosis 40:245–255

Franceschini G, Bondioli A, Mantero M, et al. (1982) Increased apoprotein B in very low density lipoproteins of patients with peripheral vascular disease. Arteriosclerosis 2:74–80

Franceschini G, Sirtori M, Gianfranceschi G, et al. (1985) Reversible increase of the apo CII/CIII-1 ratio in the very low density lipoproteins after procetofen treatment in hypertriglyceridemic patients. Artery 12:363–381

Gaddi A, Descovich GC, Noseda S, et al. (1984) Controlled evaluation of pantethine, a natural hypolipidemic compound in patients with different forms of hyperlipoproteinemia. Atherosclerosis 50:73–83

Gibson JC, Rubinstein A, Brown WV (1984) Precipitation of apo E-containing lipoprotein by precipitation reagents for apolipoprotein B. Clin Chem 30:1784–1788

Gidez LI, Miller GJ, Burstein M, Eder HA (1979) Analysis of plasma HDL subclasses by a precipitation procedure. In: Rep HDL Methodology Worksh, NIH Publ 82–1661, Bethesda, MD, p 328

Gugler R (1978) Clinical pharmacokinetics of hypolipidemic drugs. Clin Pharmacokinet 3:425–439

Kane JP, Sata T, Hamilton RL, et al. (1975) Apoprotein composition of very low density lipoproteins of human serum. J Clin Invest 56:1622–1634

Knopp RH, Ginsberg J, Albert JJ, et al. (1985) Contrasting effects of unmodified and time-release forms of niacin on lipoproteins in hyperlipidemic subjects: clues to mechanism of action of niacin. Metabolism 34:642–650

Kritchevsky D, Tepper SA, Czarnecky SK, Story JA (1979) Influence of tiadenol, bis (hydroxyethylthio) 1.10 decane, on cholesterol metabolism in rats. Pharmacol Res Commun 11:475–482

The Lipid Research Clinics Coronary Primary Prevention Trial Results (1984) J Am Med Assoc 251:351–374

Mahley RW (1983 Apolipoprotein E and cholesterol metabolism. Klin Wochenschr 61:225–232

Mattson FM, Grundy SM (1985) Comparison of effects of dietary saturated, monounsaturated and polyunsaturated fatty acids on plasma lipids and lipoproteins in man. J Lipid Res 26:199–202

Nikkilä EA, Huttunen JK, Ehnholm C (1977) Effect of clofibrate on post-heparin plasma triglyceride lipase activities in patients with hypertriglyceridemia. Metabolism 26:179–186

Pan HY, Willard DA, Funker PT, McKinstry DN (1986) The clinical pharmacology of SQ 31,000 (CS-514) in healthy subjects. IX Int Symp Drugs Affecting Lipid Metabolism. Florence, Italy, October 22–25

Saku K, Gartside PS, Hynd A, Kashyap ML (1986) Mechanism of action of gemfibrozil on lipoprotein metabolism. J Clin Invest 75:1702–1712

Schmidt SB, Wasserman AG, Muesing RA, et al. (1985) Lipoprotein and apolipoprotein levels in angiographically defined coronary atherosclerosis. Am J Cardiol 55:1459–1462

Schmitz G, Assmann G (1982) Isolation of human serum HDL_1 by zonal ultracentrifugation. J Lipid Res 23:903–910

Schonfeld G, George PK, Miller J, Reilly P, Witztum J (1979) Apolipoprotein C-II and C-III levels in hyperlipoproteinemia. Metabolism 28:1001–1010

Simpson HS, Ballantyne FC, Packard CJ, Morgan HG, Shepherd J (1982) High-density lipoprotein subfractions as measured by differential polyanionic precipitation and rate zonal ultracentrifugation. Clin Chem 28:2040–2043

Sirtori CR, Franceschini G (1984) Drug effects on HDL. In: Miller NE, Miller GJ (eds) Clinical and metabolic aspects of high-density lipoproteins. Elsevier, Amsterdam, p 341

Sirtori CR, Montanari G, Gianfranceschi G, et al. (1985) Correlation between plasma levels of fenofibrate and lipoprotein changes in hyperlipidaemic patients. Eur J Clin Pharmacol 28:619–629

Sirtori CR, Tremoli E, Gatti E, et al. (1986) Controlled evaluation of fat intake in the Mediterranean diet: comparative activities of olive oil and corn oil on plasma lipids and platelets in high-risk patients. Am J Clin Nutrit 44:635–642

Sirtori CR, Franceschini G, Gianfranceschi G, et al. (1987) Effects of gemfibrozil on plasma lipoprotein/apolipoprotein distribution and platelet reactivity in hypertriglyceridemic patients. J Lab Clin Med (in press)

Sniderman A, Shapiro S, Marpole D, et al. (1980) Association of coronary atherosclerosis with hyperapobetalipoproteinemia (increased protein but normal cholesterol levels in human plasma low density (β) lipoproteins) Proc Natl Acat Sci USA 77:604–608

Wallenstein S, Fischer AC (1977) The analysis of the two-period repeated measurements cross-over design with application to clinical trials. Biometrics 33:261–272

Warhick GR, Albers JJ (1978) A comprehensive evaluation of the heparin-manganese precipitation procedure for estimating high density lipoprotein cholesterol. J Lipid Res 19:65–76

Warnick GR, Cheung MC, Albers JJ (1979) Comparison of current methods for high-density lipoprotein cholesterol quantitation. Clin Chem 25:596–604

Warnick GR, Albers JJ, Leary ET (1980) HDL cholesterol: results of interlaboratory proficiency test. Clin Chem 26:169–170

Warnick GR, Benderson JM, Albers JJ (1983) Interlaboratory proficiency survey of high-density lipoprotein cholesterol measurement. Clin Chem 29:516–519

Witztum JL, Dillingham MA, Giese W, et al. (1980) Normalization of triglycerides in type IV hyperlipoproteinemia fails to correct low levels of high-density-lipoprotein cholesterol. N Engl J Med 30:907–914

Metabolic Studies with Lovastatin in Patients with Primary Hypercholesteremia

D. B. HUNNINGHAKE, D. M. HIBBARD, W. C. DUANE, M. L. FREEMAN, W. F. PRIGGE, K. J. GRAHAM, and R. L. GEBHARD [1]

1 Introduction

Lovastatin (mevinolin) is a competitive inhibitor of HMG CoA (hydroxymethyl-gluturate coenzyme A) reductase. Lovastatin decreases cholesterol synthesis and increases the number of LDL (low density lipoprotein) receptors and the catabolism of LDL (1,2). A variety of clinical trials demonstrate that lovastatin administration in humans is associated with significant reductions in LDL cholesterol up to 40% (3). The efficacy of lovastatin in humans appears to be established, but additional studies in humans related to long-term safety and mechanism of action are still in progress. This report addresses several of these issues in a preliminary fashion.

Many of the patients who are or will be receiving lovastatin have manifestations of coronary heart disease and are receiving many other drugs for a variety of therapeutic reasons. Lovastatin is metabolized by the liver and it is desirable to know if the disposition of other drugs which are also metabolized by the liver would be affected by lovastatin administration. Antipyrine, which undergoes biotransformation by at least three independent hepatic oxidative pathways involving N-demethylation and aliphatic hydroxylation, is frequently used as an indicator of hepatic drug-metabolizing activity in vivo (4). The purpose of the current study was to determine the effect of lovastatin on the biotransformation and elimination of antipyrine.

A second major area of investigation involved a study of the effects of lovastatin on jejunal HMG CoA reductase activity in humans. The liver and the gastrointestinal tract are the two major sites for cholesterol synthesis. The classic plasma pharmacokinetic data for lovastatin are not yet available to permit one to predict duration of effect and appropriate dosing schedules. The availability of an assay of HMG CoA reductase activity in the jejunum provided us with the opportunity to assess the degree of inhibition of the activity of this enzyme following lovastatin administration and to obtain some estimate of the duration of effect of lovastatin (5).

Other hypolipidemic agents, such as clofibrate, have been demonstrated to either increase the lithogenicity index or incidence of cholelithiasis. Since lovastatin significantly decreases blood cholesterol levels and was postulated to decrease mucosal

---

1 Department of Medicine and Pharmacology, University of Minnesota, Minneapolis, MN 55455, USA

Drugs Affecting Lipid Metabolism
Ed. by R. Paoletti et al.
© Springer-Verlag Berlin Heidelberg 1987

reductase activity, it was felt desirable to assess the gallbladder bile cholesterol saturation index (lithogenicity index) (6,7).

2 Methods

Antipyrine. Twelve healthy subjects, nine men and three women, with type II hyperlipoproteinemia (LDL-C $\geq$ 190 mg dl^{-1}, triglycerides less than 250 mg dl^{-1}) between the ages of 21 and 59 volunteered for the study. All were nonsmokers, abstained from alcohol and all other drugs, and were on a cholesterol-lowering diet and limited caffeine intake throughout the study. Each subject gave written consent and was admitted to General Clinical Research Center (GCRC) on four separate occasions. At time zero for each visit the subject received 18 mg kg^{-1} antipyrine in 200 ml water orally. The four studies were initiated in the following sequence in relationship to lovastatin administration: (1) baseline (placebo); (2) acute (lovastatin 40 mg twice daily for 6 days); (3) chronic (lovastatin 40 mg twice daily for 26 days); (4) retest (placebo for 14 days).

Plasma samples were collected in heparinized tubes prior to zero, 0.25,0.5,1,2,3,4, 6,8,12,24, and 48 h postantipyrine administration. The samples were spun immediately and stored at -20°C until analyzed. All plasma samples were analyzed by the previously described HPLC method within 1 month of collection (8). The following pharmacokinetic parameters were calculated from the data obtained from each subject after administration of single doses of antipyrine: Area under the Plasma Concentration vs time curve (AUC) by the trapezoidal method for the early part of the curve, and the log-trapezoidal method extrapolated to infinite time for that portion of the curve which appears linear in a semilog plot. The negative of the slope of this line gave the terminal rate of elimiation of antipyrine from plasma (λ_z = terminal slope of the curve; concentration vs time). The time to peak Tmax (time of maximum measured concentration) and maximum peak concentrations Cmax were also calculated. The pharmacokinetic parameters calculated for acute, chronic, and retest were each compared to those calculated for baseline using a paired t-test. The cumulative percent of dose recovered for the urinary metabolites (3-hydroxymethylantipyrine, n-demethylantipyrine, and 4-hydroxyantipyrine) and antipyrine were calculated.

Gastrointestinal Studies. Patients with type II hyperlipoproteinemia who had given informed consent were used in this study. All patients were on a cholesterol-lowering diet and were asked to fast for at least 12 h before intubation at 0800 h. After cetacaine spray, a Carey capsule small intestinal biopsy tube was passed into the duodenum. Gallbladder bile was collected on ice by siphonage following administration of CCK-A (Kinevac 0.02 μg kg^{-1}) IV. A small bowel biopsy was obtained by suction. The patients described in this study then had a second procedure after 12 weeks of treatment with lovastatin. Bile samples were analyzed for bile acids (hydroxy-steroid dehydrogenase method) lecithin (colorimetry), and cholesterol (extracation and GLC) (7). Mucosal tissue was homogenized and assayed for total activity for HMG CoA reductase (5).

3   Results

The plasma antipyrine levels during the four treatment periods (two with placebo and two with lovastatin) are illustrated in Fig. 1. No significant differences between the plasma antipyrine levels for the various time periods are noted for any treatment period.

Some of the classic pharmacokinetic parameters for the plasma antipyrine study are illustrated in Table 1. There were no consistent differences during any of the treatment periods for the AUC, terminal rate elimination constant, Tmax or Cmax.

The cumulative urinary excretion of antipyrine and the three metabolites as percent (mean ± SD) of the administered doese for the treatment periods were as follows: baseline 56.2 ± 7.3; acute 57.8 ± 8.1; chronic 56.0 ± 11.2; and retest 58.6 ± 14.7. There were no significant differences within the treatment groups.

The levels of HMG CoA reductase before and 12 weeks after lovastatin administration are illustrated in Fig. 2. There was a mean reduction of 60% in the mucosal reductase activity (9.4 ± 3.3 vs 3.8 ± 0.7 in five patients, $p < 0.05$).

The biliary saturation index was measured in five patients before and 12 weeks after lovastatin therapy. The mean LDL-cholesterol was lowered from 253 to 149 mg dl^{-1}. The mean gallbladder bile saturation index was significantly altered (1.04 ± 0.1 vs 0.89 ± 0.2), and all patients showed some reduction in saturation index.

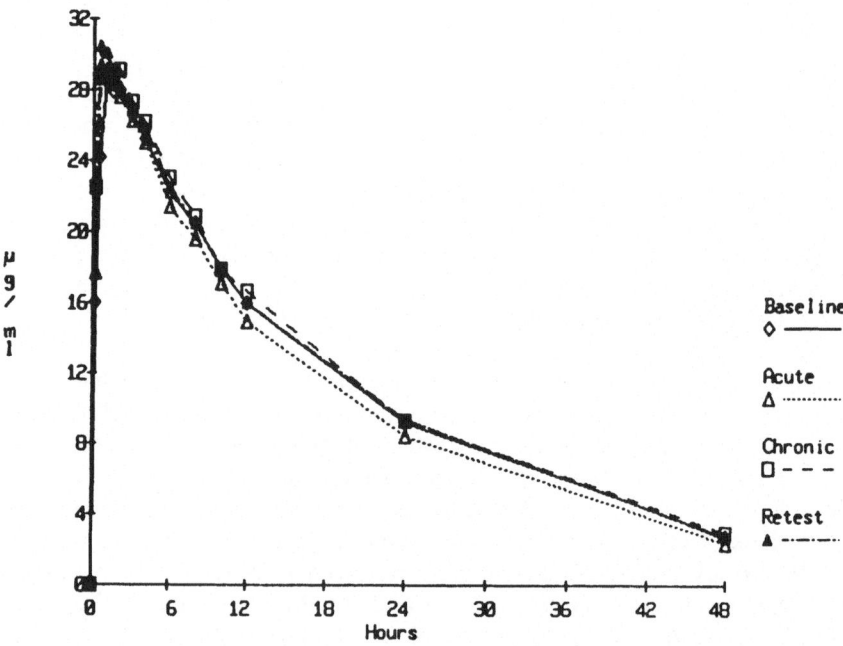

Fig. 1. The concentration of plasma antipyrine at the various time intervals for the four treatment groups

Table 1. Mean values and standard deviations of the pharmacokinetic parameters for the disposition of antipyrine before, during, and after administration of lovastatin

	Baseline		Acute		Chronic		Retest	
	Mean	SD	Mean	SD	Mean	SD	Mean	SD
Cmax (μg ml^{-1})	30.3	4.3	32.2	4.8	32.1	5.4	34.2	7.7
Tmax (h)	1.1	0.5	0.8	0.4	1.0	0.8	0.9	0.7
AUC (μg h^{-1} ml^{-1}) $(0 \rightarrow \infty)$	622	144	587	84	615	155	599	126
λ_z (h^{-1} x 1000)	51.8	8.4	54.2*	9.6	50.8	10.3	51.5	8.8

* $p < 0.05$ as compared to baseline.

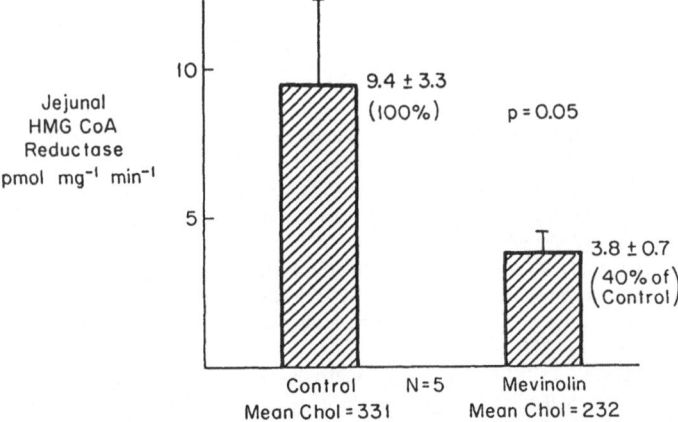

Fig. 2. The level of HMG CoA reductase activity in the jejunum before and after lovastatin therapy

4 Discussion

The results of this study indicate that lovastatin had little effect, if any on the metabolism of antipyrine either after short-term administration (approximately 1 week) or long-term administration (approximately 4 weeks). In addition, there was no apparent effect on absorption or elimination as measured by AUC$_{(0\rightarrow\infty)}$ and λ_z. The urinary data validates this by showing similar recoveries of the oral dose in each treatment group. The results of this study suggest that the potential for clinically significant oxidative metabolic drug interactions with mevinolin is minimal.

Lovastatin inhibited the activty of intestinal HMG CoA reductase activity for at least 12 h following the last dose of medication. This study provides in vivo confirmation of the drug's pharmacologic affect and demonstrates that there is a chronic suppression (12 weeks) of HMG CoA reductase activity in the gut. In rodent studies, lovastatin rapidly increases the the total quantity of reductase enzyme in cells by inhibiting its degradation (9). This study in humans indicates that there is an inhibition of enzyme activity, and the chronic reduction of reductase activity correlates with the persistent LDL-lowering affect of lovastatin in man.

Lovastatin could influence the secretion of cholesterol into the bile or conversion of cholesterol into bile acids and thus influence the formation of cholesterol gall-stones. There was no increase in the lithogenicity index of gallbladder bile following lovastatin treatment. This finding suggests that lovastatin should not increase the risk for a cholesterol gallstone in humans.

The preliminary findings reported from these studies suggests that lovastatin will be an acceptable drug for long-term administration in man. It is unlikely that it will interfere with the hepatic metabolism of co-administered drugs and it does not increase lithogenicity of gallbladder bile. Additionally, a chronic and prolonged suppression ($>$ 12 weeks) of HMG CoA reductase activity following the last dose of lovastatin was demonstrated.

References

1. Tobert JA et al. (1982) Cholesterol lowering effect of mevinolin an inhibitor of 3-hydroxy-3-methylglutaryl coenzyme A reductase in healthy volunteers. J Clin Invest 69:913–919
2. Bilheimer DW, Grundy SM, Brown MS, Goldstein JL (1983) Mevinolin stimulates receptor-mediated clearance of low density lipoprotein from plasma in familial hypercholesterolemia hepterozygotes. Trans Assoc Am Phys 96:1–9
3. Illingworth DR, Sexton GJ (1984) Hypocholesterolemic effects of mevinolin in patients with heterozygous familial hypercholesterolemia. J Clin Invest 74:1972–1978
4. Vesell ES (1979) The antipyrine test in clinical pharmacology conceptions and misconceptions. Clin Pharmacol Ther 26:275–286
5. Gebhard RL, Stone BG, Prigge WF (1985) 3-hydroxy-3-methylglutaryl coenzyme A reductase activity in the human gastrointestinal tract. L Lipid Res 26:47–53
6. Carey MC, Small DM (1978) The physical chemistry of cholesterol solubility in bile. J Clin Invest 61:998–1026
7. Duane WC (1978) Simulation of the defect of bile acid metabolism associated with cholesterol cholelithiasis by sorbitol ingestion in man. J Lab Clin Med 91:969–978
8. DeVries JX, Staiger C, Wang NS, Schlicht F, Oroknay B (1984) Simultaneous determination of antipyrine and metabolites in human plasma and urine by high-performance liquid chromatography. J Chromatol 309:219–221
9. Brown MS, Goldstein JL, Dietsch JM (1979) Active and inactive forms of 3-hydroxy-3-methylglutaryl coenzyme A reductase in the liver of the rat. J Biol Chem 254:5144–5149

Reduction of Plasma Lipids and Lipoproteins by Dietary Fish Oils in Patients with Hypertriglyceridemia

B. E. Phillipson, W. E. Connor, W. S. Harris, and D. R. Illingworth [1]

1 Introduction

The relative infrequency of coronary heart disease (CHD) and thrombosis among Eskimos has raised the possibility that their unusual diet might be protective (1–5). Most Eskimos and Northwest Coast Indians traditionally derived the bulk of their food from the sea. Although some marine animals, fish, and shellfish do have a high cholesterol and fat content, their fat is unique in that it contains large quantities of long-chain, highly polyunsaturated omega-3 fatty acids (5).

Omega-3 fatty acids have a unique structure and cannot be synthesized by animals. The dietary sources of omega-3 fatty acids are some vegetable oils, leafy vegetables, and, in particular, fish and shellfish. Linolenic fatty acid, (C18:3), is obtained from vegetable products. Eicosapentaenoic acid, EPA, (C20:5), and docosahexaenoic acid, DHA, (C22:6) are derived from fish and shellfish and phytoplankton (the plants of the ocean) and are highly concentrated in fish oils. Although fish oils have been found to have a hypolipidemic effect (6), metabolic differences between dietary fish oils and vegetable oils have not been appreciated until recently (7,8). The present chapter discusses the efficacy of dietary fish oils as hypotriglyceridemic agents in patients with moderate to severe hypertriglyceridemia.

2 Methods

Subjects and Study Design. Twenty hypertriglyceridemic patients volunteered for the study (9) and informed consent was obtained from each patient. All patients had normal thyroid, renal, and hepatic function and were in a steady state before entry into the study. Two different control diets were used for the two groups of patients, depending on the phenotype of hyperlipidemia. The type IIb group received their usual low-cholesterol (150 mg), low-fat (20 to 30% of total calories) diet, followed by a fish-oil diet for 4 weeks, which was followed, in some patients, by a 4-week period of a diet high in a vegetable oil rich in omega-6 fatty acids. Both diets were

1 The Oregon Health Sciences University, Portland, OR 97201, USA

Drugs Affecting Lipid Metabolism
Ed. by R. Paoletti et al.
© Springer-Verlag Berlin Heidelberg 1987

balanced for cholesterol (about 325 mg day^{-1}) and contained 30% of total calories as fat. The diets in all periods were eucaloric.

For the type V group, the control diet was very low in fat (5–10%) with a virtual absence of cholesterol. The second diet contained fish oil, which accounted for 20 or 30% of total calories, and 350 mg cholesterol day^{-1}. The third diet was similar, but contained vegetable oil as the fat source.

The experimental diets differend mainly in their fatty acid composition. The two control diets contained relatively similar amounts of saturated, monounsaturated, and diunsaturated fatty acids, but provided only traces of omega-3 fatty acids, which accounted for 22 to 35% of total fatty acids in the fish-oil diets.

Laboratory Procedures. The fatty acid composition of the diets was determined by GLC methods and total cholesterol and triglyceride levels of plasma and lipoprotein fractions were measured with an Autoanalyzer II (Technicon Instruments). Apolipoproteins A-1,B,C-III, and E were determined in whole plasma by electroimmunoassay.

3 Results

Dietary fish oil had a profound hypolipidemic effect in each of the 20 patients with hypertriglyceridemia and total plasma cholesterol and triglyceride levels were consistently lowered. Body weights remained constant and the diets were well tolerated.

In patients with phenotypic type IIb hyperlipidemia, the fish-oil diet decreased the plasma cholesterol levels by 27%. The decline in VLDL cholesterol was most striking, but LDL and HDL cholesterol also decreased. The plasma triglyceride changes were even greater than the cholesterol changes with the fish-oil diet (−64%), largely because of the reduction in VLDL triglyceride levels which fell from 216 to 55 mg dl^{-1} (2.44 to 0.62 mmol l^{-1}). The highly polyunsaturated vegetable-oil diet led to less marked decreases in plasma cholesterol and triglyceride levels in the type II-b group as compared with the effects of fish oil because the vegetable-oil diet had a much smaller effect on VLDL cholesterol and triglyceride. LDL values were similar, but HDL cholesterol levels were higher after the vegetable-oil diet.

The effects of the fish-oil diet were even more striking in the patients with the type V phenotype. During the fish-oil diet, the total plasma triglyceride decreased from a control value of 1353 to 281 mg dl^{-1} (15.29 to 3.18 mmol l^{-1}), a drop of 79%. The VLDL triglyceride levels decreased similarly, from 1087 to 167 mg dl^{-1} (12.28 to 1.89 mmol l^{-1}). Plasma cholesterol levels dropped to the normal range after the fish-oil diet, from 373 to 207 mg dl^{-1}). Most of this was due to decreases in the level of VLDL cholesterol, which fell from 270 to 70 mg dl^{-1} (7.02 to 1.82 mmol l^{-1}). LDL levels rose from their initially low levels of cholesterol, 84 to 125 mg dl^{-1} (2.18 to 3.25 mmol l^{-1}). Plasma apoprotein changes reflected the changes in lipoprotein lipids and significant reductions occurred in the concentrations of apoproteins B and C-III during the fish-oil diet.

When the omega-6-rich vegetable oil replaced the fish oil in the diets of eight patients, all had increases in plasma triglyceride levels within 3 to 4 days. After 10 to 14 days of the vegetable-oil diet, the mean plasma triglyceride value rose by 198% and the VLDL triglyceride value increased from 171 to 550 mg dl^{-1} (1.93 to 6.22 mmol l^{-1}). The plasma cholesterol level also increased, from 195 to 264 mg dl^{-1} (5.07 to 6.86 mmol l^{-1}). Because of worsening hypertriglyceridemia and the risk of abdominal pain typical of the type V disorder, the vegetable-oil diet was discontinued prematurely in all patients in the type V group.

4 Discussion

In the 20 patients with hypertriglyceridemia described in the present study, fish oil in the diet led to an even more profound hypolipidemic effect than previously observed in normal subjects. There are number of possible mechanisms by which dietary fish oils could exert their hypotriglyceridemic action. The most likely mechanism is the reduction of VLDL synthesis in the liver. Dietary fish oil has been shown to block the usual induction of hypertriglyceridemia that an abrupt increase in dietary carbohydrate causes. In turnover studies, fish oil inhibited VLDL apoprotein B synthesis and VLDL triglyceride synthesis. Furthermore, fatty acid synthesis in rat liver has been inhibited by the omega-3 fatty acids of fish oil, which may result in reduced VLDL synthesis. Our results indicate that consumption of fish and fish oils have therapeutic benefit in patients with hypertriglyceridemia. In our instructions to patients, we stress the use of fish, properly prepared without added fat in place of meat and high-fat dairy products. We suggest up to 6 oz day^{-1} of either high-fat or low-fat fish. We also allow small quantities (i.e., 2 to 3 oz) of shellfish as these also contain omega-3 fatty acids (EPA and DHA). In addition, fish oils may be used therapeutically in certain patients.

References

1. Dyerberg J, Bang HD, Stofferson E, Moncada S, Vane JR (1978) Eicosapentaenoic acid and prevention of thrombosis and atherosclerosis. Lancet 2:117–119
2. Dyerberg J, Bang HO (1979) Hemostatic function und platelet polyunsaturated fatty acids in Eskimos. Lancet 2:433–435
3. Bang HO, Dyerberg J. Plasma lipids and lipoproteins in Greenlandic West Coast Eskimos (1972) Acta Med Scand 192:85–94
4. Dyerberg J, Bang HO, Hjorne N (1975) Fatty acid composition of the plasma lipids in Greenland Eskimos. Am J Clin Nutrit 28:958–966
5. Bang HO, Dyerberg J, Hjorne N (1976) The composition of food consumed by Greenlandic Eskimos. Acta Med Scand 200:69–73
6. Harries WS, Connor WE (1980) The effects of salmon oil upon plasma lipids, lipoproteins and triglyceride clearance. Trans Assoc Am Phys 43:148–155
7. Goodnight SH, Harris WS, Connor WE, Illingworth DR (1982) Polyunsaturated fatty acids, hyperlipidemia, and thrombosis. Arteriosclerosis 2:87–113
8. Harris WS, Connor WE, McMurry MP (1983) The comparative reductions in the plasma lipids and lipoproteins by dietary polyunsaturated fats: salmon oil vs. vegetable oils. Metabolism 32:179–184
9. Phillipson BE, Rothtrock DW, Connor WE, Harris WS, Illingworth DR (1985) Reduction of plasma lipids, lipoproteins and apoproteins by dietary fish oils in patients with hyperglyceridemia. N Engl J Med 312:1210–1216

Effects of Monounsaturated Fatty Acids Versus Complex Carbohydrates on Serum Lipoprotein Subfractions and Apolipoproteins in Healthy Men and Women

R. P. Mensink and M. B. Katan [1]

1 Introduction

Recently, we have reported that an olive oil-rich diet and a high-carbohydrate, high-fibre diet lower serum total cholesterol levels to the same extent relative to a Western-type diet, high in total and saturated fat. High-density lipoproteins (HDL) cholesterol concentrations did not change on the olive oil-rich diet, but decreased on the high-carbohydrate diet (1). It was suggested that in consideration of the supposed anti-atherogenic effect of HDL (2,3), reducing total fat intake per se might not be optimal for the prevention of coronary heart disease. However, there are indications that the protective role of HDL is confined to the HDL_2 fraction (4) or that apolipoprotein concentrations are better indicators of the risk for atherosclerosis than the lipid components of lipoproteins (5,6). Here, we report the effects of an olive oil-rich and a high-carbohydrate diet on the distribution of cholesterol over the various lipoproteins and on apolipoproteins A_I and B levels.

2 Methods

Forty-eigth healty men and women first consumed for 17 days a Western-type diet, high in saturated (20 en%) and total fat (38 en%). Twenty-four of the subjects then consumed for 32 days a high-carbohydrate, high-fibre diet (total fat 22 en%) and the other 24 an olive oil-rich diet (oleic acid 24 en%, total fat 41 en%). The amounts of protein (12—14 en%), polyunsaturated fat (4—5 en%) and cholesterol (31—35 mg MJ^{-1}) were similar in all three diets. Each diet consisted of conventional, mixed solid foods and menus were changed daily. All foodstuffs were supplied individually according to each person's energy need. During the study changes in weight did not exceed 2.5 kg.

Equal amounts of sera from each subjects obtained on days 14 and 17 (control period) were pooled and thus aliquots were obtained on days 46 and 49 (test period).

---

1 Department of Human Nutrition, Agricultural University, De Dreijen 12, 6703 BC Wageningen, The Netherlands

Drugs Affecting Lipid Metabolism
Ed. by R. Paoletti et al.
© Springer-Verlag Berlin Heidelberg 1987

Table 1. Effects of a high-carbohydrate, high-fibre diet and an olive oil-rich diet on cholesterol concentrations in the various lipoproteins, and on apolipoproteins A_I and B levels and their ratio (mean ± SD) relative to levels on a high saturated-fat control diet period

	High-carbohydrate (n=24)	Olive oil-rich (n=24)
VLDL (mmol l^{-1})		
Saturated-fat control diet	0.34 ± 0.21	0.36 ± 0.34
Change on test diet	+0.08 ± 0.15	−0.08 ± 0.25 [b]
IDL (mmol l^{-1})		
Saturated-fat control diet	0.24 ± 0.16	0.21 ± 0.10
Change on test diet	+0.01 ± 0.19	+0.03 ± 0.10
LDL (mmol l^{-1})		
Saturated-fat control diet	2.57 ± 0.67	2.54 ± 0.55
Change on test diet	−0.45 ± 0.67	−0.50 ± 0.44
HDL_1 plus Lp(a) (mmol l^{-1})		
Saturated-fat control diet	0.24 ± 0.11	0.24 ± 0.09
Change on test diet	−0.01 ± 0.07	+0.01 ± 0.06
HDL_2 (mmol l^{-1})		
Saturated control diet	0.47 ± 0.22	0.50 ± 0.24
Change on test diet	−0.08 ± 0.17	−0.03 ± 0.10
HDL_3 (mmol l^{-1})		
Saturated control diet	0.87 ± 0.16	0.87 ± 0.15
Change on test diet	−0.09 ± 0.15	+0.01 ± 0.13 [b]
Bottom fraction (mmol l^{-1})		
Saturated control diet	0.07 ± 0.07	0.05 ± 0.06
Change on test diet	+0.08 ± 0.08	+0.11 ± 0.07
Apolipoprotein A_I (mg l^{-1})		
Saturated-fat control diet	1816 ± 248	1703 ± 250
Change on test diet	−111 ± 221	+ 47 ± 170 [c]
Apolipoprotein B (mg l^{-1})		
Saturated-fat control diet	938 ± 206	952 ± 226
Change on test diet	− 8 ± 160	− 80 ± 182 [a]
Apo-A_I/apo-B		
Saturated-fat control diet	2.01 ± 0.44	1.91 ± 0.64
Change on test diet	−0.12 ± 0.29	+0.30 ± 0.50 [d]

Statistical comparison between diets:
[a] $p < 0.1$;
[b] $p < 0.05$;
[c] $p < 0.01$;
[d] $p < 0.001$.

Lipoproteins were fractionated by density ultracentrifugation (7). The following density classes (d in mg ml^{-1}) were isolated: VLDL (d < 1.010), IDL (1.010 < d < 1.019), LDL (1.019 < d < 1.055), HDL$_1$ plus lipoprotein(a) (1.055 < d < 1.075), HDL$_2$ (1.075 < d < 1.100), HDL$_3$ (1.100 < d < 1.180) and a bottom fraction (d > 1.180). The mean recovery of cholesterol in the lipoprotein fractions was 93.0 ± 5.3% for the control and 93.4 ± 6.5% and for the test period. All samples of one subject were analyzed within one run. Apolipoprotein A$_I$ (apo-A$_I$) and apolipoprotein B (apo-B) were measured in whole serum by radial immunodiffussion. For one subject the samples obtained at the end of the control period and at the end of the test period were analyzed in duplicate on one plate.

The response to the test diet was calculated per subject as the change from the end of the control period to the end of the test period.

3 Results

Table 1 shows that VLDL-cholesterol increased by 0.08 mmol l^{-1} on the high-carbo-hydrate, high-fibre diet and fell by 0.08 mmol l^{-1} on the olive oil-rich diet ($p < 0.05$ for difference between diets). HDL$_3$-cholesterol fell by 0.09 mmol l^{-1} on the high-carbohydrate diet and increased by 0.01 mmol l^{-1} on the olive oil-rich diet ($p < 0.05$). HDL$_2$-cholesterol fell to a greater extent on the carbohydrate diet (0.08 mmol l^{-1}) than on the olive oil diet (0.03 mmol l^{-1}), although the difference between the diets was not statistically significant. Apo-A$_I$ levels decreased by −111 mg l^{-1} and apo-B levels by −8 mg l^{-1} on the high-carbohydrate diet. These values were +47 mg l^{-1} and −80 mg l^{-1} on the olive oil-rich diet respectively. Differences between diets were statistically significant for both A$_I$ ($p < 0.01$) and B ($p < 0.1$). The apo-A$_I$ to apo-B ratio decreased by 0.12 or 6% on the carbohydrate diet and increased by 0.30 or 16% on the olive oil-rich diet ($p < 0.001$).

4 Discussion

We have reported that total serum cholesterol concentrations fell to the same extent on a high-carbohydrate, high-fibre diet and on an olive oil-rich diet as compared to a Western-type diet. In addition, the olive oil-rich diet, unlike the high-carbohydrate diet, caused a specific fall in non-HDL-cholesterol (1). We now have evaluated the effects of both diets on cholesterol concentrations in the seven lipoprotein subfractions and on apolipoprotein A$_I$ and apolipoprotein B levels.

It has been suggested that variations in HDL are mostly located in the HDL$_2$ fraction (8). This is not confirmed by our results. The decrease in HDL-cholesterol on the high-carbohydrate diet was located both in HDL$_2$ and HDL$_3$, although the former was not statistically significant from the change on the olive oil-rich diet. Differences in the non-HDL fractions between the high-carbohydrate and the olive oil-rich diet were mainly located in the VLDL fraction: VDL-cholesterol increased on the high-carbohydrate diet and decreased on the olive oil-rich diet. Thus, the carbohydrate diet caused an unfavourable redistribution of cholesterol from HDL_2 and HDL_3 to VLDL.

It has been suggested that the apo-A_I to apo-B ratio is superior to the more widely used HDL-cholesterol to LDL tor total cholesterol ratio as an indicator for atherosclerotic risk (5). We found a marked increase in this ratio when saturated fatty acids were replaced by monounsaturated fatty acids, but a decrease when saturates were replaced by carbohydrates. Hollenbeck et al. (9) showed a decrease in both apo-A_I and apo-B on a low-fat diet in insulin-dependent women, but no change in their ratio. However, we have compared monounsaturated fatty acids with complex carbohydrates, while Hollenbeck et al. (9) have compared saturated and monounsaturated fatty acids with carbohydrates.

Thus, indices proposed as predictors of the risk for coronary heart disease were more favourably affected by an olive oil-rich diet than by a high-carbohydrate, high-fibre diet.

Acknowledgements. This study was supported by a grant from the Commission of the European Communities. We are grateful to the technical and dietary staff of the Department of Human Nutrition for their assistance; to Mrs. P. van Heijst for the determinations of the apolipoproteins; and to the volunteers for their cooperation and interest.

References

1. Mensink RP, Katan MB (1987) Effect of monounsaturated fatty acids versus complex carbohydrates on high density lipoproteins in healthy men and women. Lancet 1:122–125
2. Schaefer EJ (1984) Clinical, biochemical and genetic features in familial disorders of high density lipoproteins deficiency. Arteriosclerosis 4:303–322
3. Brensike JF, Levy RI, Kelsey SF, et al. (1984) Effects of therapy with cholestyramine on progression of coronary arteriosclerosis: results of the NHLBI type II coronary intervention study. Circulation 69:313–324
4. Gofman JW, Young W, Tandy R (1966) Ischemic heart disease, atherosclerosis and longevity. Circulation 34:679–697
5. Avogaro P, Bittolo Bon G, Cazzolato G, Quinci GB (1979) Are lipoproteins better discriminators than lipids for atherosclerosis? Lancet i:901–903
6. Maciejko JJ, Holmes DR, Kottke BA, Zinsmeister AR, Dinh DM, Mao SJT (1983) Apolipoprotein-AI as a marker of angiographically assessed coronary-artery disease. N Engl J Med 309:385–389
7. Terpstra AHM, Woodward JH, Sanchez-Muniz FJ (1981) Improved techniques for the separation of serum lipoproteins by density gradient ultracentrifugation – visualization by prestaining and rapid separation of serum lipoproteins from small volumes of serum. Anal Biochem 111: 149–157
8. Anderson DW (1978) H.D.L. cholesterol: the variable components. Lancet i:819–820
9. Hollenbeck CB, Connor WE, Riddle MC, Alaupovic P, Leklem JE (1985) The effects of a high-carbohydrate low-fat cholesterol-restricted diet on plasma lipid, lipoprotein, and apoprotein concentrations in insulin-dependent (type I) diabetes mellitus. Metabolism 34:559–566

Changes of Eicosanoid Formation in Relation to Dietary Lipids and in Hyperlipaemia

C. Galli, C. Mosconi, L. Medini, and E. Tremoli[1]

1 Introduction

The composition of cellular lipids (e.g. platelets, leukocytes) is affected by manipulations of dietary fat and/or pathological states, such as hyperlipidemias or other lipid metabolic disorders. As a result, functional parameters are modified, possibly as a consequence of altered eicosanoid production.

2 Dietary Fatty Acids and the Arachidonic Acid Cascade

The precursor-product relationship between 20 C-polyunsaturated fatty acids and the eicosanoids represents the obvious metabolic link between dietary fatty acids and these biologically active products. Studies based on manipulations of dietary fats, e.g. in experimental animals, have indeed shown that there is generally a good correlation between levels of eicosanoid precursor(s) fatty acid(s) in cellular membranes and eicosanoid production after stimulation, e.g. in platelets, in in vitro systems. However, the accumulation of E-precursor fatty acids in cellular lipids is dependent upon complex processes, such as conversion of short-chain PUFA (SCP) derived from the diet to the long-chain PUFA (LCP) found in tissues, and/or interactions between fatty acids of different series in their metabolic conversion and incorporation in cellular lipids.

Thus, as a consequence of the multiple steps between the intake of dietary fatty acids and the incorporation of LCP in cellular lipids, complex relationships between dietary fatty acids and E-formation are observed.

Also, considerable species differences appear to exist in LCP metabolism, and this is reflected in appreciable interspecies variations in the levels of LCP in the plasma compartment, as indicated in Table 1. The data suggest that the overall conversion of LA to AA is more efficient in rats than in the other species, and that LCP metabolism in rabbits is not very active.

---

1 Institute of Pharmacological Sciences, University of Milan, Italy

Drugs Affecting Lipid Metabolism
Ed. by R. Paoletti et al.
© Springer-Verlag Berlin Heidelberg 1987

Table 1. Levels (weight percentages) of linoleic (LA) and arachidonic (AA) acids in plasma phospholipids of different species[a]

Fatty acid	Rat (Socini et al. 1983)	Rabbit (Masi et al. 1986)	Man (Dougherty et al. 1987)
LA	11.8 ± 3.4	25.8 ± 5.3	24.7 ± 3.7
AA	8.9 ± 2.8	2.2 ± 0.2	12.5 ± 2.3
AA/LA	0.75	0.08	0.51

[a] Values are percentage levels and represent the average ± SD of at least 10 determinations. Rats and rabbits were fed semi-synthetic diets containing 10% by weight corn oil, and human subjects consumed a diet containing 4.5 en% LA.

As a consequence of the different rates of PUFA metabolism in different animal species, the effects of enhanced dietary levels of linoleic acid on the levels of arachidonic acid in plasma and tissue lipids are quite different. In humans, e.g. the administration of diets enriched in LA does not result in elevation of AA levels in plasma phospholipids (PL) (Lasserre et al. 1985). Also, in both humans and rabbits, elevation of dietary LA resulted in elevation of LA and reduction of AA and of the AA/LA ratio in platelet PL (Galli et al. 1981; Tremoli et al. 1986). These fatty acid changes may be responsible for the resulting decreased formation of platelet TxB_2 and platelet aggregation. In contrast, in rats, animal species with a high rate of metabolic conversion of LA to AA, when LA was increased in the diet of about 3 en% above the levels present in the standard diet, AA levels in plasma and platelet lipids (Giani et al. 1984) were significantly elevated.

A number of elegant and important studies in humans have shown that the administraion of eicosapentaenoic acid (20:5 n-3, EPA) results in incorporation of this fatty acid in lipids of various cells and tissues, including platelets and leukocytes. Functional parameters are hence modified, possibly as a result of the formation of thromboxane and leukotriene products derived from this fatty acid (Fischer and Weber 1983, 1984).

EPA administration affects eicosanoid production, however, also through additional mechanisms besides incorporation of the fatty acid in cellular lipids. This occurs in situations when interactions between dietary w6 and w3 fatty acids prevent a significant incorporation of EPA in cellular compartments (Socini et al. 1983). Tx production was depressed in platelets from animals fed a combination of corn oil + fish oil, rich in EPA, in spite of a lack of accumulation of EPA in these cells, in a proportion similar to that observed in animals fed fish oil alone, which results in accumulation of EPA in platelets.

This type of study was further extended by comparatively evaluating the effects of the administration of an EPA-enriched oil (MaxEPA) vs a "neutral" oil, such as olive oil, on various platelet and leukocyte parameters in the rat. Both oils were administered by gastric intubation for 8 weeks to rats on a standard diet. Thus, the same amount of linoleic acid was supplied to both groups of animals, but EPA or oleic acid respectively, were the main supplements of each treatment. EPA did ac-

Table 2. Levels of LA and AA in plasma, platelet, red blood cell and PMNL PL in MaxEPA (EPA) and olive oil (00) treated rats[a]

Fatty acid	Plasma 00	Plasma EPA	Platelet 00	Platelet EPA	Red Cells 00	Red Cells EPA	PMNL 00	PMNL EPA
LA	*23.6*	*26.5*	*8.2*	*11.9*	*11.1*	*12.9*	*8.0*	*11.1*
AA	*14.9*	*9.8*	*25.5*	*20.9*	*23.8*	*20.0*	*21.3*	*12.4*
EPA	*0.5*	*0.7*	*1.0*	*4.3*	*1.2*	*2.1*	*1.6*	*2.3*
AA/LA	0.63	0.37	3.17	1.76	2.14	1.55	2.66	1.11

[a]   Values are the mean percentage levels obtained in determinations carried out on 6 samples per group of animals. Values in *italics* are significantly different from each other.

Table 3. Eicosanoid production in EPA and olive oil (00) treated rats[a]

Group	Platelet TxB$_2$	Aortic 6-keto-PGF$_{1\alpha}$	PMNL TxB$_2$	PMNL LTB$_4$	PMNL LTC$_4$
00	31.5	9.6	12.5	114.2	3.4 n.s.
EPA	·15.9	6.2	7.8	84.2	2.5

[a]   Values are the average of determinations carried out on 6 or more samples. All differences, except those for PMNL LTC$_4$ are statistically different. All compounds were measured by highly specific RIAs. Platelet TxB$_2$, ng ml^{-1} PRP (400,000 μl^{-1}) 1 min after 5 I.U. thrombin. Aortic 6-keto-PGF$_1$ alfa, pg μl^{-1} perfused buffer (0.3 ml min^{-1}). PMNL eicosanoids, μg/ 5,000,000 cells 7 min after incubation with A 23187.

cumulate in cellular PL, but also, a significant reduction of the 20:4/18:2 ratio was observed suggesting that the conversion of LA to AA was depressed in the presence of EPA (Tables 2 and 3).

TxB$_2$ formation in stimulated platelets, and prostacyclin release from perfused aortas measured as the stable metabolite 6-keto-PGF$_{1\alpha}$ alfa, were both depressed by MaxEPA (Table 3). In stimulated PMNL, differences were observed for TxB$_2$ and LTB$_4$ formation.

These studies indicate that changes of cellular eicosanoids and of functional parameters after dietary EPA are mediated by several processes, such as, not only incorporation of EPA, but also inhibition of the conversion of LA to AA and, in cellular membranes, elevation of LA, which has been shown to interfere with E production in "in vitro" systems.

3 Plasma Cholesterol and Platelet Eicosanoids

In addition to modifications of cellular lipids as a consequence of manipulations of dietary lipids, complex modifications of membrane lipid composition and metabolism may derive from generalized alterations of lipid metabolism, such as hyperlipaemia and atherosclerosis, and these could, in turn, influence E production by affecting the precursor pools or their utilization. Several reports, including date from our Institute (Tremoli et al. 1984), appeared in the literature indicating that platelets of hyperlipaemic subjects are hyper-reactive to aggregating agents and produce more thromboxane after stimulation.

A correlation between levels of serum cholesterol and platelets TxB_2 formation has been observed also in populations of normal subjects subjected to dietary interventions. Two studies were carried out a few years ago in North Karelia and South Italy, respectively aimed to evaluate the effects of changing the composition of dietary fatty acids on several parameters, including platelet Tx formation in normal couples living in rural areas and consuming the typical diets (Enholm et al. 1982; Tremoli et al. 1986). After a period of 2 weeks on their typical diets, the Finnish couples had a diet with low fat, low saturates, high polyunsaturates for a period of 6 weeks, subsequently returning to their diets for a further period of 6 weeks. In the Italian study, on opposite scheme was followed. In the Finnish male subjects after inter-, vention, a drop of about 25% was observed in both serum cholesterol and platelet TxB_2 formation after stimulation (Tremoli et al. 1986), and an opposite effect was observed in the Italian study.

In order to further explore whether changes of platelet eicosanoid production under conditions of enhanced plasma cholesterol levels could be associated with modifications of the levels or utilization of the fatty acid precursor, we have recently evaluated the distribution of the arachidonic acid pool in various phospholipids of total platelet membranes from type IIa hypercholesterolemic subjects (serum cholesterol ≥ 300 mg dl^{-1}). Although only a trend towards greater cholesterol and phospholipid levels was observed in the hypercholesterolemic group, levels of AA (ex-

Table 4. Levels of AA in major phospholipid classes of platelets from control and hypercholesterolemic subjects[a]

PL classes	Control (n=7)	HC (n=13)
PC	16.1	30.2
PE	53.0	89.1
PS	9.3	16.0
PI	6.0	18.0

[a] Values are mg^{-1} protein and represent the average from the number of analyses indicated in parentheses. Standard deviations were within 10% of the values. All differences are statistically significant (p <0.01), the most relevant being found in PI-associated AA.

pressed on the basis of the protein content) appeared significantly higher in the PI pool of the HC group than in controls, as shown in Table 4.

Although PL classes other than PI may represent a greater overall store of precursor(s) for eicosanoid formation, the inositol phosphatides appear to be involved in the first steps of cell activation. The observation of an enhanced AA pool in PI of hyperlipidemic platelets may thus play a role in the increased platelet Tx production reported in these patients.

In conclusion, changes of cellular lipids, induced by manipulations of dietary fatty acids, or resulting from generalized alterations of lipid metabolism, may affect the size of the E-precursor pool or the processes involved in their utilization, thus resulting in alterations of E production.

Further research should be aimed to elucidate the mechanisms of these effects.

Acknowledgement. This work was supported in part by contract 85-01654 of the Italian Research Council, CNR.

References

Dougherty RM, Galli C, Ferro-Luzzi A, Iacono JM (1987) The lipid and phospholipid fatty acid composition of plasma, red blood cells and platelets and how they are affected by dietary lipids: a study of normal subjects from Italy, Finland and the U.S.A. Am J Clin Nutrit 45: 443–455

Enholm E, Huttunen JK, Pietinen P, Leino U, Mutanen M, Kostiainen E, Pillarainen J, Dougherty R, Iacono J, Puska P (1982) Effect of diet on serum lipoproteins in a population with a high risk of coronary heart diease. N Engl J Med 307:850–862

Fischer S, Weber PC (1983) Thromboxane A_3 (TxB$_3$) is formed in human platelets after dietary eicosapentaenoic acid (C20:5 n-3). Biochem Biophys Res Commun 116:1091–1094

Fischer S, Weber PC (1984) Prostaglandin in I_3 is formed in vivo in man after dietary eicosapentaenoic acid (C20:5 n-3). Nature (London) 307:165–167

Galli C, Agradi E, Petroni A, Tremoli E (1981) Differential effects of dietary fatty acids on the accumulation of arachidonic acid and its metabolic conversion through the cyclooxygenase and the lipoxygenase in platelets and vascular tissue. Lipids 16:165–172

Giani E, Masi I, Colombo C, Galli C (1984) Sex differences in platelet thromboxane and arterial prostacyclin production in control and n-6 fatty acid supplemented rats. Prostaglandins 28: 573–583

Lasserre M, Mendy F, Spielmann D, Jacotot B (1985) Effects of different dietary intake of essential fatty acids on C20:3 w6 and C20:4 w6 serum levels in human adults. Lipids 20:227–233

Masi I, Giani E, Galli C, Tremoli E, Sirtori CR (1986) Diets rich in saturates, monounsaturates and polyunsaturates fatty acids differently affect plasma lipids, platelet and arterial wall eicosanoids in rabbit. Ann Nutrit Metab 30:66–72

Socini A, Galli C, Colombo C, Tremoli E (1983) Fish oil administration as a supplement to a corn oil containing diet affects arterial prostacyclin production more than platelet thromboxane formation in the rat. Prostaglandins 25:693–710

Tremoli E, Maderna P, Colli S, Morazzoni G, Sirtori M, Sirtori CR (1984) Increased platelet sensitivity and thromboxane B_2 formation in type II hyperlipoproteinemic patients. Eur J Clin Invest 14:329–346

Tremoli E, Petroni A, Socini A, Maderna P, Colli S, Paoetti R, Galli C, Ferro-Luzzi A, Strazzullo P, Mancini M, Puska P, Iacono J, Dougherty R (1986) Dietart interventions in North Karelia, Finland and South Italy. Modification of thromboxane B_2 formation in platelets of male subjects only. Atherosclerosis 59:101–111

Effects of Omega-3 Polyunsaturated Fatty Acids on Human Leukocyte Function and Biochemistry

R. I. SPERLING and K. F. AUSTEN [1]

1 Introduction

The 5-lipoxygenase pathway, which generates monohydroxy-fatty acids and leuko-trines via metabolism of arachidonic acid, is distinguished in a number of ways from the better-known cyclooxygenase pathway, which produces prostaglandins, throm-boxanes, and heptadecatrienoic acid. The 5-lipoxygenase pathway has been identified in a limited number of cell types, the majority of which are of bone marrow origin; its expression requires activation of the initial enzyme in the cascade (the 5-lipoxygenase enzyme) as well as substrate availiability (Borgeat and Samuelsson 1979) for synthesis of products. In contrast, the cyclooxygenase pathway is virtually ubiquitous among nucleated cells, merely requiring substrate availability for product generation. Arachidonic acid (AA, 20:4) is the major substrate for both pathways, and is an ω-6 polyunsaturated fatty acid (i.e., the terminal unsaturation is six carbons from the ω-methyl group); it is derived from ω-6 polyunsaturated fatty acids of land-based plants and animals. Eicosapentaenoic acid (EPA; 20:5, ω-3) and docosahexaenoic acid (DHA; 22:6, ω-3) are both prominent in marine organisms and especially as esterified components of fish oil triglycerides. Both EPA and DHA inhibit the cyclooxygenase in a cell-free preparation and EPA has been shown to inhibit this enzyme in intact cells (Needleman et al. 1979; Corey et al. 1983). The effects of the ω-3 polyunsaturated fatty acids on the 5-lipoxygenase pathway in human cells both from healthy volunteers and patients with defined disease states are being studied.

2 Metabolism of ω-6 and ω-3 Fatty Acids via the 5-Lipoxygenase Pathway

Arachidonic acid is metabolized by 5-lipoxygenase, the first enzyme of this metab-olic pathway, to generate 5S-hydroperoxy-6-*trans*-8,11,14-*cis*-eicosatetraenoic acid (5-HPETE), an unstable intermediate in the cascade. Likewise, EPA is converted to

---

1 Departments of Medicine, Harvard Medical School and of Rheumatology and Immunology, Brigham and Women's Hospital, Boston, MA 02115, USA

Drugs Affecting Lipid Metabolism
Ed. by R. Paoletti et al.
© Springer-Verlag Berlin Heidelberg 1987

5-hydroperoxy-eicosapentaenoic aicd (5-HPEPE), with the same stereochemistry, and DHA is converted to a much lesser extent, to its analogous products (4- and 7-HPDHA). Each hydroperoxy-fatty acid is reduced, at least in part, to a corresponding secondary alcohol, 5S-hydroxy-6-*trans*-8,11,14-*cis*-eicosatetraenoic acid (5-HETE), 5-HEPE, and 4- and 7-HDHA, respectively. However, the reduction to the mono-hydroxy-metabolite is the obligatory final step in processing of DHA by the 5-lipo-oxygenase pathway, whereas the hydroperoxy-fatty acids derived from arachidonic acid and EPA may also be dehydrated to form the respective epoxide leukotrienes, LTA$_4$ (5,6-*trans*-oxido-7,9-*trans*-11,14-*cis*-eicosatetraenoic acid) and the corresponding LTA$_5$ by the hydroperoxy-fatty acid dehydrase activity of the 5-lipoxygenase enzyme (Rouzer et al. 1986).

The epoxide leukotrienes are unstable in aqueous environments and are non-enzymatically hydrolyzed to dihydroxy-leukotrienes; in the case of arachidonic acid-derived LTA$_4$, the major products are 5S,12R- and 5S,12S-dihydroxy-6,8,10-*trans*-14-*cis*-eicosatetraenoic acids (6-*trans*-LTB$_4$ diastereoisomers), and for EPA-derived LTA$_5$, the 6-*trans*-LTB$_5$ diastereoisomers with analogous stereochemistry. Some human cells, such as the polymorphonuclear neutrophilic leukocyte (PMN), monocyte, and pulmonary alveolar macrophage possess a potent catalytic activity for conversion of epoxide leukotrienes to biologically active dihydroxy-leukotrienes, 5S,12R-dihydroxy-6,14-*cis*-8,10-*trans*-eicosatetraenoic acid (LTB$_4$) from the arachidonic acid intermediary metabolite LTA$_4$ (Borgeat and Samuelsson 1979; Fels et'al. 1982; Godard et al. 1983; Williams et al. 1984), and analogously, LTB$_5$ from EPA via LTA$_5$ (Lee et al. 1984a). The enzyme, an epoxide hydrolase, which is a cytosolic component of human PMN (Rådmark et al. 1984), is also found in erythrocytes and blood plasma (Fitzpatrick et al. 1983, 1984), and may be the 5-lipoxygenase pathway enzyme with the broadest distribution among cell types. In the monocyte, another enzyme for further metabolism of expoxide leukotrienes is a glutathione-S-transferase isoenzyme, named LTC$_4$ synthase. This enzyme is most likely microsomal, by analogy with the rat basophil leukemia enzyme (Bach et al. 1984; Yashimoto et al. 1985), and converts LTA$_4$ to 5S-hydroxy-6R-S-glutathionyl-7,9-*trans*-11,14-*cis*-eicosate-traenoic acid (LTC$_4$) and LTA$_5$ to LTC$_5$. Further metabolism of these peptido-leukotrienes to other biologically active molecules occurs via the sequential cleavage of the glutathionyl side chain to produce 5S-hydroxy-6R-S-cysteinyl-glycyl-7,9-*trans*-11,14-*cis*-eicosatetraenoic acid (LTD$_4$) and 5-S-hydroxy-6R-S-cysteinyl-7,9-*trans*-11, 14-*cis*-eicosatetraenoic acid (LTE$_4$) from LTC$_4$ and, similarly, LTD$_5$ and LTE$_5$ from LTC$_5$.

3 ω-3 Fatty Acids and the Human 5-Lipoxygenase Pathway

The most relevant assessment of the effect of ω-3 fatty acids on arachidonic acid metabolism via the 5-lipoxygenase pathway will ultimately derive from measurements in body fluids of products generated in vivo by stimulus-specific host responses.

Limitations of the current assays for LTC_4, LTD_4, and LTE_4 due to unacceptable background levels in blood plasma, and for LTB_4, due to rapid and almost total in vivo degradation (Serafin et al. 1984), led to recent studies in which product generation, from isolated cells after in vitro stimulation, was analyzed. Methodology includes: ex vivo leukocyte activation after dietary manipulation to study the effect of EPA incorporated into cellular lipids; in vitro addition of EPA or DHA to leukocytes before cell activation, to model the effects of unesterified polyunsaturated fatty acids in plasma (Dole 1956; Waki and Lands 1968); and in vitro incorporation of the ω-3 fatty acids into cells prior cell activation.

In a study of the effects of incorporating EPA into cellular lipids of PMN and monocytes, seven healthy human subjects supplemented their usual diet with fish oil triglycerides (18 g daily), providing 3.2 g EPA and 2.2 g DHA daily for 6 weeks. Products of the 5-lipoxygenase pathway from both arachidonic acid and the ω-3 fatty acids were generated in response to ex vivo activation of PMN suspensions and monocyte monolayers with the calcium ionophore A23187, in dose- and time-dependent protocols. The products, resolved by reverse phase-high performance liquid chromatography, were quantitated by integrated optical density and/or radio-immunoassays. In separate experiments, [^{3}H]arachidonic acid was preincorporated into cellular lipid pools to assess the effects of ω-3 fatty acids on hydrolysis of arachidonyl esters (Lee et al. 1985). In a second study, analogous assessments were carried out for the products of PMN from healthy donors on their usual diet; the cells, in suspension, were activated with 10 μM calcium ionophore in vitro for 5 min in the presence of 0–40 μg ml^{-1} EPA or DHA (Lee et al. 1984b).

The only significant changes in the relative content of cellular fatty acids after 6 weeks of fish oil dietary supplementation, were increases in EPA of more than 7-fold in PMN and 15-fold in monocytes. The capacity of A23187 to release pre-incorporated [^{3}H]arachidonic acid from its esterified pools in PMN and monocytes was inhibited by 37% and 39% at 6 weeks, respectively, as compared with release from cells obtained prior to dietary supplementation (Lee et al. 1985). In the second study, exogenous EPA and DHA did not affect arachidonyl ester hydrolysis in PMN (Lee et al. 1984b). In addition to the possibility that esterified EPA could directly inhibit a critical fatty acyl hydrolase, it could also have an indirect effect by inhibiting the subsequent biosynthesis of 5-HETE and LTB_4, which have been noted to provide endogenous augmentation of phospholipase activity (Billah et al. 1985). At the next level of the cascade, 6 weeks of dietary supplementation resulted in inhibition of the total generation of 5-lipoxygenease products by >50% and >30% in PMN and mono-cytes, respectively, indicating a possible inhibitory effect on the 5-lipoxygenase of the PMN (Lee et al. 1985). In contrast, in vitro addition of EPA at 5 and 10 μg ml^{-1} (the lower concentrations tested) enhanced 5-lipoxygenase activity in A23187-activated PMN from untreated donors; nonesterified DHA was notably without effect at this and all subsequent steps in the cascade (Lee et al. 1984b).

Neither exogenous nor incorporated EPA have an effect on the dehydrase activity which generates epoxide leukotrienes from the hydroperoxy-fatty acids, and incorporated EPA does not modulate the function of the epoxide hydrolase in either PMN or monocytes. However, when presented in vitro, EPA caused a dose-dependent inhibition of the epoxide hydrolase in calcium ionophore-stimulated PMN (Lee et al.

1984b). Although the mechanism for this finding is yet to be elucidated, it is possible that LTA$_5$ is acting as an inhibitor as well as a less preferred substrate (Nathaniel et al. 1985). Exogenous, nonesterified EPA did not affect the degradation of LTB$_4$ by ω-oxidation in calcium ionophore-stimulated PMN.

In addition to inhibiting 5-lipoxygenase pathway product generation, 6 weeks of fish oil dietary supplementation decreased the maximal LTB$_4$-mediated chemotactic responsiveness of the PMN by a mean of 70%, reduced the chemotactic potency of the agonist by 100-fold, and eliminated the LTB$_4$-mediated enhancement of PMN adherence to endothelial monolayers (Lee et al. 1984a, 1985; Hoover et al. 1984). Thus, the dietary supplement might be expected to decrease both margination and diapedesis of PMN into an inflammatory site in response to this potent, naturally occurring mediator. Both the effects on the 5-lipoxygenase pathway and the alterations of PMN functional responses a reverted to the prediet levels 6 weeks after discontinuation of the fish oil supplement, supporting a causal relationship to the dietary alteration.

4 The Effects of ω-3 Polyunsaturated Fatty Acids in Rheumatoid Arthritis

In a recent study, 12 patients with active rheumatoid arthritis and on stable regimens of nonsteroidal anti-inflammatory drugs (NSAIDs) received a fish oil dietary supplement for 6 weeks. After 6 weeks of dietary supplementation with fish oil, the ratio of arachidonic acid to EPA in neutrophil cellular lipids decreased from 81:1 to 2.7:1, and calcium ionophore stimulation of PMN and monocytes generated both tetraene and pentaene products of the 5-lipoxygenease pathway. Decreases in the joint-pain index and the patient's assessment of disease activity reached statistical significance at the 95% confidence limit at 6 weeks as compared to the prediet period, and six of the seven other measures of disease activity also tended toward improvement. The generation of LTB$_4$ by calcium ionophore-stimulated neutrophils declined significantly at the 6 week evaluation compared to prediet values, without concomitant statistically significant changes in the generation of the other 5-lipoxygenase pathway products, thus indicating inhibition at the level of the epoxide hydrolase step (Sperling et al. 1987a). The generation of 5-lipoxygenase products by calcium ionophore-stimulated monocytes was not suppressed, but a significant decline in the generation of platelet-activating factor was demonstrated at week 6, suggesting inhibiton at the level of phospholipase A$_2$ without concomitant limitation of arachidonic acid availability sufficient to affect the generation of its metabolites via the 5-lipoxygenase pathway. Suppression of platelet-activating factor generation by monocytes was also seen in the study of healthy volunteers at week 6, however, in that study phospholipase A$_2$ inhibition was also evidenced by concomitant suppression of arachidonic acid release and 5-lipoxygenase pathway product generation (Lee et al. 1985; Sperling et al. 1987b). Neutrophil chemotaxis to LTB$_4$ and N-formyl-methionyl-leucyl-phenyl-alanine increased significantly to both agonists by week 6 from prediet levels which were lower than those seen in healthy donors. The contrasting effects of fish oil on

LTB$_4$-mediated PMN chemotaxis may relate to the underlying disease state, background medication, or to the level of chemotactic responsiveness prior to the initiation of dietary modification.

The evidence for some clinical improvement at 6 weeks, when treatment with both dietary fish oil supplementation and nonsteroidal antiinflammatory drugs was employed, may indicate that attenuation of arachidonic acid metabolism by both the cyclooxygenase and 5-lipoxygenase pathways represents and advantageous therapeutic goal.

Despite the suggestion of an anti-inflammatory effect of ω-3 fatty acids, it should be emphasized that the expectation of down-regulating pro-inflammatory events by ω-3 fatty acid administration is still inferential from ex vivo and in vitro experimental studies. The effect of the fatty acids on each 5-lipoxygenase-containing cell type using transmembrane activators, as well as the calcium ionophore, needs to be assessed both in healthy subjects and in patients with active inflammatory diseases. Large, double-blind, placebo-controlled clinical trials will be necessary to evaluate the role of ω-3 fatty acids in the therapy of human inflammatory disease.

5 Conclusion

The myriad of effects on chemotaxis, leukotriene generation and platelet activating factor formation induced by dietary fish oil supplementation demonstrated in these several studies needs to be related mechanistically to lipid biochemistry and metabolism. A mechanism in which the ω-3 fatty acids exert a primary effect on one site of action which either (1) results in a sequential series of effects, or (2) which then modulates several additional processes in an independent, parallel fashion would be most satisfying; however, (3) a primary effect at multiple sites of action clearly is an additional possibility. Although these studies do not directly address the issue of the mechanism of fish oil action, some insights may be deduced. In both the study of healthy subjects and of patients with rheumatoid arthritis, modulation of LTB$_4$-mediated and calcium ionophore-mediated neutrophil stimulation was demonstrated, suggesting multiple sites of action. The selective down-regulation of neutrophil LTB$_4$ synthesis, both by exogenous EPA presented in vitro to PMN from healthy volunteers and by dietary supplementation in patients with rheumatoid arthritis, appears to be due to inhibition of epoxide hydrolase activity and is apparently unrelated to other effects seen in these studies. This also supports the multiple sites of action hypothesis. The opposing effects of dietary fish oil supplementation on LTB$_4$-mediated PMN chemotaxis in healthy volunteers and patients with rheumatoid arthritis is difficult to explain by a single site of action; these data also indicate multiple sites of action of the ω-3 polyunsaturated fatty acids. The contrasting results seen in these studies may represent the summation of the individual effects of ω-3 fatty acids on each of the sites of action as modulated by underlying disease states, activity of disease, background drug therapy, and the prediet baseline in the study population.

References

Bach MK, Brashler JR, Morton DR Jr (1984) Solubilization and characterization of the leukotriene C_4 synthetase of rat basophil leukemia cells: a novel, particulate glutathione S-transferase. Arch Biochem Biophys 230:455–465

Billah MM, Bryant RW, Siegel MI (1985) Lipoxygenase products of arachidonic acid modulate biosynthesis of platelet-activating factor (1-0-alkyl-2-acetyl-*sn*-glycero-3-phosphocholine) by human neutrophils via phospholipase A_2. J Biol Chem 260:6899–6906

Borgeat P, Samuelsson B (1979) Arachidonic acid metabolism in polymorphonuclear leukocytes: effects of ionophore A23187. Proc Natl Acad Sci USA 76:2148–2152

Corey EJ, Shih C, Cashman JR (1983) Docosahexaenoic acid is a strong inhibitor of prostaglandin but not leukotriene biosynthesis. Proc Natl Acad Sci USA 80:3581–3584

Dole VP (1956) A relation between non-esterified fatty acids in plasma and the metabolism of glucose. J Clin Invest 35:150–154

Fels AO, Pawlowski NA, Cramer EB, King TKC, Cohn ZA, Scott WA (1982) Human alveolar macrophages produce leukotriene B_4. Proc Natl Acad Sci USA 79:7866–7870

Fitzpatrick F, Haeggström J, Granström E, Samuelsson B (1983) Metabolism of leukotriene A_4 by an enzyme in blood plasma: a possible leukotactic mechanism. Proc Natl Acad Sci USA 80:5425–5429

Fitzpatrick F, Liggett W, McGee J, Bunting S, Morton D, Samuelsson B (1984) Metabolism of leukotriene A_4 by human erythrocytes. A novel cellular source of leukotriene B_4. J Biol Chem 259:11403–11407

Hoover RL, Karnovsky MJ, Corey EJ, Austen KF, Lewis RA (1984) Leukotriene B_4 action on endothelium mediates augmented neutrophil/endothelial adhesion. Proc Natl Acad Sci USA 81:2191–2193

Lee TH, Mencia-Huerta J-M, Shih C, Corey EJ, Lewis RA, Austen KF (1984a) Characterization and biologic properties of 5,12-dihydroxy derivatives of eicosapentaenoic acid, including leukotriene B_5 and the double lipoxygenase product. J Biol Chem 259:2383–2389

Lee TH, Mencia-Huerta J-M, Shih C, Corey EJ, Lewis RA, Austen KF (1984b) Effects of exogenous arachidonic, eicosapentaenoic, and docosahexaenoic acids on the generation of 5-lipoxygenase pathway products by ionophore-activated human neutrophils. J Clin Invest 74:1922–1933

Lee TH, Hoover RL, Williams JD, Sperling RI, Ravalese JR III, Spur BW, Robinson DR, Corey EJ, Lewis RA, Austen KF (1985) Effect of dietary enrichment with eicosapentaenoic and docosa-hexaenoic acids on in vitro neutrophil and monocyte leukotriene generation and neutrophil function. N Engl J Med 312:1217–1223

Nathaniel DJ, Evans JF, Leblanc Y, Léveillé C, Fitzsimmons BJ, Ford-Hutchinson AW (1985) Leukotriene A_5 is a substrate and an inhibitor of rat and human neutrophil LTA_4 hydrolase. Biochem Biophys Res Commun 131:827–835

Needleman P, Raz A, Minkes MS, Ferendelli JA, Sprecher H (1979) Proc Natl Acad Sci USA 76:944–948

Rådmark O, Shimizu T, Jörnvall H, Samuelsson B (1984) Leukotriene A_4 hydrolase in human leukocytes. Purification and properties. J Biol Chem 259:12339–12345

Rouzer CA, Matsumoto T, Samuelsson B (1986) Single protein from luman leukocytes possesses 5-lipoxygenase and leukotriene A_4 synthease activities. Proc Natl Acad Sci USA 83:857–861

Serafin WE, Oates JA, Hubbard WC (1984) Metabolism of leukotriene B_4 in the monkey. Identification of the principal nonvolatile metabolite in the urine. Prostaglandins 27:899–911

Sperling RI, Weinblatt M, Robin J-L, Ravalese J III, Hoover RL, House F, Coblyn JS, Fraser PA, Spur BW, Robinson DR, Lewis RA, Austen KF (1987a) Effects of dietary supplementation with marine fish oil on leukocyte lipid mediator generation and function in rheumatoid arthritis. Arth Rheum 30:987–998

Sperling RI, Robin J-L, Kylander KA, Lee TH, Lewis RA, Austen KF (1987b) The effects of N-3 Polyunsaturated fatty acids on the generation of PAF-acether by human monocytes. J Immunol (in press)

Waki K, Lands WEM (1968) Control of lecitin biosynthesis in erythrocyte membranes. J Lipid Res 9:12–18

Williams JD, Czop JK, Austen KF (1984) Release of leukotrienes by human monocytes on stimulation of their phagocytic receptor for particulate activators. J Immunol 132:3034–3040

Yashimoto T, Soberman RJ, Lewis RA, Austen KF (1985) Isolation and characterization of leukotriene C_4 synthetase of rat basophilic leukemia cells. Proc Natl Acad Sci USA 82:8399–8403

Dietary and Biochemical Studies with Textured Soy Proteins

C. R. SIRTORI, A. CANAVESI, C. MANZONI, and M. R. LOVATI [1]

1 Introduction

The hypocholesterolemic activity of vegetable proteins, particularly from soybean, was suggested early this century from animal studies (Ignatowski 1909; Meeker and Kesten 1940). In more recent years, several authors have demonstrated that the substitution of animal proteins with soybean proteins in the diet significantly reduces plasma cholesterol levels both in experimental animals (Terpstra et al. 1984) and in man (Sirtori et al. 1977). The soybean protein diet appears to be an effective alternative to drug treatments in adult type II hypercholesterolemic patients (Sirtori et al. 1977; Descovich et al. 1980), the activity being independent of dietary cholesterol and only partially influenced by the P/S ratio (Sirtori et al. 1979).

Unfortunately, the mode of action of textured vegetable proteins (TVP) from soy has remained quite elusive. Part of the difficulty in the clarification of the mechanisms probably lies in the selective activity of the diet in spontaneously hypercholesterolemic patients and in diet-induced hypercholesterolemic animals (Kim et al. 1978; Terpstra et al. 1982), with little or no activity in normolipidemic animals and humans (Van Raaij et al. 1982). Moreover, significant discrepancies exist between the results of animal and human studies: in fat-fed animals the introduction of vegetable proteins in the diet leads to increased fecal steroid excretion, as well as to an increased turnover of circulating cholesterol (Fumagalli et al. 1978; Huff and Carroll 1980). This is not the case of type II patients treated with this regimen, where in spite of the dramatic plasma cholesterol reduction, no increase of fecal neutral or acidic steroid excretion, or accelerated turnover of cholesterol can be ascertained (Fumagalli et al. 1982).

In view of the significant activity of the diet in type II patients and of the uncertainties about the mode of action, interest is currently very high on this topic (Gibney and Kritchevsky 1983). The soybean diet may, in fact, represent a means of reducing the diet-linked atherogenic risk. The object of this presentation will be to review some recent studies on: (1) a new formulation of TVP which may favorably affect high density lipoprotein (HDL)-cholesterolemia; (2) a study in children, where the soybean diet may represent a choice treatment for hypercholesterolemia; (3) data

1 Center E. Grossi Paoletti, Institute of Pharmacological Sciences, University of Milano and Divisione Medica Vergani, Ospedale di Niguarda Ca' Granda, Milano, Italy

Drugs Affecting Lipid Metabolism
Ed. by R. Paoletti et al.
© Springer-Verlag Berlin Heidelberg 1987

from animal studies, as well as preliminary data from humans, on the regulation of low density lipoprotein (LDL) receptors following this regimen.

2 Activity of TVP Added with Lecithin (L-TVP)

A new TVP preparation containing 6% by weight lecithin (L-TVP) was recently tested in 65 patients with stable type II hyperlipoproteinemia. The study was divided into four phases, i.e. phase I: 4 weeks of traditional low lipid/low cholesterol diet; phase II: 4 weeks of total substitution of animal proteins with L-TVP; phase III: 4 weeks of standard low lipid/low cholesterol diet; phase IV: 4 weeks with partial (50%) substitution of animal proteins with L-TVP.

The diets were well tolerated with negligible changes in body weight. Monitoring of plasma lipids showed, during phase I, only very small changes, in view of the strict dietary regimen already followed by the patients. Substitution of animal proteins with L-TVP dramatically reduced plasma cholesterol levels, both in males (-20.3%, $p < 0.001$) and in females (-17.3%, $p < 0.001$). These lipoprotein changes were fully reversible during the ensuing 4 weeks (phase III), whereas during phase IV, of partial substitution, a new reduction of cholesterolemia was recorded (-14.4% in males and -12.2% in females, both $p < 0.01$, compared to baseline) (Sirtori et al. 1985).

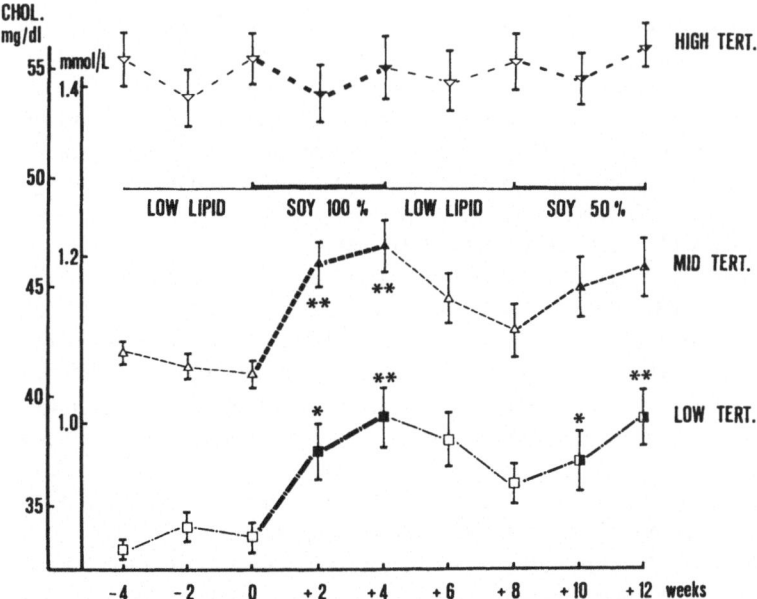

Fig. 1. HDL-cholesterol levels during a controlled evaluation of the lecithinated soybean protein diet given as a total (100%), or partial (50%) substitution of animal proteins to type II patients. When patients are stratified into tertiles according to the starting HDL levels, it becomes clear that patients in the mid and low tertiles responded with an HDL-cholesterol rise during the new diet (Sirtori et al. 1985)

Changes of LDL essentially reflected the reduction of total cholesterolemia. However, the most interesting finding was related to HDL-cholesterol (HDL-C) levels. There was a clear tendency for a rise of about 8% (p:ns) both during the total and partial substitution with L-TVP (phases II and IV). When, however, the starting HDL-C levels were stratified into tertiles, it became clearly apparent that HDL-C had significantly risen only in the patients within the two lower tertiles (Fig. 1) (correlation coefficient between baseline HDL-C and rise during L-TVP; $r = 0.371, p < 0.01$).

It is possible that the addition of lecithin to TVP may, in some way, affect the capacity of HDL to incorporate cholesterol. This mechanism, already supported by in vitro findings (Zierenberg et al. 1981) may become operant in cases where the transport of cholesterol in HDL, because of congenital or acquired dysfunctions, is less efficient. Based on these findings, the L-TVP diet may provide an effective therapeutic tool for the type II patients with reduced HDL-cholesterol.

3   Evaluation of the L-TVP Diet in Type II Children

Hypercholesterolemia poses a serious therapeutic problem in childhood, since the physician may not be willing to prescribe long-term drug therapies in pre-pubertal children. We, therefore, elected to evaluate the soybean diet regimen in a group of 16 children with familial type II disease, within a multi-center study (Gaddi et al. 1987).

After 4 weeks of a traditional low lipid-low cholesterol diet, the L-TVP diet was prescribed to the children within a program, allowing an adequate calorie intake for normal growth. All 16 children and their families started the protocol with good compliance. Unfortunately, due to intercurrent illness or movement out of town, only 12 children completed 2 months or more of treatment. The results (Fig. 2) clearly show that the initial low lipid diet did not achieve any remarkable change in the plasma lipid/lipoprotein levels. A significant reduction of total cholesterol was, instead, recorded already after 2 weeks of the L-TVP diet.

At the fourth week, the mean cholesterol in all 16 subjects was reduced to -19.2% vs baseline ($p < 0.001$); LDL-cholesterol levels were also reduced by 19.7% ($p < 0.001$), triglyceridemia increased slightly, as also did HDL-C (from 35.6 ± 6.4 to 36.8 ± 6.5 mg dl^{-1}). After 18 weeks, when the study was considered as completed, the 12 children still participating showed the stabilization of their cholesterolemia, with a mean reduction of 21.6%, while a further rise of HDL-C could be recorded (to 37.0 ± 8.6 mg dl^{-1} at the end of the treatment, p:ns vs baseline). Most of the children have followed the regimen for more prolonged periods of time (up to 4 years), with complete maintenance of their biochemical improvement and normal growth.

The soybean diet may, therefore, offer a satisfactory alternative to drug treatments in pre-pubertal children. Similar findings have been recently reported by Widhalm (1986) who tested the regimen in 11 children with familial disease, starting at the age of 2 years. This author reported a reduction of 32% in total cholesterol and 37% in LDL-cholesterol levels vs baseline (on a free diet).

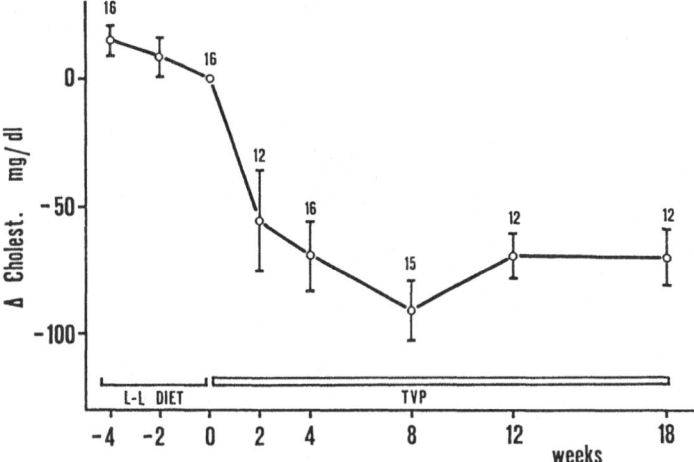

Fig. 2. Changes of total cholesterol during the soybean protein diet in a sample of 16 children with familial hypercholesterolemia (mean plasma cholesterol levels at time *0*: 360.3 ± 27.8 mg dl^{-1}). A mean reduction of cholesterolemia of 21.8% was recorded after 8 weeks of dietary treatment. *Figures* above the means ± SD represent the number of participating patients (Gaddi et al. 1987)

4 Activity of the Soybean Protein Diet on Lipoprotein Receptors

The discrepant results between animal and human studies in terms of fecal steroid excretion (increased excretion in animals, no change in humans) have raised interest in the receptor regulation of lipoprotein metabolism during different diets. This topic is of particular significance after a recent study indicating that the addition of cholesterol to a human diet is hypercholesterolemic only when the diet contains animals proteins, not when dietary proteins are represented by soy (Meinertz et al. 1984).

We evaluated in two different studies, the activity of liver lipoprotein receptors in rats, when exposed to cholesterol-rich regimens with two different dietary proteins. In the *first study,* two groups of female rats were fed on a 2% cholesterol diet with either casein or soy protein. There was, as expected (Terpstra et al. 1982), a clear difference between the cholesterol response to the two diets, with an approximate doubling of total cholesterol and a 12-fold increase of very low density lipoprotein (VLDL)-associated cholesterol.

In this experiment, the binding of lipoprotein to liver cells was tested by means of the separated β-VLDL fraction from blood of male rabbits fed for 4 weeks a 2% cholesterol regimen. The receptorial activity was determined by incubating rat liver membrane preparations (Kovanen et al. 1979) in the presence of ^{125}I-labelled rabbit β-VLDL.

In spite of the significant cholesterol load administered to these animals, the binding of β-VLDL to liver membranes from soybean-fed rats did not differ from that of control animals (Fig. 3), whereas it was markedly reduced in the case of animals receiving casein as the major dietary protein (Sirtori et al. 1984). The evalua-

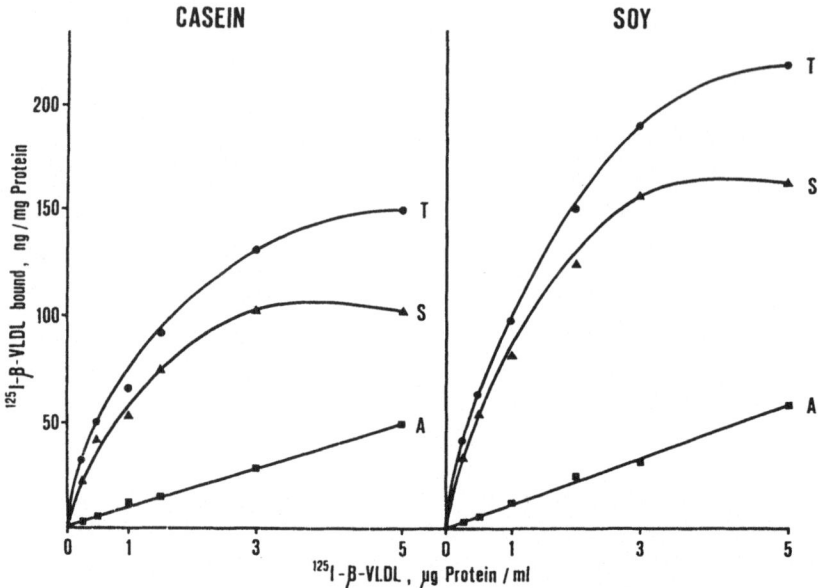

Fig. 3. Binding of cholesterol-rich β-VLDL to liver membranes from rats treated with a choles-teremic regimen containing either casein or soy protein. Data are expressed as total (*T*), aspecific (*A*) or specific (*S*) binding. A clear reduction of specific binding is noted in membranes from casein+cholesterol-fed rats, whereas in the soy+cholesterol-fed animals data to not differ from those of animals on a standard diet (Sirtori et al. 1984)

tion of the enzymes involved in the receptor pathway provided data consistent with a moderately reduced receptor activity in the case of the soy-cholesterol fed rats and in a total suppression for the casein-cholesterol group.

In the *second study,* a turnover experiment with autologous β-VLDL from rats was carried out. Recipient animals had received, as before, soy+cholesterol and casein+ cholesterol. There was clear evidence, in this second experiment, of an accelerated catabolism of β-VLDL after the soy protein regimen (Lovati et al. 1985). This finding complements data from the previous experiment and indicates that some component(s) of soy protein may modulate liver receptor activity, when this is suppressed by cholesterol (Redgrave 1984).

Recently, in a clinical study, the LDL receptor activity was examined in freshly isolated mononuclear cells from type II patients, receiving either a low lipid diet with animal proteins or an identical diet (in terms of cholesterol and lipid content) with TVP. The results of this study, which are currently being evaluated, definitely suggest that the receptor activity is dramatically increased after the soybean protein regimen in these patients with type II hyperlipoproteinemia.

5 Conclusions

The experience gained in the past 14 years clearly indicates that the soybean protein regimen is therapeutically very effective in patients with type II hyperlipoproteinemia. A similar activity is exerted in animal models with diet-induced hypercholesterolemia. Apparently, the available data suggest a different regulation of the high affinity receptors for lipoproteins, when the soybean diet is either administered in the presence of high cholesterol loads (e.g. animal studies) or per se to type II patients. The mechanism of the change in the lipoprotein receptor regulation needs further study.

Acknowledgements. These studies were supported by the Consiglio Nazionale delle Ricerche of Italy and by a grant-in-aid from Gipharmex SPA, Milano, Italy

References

Descovich GC, Ceredi C, Gaddi et al. (1980) Multicenter study of soybean protein diet for outpatient hypercholesterolaemic patients. Lancet ii:709–712

Fumagalli R, Paoletti R, Howard AN (1978) Hypocholesterolemic effect of soya. Life Sci 22: 947–952

Fumagalli R, Soleri L, Farina R et al. (1982) Fecal colesterol excretion studies in type II hypercholesterolemic patients treated with the soybean protein diet. Atherosclerosis 43:341–352

Gaddi A, Descovich GC, Noseda G et al. (1987) Hypercholesterolaemia treated by soybean protein diet. Arch Dis Childh 62:274–278

Gibney MJ, Kritchevsky D (1983) Animal and vegetable proteins in lipid metabolism and atherosclerosis. Liss, New York

Huff MW, Carroll KK (1980) Effect of the dietary protein on turnover, oxidation and absorption of cholesterol and on steroid oxidation in rabbits. J Lipid Res 21:546–558

Ignatowski A (1909) Über die Wirkung des tierischen Eiweißes auf die Aorta und die parenchymatosen Organe der Kaninchen. Virchows Arch 198:248–270

Kim DN, Lee KT, Reiner JM et al. (1978) Effect of a soy product on serum and tissue cholesterol concentrations in swine fed high-fat, high-cholesterol diets. Exp Mol Pathol 29:385–399

Kovanen PT, Brown MS, Goldstein JL (1979) Increased binding of low density lipoprotein to liver membranes from rats treated with 17 α-ethinyl estradiol. J Biol Chem 254:11367–11373

Lovati MR, Allievi L, Sirtori CR (1985) Accelerated early catabolism of very low density lipoproteins in rats after dietary soy proteins. Atherosclerosis 56:243–246

Meeker DR, Kesten HD (1940) Experimental atherosclerosis and high protein diets. Proc Soc Exp Biol Med 45:543–545

Meinertz H, Nilausen K, Faergeman O (1984) Effects of dietary soy protein and casein on plasma lipoproteins in normolipidemic subjects. Circulation 70:1161 (Abstr)

Redgrave TG (1984) Dietary proteins and atherosclerosis. Atherosclerosis 52:349

Sirtori CR, Agradi E, Conti F et al. (1977) Soybean protein diet in the treatment of type II hyperlipoproteinaemia. Lancet i:275–277

Sirtori CR, Gatti E, Mantero O et al. (1979) Clinical experience with the soybean protein diet in the treatment of hypercholesterolemia. Am J Clin Nutrit 92:1645–1658

Sirtori CR, Galli G, Lovati MR et al. (1984) Effects of dietary proteins on the regulation of liver lipoprotein receptors in rats. J Nutrit 114:1493–1500

Sirtori CR, Zucchi-Dentone C, Sirtori M et al. (1985) Cholesterol-lowering and HDL-raising properties of lecithinated soy proteins in type II hyperlipidemic patients. Ann Nutrit Metab 29:348–357

Terpstra AHM, Tintelen G van, West CE (1982) The hypocholesterolemic effect of dietary soy protein in rats. J Nutrit 112:810–817

Terpstra AHM, West CE, Fennis JTCM et al. (1984) Hypocholesterolemic effect of dietary soy protein versus casein in rhesus monkeys. Am J Clin Nutrit 39:1–7

Van Raaij JAM, Katan MB, West CE et al. (1982) Influence of diets containing casein, soy isolate and concentrate on serum cholesterol and lipoproteins in middle-aged volunteers. Am J Clin Nutrit 35:925–934

Widhalm K (1986) Effect of diet on serum lipids and lipoprotein in hyperlipoproteinemic children. In: Beynen AC (ed) Nutritional effects on cholesterol metabolism. Transmondial, Voorthuizen, pp 133–140

Zierenberg O, Assmann G, Schmitz G et al. (1981) Effect of polyenephosphatidylcholine on cholesterol uptake by human high density lipoprotein. Atherosclerosis 39:527–542

Detection of Reactive Free Radicals in Livers of Ethanol-Fed Rats: Potentiating Effect of High Fat Diets

P. B. McCay and L. A. Reinke [1]

1 Introduction

Lipid peroxidation has been associated with alcoholic liver injury as a result of investigations by Di Luzio and co-workers (1,2) who reported that antioxidants prevented the development of fatty liver after a large acute dose of ethanol. Other investigators have subsequently reported that lipid peroxidation appears to occur in the liver of animals given an acute dose of alcohol. The evidence for this was the detection of malondialdehyde by use of the thiobarbituric acid assay (3,4). Chronic, excessive use of alcohol by humans results in liver injury characterized by fatty infiltration which can lead to fibrotic degeneration and necrosis (5,6), but evidence that these effects are linked to the peroxidative phenomenon is still controversial. In this report, we present evidence that reactive free radicals, capable not only of initiating lipid peroxidation, but also of reacting with other cell components, are generated in the hepatic endoplasmic reticulum as a consequence of ethanol consumption.

Using the spin-trapping procedure (7), we have previously demonstrated that it is possible to trap free radicals in vivo at the site of their formation in various organs under conditions which are believed to promote the formation of free radicals, such as exposure to carbon tetrachloride (8) or gamma radiation (9). This spin-trapping procedure has been employed in the investigations described in this report to demonstrate that carbon-centered lipid free radicals are generated in the hepatic endoplasmic reticulum of animals exposed to ethanol. The results indicate that lipid peroxidation occurs as a very early event in the liver following ethanol intake, and that the intensity of free radical formation is enhanced by consumption of high levels of dietary fat.

2 Methods

Young adult female Sprague-Dawley rats, weighing between 140–150 g, were fed either a high-fat (fat provided 35% of calories) or a low-fat (fat provided 12% of calories) diet containing ethanol as 36% of the total caloric content (28). Control animals

---

1 Oklahoma Medical Research Foundation and Department of Pharmacology,
   University of Oklahoma Health Sciences Center, Oklahoma City, OK 73104, USA

Drugs Affecting Lipid Metabolism
Ed. by R. Paoletti et al.
© Springer-Verlag Berlin Heidelberg 1987

Table 1. Composition of liquid diets

	High fat diet		Low fat diet	
	Ethanol	Control	Ethanol	Control
			(gl^{-1})	
Casein	41.4	41.4	41.4	41.4
L-cysteine	0.5	0.5	0.5	0.5
DL-methionine	0.3	0.3	0.3	0.3
Corn oil	8.5	8.5	2.5	2.5
Olive oil	28.4	28.4	8.5	8.5
Safflower oil	2.7	2.7	2.7	2.7
Choline bitartrate	0.53	0.53	0.53	0.53
Fiber	10.0	10.0	10.0	10.0
Dextrin-maltose	25.6	115.2	83.6	173.3
Ethanol	50.0	0.0	50.0	0.0
Suspending agent K [a]	2.0	2.0	4.0	4.0

[a] Suspending agent K is a nonnutritive suspending agent used by Bioserve, Inc. in place of sodium carrageenate (28). Also included in the diet formulation are AIN 1977 recommendations of vitamins and minerals (29).

were fed the same diets except that the dextrin-maltose was substituted for ethanol to provide an isocaloric control diet (see Table 1), and these animals were pair-fed with the rats receiving the ethanol-containing diet. The animals were fed the diets for at least 2 weeks before the experimental procedures were carried out. The animals grew well on these diets and appeared healthy.

At the beginning of the experiment, animals from each dietary group were administered 1.0 ml of a 0.05 M solution of the spin-trapping agent, trimethoxyphenyl-t-butyl nitrone intragastrically. After 15 min, the animals were decapitated and the livers were quickly weighed and then homogenized in 15 vol chloroform-methanol (2:1). One-fifth volume of 0.9% NaCl solution was added to bring about phase separation, and the chloroform layer recovered. Evaporation of the chloroform layer yielded the total lipids of the liver. Samples (0.2 ml) of the lipid extracts were placed in Pasteur pipettes which had the capillary end sealed. The lipid extracts were forced into the capillary portion of the pipettes by centrifugation. The pipettes were then assayed in a Varian E-9 electron spin resonance spectrometer for the presence of an EPR signal. The capillary bores of the pipettes were sufficiently uniform to provide little difference in the intensity of an EPR standard from pipette to pipette.

The intensity of the EPR signals of the spin adducts obtained from the various animals could be compared by dividing the height of the signal by the number of grams of liver extracted. The signal intensity can be expressed as signal height (mm)/ gram liver.

In some studies, in order to determine if there was a specific intracellular localization of the any observed EPR signal, the liver was homogenized and fractionated into subcellular components before extraction of lipids from the different subfractions (microsomes, mitochondria, plasma-membrane nuclear pellet, and the cytosol).

When rats or mice were subjected to an acute dose of ethanol, the animals were fed commercial laboratory rations ad libitum, the ethanol was given (5.0 g kg^{-1} body wt.) with the spin-trapping agent intraperitoneally.

The spectrometer settings were: microwave power, 25 mw; modulation amplitude, 1 G; time constant, 10 s; scan range, 100 G; scan time, 16 or 30 min; temperature, 25°.

3 Results

The lower EPR scan shown in Fig. 1 is typical of that observed in liver lipid extracts of rats which had been fed the high fat diet containing ethanol. The nitrogen and beta-hydrogen splitting constants of the signal (a_N = 14.5 g, a^H_β = 3.25 G) are characteristic of those of a lipid radical adduct of trimethoxyphenyl-N-t-butyl nitrone (8). Liver lipid extracts of rats which were fed the control high fat diet isocalorically show no indication of such radical generation in the liver (Fig. 1.). Additional control studies were performed in which the livers of ethanol-fed animals were homogenized in the chloroform-methanol solvent which also contained 0.025 M trimethoxyphenyl-N-t-butyl nitrone, followed by recovery of the total lipids extracted, which were then scanned in the EPR spectrometer. This was done to demonstrate that the formation of the spin adduct was not an artifact of the extraction process. However, none of these control extractions of livers of ethanol-fed rats exhibited an EPR signal.

Rats fed the *low fat diet* containing an equivalent amount of ethanol also showed the same type of radical generation (lower scan, Fig. 2), but the intensity of the signal was always considerably less than that observed in animals fed the high fat diet containing ethanol. Table 2 shows the relative intensities of lipid radical generation in the livers of animals fed the different diets. The intensity of radical generation in the livers of animals given ethanol in the low fat diet was consistently only about one-half

Table 2. Lipid radical signal intensity in liver of rats fed ethanol with either a high or a low fat diet[a]

Diet	Signal intensity (Millimeter signal height/gram liver)
High fat + ethanol	14.31
High fat control (no ethanol)	0.00
Low fat + ethanol	6.34
Low fat control (no ethanol)	0.00

[a] Results are averages of three experiments.

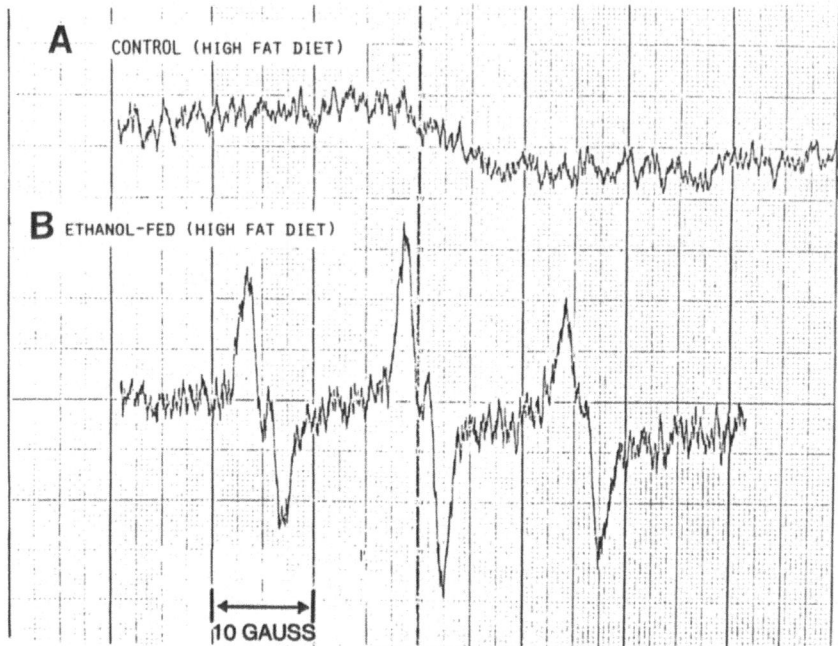

Fig. 1A, B. EPR spectra of liver lipid extracts from rats fed a high fat diet for 2 weeks. **A** Control;
B ethanol-fed. The animals were administered the spin trap, trimethoxyphenyl-t-butyl nitrone,
15 min before removing the liver for immediate extraction of the liver

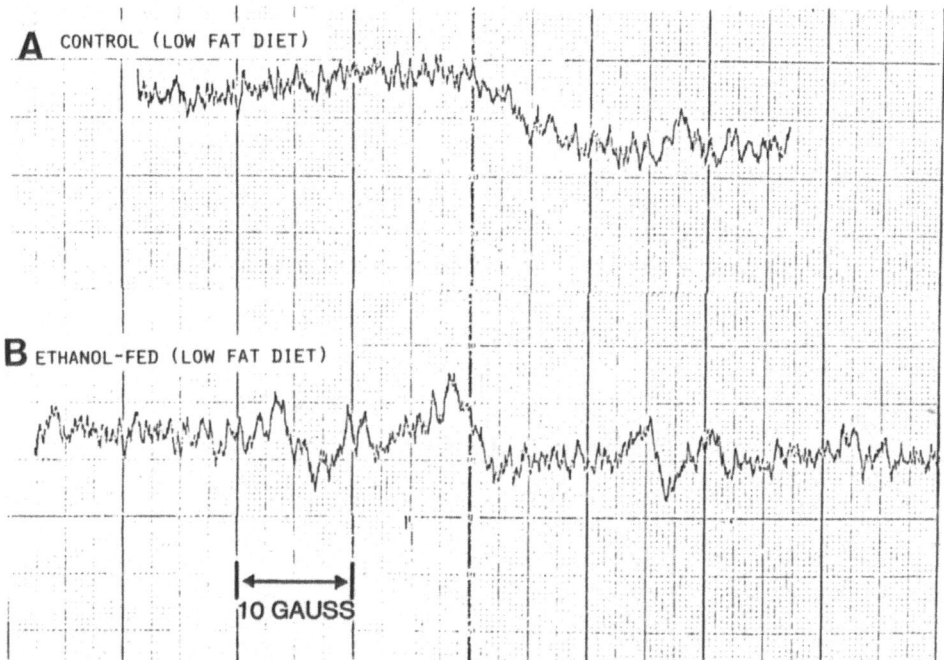

Fig. 2A, B. EPR spectra of liver lipid extracts from rats fed a low fat diet for 2 weeks. **A** Control;
B ethanol-fed. The animals were administered the spin trap as in Fig. 1

that observed in animals fed ethanol in the high fat diet. These results indicated that the higher level of fat consumption per se appears to augment free radical formation in the liver since the caloric intake of both the high and the low fat dietary groups was the same. It should be noted that a comparable intake of ethanol ($14-16$ g kg^{-1} body wt. day^{-1}) was achieved with both the high fat and low fat diets.

In order to determine whether the spin adduct of the lipid radical was localized within any particular cell organelle, the following experiment was done. Rats were fed the high fat diet containing ethanol (as 35% of calories) for 2 weeks and then they were given the spin-trapping agent (trimethoxyphenyl-N-t-butyl nitrone). After 30 min, the livers were removed, homogenized, and fractionated by differential centrifugation in 0.25 M sucrose. Each subcellular fraction was then extracted with chloroform:

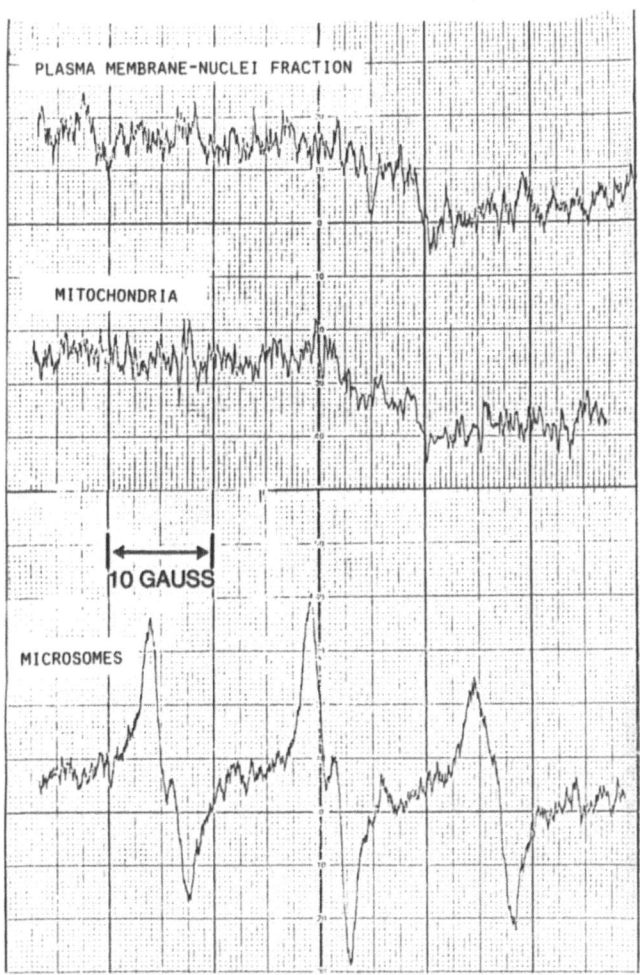

Fig. 3. EPR spectra of lipid extracts from subcellular fractions of rat livers of animals fed a high fat diet containing ethanol for 2 weeks. The spin-trapping agent was administered to the animals as described in Fig. 1

methanol and the total lipids recovered. EPR scans of each of the lipid extracts were
made and an example of these scans is shown in Fig. 3. Only the microsomal fraction
contained the spin adduct. Neither the mitochondrial nor the plasma membrane-
nuclear fraction had a detectable signal. In addition, the lipid extract of the cytosolic
fraction also had no signal (data not shown).

Investigations in our laboratory had established earlier that the spin-trapping
agent, trimethoxyphenyl-N-t-butyl nitrone was able to penetrate into cells of all
organs which were tested in the intact animal (brain, spleen, heart, kidney) (9). In
view of this widespread distribution, a search for evidence of free-radical generation
in other organs in ethanol-fed rats was undertaken. Figure 4 shows EPR scans of
lipid extracts of tissues of rats which had been fed ethanol for 2 weeks and then given
the spin trap 30 min prior to preparing the lipid extracts. These scans show that
no radical adducts were apparent in brain, spleen, or kidney of these rats, but that
a strong signal was observed in the liver extracts. These results indicate that the liver
is probably the major site of free-radical generation in the animal.

In order to determine whether or not the formation of the lipid radicals in the liver
of ethanol-treated animals was a short-term, transient phenomenon, rats which had
been fed the ethanol-supplement diet for at least 2 weeks were transferred to the
control diet containing no ethanol. The animals were assayed for free-radical genera-

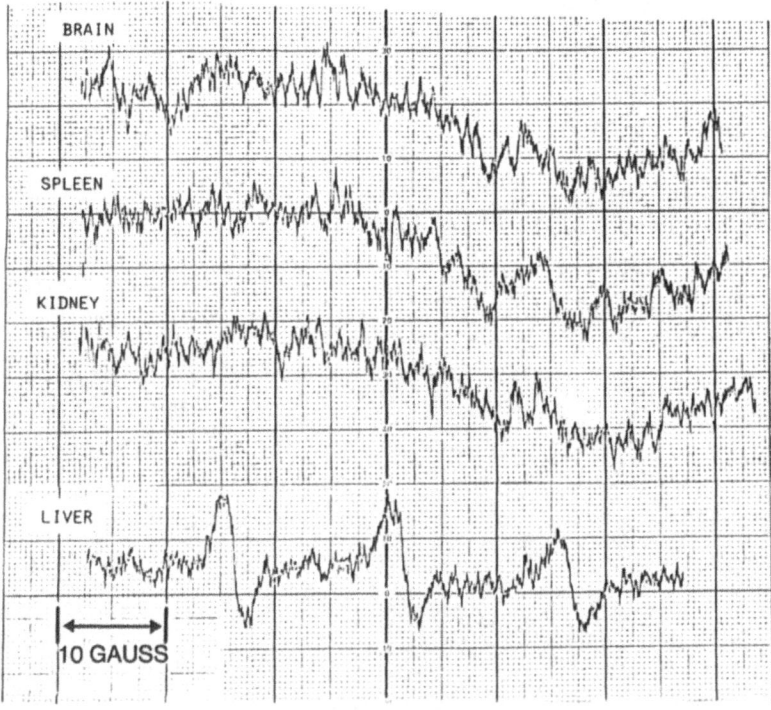

Fig. 4. EPR spectra of lipid extracts from various organs of rat which had been fed a high fat diet
containing ethanol for 2 weeks. The spin-trapping agent was administered to these animals 30 min
prior to removal of the organs for extraction

Table 3. Signal strength of ethanol-dependent lipid free radicals generated in vivo in liver as a function of time following withdrawal of ethanol feeding[a]

Time following[b] ethanol withdrawal (h)	Signal intensity (Millimeters signal height/gram liver)
0 (Time of diet transfer)	16.14
20	6.73
40	6.36
70	6.83
90	3.24

a   Rats were fed the ethanol-containing high fat diet for 2 weeks and then transferred to an isocaloric high fat diet containing no ethanol. Spin trapping was performed on the animals at the various times subsequent to withdrawal, as indicated, to determine the intensity of lipid radical generation at those times. Animals were administered the spin trap, $(MO)_3 PBN$, intraperitoneally and the livers removed for adduct extraction 30 min later.

b   One animal per time point.

tion in the liver at 0 h (immediately after the diet transfer), and at intervals between 20 and 90 h (Table 3). The results show that lipid free radicals continued to be generated in the liver throughout the entire period even though the blood alcohol level had returned to zero within the first 12 h. In view of the fact that lipid free radicals of this type are very reactive and possess a very short half-life, the data obtained demonstrate that the radicals are being continuously generated for at least 3.5 days after withdrawal of ethanol feeding at a time when the tissue content of ethanol would be expected to be undetectable.

All of the studies described above were done with rats which were fed ethanol for 2 weeks. In order to determine if the ethanol must be consumed for a period of time before the free radical phenomenon was initiated, rats and mice were given acute doses of ethanol (5 g kg^{-1} body wt.) together with the spin-trapping agent as an intraperitoneal injection. The livers of these animals were extracted 15 min after the injection. In experiments with rats the liver extracts showed evidence of carbon-centered lipid free radical production within that short time period (Fig. 5). Results with mice produced similar results.

It is known that prior treatment with ethanol, administered either chronically or acutely, increases the hepatotoxicity of carbon tetrachloride (10,11). This is apparently due to an induction by ethanol of a form of cytochrome P_{450} which is much more effective in metabolizing carbon tetrachloride than the forms of P_{450} cytochrome P_{450} induced by phenobarbital (12). Carbon tetrachloride is known to be metabolized to the trichloromethyl radical in the liver, and these radicals have been trapped in the hepatic endoplasmic reticulum in vivo (13). The metabolism of this compound is associated with the production of lipid peroxidation and gives rise to

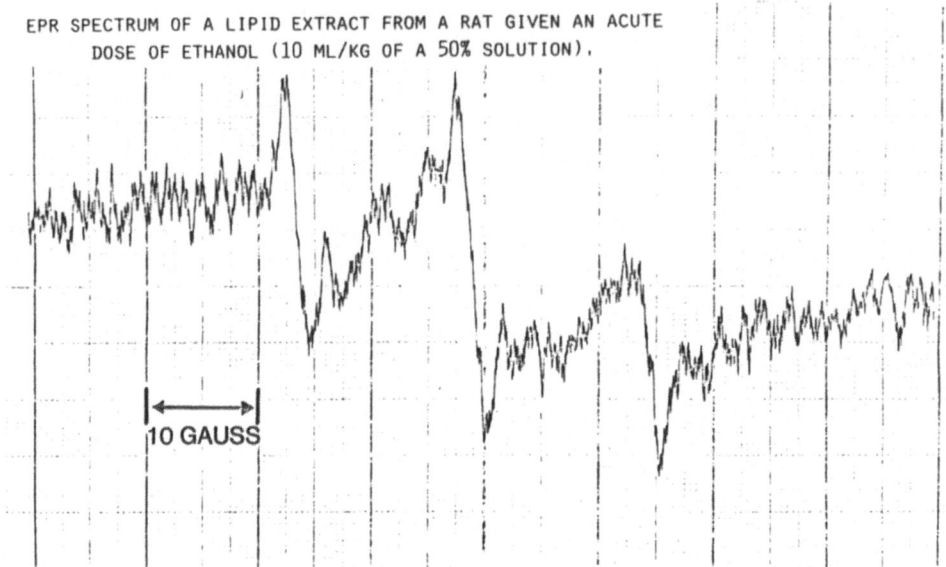

EPR SPECTRUM OF A LIPID EXTRACT FROM A RAT GIVEN AN ACUTE
DOSE OF ETHANOL (10 ML/KG OF A 50% SOLUTION).

10 GAUSS

Fig. 5. EPR spectrum of liver lipid extract from at rat given an acute dose of ethanol together with the spin-trapping agent. The liver was extracted 15 min after administration

Table 4. Enhancement of trichloromethyl radical generation in vivo in livers of ethanol-fed and rats which were exposed to carbon tetrachloride[a]

Diet	Signal strength[b] (Millimeters signal height/gram liver)
High fat + ethanol	25.78
High fat control	14.35
Low fat + ethanol	17.20
Low fat control	12.97

[a] CCl_4 (0.08 ml/100) was dissolved in 0.3 ml corn oil which was then homogenized with 1.0 ml 0.1 M phenyl-t-butyl nitrone. The mixture was administered to rat by stomach tube. 30 min later, the livers were removed and the total lipids extracted for EPR scanning.

[b] Results of three experiments.

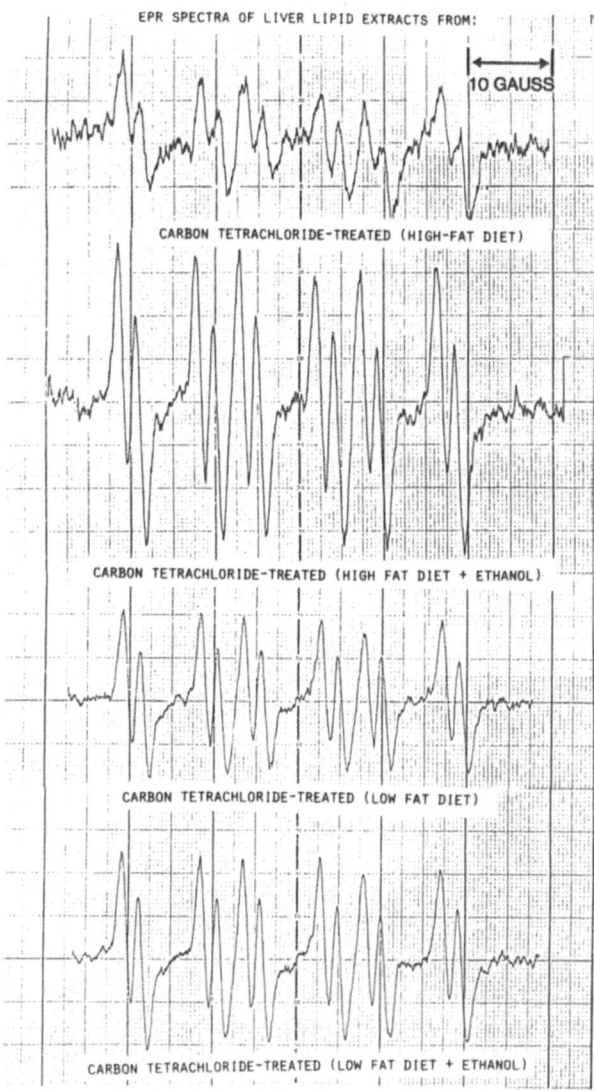

Fig. 6. EPR spectra of liver lipid extracts from rats fed either a high or low diet, with or without ethanol, and then given a dose of ^{13}C-carbon tetrachloride and the spin-trapping agent 30 min before removal of the liver for extraction of lipids

carbon-centered lipid radicals which have also been spin-trapped in vivo (8). Rats isocalorically pair-fed the high fat control diet were compared with rats fed the ethanol-containing high fat diet for a period of 2 weeks. At the end of the 2-week feeding period, all animals were given a dose of carbon tetrachloride along with the spin-trapping agent which traps trichloromethyl radicals (phenyl-t-butyl nitrone). After 30 min, the livers were removed and the total lipids extracted. An EPR extract of the total lipids showed the presence of the trichloromethyl radical adducts in animals fed the high fat diet without alcohol (Fig. 6). The lipids from the animals fed the ethanol-containing diet had more intense EPR signals (second scan), indicating that these animals experience approximately twice the number of trichloromethyl radicals generated as did the controls in the same amount of liver tissue (Table 4). The same type of experimental was performed with other animals using the low fat diet. In this case, the effect of ethanol feeding did not significantly affect the intensity of trichloromethyl radical generation (Fig. 6 and Table 4). Indicating that induction of cytochrome P_{450} by ethanol in animals fed low amounts of fat may not occur.

4 Discussion

These findings indicate that the lipid radicals are generated in the hepatic endoplasmic reticulum as a consequence of ethanol consumption. Radicals of this type represent the first step in the process of lipid peroxidation and constitute a confirmation of the view held by many investigators (3,4,14—16) that lipid peroxidation is involved in the pathological effects of ethanol on the liver following both acute (16—18) and chronic (18—20) doses. Evidence is accumulating that the metabolism of ethanol in hepatic microsomes may be mediated by the ability of the microsomes to generate hydroxy radicals ($\cdot OH$) (21—23). Assuming that this mechanism occurs in the hepatic endo-plasmic reticulum of the intact animal, the extreme reactivity of hydroxy radicals could be expected to initiate lipid peroxidation in that organelle. Other investigations give further support to this concept. For example, some free radical scavengers (antioxidants) have been shown to inhibit ethanol-induced liver and cardiac damage caused by ethanol consumption (1,2,24). Whether or not lipid peroxidation per se is involved in the tissue injury caused by ethanol, the detection of lipid radicals in the hepatic endoplasmic reticulum indicates that a highly reactive species with properties resembling those of hydroxy radicals is produced in vivo in that organelle in ethanol-treated animals. Furthermore, the feeding of high fat diets to these animals markedly enhanced the intensity of generation of these radicals in comparison to animals consuming alcohol with a low fat diet. The reason for this is not clear, but it could be related to the "permissive" effect of dietary fat on the induction of some enzymatic activities described by Century (25) and by Marshall and McLean (26). It has been shown that ethanol consumption leads to ethanol tolerance, increased metabolism of the alcohol to acetaldehyde, and the induction of a cytochrome P_{450} 3a which may be responsible for this increased metabolism (27).

Shaw et al. observed that long-term feeding of ethanol potentiated lipid peroxidation in liver (18), possibly because of an induction effect. The smaller signal obtained in our studies in animals which have been given an acute dose of ethanol may be explained by the fact that induction of cytochrome P_{450} 3a by ethanol would not have occurred in these animals. Assuming that this cytochrome, in the presence of ethanol, is responsible for generating a highly reactive chemical species in the endoplasmic reticulum, the latter would react with membrane lipids to produce the carbon-centered lipid radicals observed by spin trapping. If the initial reactive species were hydroxy radicals, they would be trapped also, but the spin adducts of hydroxy radicals are not stable and would quickly disappear before they could be observed in the EPR spectrometer. The lipid radicals trapped by trimethoxyphenyl-N-t-butyl nitrone, however, are quite stable.

The conclusions derived from this work include confirmation that lipid peroxidation is initiated in the hepatic endoplasmic reticulum in vivo during the metabolism of ethanol. Also, diets containing fat at a level similar to that of the average diet in the US significantly enhances the intensity of radical production. In addition, the lipid free radicals generated as a result of ethanol consumption continue to be produced long after the blood level of alcohol returns to zero, suggesting that a self-sustaining free radical chain reaction may be occurring in the endoplasmic reticulum. Investigations currently in progress are examining the effects of antioxidants and of different types of fat on the production of the lipid free radicals in vivo.

References

1. Di Luzio NR, Costales F (1965) Inhibition of the ethanol and carbon tetrachloride-induced fatty liver by antioxidants. Exp Mol Pathol 4:141–154
2. Di Luzio NR (1966) Mechanism of the acute ethanol-induced fatty liver and modification of liver injury by antioxidants. Lab Invest 15:50–63
3. Videla LA, Fernandez V, Ugarte G, Valenzuela A (1980) Effect of acute ethanol intoxication on the content of reduced glutathione of the liver in relation to its lipoperoxidative capacity in the rat. Febs Lett 111:6–10
4. Videla LA, Valenzuela A (1982) Alcohol ingestion, liver glutathione and lipoperoxidation: metabolic interrelations and pathological implications. Life Sci 31:2395–2407
5. Lieber CS (1985) Alcohol and liver: metabolism of ethanol, metabolic effects and pathogenesis of injury. Acta Med Scand Suppl 703:11–55
6. Lieber CS (1984) Alcohol and the liver – 1984 update. Hepatology 4:1243–1260
7. Janzen EG, Stronks HJ, Dubose CM, Poyer JL, McCay PB (1985) Chemistry and biology of spin-trapping radicals associated with halocarbon metabolism in vitro and in vivo. Environ Health Perspect 64:151–170
8. McCay PB, Lai EK, Poyer JL, Dubose CM, Janzen RG (1984) Oxygen-and-carbon-centered free radical formation during carbon tetrachloride metabolism. J Biol Chem 259:2135–2143
9. Lai EK, Crossely C, Sridhar R, Misra HP, Janzen EG, McCay PB (1986) In vivo spin trapping of free radicals generated in brain, spleen, and liver during α radiation of mice. Arch Biochem Biophys 244:156–160
10. Maling HM, Strpp B, Sipes IG, Highman B, Saul W, Williams MA (1975) Enhanced hepatotoxicity of carbon tetrachloride, thioacetamide and dimethylinitrosamine by pretreatment of rats with ethanol and some comparison with potentiation by isopropranol. Toxicol Appl Pharmacol 33:291–308

11. Traiger CJ, Plaa GL (1971) Differences in the potentiation of carbon tetrachloride in rats by ethanol and isopropranolol pretreatment. Toxicol Appl Pharmacol 20:105–112
12. Johansson I, Ingelman-Sundberg M (1985) Carbon Tetrachloride-induced lipid peroxidation dependent on an ethanol-inducible form of rabbit liver microsomal cytochrome P-450. Febs Lett 183:265–269
13. Poyer JL, McCay PB, Lai EK, Janzen EG, Davis ER (1980) Confirmation of assignment of the trichloromethyl radical spin adduct detected by spin trapping during ^{13}C-Carbon tetrachloride metabolism in vitro and in vivo. Biochem Biophys Res Commun 94:1154–1160
14. Videla LA, Fernandez V, Valenzuela A, Ugarte G (1981) Effect of (+)-cyanidanol-3 on the changes in liver glutathione content and lipoperoxidation induced by acute ethanol administration in the rat. Pharmacology 22:343–348
15. Koster U, Albrecht D, Kappus H (1977) Evidence for carbon tetrachloride- and ethanol-induced lipid peroxidation in vivo demonstrated by ethane production in mice and rats. Toxicol Appl Pharmacol 41:639–648
16. Litov RE, Irving DH, Downey JE, Tappel AL (1978) Lipid Peroxidation: A mechanism involved in acute ethanol toxicity as demonstrated by in vivo pentane production in the rat. Lipids 13:305–307
17. Comporti M, Benedetti A, Chieli E (1973) Studies on in vitro peroxidation of liver lipids in ethanol-treated rats. Lipid 8:498–502
18. Shaw S, Jayatilleke E, Ross WA, Gordon ER, Lieber CS (1981) Ethanol-induced lipid peroxidation – Potentiation by long-term alcohol feeding and attenuation by methionine. J Lab Clin Med 98:417–424
19. Reitz RC (1975) A possible mechanism for the peroxidation of lipids due to chronic ethanol ingestion. Biochim Biophys Acta 380:145–154
20. Koes M, Ward T, Pennington S (1974) Lipid peroxidation in chronic ethanol treated rats: in vitro uncoupling of peroxidation from reduced nicotine adenosine dinucleotide phosphate oxidation. Lipids 9:899–904
21. Feierman DE, Winston GW, Cederbaum AI (1985) Hydroxyl radicals – role of iron chelates superoxide, and hydrogen-peroxide. Alc Clin Exp Res 9:95–102
22. Winston GW, Cederbaum AI (1983) NADPH-dependent production of oxy radicals by purified components of the rat liver mixed function oxidase system. I. Oxidation of hydroxyl radical scavenging agents. J Biol Chem 258:1508–1513
24. Redetzki JE, Griswold KE, Nopajaroonsri C, Redetzki HM (1983) Amelioration of cardiotoxic effects of alcohol by vitamin-E. J Tox Clin 20:319–331
25. Century B (1973) A role of the dietary lipid in the ability of phenobarbital to stimulate drug detoxification. J Pharmacol Exp Ther 185:185–194
26. Marshall WJ, McLean AEM (1971) The effect of linoleic acid peroxidation and antioxidants on induction of cytochrome P-450 in rat liver. Biochem J 123:28
27. Koop DR, Morgan ET, Tarr GE, Coon MJ (1982) Purification and characterization of a unique isozyme of cytochrome P-450 from liver microsomes of ethanol-treated rabbits. J Biol Chem 257:8472–8480
28. Lieber CS, De Carli LM (1982) The feeding of alcohol in liquid diets – 2 decades of application and 1982 update. Alcoholism. Clin Exp Res 6:523–531
29. Bieri JG, Stoewsand GS, Briggs GM, Phillips RW, Woodard JC, Kanpka JJ (1977) Report of the American institute of nutrition ad hoc commitee on standards for nutritional studies. J Nutrit 107:1340–1348

The Role of the Antioxidant Tocopherol in Lipid and Lipoprotein Metabolism

H. J. KAYDEN and M. G. TRABER [1]

Vitamin E, tocopherol, was demonstrated to be an essential nutrient in several animal species some 6 decades ago, but it has only been in the past 10 years that vitamin E deficiency in humans, resulting in neurologic abnormalities (spinocerebellar ataxia and areflexia), has been generally recognized to occur as a result of lipid malabsorption syndromes. In this chapter the transport mechanisms for tocopherol will be reviewed and related to lipid malabsorption syndromes and abnormalities in lipid and lipoprotein metabolism. The symptomatology resulting from vitamin E deficiency and the evidence for in vivo lipid peroxidation will be discussed.

Tocopherol is present in fats and oils in the diet. In order for dietary fats to be absorbed, bile acids, which emulsify fats and aid in the formation of micelles, must be secreted into the intestinal lumen. The absolute requirement for bile acids for tocopherol absorption is amply demonstrated in children with cholestatic liver disease; those children who do not secrete bile acids into the intestinal lumen do not absorb vitamin E and within a period of a few years develop significant neurologic abnormalities, primarily areflexia and ataxia, requiring parenteral administration of tocopherol to control progression of the disorder, as described by Sokol et al. (1983). Traber et al. (1986) have demonstrated that a child with cholestasis could absorb tocopherol when it was provided as TPGS (tocopheryl polyethylene glycol 1000 succinate), in which tocopherol is esterified to an emulsifying agent and forms a micellar solution in the absence of bile or other lipids.

In the intestinal cell, tocopherol is incorporated into chylomicrons, the primary vehicle for transport of dietary lipids from the intestine to the circulation via the lymphatic system. As discussed by Kayden and Traber (1986), a striking example of the necessity of chylomicron formation for the absorption of tocopherol is demonstrated in patients with Abetalipoproteinemia. These patients do not synthesize apolipoprotein B, and thus do not secrete chylomicrons from the intestine, nor very low density or low density lipoproteins (VLDL or LDL, respectively) from the liver. In the absence of chylomicron secretion, unless these patients are supplemented with 100 mg tocopherol kg^{-1} day^{-1} (10 g day^{-1} for an adult), by the second decade of life they develop severe, disabling, neurologic abnormalities.

---

1 Department of Medicine, New York University School of Medicine, New York, NY 10016, USA

Drugs Affecting Lipid Metabolism
Ed. by R. Paoletti et al.
© Springer-Verlag Berlin Heidelberg 1987

Some of the tocopherol in chylomicrons is transferred by lipoprotein lipase to a variety of tissues, such as adipose tissue, brain, mammary gland, and muscles. This hypothesis is supported by two lines of evidence. As shown by Kayden (1983), type I hyperlipidemic patients, who lack lipoprotein lipase, have markedly elevated plasma triglyceride (in excess of 1000 mg dl^{-1}) and tocopherol levels (about 20 $\mu g\ ml^{-1}$), of which more than 80% of both the circulating triglyceride and tocopherol are contained in lipoproteins with densities <1.006 (chylomicrons and VLDL). Furthermore, Traber et al. (1985) have demonstrated that lipoprotein lipase in vitro transfers tocopherol during triglyceride hydrolysis to cells.

Following the action of lipoprotein lipase, chylomicron remnants, which contain tocopherol along with other dietary lipids, are rapidly taken up by the liver via the receptor for apolipoprotein E. In studies in progress, we have noted that a type III hyperlipidemia patient, with elevated BVLDL, has more than 50% of the circulating tocopherol in the BVLDL fraction; again demonstrating that tocopherol is an integral part of lipoproteins, and when there is an impairment of lipoprotein catabolism, which results in elevated plasma levels of lipoproteins, then tocopherol is also elevated, suggesting that the defect in lipoprotein catabolism prevents the normal transfer of tocopherol to tissues.

Following the uptake of chylomicron remnants by the liver, tocopherol is secreted within VLDL; the action of lipoprotein lipase on VLDL results in the formation of LDL. Traber and Kayden (1984) have demonstrated in vitro that tocopherol is delivered to cells by the uptake of LDL via the LDL receptor. However, patients with the homozygous form of familial hypercholesterolemia do not become vitamin E deficient, probably because tocopherol can be delivered via the lipoprotein lipase mechanism; additionally, as shown by Bjornson et al. (1976), tocopherol present in both LDL and HDL are readily exchangeable with membranes. Thus, delivery of tocopherol to tissues may also take place by unidirectional transfer from tocopherol-rich lipoproteins to tocopherol-poor membranes. Another demonstration of this latter mechanism is patients with Abetalipoproteinemia, who do not have detectable levels of VLDL or LDL in their plasma, and even when supplemented with vitamin E, seldom have plasma levels of tocopherol that reach one-tenth of normal. Nonetheless, the absorption of tocopherol from large dietary supplements provides adequate amounts of tocopherol to maintain normal nerve function and to raise adipose tissue levels of tocopherol to normal levels.

As vitamin E deficiency is associated with a dying back of the peripheral nerves, we are measuring the tocopherol content of biopsies of sural nerve taken from vitamin E deficient subjects with neurologic disease and from patients with peripheral neurologic disease without vitamin E deficiency. Our preliminary data suggests that the tocopherol content of sural nerves from vitamin E deficient subjects is strikingly reduced compared to that of the controls; whether this reduction in the tocopherol content is the *cause* or a *result* of the dying back of the peripheral nerves is under investigation. One of our hypotheses is that vitamin E deficiency results in a reduction of the content of the antioxidant tocopherol in the nerves, thus allowing damage by free radicals to take place.

Extensive evidence of free-radical damage should be demonstrable in humans with vitamin E deficiency. In one Abetalipoproteinemic patient, who died of cardiac arrhythmias and came to autopsy, Kayden (1972) described that extensive amounts of ceroid pigment, which represents peroxidative changes of lipids, were present in the heart muscle, especially along the conducting system, presumably contributing to the arrhythmias. As discussed by Brin and Traber (1985), characteristic neuromuscular abnormalities have been studied in experimental vitamin E deficiency in monkeys; similar lesions have been noted in case reports of patients who develop vitamin E deficiency, usually in association with malabsorption; e.g., blind loop (intestinal) syndrome, short bowel syndrome, cystic fibrosis, intestinal lymphangiectasia, and cholestasis.

While there is unequivocal evidence that tocopherol functions as a potent antioxidant and free-radical scavenger in vitro in a number of systems, the documentation for this activity in vivo is much more difficult to obtain. The possibility exists that this role is so swiftly and efficiently carried out, and that there is even regeneration of the partially oxidized tocopherol back to its unoxidized state, so that the failure to demonstrate any alterations in tocopherol can as easily be taken as proof of antioxidant activity, rather than evidence against the occurrence. We have therefore studied patients with documented vitamin E deficiency, as evidenced by neurologic defect in association with diminished plasma and tissue levels of tocopherol for flagrant evidence of altered plasma or adipose tissue lipids, and have found none.

One group of patients who particularly deserve study are those with familial vitamin E deficiency. These subjects present neurologic disease of differing degrees of severity, but mostly representative of spinocerebellar disease. Our preliminary studies of these subjects demonstrate that they have exquisitely low plasma vitamin E levels, and reduced adipose tissue levels, and reduced sural nerve concentrations of tocopherol. Studies of gastrointestinal function including fat absorption, etc. have all been normal. Laboratory examination of the usual clinical parameters have not been abnormal. In view of the reports by several laboratories of abnormalities in the composition of LDL after incubation with endothelial cells in tissue culture, which could be prevented by addition of vitamin E to the medium, we have studied the lipoproteins and apolipoproteins of four patients with this disorder. No abnormality was noted in any of our four patients, nor in the other four reported, except for one female patient of 23 who has heterozygous familial hypercholesterolemia. Admittedly, the altered state of LDL may have been such that the scavenger activity of macrophages may have eliminated them from the circulation; or the concentration of abnormal LDL may have been so small so as to escape detection by the relatively crude analysis of differences in electrophoretic migration, antibody responses, etc.

There are multiple systems in the body that protect and defend against free-radical injury. Tocopherol is a potent antioxidant, but it is not the only antioxidant system; however, there may be tissues in which it is the dominant or perhaps only antioxidant — especially in certan lipid membranes. One example is the sensitivity of erythrocytes deficient in vitamin E to peroxide hemolysis, especially since tocopherol may be located in a particular spacial relation to the unsaturated fatty acids of phospholipids in these membranes. Studies by Dodge et al. (1967) on peroxide hemolysis of human red blood cells deficient in tocopherol showed that oxidant changes in the fatty acids preceded the rupture of the membranes with leakage of hemoglobin and hemolysis. Whether a similar specific functional role for tocopherol exists in the nervous system (or perhaps the retina of the premature newborn baby) remains to be established. Within the circulating lipoproteins there is a presumed surfeit of tocopherol; but in large lipid-rich particles (chylomicrons, VLDL, and perhaps LDL), the tocopherol may equilibrate more rapidly with the core of the circulating moiety than with the surface components and therefore not be readily accessible as an antioxidant for either the surface proteins or phospholipids. Further studies on the nature of the toxicity of lipids and proteins to cells in culture that occur as a result of oxidation or free-radical injury to LDL, are being carried out with specific reference to the role of tocopherol in protecting against such toxicity.

References

Bjornson LK, Kayden HJ, Miller E, Moshell AN (1976) Transport of tocopherol and carotene in human blood. J Lipid Res 17:343–352

Brin MF, Traber MG (1985) Vitamin E deficiency and human neurologic disease. Lab Manag 23:57–67

Dodge JT, Cohen G, Kayden HJ, Phillips GB (1967) Peroxidative hemolysis of red blood cells from patients with abetalipoproteinemia. J Clin Invest 46:357–368

Kayden HJ (1972) Abetailipoproteinemia. Annu Rev Med 23:285–296

Kayden HJ (1983) Tocopherol content of adipose tissue from vitamin E deficient humans. In: Biology of vitamin E. Ciba Found Symp, New York, 101:70–85

Kayden HJ, Traber MG (1986) Clinical, nutritional and biochemical consequences of apo B deficiency. In: Angel A, Frohlich J (eds) Lipoprotein deficiency syndromes. Advances in experimental medicine and biology, vol 201. Plenum, New York, pp 67–81

Sokol RF, Heubi JE, Iannaccone ST, Bove KE, Balistreri WF (1983) Mechanism causing vitamin E deficiency during chronic childhood cholestasis. Gastroenterology 85:1172–1182

Traber MG, Kayden HJ (1984) Vitamin E is delivered to cells via the high affinity receptor for LDL. Am J Clin Nutrit 40:747–751

Traber MG, Olivecrona T, Kayden HJ (1985) Bovine milk lipoprotein lipase transfers tocopherol to human fibroblasts during triglyceride hydrolysis in vitro. J Clin Invest 75:1729–1734

Traber MG, Kayden HJ, Green JB, Green MH (1986) Absorption of water-miscible forms of vitamin E in a patient with cholestasis and in thoracic duct cannulated-rats. Am J Clin Nutrit 44:914–923

Newer Aspects of Fats and Lipid Metabolism

D. Kritchevsky [1]

It is generally regarded that the cholesterolemic or atherogenic effects of a fat are a function of its iodine value. However, there are some fats which, because of triglyceride structure or fatty acid content, do not follow this pattern.

Peanut oil is one such anomalous fat. Despite its iodine value (90 ± 5), peanut oil is atherogenic for rats, rabbits, and rhesus and vervet monkeys whether fed with cholesterol or as part of a semipurified, cholesterol-free diet (Kritchevsky 1983). Peanut oil contains 4–6% long-chain, saturated fatty acids (arachidic, behenic, lignoceric) and these were, at one time, thought to be responsible for its atherogenicity. A fat mixture resembling in composition peanut oil minus its component long-chain fatty acids was found to be less atherogenic for rabbits than peanut oil, but when arachidic and behenic acids were randomized into the mixture, its atherogenicity was unchanged. Randomization of peanut oil, which changes the triglyceride structure, but not the fatty acid composition, has been shown to reduce its atherogenicity.

Comparison of peanut oils obtained from the US, South America, or Africa, showed the South American oil to be the most atherogenic. The computed iodine values of the three fats were: 100, 110, and 94, respectively; and their ratios of 18:1/18:2 were: 1.65, 0.89, and 2.70.

To test the possibility that the ratio of 18:1/18:2 might exert an independent atherogenic effect, we tested oils from two special cultivars of peanut. India White (iodine value, 103; 18:1/18:2 = 1.00) and Jenkins Jumbo (iodine value, 88; 18:1/18:2 = 5.28). The results of two experiments are summarized in Table 1. The commercial peanut oil was more atherogenic than either of the two special cultivars. There was no relation of atherogenicity to either iodine value or ratio of 18:1/18:2. We are still seeking the feature of peanut oil which affects its atherogenicity. Presence of lectin is a possibility under test.

Cocoa butter is another fat whose atherogenicity is not related to its iodine value (30 ± 5). The atherogenic and cholesterolemic effects of cocoa butter have been compared in rabbits with those of palm, corn, or coconut oils. When the diet contained 2% cholesterol and the fat was fed as 6% of the diet, cocoa butter was 19% less atherogenic than palm or coconut oils, but more atherogenic than corn oil. When fed as part (14%) of a semipurified diet, cocoa butter was 67% less atherogenic than

---

1 The Wistar Institute of Anatomy and Biology, 3601 Spruce Street, Philadelphia, PA 19104, USA

Drugs Affecting Lipid Metabolism
Ed. by R. Paoletti et al.
© Springer-Verlag Berlin Heidelberg 1987

Table 1. Influence of different peanut oils on atherosclerosis in rabbits[a]

Peanut oil	No.	Cholesterol		Average atherosclerosis (arch + thoracic/2)
		Serum (mg dl^{-1})	Liver (g 100g^{-1})	
Commercial	17/18	1521	1.67	1.81
Jumbo Jenkins	17/18	1734	1.53	1.35
India White	18/18	1992	1.57	1.27

[a] Rabbits fed 1% cholesterol and 4% oil for 60 days.

coconut oil, 59% less atherogenic than palm kernel oil. The iodine values of coconut oil and palm oil are about 10 and 40, respectively (Kritchevsky et al. 1982). When rats were fed diets containing 1% cholesterol and 10% cocoa butter, corn, palm, or coconut oil, serum cholesterol levels were significantly elevated only in the palm and coconut oil groups (Table 2).

The high stearic acid content of cocoa butter (35%) has been cited as the reason that this fat does not affect lipidemia. We have recently compared the effects on lipid metabolism of a stearic acid-rich variant of soybean oil (A6) with those of commercial soybean and peanut oil (Kritchevsky et al. 1986). The calculated iodine values of the three fats were 100, 125, and 100, respectively. Contents (%) of stearic, oleic, and linoleic acids were: A6: 28.1, 19.8, and 35.5; soybean oil: 4.4, 42.8, and

Table 2. Serum and liver lipids in rats fed 1% cholesterol and 10% cocoa butter, corn, coconut, or palm oil

	Group[a]			
	Cocoa butter	Corn oil	Palm oil	Coconut oil
Weight gain (g)	141 ± 5	156 ± 6	144 ± 3	141 ± 5
Liver weight (g)	11.8 ± 0.30a	14.3 ± 0.22ab	12.5 ± 0.54b	12.8 ± 0.84
Relative liver wt.	3.66 ± 0.07a	4.28 ± 0.14ab	3.86 ± 0.12b	3.97 ± 0.22
Serum (mg dl^{-1})				
Cholesterol	123 ± 8ab	120 ± 7 cd	160 ± 9ac	153 ± 12bd
Triglycerides	110 ± 6	107 ± 11	133 ± 15	119 ± 9
Liver (g 100 g^{-1})				
Cholesterol	1.36 ± 0.08abc	2.84 ± 0.20ad	2.48 ± 0.12b	2.22 ± 0.20cd
Triglyceride	1.92 ± 0.18abc	5.89 ± 0.35ade	2.68 ± 0.28 bd	3.51 ± 0.46cd

[a] Values in horizontal row bearing same letter are significantly ($p <0.05$) different.

Table 3. Influence of three fats on lipids in Wistar rats

	Fat [a]		
	A6	Soybean Oil	Peanut Oil
Weight gain (g)	24 ± 4 ab	81 ± 6 ac	58 ± 6 bc
Liver weight (g)	5.7 ± 0.22 ab	8.9 ± 0.22 ac	7.5 ± 0.43 bc
Relative liver wt.	2.46 ± 0.06 ab	3.10 ± 0.06 a	2.84 ± 0.13
Serum lipids (mg dl^{-1})			
Cholesterol	54 ± 3 ab	83 ± 6 a	75 ± 3 b
Triglycerides	31 ± 1 ab	38 ± 2 a	42 ± 2 b
Phospholipids	49 ± 3 ab	63 ± 5 a	65 ± 3 b
Liver lipids (g 100 g^{-1})			
Cholesterol	0.20 ± 0.02	0.20 ± 0.01	0.22 ± 0.01
% Ester	12.4 ± 1.3 ab	23.4 ± 1.4 a	19.1 ± 1.6 b
Triglycerides	0.47 ± 0.11 a	0.90 ± 0.07 a	0.70 ± 0.07
Phospholipids	3.07 ± 0.13	3.03 ± 0.15	3.01 ± 0.06

[a] Values in horizontal row bearing same letter are significantly ($p < 0.05$) different.

36.7; and peanut oil: 1.9, 46.1, and 35.6. Rats were fed a semipurified diet containing 14% of these fats. Rats fed A6 showed lower weight gain and liver weight. Rats fed A6 also exhibited significantly lower serum lipids and lower levels of liver triglycerides and cholesteryl esters (Table 3).

Generally, unsaturated fats are hypocholesterolemic relative to saturated fat. However, not all fats conform to this generalization; the recent observations that monounsaturated fat may specifically lower LDL cholesterol is a case in point (Grundy, 1986). Investigation into the mechanism responsible for the apparently aberrant behavior of certain fats may provide clues to roles played by specific fatty acids or specific triglyceride conformation in lipid metabolism. This new knowledge may lead to increased understanding of fat metabolism and atherosclerosis and may offer clues to new and useful fat formulations.

Acknowledgment. This work was supported, in part, by a grant (HL-03299) and a Research Career Award (HL-00734) from the National Institutes of Health and by funds from the Commonwealth of Pennsylvania.

References

Ahrens EH, Jr (1957) Nutritional factors and serum lipid levels. Am J Med 23:928–952
Grundy SM (1986) Comparison of monounsaturated fatty acids and carbohydrates for lowering plasma cholesterol. New Engl J Med 314:745–748
Kritchevsky D (1983) Dietary influence on lipids and lipoprotein levels in animals and atherosclerosis. Progr Biochem Pharmacol 19:151–165
Kritchevsky D, Tepper SA, Bises G, Klurfeld DM (1982) Experimental atherosclerosis in rabbits fed cholesterol-free diets. 10. Cocoa butter and palm oil. Atherosclerosis 41:279–284
Kritchevsky D, Tepper SA, Klurfeld DM, Fehr WR, Hammond EG (1986) Influence of A6 soybean oil on cholesterol metabolism in rats. Nutrit Rep Int 35:265–268

Can We Retard Atherogenesis by Modifying High-Density Lipoprotein Metabolism?

N. E. MILLER [1]

1 Introduction

During the past 10 years plasma high-density lipoprotein (HDL) cholesterol concentration has emerged as a major risk factor for coronary heart disease (CHD) in several westernized societies. The association between HDL and CHD has been found in both sexes, is independent of other lipoproteins, and is at least as strong as that between CHD and low-density lipoprotein (LDL) concentration. Angiographic and autopsy studies have indicated that the severity of coronary atherosclerosis is also a negative function of HDL. The epidemiology of HDL in relation to CHD has recently been reviewed.

At present the mechanism of the association between HDL and atherogenesis is uncertain. The original proposal that it reflects the participation of HDL in reverse-cholesterol transport remains plausible, but alternative mechanisms, both direct and indirect, have been suggested. Much more experimental work is needed in this area. Nevertheless, it is my belief that we have already reached the stage at which modification of HDL metabolism warrants serious consideration as a possible future adjunct to LDL-lowering therapy in the prevention of atherosclerosis. This chapter summarizes the evidence upon which this conviction is based.

2 Do We Need to Consider Modifying HDL now that There is Good Evidence that Lowering LDL Reduces Risk of CHD?

The reduction of CHD incidence in hypercholesterolemic patients achieved with cholestyramine in the LRC-CPPT (LRC Program 1984) was of great importance as a test of the "lipid hypothesis". However, the overall reductions of LDL (12%) and CHD (19%) were modest. Clearly, LDL-lowering drugs that are better tolerated and more effective are urgently needed, and no doubt will be found. Even when such agents are available though, it is questionable whether they will be sufficient on their

---

1 Department of Chemical Pathology and Metabolic Disorders, St Thomas' Hospital Medical School, London, Great Britain

Drugs Affecting Lipid Metabolism
Ed. by R. Paoletti et al.
© Springer-Verlag Berlin Heidelberg 1987

own to reduce the LDL/HDL ratio adequately in all hypercholesterolemic subjects. Figure 1 shows the relationship between progression of femoral atherosclerosis over 2 years and the LDL/HDL ratio in 24 subjects (Duffield et al. 1983). It can be seen that zero growth occurred at a ratio of about 2.0. A rather similar result was obtained by Arntzenius (1986) for coronary lesions. In middle-aged patients with heterozygous familial hypercholesterolemia (FH), in which HDL has been shown to retain its association with CHD (Streja et al. 1978), the mean LDL/HDL ratio is approximately 7.0 at a mean HDL cholesterol level of 1.2 mmol l^{-1} (Fredrickson and Levy 1972). Thus, it can be calculated that on average a 71% reduction of LDL would be required in FH patients to lower the LDL/HDL ratio to the value at which there is zero growth of disease, in the absence of a rise in HDL cholesterol.

Of greater significance, however, is the fact that most subjects who develop CHD have total and LDL cholesterol levels that are below the 95th centile (Castelli and Anderson 1986). The predictive power of HDL for CHD in normocholesterolemic subjects is as strong as it is in hyperlipidemic subjects (Gordon et al. 1977; Goldbourt and Medalie 1979; Jacobs 1985; Assmann et al. 1986; MRFIT Research Group 1986; Reed et al. 1986). Moreover, in some communities a low HDL level appears to be a more common 'cause' of CHD than a high LDL level. Maciejko et al. (1983), for example, found that approximately 50% of subjects with angiographically defined

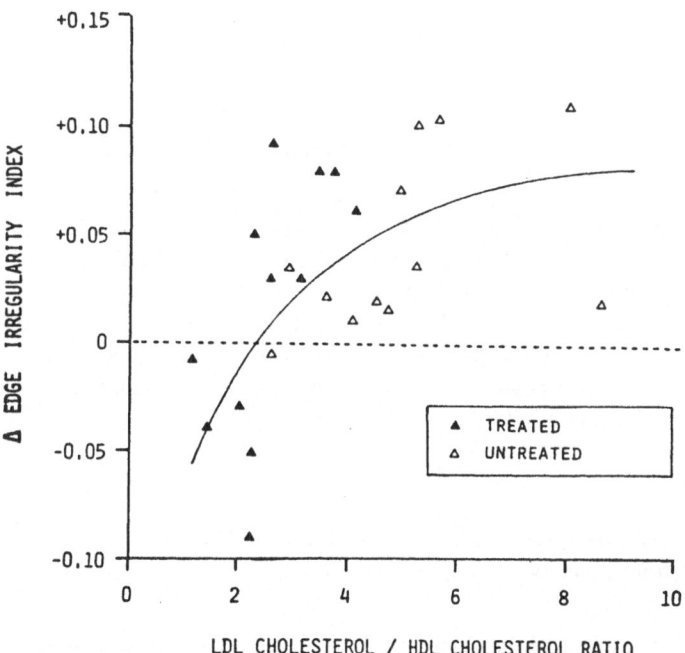

Fig. 1. Relationship between change in extent of femoral atherosclerosis over 2 years (positive = progression; negative = regression), as quantified by computerized image analysis, and the mean LDL cholesterol/HDL cholesterol ratio during the follow-up period in 24 hyperlipidaemic subjects. (Data from Duffield et al. 1983)

coronary disease had an HDL cholesterol level below the 10th centile for healthy individuals, whereas total cholesterol and triglyceride levels were similar in cases and controls. Ordovas et al. (1986) obtained a similar result, and provided evidence that an allele of the gene for the apo AI of HDL may be the most common inherited lipo-protein disorder leading to premature coronary disease in the USA. Occurring with a frequency of 4% in the general population, possession of the allele lowered HDL cholesterol by 36% on average, and appeared to increase the risk of premature coronary atherosclerosis by ten-fold. Thus, in a substantial number of North Americans ap-propriate modification of HDL metabolism might have a greater impact on CHD risk than reduction of LDL concentration.

3   Does HDL Remain a Predictor of Coronary Disease in Subjects Given a Lipid-Lowering Diet or Drug?

Evidence has been provided by the Oslo Study (Holme 1982), the LRC-CPPT (LRC Program 1984) and the MRFIT (MRFIT Research Group 1986) that HDL retains its association with CHD risk when LDL is lowered by diet or cholestyramine. In two trials that examined the progression of coronary lesions by angiography in subjects receiving lipid-lowering therapy by drugs and/or diet (Nikkilä et al. 1984; Arntzenius 1986), the negative association of lesion growth with HDL cholesterol was stronger than its positive association with total or LDL cholesterol.

4   Is There any Evidence that Raising HDL can Reduce CHD Risk?

In the LRC-CPPT (LRC Program 1984) the small increment in HDL cholesterol produced by cholestyramine made a contribution to the reduction of CHD incidence that was statistically independent of that made by the decrease in LDL concentration. More recently, long-term treatment of epileptics with anticonvulsants, which are known to raise HDL cholesterol without substantially lowering the concentration of LDL, was recently found in Finland to be associated with a 29% decrease in mortality from CHD (Muuronen et al. 1985).

5   Is There any Evidence that Raising HDL can be Associated with a Decrease in Atherogenesis?

In the NHLBI type II intervention study (Levy et al. 1984) the magnitude of the rise in HDL cholesterol which occurred during cholestyramine therapy was more reliable in predicting the extent of coronary lesion progression over 5 years than was the magnitude of the decrease in LDL concentration.

In animal studies three classes of compound which raise HDL cholesterol have also been reported to reduce atherogenesis, ethanol (Rudel et al. 1981), oestrogens (Vesselinovitch et al. 1974) and monoterpenes (Benko et al. 1961; Bell et al. 1980), although the interpretation of the first two is complicated by coincident changes in other lipoproteins. There have been few successful attempts to synthesize novel compounds with selective HDL-raising activity. One drug of considerable interest in this context, however, is BRL26314, which has been found to raise HDL without significantly lowering other lipoproteins, and to inhibit atherogenesis in cholesterol-fed rabbits (Fears et al. 1984). Its mode of action is not known.

6 Conclusion

Thus, there is a growing body of evidence supporting the notion that appropriate modification of lipoprotein metabolism to raise HDL cholesterol might be of value in the prevention of atherosclerosis in at least two situations: as an adjunct to LDL-lowering therapy in certain severely hypercholesterolemic subjects, and in subjects who have normal LDL levels but who are at increased risk of CHD on account of genetically determined low levels of HDL.

How should this possibility be further pursued? Clearly, we need urgently to identify the mechanism of the association between HDL and atherogenesis. Is it direct or indirect? If the former, is it related to reverse cholesterol transport or to effects on, for example, prostacyclin synthesis or fibrinolysis? What are the critical components of HDL metabolism? HDL levels have been found in different studies to be correlated positively with the synthetic rates of apos AI and AII, with lipoprotein lipase activity and with LCAT concentration, and to be correlated negatively with the fractional catabolic rates of the apos and with hepatic lipase activity. Which of these are true determinants of HDL cholesterol, and which, if any, would make effective targets for pharmacological intervention? These and other questions warrant immediate debate and investigation, involving the combined efforts and close liason of academic institutions and the pharmaceutical industry.

References

Arntzenius AC (1986) Diet, lipoproteins and the progression of coronary atherosclerosis. Drugs 31 (Suppl 1): 61–65

Assmann G, Schulte H, Oberwittler W, Hauss W (1986) New aspects in the prediction of coronary heart disease: the PROCAM study. In: Fidge NH, Nestel PJ (eds) Atherosclerosis VII. Elsevier, Amsterdam, p 19

Bell GD, Bradshaw JP, Burgess A, et al. (1980) Elevation of serum HDL cholesterol by Rowachol, a proprietary mixture of six pure monoterpenes. Atherosclerosis 36:47–54

Benko S, Macher A, Szarvas F, Tiboldi T (1961) Effect of essential oils on atherosclerosis of cholesterol-fed rabbits. Nature (London) 190:731–732

Castelli WP, Anderson K (1986) A population at risk. Prevalence of high cholesterol levels in hypertensive patients in the Framingham study. Am J Med 80(2A): 23–32

Duffield RGM, Lewis B, Miller NE et al. (1983) Treatment of hyperlipidaemia retards progression of symptomatic femoral atherosclerosis. A randomized controlled trial. Lancet ii:639–642

Fears R, Esmail A, Walker P, Rush WR, Ferres H (1984) Hyperalphalipoproteinaemic activity of BRL26314 – II. Inhibition of atherosclerosis in rabbits. Biochem Pharmacol 33:219–228

Fredrickson DS, Levy RI (1972) Familial hyperlipoproteinemia. In: Stanbury JB, Wyngaarden JB, Fredrickson DS (eds) The metabolic basis of inherited disease. McGraw-Hill, New York, p 545

Goldbourt U, Medalie JH (1979) High density lipoprotein cholesterol and incidence of coronary heart disease – the Israeli ischemic heart disease study. Am J Epidemiol 109:296–308

Gordon T, Castelli WP, Hjortland MC et al. (1977) HDL cholesterol and CHD risk – the Framingham study. Am J Med 62:707–714

Holme I (1982) On the separation of the intervention effects of diet and antismoking advice on the incidence of major coronary events in coronary high risk men. The Oslo study. J Oslo City Hosp 32:31–54

Jacobs DR (1985) HDL cholesterol and coronary heart disease, cardio-vascular disease and all cause mortality. Circulation 72 (III): 185

Levy RI, Brensike JF, Epstein SE et al. (1984) The influence of changes in lipid values induced by cholestyramine and diet on progression of coronary disease: results of the NHLBI type II coronary intervention study. Circulation 69:325–337

LRC Program (1984) The lipid research clinics coronary primary prevention trial results. II. J Am Med Assoc 251:365–374

Maciejko JJ, Holmes DR, Kottke BA et al. (1983) Apolipoprotein AI as a marker of angiographically assessed coronary-artery disease. N Engl J Med 309:385–389

MRFIT Research Group (1986) Relationship between baseline risk factors and coronary heart disease and total mortality in the multiple risk factor intervention trial. Prev Med 15:254–273

Muuronen A, Kaste M, Nikkilä E, Tolppanen E-M (1985) Mortality from ischaemic heart disease among patients using anticonvulsant drugs: a case-control study. Br Med J 291:1481–1483

Nikkilä E, Viikinkoski P, Valle M, Frick MH (1984) Prevention of progression of coronary atherosclerosis by treatment of hyperlipidaemia. Br Med J 289:220–223

Ordovas JM, Schaefer EJ, Salem D et al. (1986) Apolipoprotein AI gene polymorphism associated with premature coronary artery disease and familial hypoalphalipoproteinemia. N Engl J Med 314:671–677

Reed D, Yano K, Kagan A (1986) Lipids and lipoproteins as predictors of coronary heart disease, stroke and cancer in the Honolulu heart program. Am J Med 80:871–878

Rudel LL, Leathers CW, Bond MG, Bullock BC (1981) Dietary ethanol-induced modifications in hyperlipoproteinemia and atherosclerosis in nonhuman primates. Arteriosclerosis 1:144–155

Streja D, Steiner G, Kwiterovich PO (1978) Plasma high density lipoproteins and ischemic heart disease. Studies in large kindred with familial hypercholesterolemia. Ann Int Med 89:871–880

Vesselinovitch D, Wissler RW, Fisher-Dzoga K et al. (1974) Regression of atherosclerosis in rabbits. Atherosclerosis 19:259–275

Effects of Lipid-Lowering Drugs on High Density Lipoprotein Structure and Metabolism

J. Shepherd and C. J. Packard [1]

1 Introduction

The inverse relationship between plasma high density lipoproteins (HDL) and cardio-vascular risk, though amply documented in the 1960s (Gofman et al. 1966), was not widely appreciated until its re-emphasis in the publication of Miller and Miller (1975). Even now, despite the weight of evidence supporting the concept, our understanding of the mechanism which underlies it remains rudimentary. The most popular view (Miller 1984) endows the lipoprotein with the capacity to accept cholesterol from parenchymal cells, possibly following its interaction with a membrane receptor which recognises the major apolipoprotein A component of the particle. Esterification of the assimilated sterol by lecithin:cholesterol acyltransferase leads to its entrapment within the hydrophobic core of the particle which expands and limits continued cholesterol uptake. At this point it is envisaged that further cycles of sterol capture depend upon transfer of the core ester into less dense lipoproteins. These apolipoprotein B-containing very low and low density lipoproteins (VLDL and LDL) are probably responsible for the ultimate delivery of sterol to the liver, so making centripetal cholesterol transport dependent on the integrated activities of all major plasma lipoprotein fractions. Thus, it is not surprising that most prescribed hypolipidaemic drugs, which were originally designed to lower plasma VLDL and LDL, should also have an impact on HDL metabolism.

Whether this perturbs the process of reverse cholesterol transport is still open to question; and indeed it is inadvisable to imply that pharmacologic manipulation of plasma HDL might alter cardiovascular risk since measurement of HDL levels alone gives no indication of the dynamics of cholesterol flux through the system. Having said that, it is noteworthy that the increase in HDL which followed administration of cholestyramine in the Lipid Research Clinics Coronary Primary Prevention Trials (1984) was independently cardioprotective to the study participants. The discussion which follows outlines our understanding of the impact of lipid-lowering drug therapy on HDL metabolism.

1 University Department of Pathological Biochemistry, Royal Infirmary, Glasgow G4 OSF, Great Britain

Drugs Affecting Lipid Metabolism
Ed. by R. Paoletti et al.
© Springer-Verlag Berlin Heidelberg 1987

2   Sequestrant Resins and HDL Metabolism

The polymeric agents cholestyramine and colestipol deplete the liver of bile acids by sequestering them in the intestine and interrupting their enterohepatic circulation (see Packard and Shepherd 1982, for a review of this topic). The liver responds by synthesising more bile acids from cholesterol which it acquires by assimilation of LDL from the bloodstream. Early studies of this process failed to detect the subtle changes which the therapy induced in HDL, largely because these were masked by major reductions of the order of 15–25% in plasma LDL. Subsequent detailed investigations, however, showed that the drugs had produced small but significant perturbations in HDL apoprotein metabolism which were accompanied by alterations in the plasma distribution of its subfractions, HDL_2 and HDL_3. When type II hyperlipoproteinaemic subjects were given moderate doses of cholestyramine (16 g day^{-1}), the plasma concentration of apolipoprotein AI (apo AI, the major protein in HDL) rose 12%; the other A protein, apo AII, showed no change. Kinetic studies indicated that the increment in apo AI resulted from augmentation of its synthesis and was accompanied by an 80% rise in the concentration of HDL_2 in the plasma. Since both the liver and intestine are capable of synthesising the protein, it could have come from either source. Experimental evidence from animal studies, however, favours the gut. Bearnot and colleagues (1982) have shown that biliary diversion in rats promotes the release of apo AI-containing HDL into the circulation. Such particles may be the progenitors (Fig. 1) of the raised HDL_2 levels which were seen in our patients.

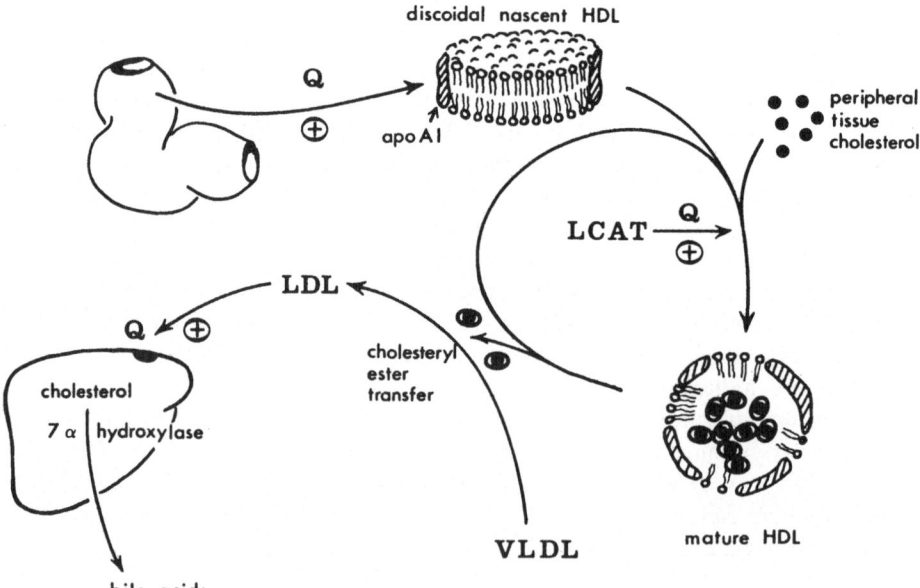

Fig. 1. Effects of cholestyramine (Q) on reverse cholesterol transport. Bile acid sequestrant resins: (1) increase synthesis of apo AI-containing HDL, probably from the intestine; (2) activate plasma lecithin:cholesterol acyltransferase; (3) increase LDL receptor activity on liver cell membranes; (4) promote bile acid production via cholesterol 7 α hydroxylase

3 Chlorophenoxyisobutyric Acid (CPIB) Derivatives and HDL Metabolism

Clofibrate derivatives are widely prescribed for the treatment of hypertriglyceridaemia. They exhibit a variety of systemic actions, notable among wich is their stimulatory effect on lipoprotein lipase. This enzyme is responsible for hydrolysis of the triglyceride core of chylomicrons and VLDL, releasing in the process coat phospholipids and apoproteins which are sequestered by HDL. The latter becomes enriched in phospholipid, increasing its flotation rate in the ultracentrifuge. Fibrate-mediated activation of the enzyme might therefore be expected to induce chronic changes in HDL structure and metabolism. But, the impact of these drugs on the lipoprotein varies from individual to individual, some patients showing significant increments in HDL cholesterol and apolipoprotein A, while in others the effect is minimal (Shepherd and Packard 1984). In general, it appears that subjects with initially high plasma triglyceride and subnormal HDL respond by reducing the former and raising the latter. In the process, the HDL core loses triglyceride at the expense of cholesteryl ester so that there is a disproportionate increment in HDL cholesterol relative to particle number. Eisenberg and colleagues (1986) have shown that these events are accompanied by an increase in particle size; and we found (Shepherd et al. 1984) that treatment with Bezafibrate expanded the circulating HDL_2 pool in the plasma without inducing significant alterations in the metabolism of either apolipoprotein AI or AII. So, these proteins do not seem to be critical regulators of the metabolic behaviour of the particles.

4 Nicotinic Acid and HDL

Nicotinic acid has a powerful effect on HDL. When administered in pharmacologic doses (3 g day^{-1}), it induces changes in the plasma concentration, apoprotein content and subfraction distribution of the lipoprotein (Shepherd and Packard 1984). Some acitons of the drug may be secondary to its triglyceride lowering effect, as observed above in the case of the fibrates, but others appear to be direct. Apolipoproteins AI and AII show divergent responses to treatment. Catabolism of the former decreases, so that its plasma concentration rises. On the other hand, both the synthesis and plasma concentration of apo AII fall. These changes are accompanied by an increment in HDL_2 and a reduction in HDL_3. The impact of the drug, particularly on the plasma concentrations of apolipoproteins AI and AII is inconsistent with the concept that the lipoprotein spectrum comprises a single population of compositionally similar particles containing both proteins; and immunological studies have shown that a least two distinct particle types exist. In normal subjects (Atmeh et al. 1983), about 55% of AI is found in association with particles that contain AII [called (AI+AII)HDL], while the remainder circulates in another species devoid of the latter [(AI)HDL]. HDL_2 is particularly enriched in (AI)HDL, while (AI+AII)HDL appears predominantly in the HDL_3 density interval. The physiological relevance of these particles is evident from their differential response to drug therapy. Nicotinic acid specifically increases the plasma concentrations of (AI)HDL, while diminishing that of (AI+AII)HDL, probably by suppressing its synthesis. Whether such changes have any influence on cardiovascular risk is not yet clear, although it is relevant to note that of all drugs employed in the Coronary Drug Project only nicotinic acid proved favourable in delaying progression of ischaemic heart disease.

5 Probucol and HDL

Probucol is a well-tolerated cholesterol-lowering agent which has little effect on triglyceride. Its hypocholesterolaemic action is not confined to LDL. Indeed, in percentage terms, HDL cholesterol is reduced more during therapy. The drug lowers the level of circulating apolipoproteins AI and AII by inhibiting their rates of synthesis (Shepherd and Packard 1984), and there is a concomitant reduction in plasma HDL_2. Examination of the changes in (AI)HDL and (AI+ALL)HDL particles gives further clues relative to the mechanism of action of the agent (Atmeh et al. 1983). In normal subjects, probucol treatment lower (AI)HDL without influencing (AI+AII)HDL. This is in distinct contrast to the action of nicotinic acid.

7 Conclusions

Lipid-lowering drugs have a wide variety of effects on plasma HDL (Table 1), but the inherent heterogeneity of the fraction makes it difficult to predit their cardioprotective potential. A number of studies, currently in progress, may soon enable us to define the extent to which drug-induced increments in plasma HDL have a moderating influence on coronary risk.

Table 1. Effect of drugs on HDL metabolism[a]

Drug	HDL cholesterol	HDL_2/HDL_3 ratio	Plasma apo AI	AI synthesis	AI catabolism
Bile acid sequestrants	↗	73% ↑	12% ↑	12% ↑	→
Nicotinic acid	23% ↑	345% ↑	7% ↑	→	8% ↓
Probucol	24% ↓	15% ↓	14% ↓	39% ↓	→
Bezafibrate	15% ↑	31% ↑	→	→	→

[a] The data in this table are culled from Shepherd and Packard (1984), Packard and Shepherd (1982) and Atmeh et al. (1983).

References

Atmeh RF, Shepherd J, Packard CJ (1983) Subpopulations of apolipoprotein AI in human high density lipoporteins. Biochim Biophys Acta 751:175–188

Bearnot HR, Glickman RM, Weinberg L, Green PHR, Tall A (1982) Effects of biliary diversion on rat mesenteric lymph apolipoprotein AI and high density lipoprotein. J Clin Invest 69:210–217

Eisenberg S, Gavish D, Kleinman Y (1986) Bezafibrate. In: Fears R, Levy RI, Shepherd J, Packard CJ, Miller NE (eds) Pharmacological control of hyperlipidaemia. Prous, Barcelona, pp 145–170

Gofman JW, Young W, Tandy R (1986) Ischaemic heart disease, atherosclerosis and longevity. Circulation 34:679–697

Miller NE (1984) Current concepts of the role of HDL in reverse cholesterol transport. In: Miller NE, Miller GJ (eds) Clinical and metabolic aspects of high density lipoproteins. Elsevier, Amsterdam, pp 187–216

Miller NE, Miller GJ (1975) Plasma high density lipoprotein concentration and the development of ischaemic heart disease. Lancet i:16–19

Packard CJ, Shepherd J (1982) The hepatobiliary axis and lipoprotein metabolism. J Lipid Res 23:1081–1098

Shepherd J, Packard CJ (1984) High density lipoprotein apoprotein metabolism. In: Miller NE, Miller GJ (eds) Clinical and metabolic aspects of high density lipoproteins. Elsevier, Amsterdam, pp 247–274

Sherpherd J, Packard CJ, Stewart JM, Atmeh RF, Clark DS, Boag DE, Carr K, Lorimer AR, Ballantyne D, Morgan HG, Lawrie TDV (1984) Apolipoprotein A and B (Sf 100–400) metabolism during Bezafibrate therapy in hypertriglyceridemic subjects. J Clin Invest 74:2164–2177

The Lipid Research Clinics Coronary Primary Prevention Trial Results (1984) 1 Reduction in incidence of coronary heart disease. Lipid Res Clin Progr. JAMA 251:351–364

Enzyme-Inducers and High-Density Lipoproteins

P. V. Luoma [1]

1 Introduction

Clinical studies orginating from the late 1970s have revealed that compounds which induce hepatic microsomal enzymes may influence hepatic lipid and protein concentrations and increase serum high-density lipoprotein (HDL) levels. They have suggested an approach to prevent the atherosclerotic vascular process. Experimentally, an enzyme inducer has retarded the cholesterol accumulation in the arterial wall and the formation of atherosclerotic plaque. In this chapter I shall concentrate on the effects of microsomal enzyme inducers on liver lipid and protein, and serum HDL-cholesterol (HDL-C) and apolipoprotein concentrations, and the manifestation of atherosclerotic disease in man.

2 Microsomal Enzyme Inducers

There are several hundred compounds which have been shown to induce hepatic microsomal enzymes. They include about 15 drugs which are inducers when given at therapeutic doses. The inducing agents have been classified into groups according to structure and the characteristic effects they have on microsomal enzymes:

1. Phenobarbital and a wide variety of structurally unrelated drugs, and environmental chemicals, e.g. DDT.
2. Polycyclic aromatic hydrocarbons, e.g. 3-methylcholanthrene.
3. Pregnenolone-16α-carbonitrile and certain other steroids.
4. Ethanol.

Clofibrate may, according to experimental studies, represent a fifth class of enzyme inducers. However, in man, it has inhibited microsomal enzymes (see Greim 1981).

---

1 Clinical Research Unit, Department of Internal Medicine, University of Oulu, SF-90220 Oulu, Finland

Drugs Affecting Lipid Metabolism
Ed. by R. Paoletti et al.
© Springer-Verlag Berlin Heidelberg 1987

3   The Effects of Inducers on Liver Lipids and Proteins

Microsomal inducers increase liver protein and phospholipid concentrations. The protein and phospholipid contents in human liver parallel the microsomal enzyme activity as assessed by hepatic cytochrome P-450 or the antipyrine kinetics (Luoma et al. 1982). Enzyme inducers may reduce triglyceride concentrations in human liver (see Luoma 1986).

Microsomal inducers stimulate hepatic HMGCoA reductase, the rate-limiting enzyme of cholesterol synthesis, and also cholesterol-7α-hydroxylase, the rate-limiting, P-450-linked enzyme of bile acid synthesis. They enhance cholesterol conversion into bile acids and their elimination into bile.

4   The Effects of Inducers on Serum High-Density Lipoproteins

Drugs and Environmental Chemicals. Phenytoin treatment has been associated with high serum HDL-C; apo A-I and total cholesterol (T-C) concentrations, and high HDL-C/T-C (see Luoma 1986) and apo A-I/A-II (Acta Pharmacol Toxicol 1986, suppl V:165A) ratios. The increase in T-C has been most pronounced at the beginning of treatment, after which a decrease to pretherapy levels has been observed.

Phenobarbital has increased serum HDL-C, LDL-C, apo B and T-C concentrations in some, but not all investigations. In one group (see Durrington 1979) the increase in HDL-C concentrations was 39% and that of LDL-C 20%.

Subjects on *carbamazepine* therapy have an increased HDL-C, and the HDL-C/T-C ratio also tends to increase (see Luoma 1986).

Anticonvulsants alone or in combination. Epileptic subjects who have been treated with phenytoin alone or combined with phenobarbital and/or other anticonvulsants, when considered as a group, have high serum HDL-C and apo A-I levels and HDL-C/T-C ratios and reduced LDL-C/HDL-C ratios (see Luoma 1986). According to some studies, T-C levels in epileptics on anticonvulsants do not deviate form those in the general population.

Other inducing drugs. Glutethimide has increased serum HDL-L and LDL-C, whereas rifampicin and antipyrine have failed to affect serum lipoprotein levels (see Luoma 1986).

An exposure to *environmental chemicals* has been associated with high serum HDL-C level (Carlson and Kolmodin-Hedman 1972).

Polycyclic Aromatic Hydrocarbons. 3-methylcholanthrene has increased serum HDL-C and the HDL-C/T-C ratio in the rat.

Pregnenolone-1 6α-Carbonitrile and Certain Other Steroids. Prednisone has been found to increase HDL-C and the subfraction HDL_2 in man (Nikkilä et al. 1987).

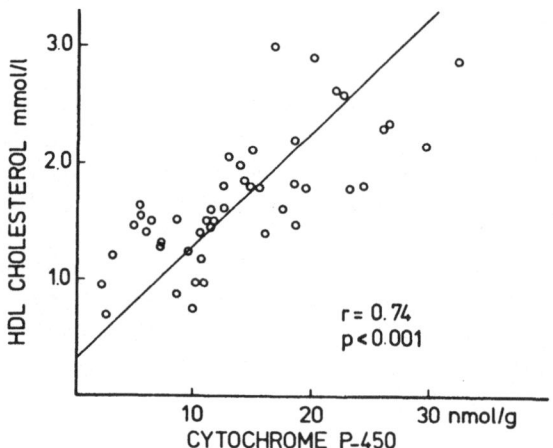

Fig. 1. Relation between plasma HDL-cholesterol and liver cytochrome P-450 concentrations in man as presented in the XVI Scandinavian Congress on Physiology and Pharmacology (Luoma et al. 1979)

Alcohol. Regular alcohol consumption induces the hepatic microsomal function and increases serum HDL-C, HDL_2, HDL_3, apo A-I and A-II, and reduces LDL-C. The effect of alcohol on lipoproteins depends on liver structure and the amount consumed. Clofibrate increases serum HDL-C in man, but possibly not through microsomal induction (see Greim 1981).

5 Induction and Liver Lipids and Proteins Related to HDL

Liver Cytochrome P-450, Lipids and Apolipoproteins. Microsomal enzyme inducers increase P-450 concentrations in human liver. Serum HDL-C and apo A-I levels (Figs. 1 and 2) and the HDL-C/LDL-C and apo A-I/A-II ratios are directly proportional (Luoma et al. 1979, 1982; Luoma 1986), and the LDL-C levels and the LDL-C/HDL-C ratios inversely proportional (see Luoma 1986) to P-450 concentrations in human liver.

Antipyrine Kinetics, Lipids and Apolipoproteins. The serum HDL-C and apo A-I levels and the HDL-C/T-C and HDL-C/LDL-C ratios are directly compatible, and the LDL-C/HDL-C ratios are indirectly compatible with the antipyrine clearance rate (see Luoma 1986). In one study a low serum LDL-C level was typical of subjects with a rapid antipyrine clearance rate, i.e. a strong induction. In another study the decrease in raised HDL-C after alcohol withdrawal paralleled the prolongation of the alcohol-caused shortening of the antipyrine half-life.

Liver Phospholipids and Proteins (Fig. 2) are directly, and liver triglycerides are indirectly proportional to HDL-C and apo A-I levels and the HDL-C/T-C ratios (Luoma et al. 1982). The apo A-I/A-II ratio also parallels liver protein concentrations.

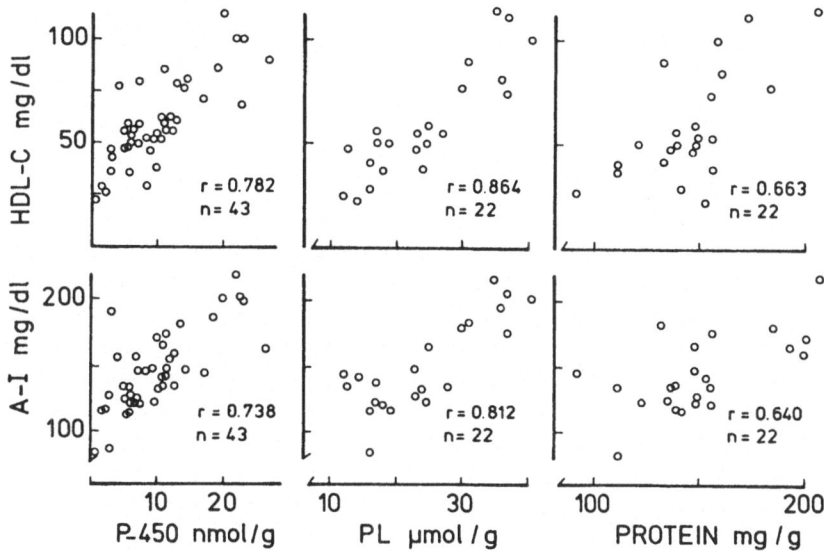

Fig. 2. Relation of plasma HDL-C and apo A-I to liver cytochrome P-450, phospholipid (PL) and protein in patients, some of whom were treated with enzyme-inducing drugs (Luoma et al. 1982)

6 Enzyme Inducers and Atherosclerosis

Experimental studies indicate that phenobarbital retards the cholesterol accumulation in the arterial wall and the formation of atherosclerotic plaque (Salvador et al. 1970). In man, a recent analysis of the causes of death of 1399 epileptic subjects who had been treated with common and anticonvulsant drugs showed that their mortality in ischemic heart diesease was about 30% lower than that or the age and sex-matched general population (Muuronen et al. 1985). In subjects regularly using alcohol, microsomal induction may play a role both in the HDL elevation, and the low incidence of coronary deaths.

7 Discussion

Enzyme inducers probably increase serum HDL levels by enhancing the production of apo A-I and HDL. This possibility is supported by their inducing effect on hepatic protein and apo A-I mRNA synthesis (Chao et al. 1985). How various microsomal inducers influence the catabolism of triglyceride-rich lipoproteins and HDL apolipoproteins, and the intestinal function contributing to serum HDL levels is not yet activity often.

The hepatic microsomal fully understood decreases and the atherogenity of serum lipoproteins (the LDL/HDL ratio) often increases with age. Enzyme inducers can prevent these phenomena, and could hence retard the atherosclerotic vascular process in man.

In addition to drugs, dietary factors and physical exercise stimulate hepatic microsomal function and may hence favourably influence the serum lipoprotein profile.

8 Summary

Drugs such as phenytoin, phenobarbital, carbamazepine and gluthethimide, and alcohol, which induce hepatic microsomal enzymes, increase serum HDL-C and apolipoprotein levels. The inducers increase the hepatic microsomal enzyme and apo A-I mRNA, and the protein and phospholipid concentrations. The increases in serum the HDL-C, apo A-I and HDL-C/LDL-C ratio parallel liver protein and phospholipid concentrations, and microsomal enzyme activity as assessed by hepatic cytochrome P-450 or the antipyrine kinetics. Experimentally, phenobarbital has inhibited the formation of atherosclerotic plaque. In man, the use of enzyme inducers, drugs or alcohol, has been associated with a low mortality rate resulting from ischemic heart disease.

References

Carlson LA, Kolmodin-Hedman B (1972) Hyper-α-lipoproteinaemia in men exposed to chlorinated hydrocarbon pesticides. Acta Med Scand 192:29–32

Chao YS, Pickett CB, Yamin TT, Guo LS, Alberts AW, Kroon PA (1985) Phenobarbital induces rat liver apolipoprotein A-I mRNA. Mol Pharmacol 27:394–398

Durrington PN (1979) Effect of phenobarbitone on plasma apolipoprotein B and plasma high-density lipoprotein cholesterol in normal subjects. Clin Sci 56:501–503

Greim HA (1981) An overview of the phenomena of enzyme induction: their relevance to drug action and interactions. In: Jenner B, Testa B (eds) Concepts in drug metabolism, Pt B. Dekker, New York, pp 219–263

Luoma PV (1986) Enzyme inducers. In: Fears R (ed) Pharmacological control of hyperlipidaemia. Prous, Barcelona, pp 365–376

Luoma PV, Pelkonen RO, Sotaniemi EA (1979) Plasma high-density lipoprotein cholesterol and hepatic drug metabolizing enzyme activity in man. Acta Physiol Scand Suppl 473:226A

Luoma PV, Sotaniemi EA, Pelkonen RO, Savolainen MJ, Ehnholm C (1982) Induction and lipoproteins. Lancet 1:625

Muuronen A, Kaste M, Nikkilä EA, Tolppanen E-M (1985) Mortality from ischaemic heart disease among patients using anticonvulsive drugs: case-control study. Br Med J 291:1481–1483

Nikkilä EA, Taskinen M-R, Jane T (1987) Plasma high-density lipoprotein concentration and subfraction distribution in relation to triglyceride metabolism. Amer Heart J 113:543–548

Salvador RA, Atkins C, Haber S, Kozma C, Conney AH (1970) Effect of phenobarbital on the development of atheromatosis in the cholesterol-fed rabbit. Biochem Pharmacol 19:1975–1981

Gonadal Steroids and Hormone Antagonists

D. Crook [1], I. F. Godsland [1], J. C. Montgomery [2], and V. Wynn [1]

1 Introduction

Continued investigation into the role of plasma lipoproteins in the development of atherosclerosis has led to increased interest in their interrelationships with gonadal hormones.

2 Oestrogens

Whereas cross-sectional studies have indicated only a weakly positive correlation between plasma oestradiol and HDL levels, there is a well-documented increase in HDL- (and in particular, HDL2-) cholesterol in response to the administration of exogenous oestrogen, both natural and synthetic. Kinetic studies of HDL metabolism in pre-menopausal women (Schaefer et al. 1983) have shown that the rise in HDL concentrations is associated with an increased rate of synthesis of its major protein component, apolipoprotein AI. In addition, post-menopausal women taking exogenous oestrogens have been shown to have higher levels of HDL-cholesterol than their age-matched non-user controls.

The actions of orally-administered oestrogens differ markedly according to their natural or synthetic origin, particularly in respect to their effects on plasma triglycerides. Synthetic oestrogens (e.g. ethinyl oestradiol) markedly increase plasma triglyceride concentrations, whereas natural oestrogens (such as 17-beta oestradiol) appear to have no effect (Buckman et al. 1980; Lobo et al. 1980).

Because changes in lipoprotein concentrations may be caused by the artificially-high intra-hepatic concentrations resulting from the oral route of administration attention has been focussed on parenterally administered oestrogen. Subcutaneous implants of natural oestradiol have been reported to increase plasma HDL concentrations in some, but not all, studies.

---

1 The Cavendish Clinic, 21 Wellington Rd, London, Great Britain
2 Dulwich Hospital Menopause Clinic, London, Great Britain

Drugs Affecting Lipid Metabolism
Ed. by R. Paoletti et al.
© Springer-Verlag Berlin Heidelberg 1987

Table 1. Effects of paranterally-administered natural oestrogen in nine premenopausal women

	Baseline	Post-implant
	mean ± SD	
Oestradiol (pmol l^{-1})	449 ± 158	859 ± 433**
SHBG[a] (nmol l^{-1})	54 ± 22	61 ± 18
Plasma chol (mg dl^{-1})	164 ± 20	178 ± 48
Plasma trig (mg dl^{-1})	55 ± 18	75 ± 21
HDL chol (mg dl^{-1})	69 ± 20	76 ± 21*
HDL2 chol (mg dl^{-1})	23 ± 12	31 ± 9*
HDL3 chol (mg dl^{-1})	46 ± 16	45 ± 16
LDL chol (mg dl^{-1})	83 ± 23	87 ± 27

[a] SHBG = Sex hormone binding globulin.
* $P < 0.05$ (Wilcoxon paired-sign ranks rest); ** $P < 0.01$.

We have recently studied a group of premenopausal women given subcutaneous
implants of 17-beta oestradiol as treatment for the premenstrual syndrome (Magos
et al. 1986). Nine women (mean age 38 years, range 31–43 years) were studied both
before and 6 weeks after implantation of a 100-mg oestradiol pellet in the lower
abdominal wall. No other hormones or drugs were administered during this period.

Plasma HDL-, HDL2- and HDL3-cholesterol concentrations were measured after
precipitation with heparin/manganese and dextran sulphate. LDL-cholesterol concentrations were estimated by the Friedewald formula. The results of this study are
summarised in Table 1.

The mean plasma oestradiol level in this group doubled in response to the implant,
while still remaining in the normal range for premenopausal women. There was no
change in SHBG concentrations, in contrast to the striking elevations seen when
oestrogen is administered by oral routes (Lobo et al. 1980). The non-significant
increases seen in the mean plasma cholesterol and triglyceride levels reflect the
moderately hyperlipidaemic response of a single subject; there were no changes in the
values obtained with all other subjects. Plasma HDL concentrations were increased
by oestrogen administration and this rise was accounted for by the HDL2 subclass
alone. In this group of healthy premenopausal women, LDL-cholesterol concentrations were unaffected by oestrogen.

3 Androgens

Weakly positive correlations between plasma testosterone and HDL levels have been
seen in some, but not all, cross-sectional studies. Conversely, male puberty is associated
with a marked decrease in HDL levels, and HDL levels may be high in castrated males.

Administration of methyl testosterone lowers HDL levels. Other steroids possessing androgenic activity (e.g. certain progestins and anabolic steroids) also lower HDL levels. However, remarkably little is known about the effects of natural testosterone on HDL metabolism in either sex.

4 Anti-Oestrogens

According to standard biological assessment techniques, many of the progestins used in the combined oral contraceptive (OC) pill possess anti-oestrogenic activity. Analysis of the effects of combined OC on HDL metabolism indicates that such progestins inhibit the rise in HDL concentrations that would occur with oestrogen alone. If the androgenic strength of the progestin is sufficient, HDL cholesterol levels may even be lowered despite the presence of oestrogen (Wynn and Nithythananthan 1982).

5 Anti-Androgens

Although the relationships between the androgenicity of a compound and its effects on serum lipoproteins are poorly characterised, it might be predicted that an anti-androgen would raise HDL- (and especially HDL2-) cholesterol concentrations and lower those of LDL. However, the results of studies performed by our group (Wynn et al. 1986) on the potent anti-androgen *cyproterone acetate* (CA) conflict with such assumptions.

We studied 63 women (mean age 27 years) who were being treated for hirsutism and severe acne. These subjects received an oral dose of CA (either 50 or 100 mg) in combination with 50 μg synthetic oestrogen (EO). Serum lipoproteins were evaluated both before and 2–6 months after commencing therapy. In order to distinguish the effects of these two steroids on lipoprotein metabolism, two further groups were similarly studied: one comprised of 13 women taking CA alone and one of 11 women taking EO alone. The results of this study are summarised in Table 2.

Triglyceride concentrations were unchanged by CA alone but were significantly increased by EO both alone and in combination with CA. Total cholesterol concentrations were reduced by CA alone but not in combination with EO or by EO alone. HDL-cholesterol concentrations were unchanged by the CA and EO combination but were increased by EO alone and decreased (although not significantly) by CA alone. These opposite effects of individually-administered CA and EO on HDL-cholesterol were restricted to the HDL2 subclass. Interestingly, HDL2-cholesterol concentrations were reduced by the CA and EO combination, although the total HDL-cholesterol concentrations were maintained by a compensatory increase in HDL3-cholesterol. LDL-cholesterol concentrations were reduced by both steroids alone and in combination.

In summary, although the anti-androgen cyproterone acetate was shown to have an oestrogen-like ability to reduce LDL-cholesterol concentrations, paradoxically it reduced HDL2-concentrations and was even able to overcome the oestrogen-related increase in this metabolically important HDL subclass.

Table 2. Serum lipoprotein responses to cyproterone acetate (CA) and synthetic oestrogen (EO)

Treatment group		Trlg	Chol	HDL	HDL2	HDL3	LDL
					$(mg\ dl^{-1})$		
CA + EO (n=63)	OFF	81	185	58	23	35	109
	ON	115	181	57	18	39	99
		***			*	*	*
CA (n=13)	OFF	86	197	57	20	37	125
	ON	88	176	51	14	37	105
			**		*		*
EO (n=11)	OFF	87	220	63	21	42	136
	ON	119	217	79	34	45	113
		*		**	**		**

* $P < 0.05$; ** $P < 0.01$; *** $P < 0.001$.

References

Buckman MT, Johnson J, Ellis H, Srivastava L, Peake G (1980) Metabolism 29:803–805
Lobo RA, March CM, Goebelsmann U, Krauss RM, Mishell DR, Jr (1980) Am J Obstet Gynecol
    138:714–719
Magos AL, Brincat M, Studd JWW (1986) Br Med J 292:1629–1633
Schaefer EJ, Foster DM, Zech LA, Lindgren FT, Brewer HB, Jr, Levy RI (1983) J Clin Endo
    Metab 57:262–267
Wynn V, Niththyananthan R (1982) Am J Obstet Gynecol 142:766–772
Wynn V, Godsland IF, Seed M, Jacobs HS (1986) Clin Endocrinol 24:183–191

The Effects of Adrenoceptor and Adrenergic Blocking Drugs on Plasma Lipoproteins

A. Lehtonen [1]

1 Adrenoceptors

The catecholamines synthesized and secreted by the adrenal medulla regulate a wide spectrum of metabolic processes. Glugoneogenesis, glycogenolysis and lipolysis, for example, are strongly influenced by catecholamines.

In human fat cells, activation of α-receptors inhibits, but activation of β-receptors stimulates, cyclic AMP accumulation and lipolysis. The overall metabolic responsiveness to catecholamines may depend not only in the regulation of the β-receptor mediated effects, but also on the degree of stimulation of the α-receptors. Under most normal conditions, however, the concomitant stimulation by catecholamines of inhibitory α-receptors seems to be of minor importance for overall lipolytic reaction.

The main function of the adipose tissue is to supply fatty acids to be utilized by different peripheral tissues, and glycerol, which is a gluconeogenic substrate in the liver. Catecholamine stimulation of hormone-sensitive lipase and lipolysis in man appears to be mediated by β_1 and β_2 adrenoceptors. Adrenoceptor blockade might therefore be expected to inhibit lipolysis and reduce free fatty acid concentrations.

Epidemiologic evidence indicates that hypertension is one of the strongest among several risk factors for coronary heart disease (CHD). A possible reason for the difficulty in demonstrating benefit with regard to CHD mortality through treatment of hypertension is that current treatment regimes aggravate one or more risk factors for CHD in the process of lowering blood pressure. The purpose of this chapter is to review the literature on the effect of adrenergic blocking drugs on plasma lipids.

1.1 Non-Selective Beta-Blockade

Total plasma cholesterol and triglycerides increased significantly during 12 months' sotalol therapy (Lehtonen and Viikari 1979). Total plasma cholesterol increased by 17% and plasma triglycerides by about 66% during 1 year. Plasma HDL-cholesterol decreased by 26% at the same time. The ratio of HDL-cholesterol to total cholesterol decreased during treatment from 0.28 to 0.18.

1 Department of Medicine, Turku City Hospital, Kunnallissairaalantie 20, SF-20700 Turku, Finland

Drugs Affecting Lipid Metabolism
Ed. by R. Paoletti et al.
© Springer-Verlag Berlin Heidelberg 1987

Most studies have been of propranolol, but they are short-term or medium-term studies from 4 weeks to 6 months. The general finding is that the treatment with propranolol increases plasma triglyceride concentration (Day et al. 1982). The changes in total plasma cholesterol and LDL-cholesterol concentrations are small, but the plasma HDL-cholesterol level seems to decrease significantly during treatment with propranolol (Day et al. 1982).

1.2 Selective Beta-Blockade

The concentrations of plasma triglycerides and total cholesterol increased slightly during 6-month atenolol treatment (Lehtonen and Marniemi 1984). The level of HDL-cholesterol during atenolol treatment was significantly lower than before it, and this decrease was especially due to the decrease of HDL_2-cholesterol concentration. The level of HDL_3-cholesterol decreased slightly.

Both metoprolol and atenolol have been reported to increase plasma triglyceride concentration, whereas plasma cholesterol levels have remained mostly unchanged. Both metroprolol and atenolol have reduced HDL-cholesterol in many but not in all studies (Välimäki et al. 1986).

1.3 Beta-Blockade and Intrinsic Sympathomimetic Activity (ISA)

The concentration of serum triglycerides and total cholesterol remaind approximately constant during 12-month treatment of hypertension with pindolol (Lehtonen et al. 1982). The concentration of HDL-cholesterol was increased after the first month of therapy. The ratio of HDL-cholesterol to total cholesterol increased from 0.18 to 0.20 during therapy.

There were no significant changes in the concentrations of total plasma cholesterol, LDL-cholesterol, VLDL-cholesterol and triglycerides during acebutolol treatment (Lehtonen 1984). The level of HDL-cholesterol decreased slightly but not significantly.

Penbutolol with partial agonist activity decreased slightly serum HDL-cholesterol, especially HDL_3-cholesterol concentration (Välimäki et al. 1986).

1.4 Combined Alpha-Beta Blocker

There is controversy about whether labetalol has any effect at all on serum lipids, but a slight fall in HDL-cholesterol during treatment with labetalol has been found in a study (Table 1).

1.5 Combination of Diuretics with Beta Blockers

The principal effect of the thiazide diuretics on lipoproteins is to increase both total plasma cholesterol and total plasma triglyceride concentrations. The possible interaction between beta blockers and diuretics are interesting because the combination

Table 1. Percentage changes in total serum cholesterol, LDL-cholesterol, HDL-cholesterol and triglycerides during β-blocker and α-blocker monotherapy according to different studies

	Duration (months)	Total cholesterol		LDL cholesterol		HDL cholesterol		Triglycerides	
β-blocker									
Non-selective	1 – 12	–9	+16	–6	+32	–17	–26	+10	+66
Selective	1 – 6	–1	+ 8	–2	+ 5	+ 1	–13	+14	+62
With ISA	1 – 12	–5	+ 5	–3	– 7	– 7	+20	– 9	+28
(pindolol, acebutolol)									
α-blocker	2 – 7	–1	– 9			– 4	+11	+ 1	–17

therapy is usual. The increase in LDL-cholesterol seen with diuretic therapy were prevented or reversed by the concomitant administration of a beta blocker (Sihiff et al. 1982). A significant increase in serum LDL-cholesterol that occurred during monotherapy with clopamide was no longer present at 4 weeks following the addition of the beta blocker pindolol.

1.6 Alpha-Blocking Agents

Alpha adrenoceptor blockade with prazosin results in a fall in total serum cholesterol and serum triglyceride concentrations and an improvement in the HDL/total cholesterol ratio (Leren et al. 1980). During the treatment with prazosin, the level of serum HDL cholesterol tends to increase slightly.

The total plasma cholesterol decreased 9% and LDL-cholesterol 17% after 20 weeks' treatment with doxazosin, a new vasodilator of the postsynaptic alpha-blocking series descended from prazosin (Lehtonen et al. 1986). Total HDL-cholesterol and HDL$_2$-cholesterol concentrations increased slightly during doxazosin treatment. The ratio HDL/total cholesterol increased significantly during doxazosin treatment.

2 Discussion

At the present stage any of the lipoprotein changes induced by antihypertensive drugs should be categorized as biochemical side effects. The changes in plasma lipoproteins, if sustained over the long term, may influence the overall cardiovascular risk of patients receiving antihypertensive care.

Little is known of how antihypertensive agents influence lipoprotein metabolism (Sacks and Dzau 1986). It is postulated that alpha-adrenergic stimulation may have an important role in suppressing adipose tissue lipase with a secondary reduction in plasma HDL-cholesterol and raising triglycerides. In most studies, however, no effect on plasma post-heparin lipase activity with beta-blocker therapy have been found. Prazosin, on the other hand, increases extrahepatic plasma lipoprotein lipase. Epinephrine increases synthesis of lipoprotein lipase by adipocytes. Therefore, adrenergic agents may influence lipoprotein lipase activity by regulating its synthesis by these tissues. Net alpha inhibition and beta stimulation of lipase could explain why prazosin lowers plasma triglyceride levels and raises high-density lipoprotein levels, whereas beta blockers produce the opposite effects.

The blood levels of glucose and insulin represent, besides free fatty acids, the most important components for hepatic lipoprotein production. The serum insulin levels have been reported to be unaltered during beta-blocker therapy by most studies, but there are reports on impaired glucose tolerance.

The unesterified high-density lipoprotein cholesterol is esterified by lecithin cholesterol acyl transferase (LCAT) and the resulting cholesterol esters move from the surface to the core of the high-density lipoprotein particle. The activity of LCAT is decreased by beta blockers without ISA, but this activity is increased by beta blockers with ISA (Lehtonen et al. 1982), which may contribute to the decrease in high-density lipoprotein found in patients who receive propranolol and to the increase in high-density lipoprotein cholesterol in patients receiving pindolol.

The role of the drug-induced effects on serum lipoproteins in the pathogenesis of coronary heart disease is unproved. Long-term studies are required in which the effect of an antihypertensive that is known not to produce deleterious changes in lipoprotein levels is compared with one that does affect lipoproteins.

References

Day JL, Metcalfe J, Simpson CN (1982) Adrenergic mechanisms in control of plasma lipid concentrations. Br Med J 284:1145–1148

Lehtonen A (1984) The effect of acebutolol on plasma lipids, blood glucose and serum insulin levels. Acta Med Scand 216:57–60

Lehtonen A, Marniemi J (1984) Effect of atenolol on plasma HDL-cholesterol subfractions. Atherosclerosis 51:335–338

Lehtonen A, Viikari J (1979) Long-term effect of sotalol on plasma lipids. Clin Sci 57 (Suppl): 405–407

Lehtonen A, Hietanen E, Marniemi J, Peltonen P, Niskanen J (1982) Effect of pinodolol on serum lipids and lipid metabolizing enzymes. Br J Clin Pharmacol 13 (Suppl 2):445–447

Lehtonen A, Himanen P, Saraste M, Niittymäki K, Marniemi J (1986) Double-blind comparison of the effects of long-term treatment with doxazosin or atenolol on serum lipoproteins. Br J Clin Pharmacol 21:77–81

Leren P, Helgeland A, Holme I, Foss OP, Hjermann I, Lund-Larsen PG (1980) Effect of propranolol and prazosin on blood lipids. The Oslo study. Lancet ii:4–6

Sacks FM, Dzau VJ (1986) Adrenergic effects on plasma lipoprotein metabolism. Speculation on mechanisms of action. Am J Med 80 (Suppl 2A):71–81

Sihiff H, Weidmann P, Mordasini R, Riesen W, Bachmann C (1982) Reversal of diuretic-induced increases in serum low-density lipoprotein cholesterol by the beta-blocker pindolol. Metabolism 31:411–416

Välimäki M, Maass L, Harno K, Nikkilä EA (1986) Lipoprotein lipids and apoproteins during beta-blocker administration: comparison of penbutolol and atenolol. Eur J Clin Pharmacol 30: 17–20

The Influence of Terpenes and Analogues on Parameters of Atherosclerosis

D. A. WHITE [1], G. D. BELL [2], F. CACCIAGUERRA [1], and B. MIDDLETON [1]

Rowachol, a proprietary mixture of the six monoterpenes menthol, menthone, borneol, cineole, camphene and pinene in olive oil, reduced the lithogenicity of bile (Doran et al. 1979) and caused dissolution of cholesterol gallstones in man (Bell and Doran 1979). Daily administration of the drug (6–9 capsules) to 16 patients for periods of 2–28 weeks gave a progressive increase in serum HDL-cholesterol concentration with no accompanying change in the concentrations of serum total cholesterol or triglyceride (Bell et al. 1980) (Fig. 1). Similar effects in humans have been shown recently by Leiss and von Bergmann (1985) who administered Rowachol (200 mg t.i.d) to young healthy volunteers and measured a significant increase in the ratio of HDL-cholesterol to total cholesterol in serum.

Such changes suggested an alteration in cholesterol metabolism and administration of menthol or cineole to rats caused a 70% inhibition of the activity of hepatic HMGCoA reductase measured 17 h after dosing in vivo. This inhibition of enzyme activity correlated well with the inhibition of C_2-flux from acetate into non-saponifiable lipid (Clegg et al. 1980). However, no direct effect on HMGCoA reductase was seen on direct addition of the terpene to the in vitro assay system. Loss of activity in vivo appeared to be due to a decrease in the amount of enzyme and possibly some "crippled enzyme" (Clegg et al. 1982).

3,3,5-Trimethylcyclohexanol (TMC) shows some structural similarities to the monoterpenes and also causes inhibition of rat hepatic HMGCoA reductase in vivo (B. Middleton et al. 1983). It is a potent choleretic at low doses, while at higher doses it causes a decrease in biliary cholesterol output and a reduction in the cholesterol saturation index of bile (Bell et al. 1984). The drug cyclandelate (Cyclospasmol: Gist Brocades nv. Delft, Holland) is the mandelic acid ester of TMC and a single dose (3 mmol kg^{-1}) caused a 50% inhibition of rat hepatic HMGCoA reductase activity (B. Middleton et al. 1983). This was confirmed as an inhibition of hepatic sterologenesis and lipogenesis in vivo in rats which had been injected with ^3H_2O after an oral dose of cyclandelate (A. Middleton et al. 1983). Administration of cyclandelate also depressed the accumulation of newly synthesized sterol into serum with a 47% decrease in the accumulation of ^3H-sterol into the LDL but not the HDL fraction.

---

1 Department of Biochemistry, University of Nottingham Medical School, Queen's Medical Centre, Nottingham NG7 2UH, England
2 Department of Medicine, The Ipswich Hospital, Ipswich, Suffolk IP4 5PD, England

Drugs Affecting Lipid Metabolism
Ed. by R. Paoletti et al.
© Springer-Verlag Berlin Heidelberg 1987

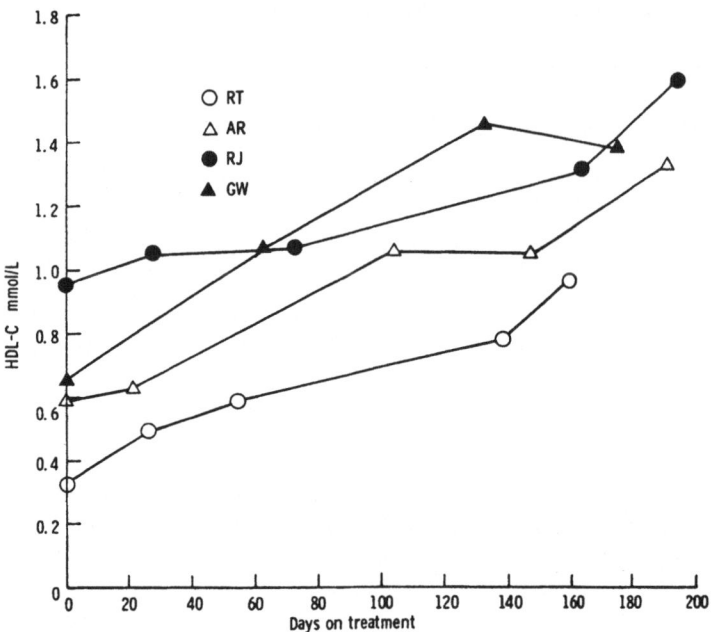

Fig. 1. The effect of Rowachol on the concentration of serum high density lipoprotein cholesterol in four individuals studied over periods in excess of 140 days (Bell et al. 1980)

In view of this inhibition of hepatic sterol synthesis and transport into serum in vivo, the effects of dietary cyclandelate on experimental atherosclerosis in rabbits were studied (Middleton et al. 1984). Rabbits were given a diet containing 1% (w/w) cholesterol for 7 weeks to initiate atherosclerosis and transferred to a low cholesterol diet with or without 0.5% (w/w) cyclandelate for 10 weeks. There was a significant increase in aortic cholesterol in all animals transferred to the low cholesterol diet but inclusion of cyclandelate decreased both the extent and severity of plaque growth. Thus, total cholesterol and calcium of aortae of cyclandelate-treated animals were decreased by 49% and 69% of control values respectively. These effects occurred in the absence of any differences from control values of total or HDL-cholesterol or in the cholesterol content of the liver and suggest a direct effect of the drug or its metabolites on the aorta.

Cyclandelate caused a dose-dependent inhibition (IC_{50} :150 μM) of ^{14}C-acetate but not mevalonate incorporation into sterol in rat hepatocytes maintained in monolayer culture, and suggested a point of inhibition at HMGCoA reductase or earlier in sterol synthesis. Similar inhibitions were noted in human monocytes and skin fibroblasts. Several other aspects of cellular cholesterol metabolism were also affected in human fibroblasts (Table 1). Thus, cholesterol esterification was inhibited by 89% within 3-h exposure to the drug with both uptake and hydrolysis of LDL-borne cholesteryl ester inhibited by 67% after 17 h. In fibroblasts, from four normal individuals, precultured with lipoprotein-deficient serum to increase surface LDL receptor number and then cultured for 17 h with LDL, cyclandelate inhibited the rate of cholesterol esterification by 93% (control 1409 ± 205 pmol h^{-1} mg^{-1}; cyclandelate treated 105 ± 80 pmol h^{-1}

Table 1. The effect of cyclandelate on cholesterol metabolism in human skin fibroblasts (Middleton et al. 1984)

Process	Control	Cyclandelate	% inhibition
		(means ± SD)	
[1–^{14}C]acetate to cholesterol (pmol h^{-1} mg^{-1})	3.4 ± 1.1	1.2 ± 0.4	65
[1–^{14}C]acetate to fatty acid	435.0 ± 166.0	391.0 ± 186.0	10
[^{3}H]rec CL-LDL[a] internalization (pmol h^{-1} mg^{-1})	1100.0 ± 180.0	310.0 ± 70.0	67
[^{3}H]rec CL-LDL[a] hydrolysis (pmol h^{-1} mg^{-1})	688.0 ± 73.0	210.0 ± 6.0	67
[^{3}H]oleate to cholesteryl ester	233.0 ± 25.0	25.0 ± 1.0	89

[a] [^{3}H]rec CL-LDL is human LDL extracted with heptane to remove endogenous cholesterol and cholesteryl ester and reconstituted with [^{3}H]cholesteryl linoleate.

mg^{-1}). Similar degrees of inhibition of cholesterol esterification rates in the presence of cyclandelate were seen in a macrophage cell line (J774), human monocytes cultured with acetyl-LDL and pig aorta slices.

These effects suggested an action of the drug on binding and internalization of LDL or a direct effect on intracellular ACAT activity. In the experiments where HMGCoA reductase was inhibited by in vivo administration of cyclandelate no reduction in ACAT activity was seen. However, addition of cyclandelate to in vitro assays of rat hepatic ACAT caused a dose-dependent and reversible inhibition of the enzyme with 70% inhibition at 100 μM (Table 2). The lack of inhibition seen after in vivo administration of the drug reflects the ease with which it is hydrolyzed in the liver and the reversibility of its action, being removed during washing of the microsomal fraction. Surprisingly TMC, the hydrolysis product, was much less inhibitory than the parent drug and the only monoterpene to show appreciable inhibition of ACAT was menthol. The ability

Table 2. The effect of terpenes and cyclandelate on rat liver microsomal ACAT activity[a]

Compound administered	ACAT activity (pmol ester min^{-1} mg^{-1} protein)	% inhibition
Control	177 ± 5	–
Menthol	121 ± 9	32
Pinene	167 ± 12	4
Menthone	190 ± 9	0
Borneol	161 ± 9	9
Camphene	168 ± 10	5
Cineole	172 ± 7	3
Rowachol	122 ± 4	31
TMC	133 ± 10	25
Cyclandelate	46 ± 8	74

[a] The assay system contained the drugs at a final concentration of 100 μM.

of cyclandelate to inhibit directly ACAT in microsomal fractions prepared from J774 cells as well suggests that this might be a general effect of the drug.

Such an inhibition of cellular ACAT would explain many of the effects previously described for terpenes and cyclandelate in particular. A direct effect on hepatic or extra-hepatic ACAT would lead to an increase in intracellular free cholesterol and decrease cholesterogenesis by inhibiting the synthesis of HMGCoA reductase. The rise in free cholesterol in fibroblasts would also lead to down-regulation of the expression of cell surface LDL receptors and decrease the uptake and hydrolysis of the LDL-borne cholesteryl ester described earlier. However, how such actions relate to the empirical observation on the rise in HDL-cholesterol following terpene therapy is unclear.

References

Bell GD, Doran J (1979) Gallstone dissolution in man using an essential oil preparation. Brit Med J 1:24

Bell GD, Bradshaw JP, Burgess A, Ellis W, Hatton J, Middleton A, Middleton B, Orchard T, White DA (1980) Elevation of serum high density lipoprotein cholesterol by Rowachol, a proprietary mixture of six pure monoterpenes. Atherosclerosis 36:47–54

Bell GD, Clegg RJ, Ellis WR, Middleton B, White DA (1984) The effects of 3,5-5-trimethylcyclohexanol on hepatic cholesterol synthesis, bile-flow and biliary lipid secretion in the rat. Br J Pharmacol 81:183–187

Clegg RJ, Middleton B, Bell GD, White DA (1980) Inhibition of hepatic cholesterol synthesis and S-3-hydroxy-3-methylglutaryl-CoA reductase by mono- and bicyclic monoterpenes administered in vivo. Biochem Pharmacol 29:2125–2127

Clegg RJ, Middleton B, Bell GD, White DA (1982) The mechanism of cyclic monoterpene inhibition of hepatic 3-hydroxy-3-methylglutaryl coenzyme A reductase in vivo in the rat. J Biol Chem 257:2294–2299

Doran J, Keighley MRB, Bell GD (1979) Rowachol – possible treatment for cholesterol gallstones. Gut 20:312–317

Leiss O, Bergmann K von (1985) Effect of Rowachol on biliary lipid secretion and serum lipids in normal volunteers. Gut 26:32–37

Middleton A, White DA, Bell GD, Middleton B (1983) In vivo inhibition of hepatic lipogenesis in the rat by cyclandelate (3,3',5-trimethylcyclohexanylmandelate). Biochem Pharmacol 32:3079–3083

Middleton B, Middleton A, Miciak A, White DA, Bell GD (1983) The inhibition of hepatic S-3-hydroxy-3-methylglutaryl-CoA reductase by 3,3,5-trimethylcyclohexanol and its mandelic acid ester, cyclandelate. Biochem Pharmacol 32:649–651

Middleton B, Middleton A, White DA, Bell GD (1984) Cyclandelate decreases pre-established atherosclerosis in the rabbit. Atherosclerosis 51:171–178

Initial Stages of Reverse Transport of Cholesterol

D. Reichl [1]

It can be estimated that in man a minimum of 1 g cholesterol day^{-1} must be removed from peripheral tissues and conveyed ultimately to the liver for processing and/or excretion. Only a small proportion of cells of extrahepatic tissues is in direct contact with blood plasma that is the main route of transport to the liver; most of those cells are separated from the bloodstream by several barriers that do not favour free flow of fluids and some of them also restrict diffusion.

Studies in vitro have shown that firstly, cholesterol leaves cells only in its unesterified form (1). The propensity of cholesterol to leave cells is apparently governed by the chemical potential of unesterified cholesterol (UC) in plasma membranes. This in turn is dictated by the balance of the reactions forming the intracellular cycle of cholesterol esterification and hydrolysis of cholesteryl esters. Our studies suggest that in vivo the balance is shifted towards production of UC (see below).

The second general conclusion that can be made from experiments in vitro is that an acceptor of UC must be present in the medium surrounding cells so that sustained mass efflux can take place; plasma HDL and reconstituted complexes of phospholipids with apolipoprotein AI have been shown to be best acceptors of cholesterol. Mass efflux of cholesterol becomes possible when the chemical potential of UC in acceptors is lower than in plasma membranes. This is so irrespective of whether the transfer of cholesterol takes place in the course of a collision of the acceptor with the cell membrane or whether UC diffuses across a layer of unstirred water. The availability of acceptor particles may be a rate-limiting factor in vivo.

Any process that reduces the occupancy of the acceptor by UC is bound to promote efflux. For instance, esterification of cholesterol may promote cholesterol efflux. Studies in man do not support the suggestion that this esterification is rate limiting in vivo (see below).

In the whole body conditions given by the interstitium are superimposed on processes that govern cholesterol efflux in vitro. Some of those properties of the interstitium that may be relevant for reverse transport should be underlined. Firstly, the interstitium is an open system. Reactants, for instance lipoproteins, enter and reaction products leave the interstitium at all times. There is also an input of free energy from outside the system. Thus, the flow of lipoproteins in the intravascular compartment as well

1 Department of Medicine and Centre for Clinical Research, St. Bartholomew's Hospital, London, Great Britain

Drugs Affecting Lipid Metabolism
Ed. by R. Paoletti et al.
© Springer-Verlag Berlin Heidelberg 1987

as in the interstitium is maintained mainly by the working heart. It affects the rate of supply of acceptors and removal of reaction products from the surface of cells in tissues.

Secondly, lipoproteins and their constituents entering tissues from plasma must cross several barriers before they reach cell surfaces and the same is true for extravascular reaction products before they reach lymph or plasma.

Thirdly, the interstitium has sieving properties (1). Consequently, smaller lipoprotein particles have a larger volume of distribution and a higher probability to reach cell surfaces than larger particles.

Lastly, lipoproteins of human extravascular extracellular fluids of peripheral tissues differ in many respects from those of plasma.

Two sources of extravascular fluids are available: Firstly, peripheral prenodal lymph that can be obtained by cannulating a collecting trunk on the dorsum of the foot. Secondly, it is possible to obtain free fluid (with virtually no contamination by blood) in the course of a palliative operation in patients with primary lymphedema of their leg.

Evidence has been presented that in many respects lipoproteins of both extravascular fluids are similar, both in terms of their concentration and composition (2). Briefly, the concentration of apoAI-containing particles in either of the fluids is approximately 20% and of apoB-containing lipoproteins approximately 10% of that in plasma (2). Those apoAI-containing lipoproteins that stain with Sudan black are mainly larger than plasma HDL. By exclusion chromatography it can also be shown that apoAI-containing lipoproteins are present in extravascular fluids that are smaller than plasma HDL isolated by ultracentrifugation.

An important feature of extravascular fluids is the high concentration of UC relative to that of cholesteryl esters. When radiolabelled cholesterol is injected intravenously to volunteers and lymph cannulated at a time interval when specific activity of tissue cholesterol is higher than in plasma, it can be shown that specific activity of lymph cholesterol esters is virtually the same as that of plasma cholesterol but lymph UC specific activity of lymph UC is much higher than that of plasma, indicating that egress of cholesterol from cells in tissues takes place at a substantially higher rate than cholesterol esterification in the interstitium (1). This is so despite the circumstance that cholesterol esterifying activity can be demonstrated in lymph under appropriate conditions. If it were LCAT, its transferase activity is suppressed but its phospholipase A-like activity need not be affected (3).

The concentration of phospholipids in human lymph and lymphedema fluid is 25% of that in plasma. A closer analysis shows that extravascularly the ratio of lecithin over sphingomyelin is close to unity. It may be that some sphingomyelin is being secreted by cells of peripheral tissues. It is also possible that phospholipase A-like activity of extravascular LCAT is underlying the observed low L/S ratio.

Our studies show that most extravascular lipoproteins have a high ratio of polar/non-polar lipids. It may be due to this circumstance that the shape of lipoproteins when viewed by electron microscopy differs from the spherical appearance of plasma lipoproteins. Most extravascular lipoproteins appear square packing. Discoidal particles are seen infrequently (2). Figure 1 summarizes our findings. Their interpretation is speculative but each proposed step is substantiated by experimental evidence. The numbers in square brackets refer to the numbers in Fig. 1.

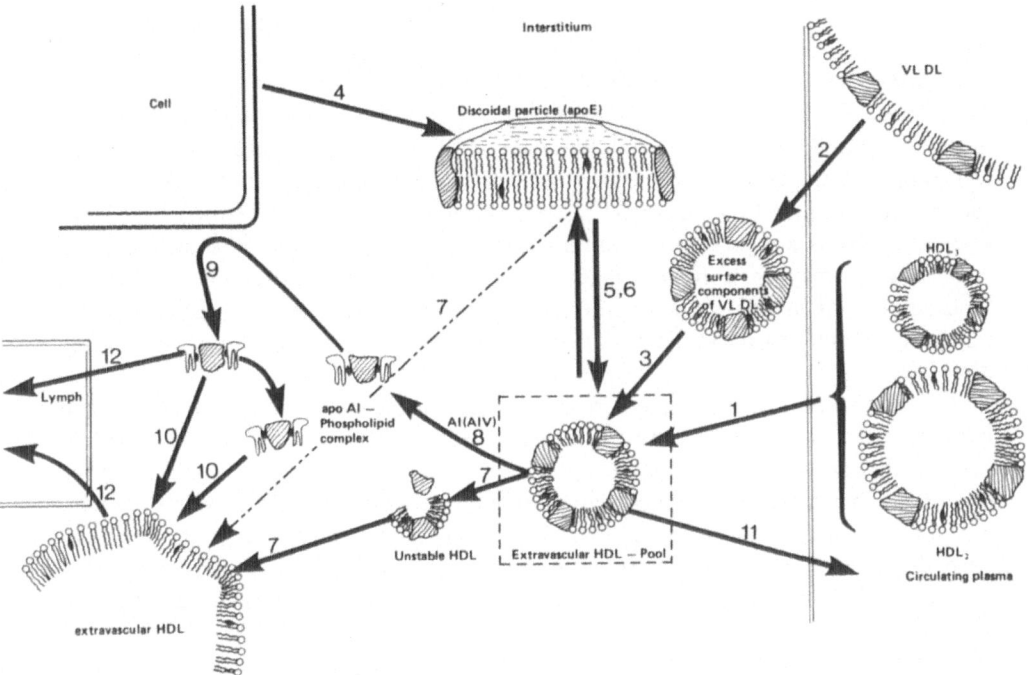

Fig. 1. Schematic representation of possible steps in extravascular metabolism of HDL. Figures refer to text below

HDL can enter the interstitium from plasma [1], possibly by vesicular transport across the endothelial cells. VLDL triglycerides are hydrolyzed at the luminal surface of endothelial cells or within transport vesicles while crossing the endothelium [2]. Excess surface components derived from VLDL may form particles in the interstitium [3]. The remaining part of VLDL including apoB contribute to IDL and LDL of extravascular fluids. Under the influence of excess phospholipid from VLDL [3] or from discoidal particles secreted by cells [4, 5] extravascular HDL may be destabilized. In man, transfer of phospholipids may be catalyzed by exchange proteins. Destabilized HDL can yield small apoAI or apoAIV phospholipid complexes [8] that can distribute in a large part of the interstitium. Due to their size, they have the highest probability to reach cell surfaces and act as primary acceptors of UC [9]. When the occupancy with cholesterol increases apoAI-phospholipid particles may aggregate. By increasing their size they decrease the probability of being at cell surfaces. They may fuse with large HDL particles [10]. The rate of supply of acceptors with low occupancy by cholesterol and that of removal of acceptors with high occupancy by cholesterol may be a rate-limiting factor for efflux of cholesterol in vivo. The remaining part of the destabilized extravascular HDL may fuse, forming the backbone of large HDL particles [7].

Complexes of apoE with phospholipids may be secreted by cells of peripheral tissues. Phospholipase A-like activity of LCAT may convert them into small spherical particles. Lipid exchange protein may transfer cholesterol esters from HDL to these particles in exchange for phospholipids. Resulting apoE-rich particles may fuse with the large HDL particles.

The large HDL are excluded from a part of the interstitium because of their sizes. The part of the interstitium in which they distribute may be characterized by relatively low resistance to fluid flow. Hence, the large particles may be carried either into lymphatic capillaries or dragged by osmotic flow through fenestrae in the wall of the venous portion of blood capillaries into the bloodstream. In terms of mass the large HDL may well represent the most efficient particles to transport unesterified cholesterol from the interstitium into flowing lymph [12] or the bloodstream [11]. Neither blood nor lymph capillaries are present in subintimal spaces of arteries. Here, the formation of large particles may be a disadvantage. Recent studies have revealed that in subendothelial spaces of arteries of experimental animals, large particles that contain predominantly UC start to accumulate soon after the onset of administration of atherogenic diets (4).

Two factors may enhance the formation of smaller from larger HDL. Firstly, when larger particles are formed in vitro from small HDL they become poorer in apoAI. It may be that an increased supply of apoAI would revert this formation. This may be one of the mechanisms by which the observed inverse relationship between the concentration in plasma of apoAI to the incidence of premature atherosclerosis may operate. Secondly, studies have shown that in the presence of excess of triglycerides and phospolipid emulsion (Intralipid) and lipid transfer protein, plasma HDL2 is converted to smaller particles. It is probably not possible to increase the availability of triglyceride-containing rich lipoproteins in subendothelial spaces but conditions might be found that would increase the local concentration of phospholipids. In view of the possible phospholipase A-like activity of LCAT it might be desirable to search for means to increase the concentration of sphingomyelin. Interestingly, one of such stimuli is the increased intracellular content of cholesterol (5).

References

1. Reichl D, Miller NE (1986) The anatomy and physiology of reverse cholesterol transport. Clin Sci 70:221–231
2. Reichl D, Forte TM, Hong J-L, Rudra DN, Pflug JJ (1985) Human lymphedema fluid lipoproteins: particle size, cholesterol and apolipoprotein distribution and electron microscopic structure. J Lipid Res 26:1399–1415
3. Nichols AV, Blanche PJ, Gong EL, Shore VG, Forte TM (1985) Molecular pathways in the transformation of model discoidal lipoprotein complexes induced by lecithin: cholesterol acyltransferase. Biochem Biophys Acta 834:285–300
4. Kruth HS (1985) Subendothelial accumulation of unesterified cholesterol – an early event in atherosclerotic lesion. Atherosclerosis 57:337–341
5. Portman OW, Alexander M (1972) Changes in arterial subfractions with aging and atherosclerosis. Biochem Biophys Acta 260:460–474

Lipolytic Enzymes and HDL: Influence of Drugs and Hormones*

M.-R. Taskinen and E. A. Nikkilä[1]

A large body of evidence indicates that both lipoprotein lipase (LPL) and hepatic lipase activities have an important role in the regulation of plasma HDL_2 and total HDL levels (1–4). During the lipolysis of triglyceride-rich particles by LPL-released surface components, i.e. phospholipids, apoproteins C's and E and free cholesterol, are transferred to HDL as demonstrated first by Patsch and co-workers (1). It is likely that both HDL_2 and HDL_3 are potential acceptors for the released surface material but the major acceptor is probably HDL_3. During this process there is a redistribution of HDL particles: the average density of HDL_3 decreases and may even reach the HDL_2 density range and particle size. If the lipolytic rate is slow as occurs in the presence of low LPL activity and/or substrate concentration, HDL_3 will incorporate the released surface components with a rise of particle density to less dense particles, i.e. light HDL_3. High LPL activity enhances the flux of the surface components to HDL_3 and consequently HDL_3 reaches the particle size of HDL_{2a} and HDL_{2b} (5). The complete conversion of HDL_3 to HDL_2 requires active LCAT reaction and an extra apo A–I molecule (5). We suggest that the degree of redistribution between HDL_3 and HDL_2 particles during lipolysis depends on the activity of the LPL system, on the pre-existing HDL particle distribution and the availability of apo A–I. Hepatic lipase is primarily a phospholipase which binds selectively to HDL_2 and hydrolyzes its phospholipids (6–8). Consequently, hepatic lipase is considered to have its major function in the catabolism of HDL_2 particles (4, 9).

Several factors alter the HDL, i.e. HDL_2 levels in consequence to changes in lipoprotein lipase and hepatic lipase activities. Examples of these are sex, physical activity, alcohol, obesity, dietary modification, drugs and hormones (4). The action of hormones on lipolytic enzymes provides a physiological model to study HDL metabolism.

We have recently studied the response of HDL subfractions and LPL during insulin treatment of type 2 diabetic patients with poor glycemic control (10). Intermittent insulin treatment aimed to normalize blood glucose and it was indeed followed by a reduction of diurnal blood glucose to near-normal levels. Elevated VLDL concentration fell dramatically during insulin therapy (the mean change being –60%). Lipoprotein lipase activity increased in the majority of the cases during insulin treatment. The mean enzyme activity was 2.3-fold higher after insulin than before it. The HDL_2 increased significant-

---

* This paper is dedicated to Esko A. Nikkilä, deceased September 21, 1986
1 Second and Third Departments of Medicine, University of Helsinki, SF-00290 Helsinki, Finland

ly during insulin therapy, whereas the HDL$_3$ fell. The opposite changes of HDL$_2$ and HDL$_3$ compensated each other and there was no change in total HDL concentration. The redistribution of particles within HDL subfractions was verified in zonal profiles of HDL subfractions. These concordant changes of LPL activity and HDL$_2$ during insulin therapy strongly imply that the observed rise in HDL$_2$ was due to the rapid intravascular lipolytic rate.

Another situation where HDL$_2$ and lipoprotein lipase activity increase concomitantly is the administration of prednisone. We have shown that the prednisone treatment to healthy men increases HDL$_2$ concentration already in 2 days. The change in HDL$_2$ was accompanied by an initial fall of HDL$_3$ level. However, the increase of HDL$_2$ was greater than the respective fall of HDL$_3$ and this resulted in a rise of total HDL level. After a few days of prednisone treatment the HDL$_3$ levels also returned to the baseline. Prednisone treatment was associated with a rise of lipoprotein lipase activity and again the change of HDL$_2$ can be explained by the rise of LPL.

We suggest that two types of changes in HDL can occur in relation to lipolysis. First, a redistribution of particles may occur within the HDL density range without any change in apo-A concentration. This process results in opposite changes in HDL subfractions as exemplified during insulin therapy. Secondly, there can be a real increase in HDL. This requires also an increase in the synthesis of apo-A. Consequently, there is a rise in both HDL subfractions. This picture is in keeping with the changes observed after prednisone administration.

Hepatic lipase activity is remarkably sex-steroid sensitive and therefore their effects on hepatic lipase provide a model to study the changes of HDL in relation to hepatic lipase activity (9, 10). Estradiol decreases the postheparin plasma hepatic lipase activity and causes an increase in HDL and in HDL$_2$ cholesterol. Levonorgestrel, a progestin with androgenic activity, increases postheparin plasma hepatic lipase activity and causes an opposite change, i.e. a fall in HDL$_2$ cholesterol. On the other hand, medroxyprogesterone acetate and desogestrel, both progestins with minor androgenic activity have no effect on either hepatic lipase activity or on HDL$_2$ cholesterol. Therefore, it is justified to suggest that the major effects of steroid hormones on plasma HDL (HDL$_2$) concentration are mediated by either induction or suppression of hepatic lipase activity (9, 10).

There is also a large body of evidence to indicate that at least clofibrate, bezafibrate and gemfibrozil increase postheparin plasma lipoprotein lipase activity (Fig. 1). Several

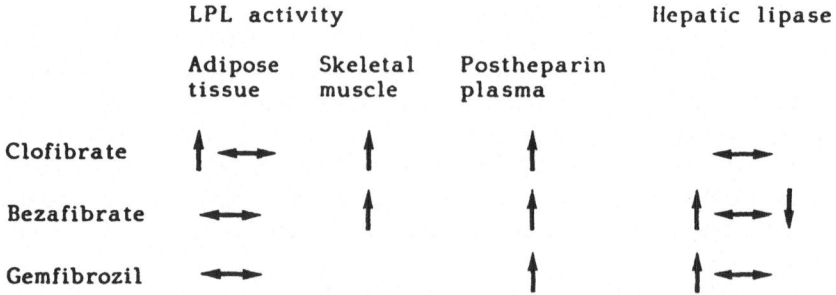

Fig. 1. Effect of fibrates on lipolytic enzymes

Table 1. Effect of bezafibrate for 2 months on HDL subfractions and lipolytic enzymes in hypertriglyceridemic men (n = 12)

	Before bezafibrate	After bezafibrate
HDL-cholesterol, mg dl^{-1}	28 ± 2	32 ± 1 [a]
HDL$_2$ mass, mg dl^{-1}	99 ± 6	105 ± 7
HDL$_3$ mass, mg dl^{-1}	148 ± 10	187 ± 8 [b]
Lipoprotein lipase activity		
Adipose tissue (μmol FFA g^{-1} h^{-1})	1.68 ± 0.36	1.33 ± 0.28
Skeletal muscle (μmol FFA g^{-1} h^{-1})	0.45 ± 0.08	0.62 ± 0.08 [b]
Postheparin plasma (μmol FFA ml^{-1} h^{-1})	22.0 ± 1.0	25.0 ± 1.2 [a]
Hepatic lipase activity (μmol FFA ml^{-1} h^{-1})	43.4 ± 4.3	37.5 ± 3.7 [b]

[a] $p < 0.05$; [b] $p < 0.01$.

studies indicate that this is due to the induction of skeletal muscle LPL, whereas adipose tissue enzyme activity is not influenced by these drugs (4, 12, 13). Consequently, the increase of HDL by these drugs has been linked at least partly to the induction of lipoprotein lipase. If this concept is valid, one would expect primarily a rise of HDL$_2$. We have recently studied the effect of bezafibrate on serum lipids and lipolytic enzymes in hypertriglyceridemic men. Bezafibrate increased significantly lipoprotein lipase activity in skeletal muscle and in postheparin plasma (Table 1). In contrast, there was no change in adipose tissue LPL activity. In addition, bezafibrate reduced hepatic lipase activity and increased significantly total HDL cholesterol. The change in total HDL seems to be mainly due to that of HDL$_3$ which increased significantly (Table 1), whereas there was no change in HDL$_2$. LPL activity in hypertriglyceridemic patients is generally low (4) and HDL particles are abnormal, representing "heavy" HDL$_3$ (14). We suggest that the modest elevation of LPL by bezafibrate may not enhance the lipolytic rate enough to induce the full conversion of HDL$_3$ to HDL$_2$. Instead, a redistribution of particles occurs within the HDL$_3$ density range resulting in the formation of "light" HDL$_3$. This concept may also explain why the data on the response of subfractions to bezafibrate have been inconsistent.

Gemfibrozil is another drug which clearly increases plasma HDL concentration. Nikkilä and co-workers (15) demonstrated several years ago that gemfibrozil increased postheparin plasma LPL activity. Therefore, one would again expect that the rise in HDL is primarily due to that of HDL$_2$. However, several reports indicate that both HDL$_2$ and HDL$_3$ subfractions increase during gemfibrozil treatment (16). Recently, Pasternack and co-workers (unpublished data) have studied the effects of gemfibrozil on HDL and lipolytic enzymes in patients with renal failure. Gemfibrozil increased significantly HDL cholesterol and postheparin plasma LPL activity. In addition, gemfibrozil therapy increased also postheparin plasma hepatic lipase activity. Saku and co-workers (17) have demonstrated that gemfibrozil stimulates also the synthesis of apo-A proteins. Therefore, the response of HDL subfractions to gemfibrozil seems to be influenced at least by three different mechanisms, which explains the rise of both HDL subfractions.

In conclusion, both lipoprotein lipase and hepatic lipase are important determinants of plasma HDL (HDL$_2$) concentration (Fig. 2). The changes of these enzyme activities

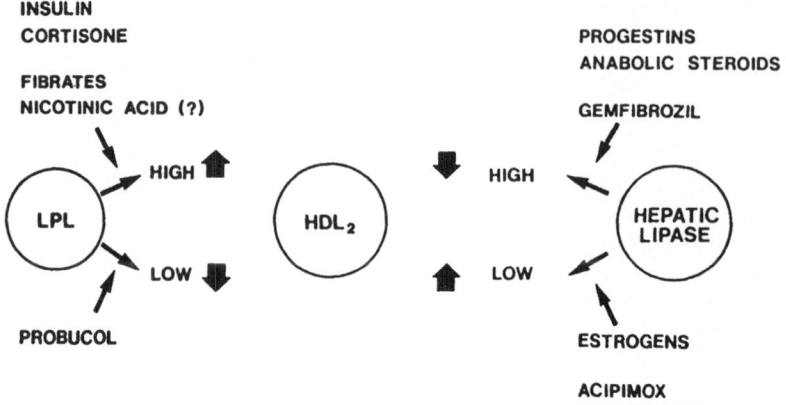

Fig. 2. Modulation of the concentration of plasma HDL_2 (HDL) by lipolytic enzymes

are followed by reciprocal changes in HDL_2 concentration. A rise of LPL is accompanied by a concordant change of HDL_2, whereas a rise of hepatic lipase decreases HDL_2. In contrast, low hepatic lipase activity is followed by a rise of HDL_2, but low LPL activity results in a fall of HDL_2. Both enzymes activities can be modified by several hormones and drugs as illustrated in Fig. 2. The action of hormones and drugs on LPL and hepatic lipase activity are accompanied by concomitant changes in HDL (HDL_2) concentration.

References

1. Patsch JR, Gotto AM Jr, Olivecrona T, Eisenberg S (1978) Formation of high density lipoprotein$_2$-like particles during lipolysis of very low density lipoproteins in vitro. Proc Natl Acad Sci USA 75:4519–4523
2. Nikkilä EA (1978) Metabolic regulation of plasma high density lipoprotein concentration. Editorial. Eur J Clin Invest 8:109–110
3. Tall AR, Small DM (1978) Plasma high-density lipoproteins. N Engl J Med 299:1232–1236
4. Nikkilä EA, Kuusi T, Taskinen M-R, Tikkanen MJ (1984) Regulation of lipoprotein metabolism by endothelial lipolytic enzymes. In: Carlson LA, Olsson AG (eds) Treatment of hyperlipoproteinemia. Raven, New York, pp 77–84
5. Nikkilä EA, Taskinen M-R, Sane T (1987) Plasma high density lipoprotein concentration and subfraction distribution in relation to triglyceride metabolism. Am Heart J 113:543–548
6. Groot PHE, Jansen H, Tol A van (1981) Selective degradation of the high density lipoprotein-2 subfraction by heparin-releasable liver lipase. FEBS Lett 129:269–272
7. Shirai K, Barnhart RL, Jackson RL (1981) Hydrolysis of human plasma high density lipoprotein$_2$ – phospholipids and triglycerides by hepatic lipase. Biochem Biophys Res Commun 100:591–599
8. Kuusi T, Nikkilä EA, Taskinen M-R, Somerharju P, Ehnholm C (1982) Human postheparin plasma hepatic lipase activity against triacylglycerol and phospholipid substrates. Clin Chim Acta 122:39–45
9. Kuusi T, Nikkilä EA, Tikkanen MJ, Taskinen M-R, Ehnholm C (1983) Function of hepatic endothelial lipase in lipoprotein metabolism. In: Schettler FG, Gotto AM, Middelhoff G,

Habenicht AJR, Jurutka KR (eds) Atherosclerosis VI. Springer, Berlin Heidelberg New York, pp 628–632

10. Taskinen M-R, Helve E, Nikkilä EA, Yki-Järvinen H (1987) Insulin therapy induces anti-atherogenic changes in lipoproteins of type 2 diabetics. Diabetes 36:81A

11. Nikkilä EA, Tikkanen MJ, Kuusi T (1983) Effects of progestins on plasma lipoproteins and heparin-releasable lipases. In: Bardin CW, Milgröm E, Mauvais-Jarvis P (eds) Progesterone and Progestins. Raven, New York, pp 411–420

12. Lithell H, Boberg J, Hellsing K, Lundqvist G, Vessby B (1978) Increase of the lipoprotein-lipase activity in human skeletal muscle during clofibrate administration. Eur J Clin Invest 8: 67–74

13. Vessby B, Lithell H, Ledermann H (1982) Elevated lipoprotein lipase activity in skeletal muscle tissue during treatment of hypertriglyceridaemic patients with bezafibrate. Atherosclerosis 44:113–118

14. Eisenberg S, Gavish D, Oschry Y, Fainaru M, Deckelbaum RJ (1984) Abnormalities in very low, low and high density lipoproteins in hypertriglycerideria. Reversal toward normal with bezafibrate treatment. J Clin Invest 74:470–482

15. Nikkilä EA, Ylikahri R, Huttunen VK (1976) Gemfibrozil: effect of serum lipids, lipoproteins, postheparin plasma lipase activities and glucose tolerance in primary hypertriglyceridemia. Proc roy Soc Med 69:58–63

16. Manninen V, Mälleönen M (1982) Effect of gemfibrozil on the blood levels of the high density lipoprotein subfractions HDL_2 and HDL_3. Research and Clinical Forums 4:77–83

17. Saku K, Gartside PS, Hynd BA, Kashyap ML (1985) Mechanism of action of gemfibrozil on lipoprotein metabolism. J Clin Invest 75:1702–1712

Human Apolipoprotein A-I and A-II Metabolism

H. B. Brewer, Jr., R. E. Gregg, S. W. Law, J. M. Hoeg, and L. A. Zech [1]

1 Introduction

Recognition of the inverse correlation between high density lipoprotein[+] cholesterol levels and the development of premature cardiovascular disease have focused a great deal of interest on HDL (Miller and Miller 1975; Rhoads et al. 1976; Castelli et al. 1977). HDL has been proposed to function in the process of "reverse cholesterol transport" by facilitating the removal of cholesterol from peripheral cells and transporting it back to the liver where it may be removed from the body (Glomset 1968). In order to gain insight into factors which modulate the plasma levels of HDL, we have systematically analyzed the metabolism of apoA-I and apoA-II, the two major apolipoproteins in HDL. This report will summarize our current studies on apoA-I and apoA-II metabolism.

2 Biosynthesis of Human ApoA-I and ApoA-II

Human apoA-I is a 267 amino acid apolipoprotein (Brewer et al. 1978) that is synthesized as a preproapolipoprotein containing an 18-amino acid prepeptide, and a 6-amino acid propeptide (Fig. 1; Cheung and Chan 1983; Gordon et al. 1983a; Karanthanasis et al. 1983; Law and Brewer 1983; Law et al. 1983; Shoulders et al. 1983). The prepeptide is cotranslationally cleaved intracellularly, and proapoA-I is secreted into plasma and lymph (Gordon et al. 1983a; Bojanovski et al. 1985). ProapoA-I undergoes posttranslational cleavage to mature apoA-I in plasma (see discussion below). ApoA-I has recently been shown to undergo a second posttranslational modification involving fatty acid acylation (Hoeg et al. 1985). Acylation is of particular interest since it dramatically increases the hydrophobicity of the apolipoprotein, thereby significantly affecting the interaction of the apolipoprotein with the hydrophobic lipids.

ApoA-II is a 154-amino acid apolipoprotein composed of two identical chains of 77 amino acid residues linked by a single disulfide bridge at position 6 in the sequence

---

1 Metabolism Disease Branch, National Heart, Lung, and Blood Institute, National Institutes of Health, Bethesda, MD 20892, USA

Drugs Affecting Lipid Metabolism
Ed. by R. Paoletti et al.
© Springer-Verlag Berlin Heidelberg 1987

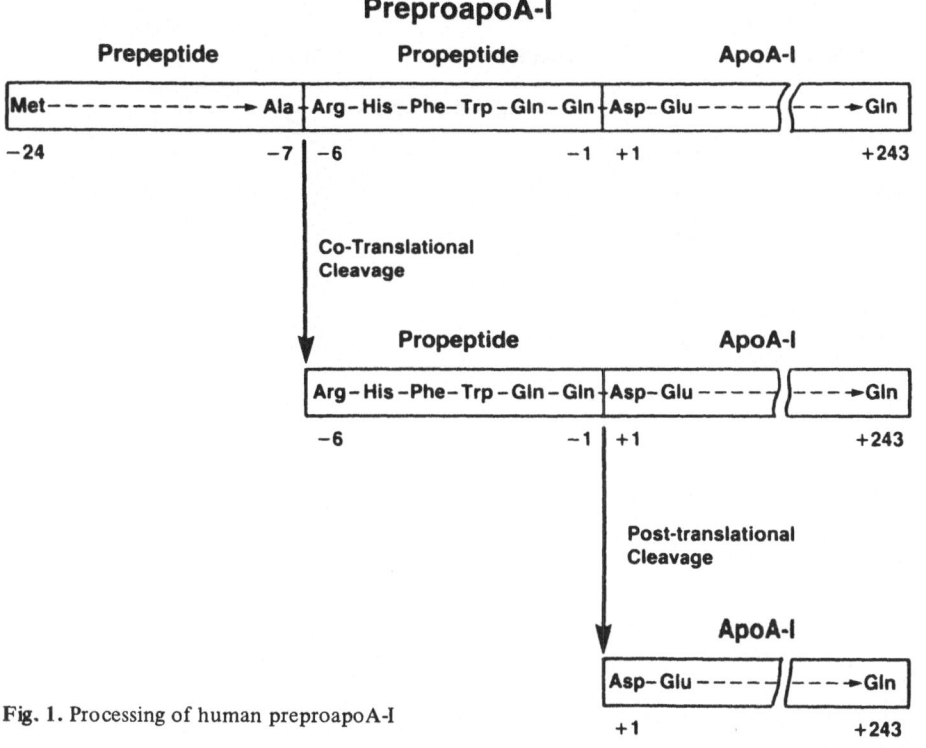

Fig. 1. Processing of human preproapoA-I

(Brewer et al. 1972). ApoA-II is a glycoprotein (Lackner et al. 1985), which is also synthesized as a preproapolipoprotein (Fig. 2; Knott et al. 1984; Lackner et al. 1984). ApoA-II is encoded as a 100-amino acid monomer containing an 18-amino acid prepeptide which is cotranslationally cleaved, and a 5-amino acid propeptide (Gordon et al. 1983b). In contrast to proapoA-I, the major apoA-II isoprotein in plasma and lymph is mature apoA-II in the form of a disulfide dimer (Brewer et al. 1972). The site of cleavage of proapoA-I terminates in two neutral residues (gln-gln), whereas proapoA-II contains two basic residues (arg-arg) at the site of cleavage to mature apoA-II. Thus, the posttranslational processing of apoA-I and apoA-II are strikingly different and appear to be under separate control.

Recently, the factors which moderate apoA-I gene expression have been analyzed in HepG-2 cells in vitro (Monge et al. 1985). The cellular level of apoA-I mRNA and the concentration of apoA-I in the tissue culture media were quantitated in cells incubated with LPDS and LPDS containing LDL (200 μg ml^{-1}). Incubation with LDL was associated with a significant increase in intracellular apoA-I mRNA and apoA-I in the media (Monge et al. 1985). These results were interpreted as indicating that the biosynthesis of apoA-I is increased following the cellular uptake of LDL by the LDL receptor pathway. The increase in biosynthesis and secretion of apoA-I following LDL uptake may function to modulate the increase in intracellular concentration of cholesterol providing a pathway for cholesterol egress from the cell.

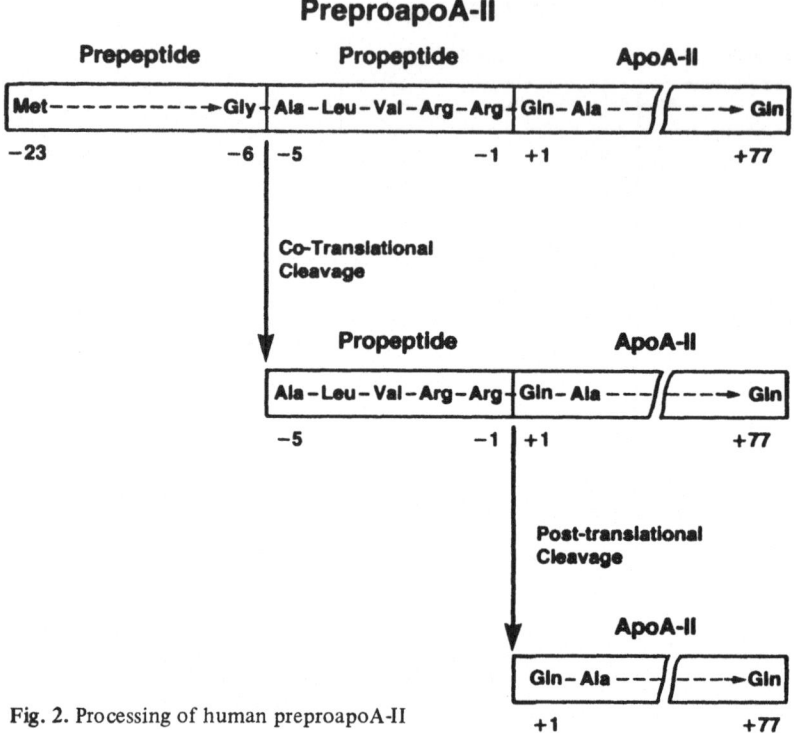

Fig. 2. Processing of human preproapoA-II

3 Human ApoA-I and ApoA-II Metabolism

The metabolism of apoA-I in man has been extensively analyzed utilizing kinetic studies of radiolabeled apoA-I (Bojanovski et al. 1985). The individual plasma isoforms of proapoA-I and mature apoA-I were purified to homogeneity from lymph VLDL and plasma HDL, respectively. The individual isoforms of proapoA-I (apoA-I$_{+2}$ isoform) and mature apoA-I (apoA-I$_{0,-1,-2}$ isoforms) were radiolabeled and injected in various combinations into normal volunteers (Bojanovski et al. 1985). The combined results from these studies established that the predominant isoprotein secreted into plasma and lymph was proapoA-I (Fig. 3). ProapoA-I was rapidly converted (residence time 3 to 4 h) to mature apoA-I (apoA-I$_0$ isoform). The apoA-I$_0$ isoform of mature apoA-I was slowly converted to the apoA-I$_{-1}$ isoform and ultimately to the apoA-I$_{-2}$ isoform. The residence times of all of the mature apoA-I isoforms were similar, and approximately 5 days in normal man. In fasting plasma approximately 4–5% of apoA-I is present as the proapoA-I isoprotein (Bojanovski et al. 1985).

The metabolism of apoA-II has also been evaluated by analysis of radiolabeled mature apoA-II. In contrast to apoA-I, there is insufficient proapoA-II present in plasma to permit isolation and kinetic analysis of the proapoA-II isoprotein (Lackner et al. 1985). Radiolabeled mature apoA-II is catabolized at a slightly faster rate (0.5 days) when compared to normal apoA-I (Zech et al. 1983). The difference in catabolism of apoA-I and apoA-II were present in both male and female patients (Zech et al. 1983).

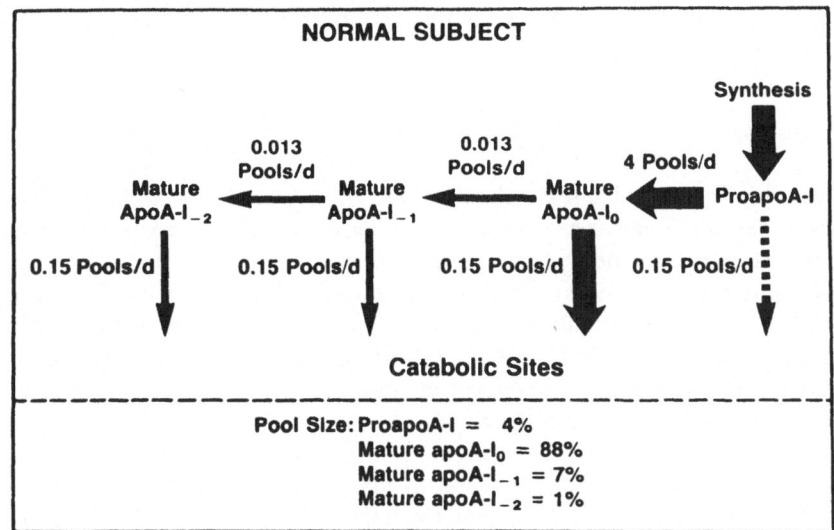

Fig. 3. Conceptual overview of human plasma ApoA-I metabolism

A kinetic model of apoA-I and apoA-II metabolism has been developed by compartmental modeling (Zech et al. 1983). In these models there is a component which is similar in both the apoA-I and apoA-II models, suggesting that apoA-I and apoA-II catabolism is linked presumably due to the presence of both apoA-I and apoA-II on the same lipoprotein particle (LpA-I,A-II). A separate component containing only apoA-I (LpA-I) without apoA-II was identified, and this apoA-I component had a residence time of 3.8 days, as compared to the residence time of 5–6 days for the LpA-I,A-II-containing lipoprotein particles (Zech et al. 1983).

The results of the compartmental analysis of apoA-I and apoA-II kinetic studies, and the presence of LpA-I,A-II as well as LpA-I and LpA-I,E particles within the lipoproteins isolated in HDL (Alaupovic 1972) have permitted the formation of a conceptual model for the role of apoA-I and apoA-II-containing lipoprotein particles in reverse cholesterol transport. The majority of the lipoprotein particles within HDL contain both apoA-I and apoA-II, and may interact with the punative HDL receptor in human liver (Hoeg et al. 1985). The lipoprotein particles containing only apoA-I may also interact with the HDL receptor. Lipoprotein particles containing both apoA-I and apoE may interact with either the HDL receptor or the apoE receptor (Hui et al. 1986). It will be of particular importance to ultimately establish the importance of each of these three pathways in the reverse cholesterol transport of cholesterol from peripheral cells back to the liver.

4 Metabolism of ApoA-I and ApoA-II in Selected Patients With Hypoalphalipoproteinemia

4.1 Tangier Disease

The most extensively studied disease characterized by hypoalphalipoproteinemia is Tangier disease. Tangier disease is a codominant familial disease characterized clinically by orange tonsils, hepatosplenomegaly, cloudy corneas, and intermittent peripheral neuropathy (Fredrickson et al. 1961). The characteristic pathological feature of Tangier disease is cholesteryl ester accumulation in the reticuloendothelial cells of the tonsil, liver, spleen, and rectal mucosa (Herbert et al. 1978). Tangier homozygotes have elevated levels of plasma triglycerides, reduced concentrations of LDL, and a striking deficiency of HDL, apoA-I, and apoA-II (Schaefer et al. 1978).

The characteristic metabolic defect in Tangier disease is a rapid catabolism of apoA-I and apoA-II leading to the marked reduction in plasma HDL (Schaefer et al. 1978; Assmann 1979). Initial studies on the rapid catabolism of the HDL apolipoproteins focused on apoA-I. Two-dimensional gel electrophoresis of apoA-I in Tangier disease revealed a relative increase of proapoA-I to mature apoA-I (Kay et al. 1981). This observation led to the proposal that the defect in Tangier disease was due to a defect in conversion of proapoA-I to mature apoA-I with rapid catabolism of proapoA-I (Gordon et al. 1983a; Zannis et al. 1983). Detailed in vitro studies (Brewer et al. 1983; Bojanovski et al. 1984a) as well as in vivo studies (Bojanovski et al. 1984b), however, established that there was no defect in the propeptide or converting enzyme, and that the rate of conversion of proapoA-I to mature apoA-I in Tangier disease was similar to normal control subjects. The relative increase in the proapoA-I isoprotein was due to a rapid catabolism of both proapoA-I and mature apoA-I, and a decreased *percent* conversion of proapoA-I to mature apoA-I (Bojanovski et al. 1984b).

The covalent structure of apoA-I in Tangier subjects has also been directly addressed. The cDNA and derived amino acid sequence of preproapoA-I was determined from a patient with Tangier disease and was shown to be similar to the structure of normal preproapoA-I (Law and Brewer 1985). Thus, the combined results of these studies have established that the apoA-I in Tangier disease is normal, and the decreased level of plasma apoA-I and HDL are due to increased catabolism. The reason for the increased rate of catabolism of the HDL apolipoproteins in Tangier disease remains elusive, and additional studies will be required to determine the molecular defect in Tangier disease. Recently, a defect in macrophages isolated from a kindred with Tangier disease was reported (Schmitz et al. 1985). It will be of interest to see if this defect will be present in other kindreds with Tangier disease.

4.2 Familial Hypoalphalipoproteinemia

Several kindreds have now been reported which are characterized by reduced levels of plasma HDL-cholesterol and an increase incidence of premature cardiovascular disease (Third et al. 1984; Ordovas et al. 1986). The molecular defect(s) in these kindreds is as yet unknown, however, the reduced levels of HDL-cholesterol have recently been investigated in the probands of three unrelated kindreds by a kinetic analysis of radio-labeled apoA-I metabolism (Roma et al. 1987). The plasma levels of apoA-I were significantly decreased, and the synthesis rate of apoA-I was similar to normal. The reduction in HDL was greater than the reduction in the plasma level of apoA-I. The combined results from these studies have been interpreted as indicating that the reduced levels of HDL-cholesterol in these three kindreds with familial hypoalphalipoproteinemia are due to an increase in apoA-I catabolism.

5 Summary

During the last decade major advances have been made in our understanding of the metabolism, function, and diseases associated with HDL. The determination of the structure and processing of preproapoA-I and preproapoA-II have provided major insights into the biosynthesis of the major HDL apolipoproteins. The metabolism of apoA-I and apoA-II in man has been analyzed, and the unique processing of proapoA-I to mature apoA-I in plasma established. The importance of the heterogeneity of the lipoprotein particles within HDL has been recognized and the different lipoproteins including LpA-I,A-II, LpA-I, and LpA-I,E in reverse cholesterol transport will be the focus of future research. Of major importance in the next decade will be the elucidation of the factors which modulate plasma HDL levels and the determination of the molecular defects in those dyslipoproteinemias which are characterized by hypoalphalipoproteinemia. The ultimate goal of these studies will be the development of improved methods to diagnose patients with hypoalphalipoproteinemia, and the development of new drugs to raise HDL levels, thereby potentially reducing the risk of the development of premature cardiovascular disease.

References

Alaupovic P (1972) Plasma lipoproteins. Conceptual development of the classification system of plasma lipoproteins. In: Peeters H (ed) Protides of the biological fluids. Pergamon, New York, pp 9–19

Assmann G (1979) Tangier disease and the possible role of high density lipoproteins in atherosclerosis. In: Gotto AM, Paoletti R (eds) Atherosclerosis reviews. Raven, New York, pp 1–28

Bojanovski D, Gregg RE, Brewer HB, Jr (1984a) Tangier disease: in vitro conversion of proapoA-I$_{Tangier}$ to mature apoA-I$_{Tangier}$. J Biol Chem 259:6049–6051

Bojanovski D, Gregg RE, Zech LA, Meng MS, Ronan R, Brewer HB, Jr (1984b) The in vivo metabolism of proapolipoprotein A-I in Tangier disease. Clin Res 32:390A

Bojanovski D, Gregg RE, Ghiselli G, Schaefer EJ, Light JA, Brewer HB, Jr (1985) Human apolipoprotein A-I isoprotein metabolism: proapoA-I conversion to mature apoA-I. J Lipid Res 26: 185–193

Brewer HB, Jr, Lux SE, Ronan R, John KM (1972) Amino acid sequence of human apoLp-Gln-II (apoA-II), an apolipoprotein isolated from the high-density lipoprotein complex. Proc Natl Acad Sci USA 69:1304–1308

Brewer HB, Jr, Fairwell T, Larue A, Ronan R, Houser A, Bronzert T (1978) The amino acid sequence of human apoA-I, an apolipoprotein isolated from high density lipoproteins. Biochem Biophys Res Commun 80:623–630

Brewer HB, Jr, Fairwell T, Meng MS, Kay L, Ronan R (1983) Human proapoA-I$_{Tangier}$: isolation of proapoA-I$_{Tangier}$ and amino acid sequence of the peptide. Biochem Biophy Res Commun 113:934–940

Castelli W, Doyle JT, Gordon T, Hames CG, Hjortland M, Hulley SB, Kagan A, Zukel WJ (1977) HDL cholesterol and other lipids in coronary heart disease: the cooperative lipoprotein phenotyping study. Circulation 55:767–772

Cheung P, Chan L (1983) Nucleotide sequence of cloned cDNA of human apolipoprotein A-I. Nucl Acid Res 11:3703–3710

Fredrickson DS, Altrocchi PH, Avioli LC (1961) Tangier disease: combined clinical staff conference at the National Institutes of Health. Ann Int Med 55:1016–1031

Glomset JA (1968) The plasma lecithin-cholesterol acyl transferase reaction. J Lipid Res 9:155–167

Gordon JI, Sims HF, Lentz SR, Edelstein C, Scanu AM, Strauss AW (1983a) Proteolytic processing of human preproapolipoprotein A-I: a proposed defect in the conversion of proA-I to A-I in Tangier's disease. J Biol Chem 258:4037–4044

Gordon JI, Bridelier KA, Sims HF, Edelstein C, Scanu AM, Strauss AW (1983b) Biosynthesis of human preproapolipoprotein A-II. J Biol Chem 258:14054–14059

Herbert PM, Gotto AM Jr, Fredrickson DS (1978) Familial lipoprotein deficiency. In: Stanbury JB, Wyngaarden JB, Fredrickson DS (eds) The metabolic basis of inherited diseases, 4th edn. McGraw-Hill, New York, pp 544–588

Hoeg JM, Demosky SJ, Jr, Edge SB, Gregg RE, Osborne JC, Jr, Brewer HB, Jr (1985) Characterization of a human hepatic receptor for high density lipoproteins. Arteriosclerosis 5:228–237

Hoeg JM, Meng MS, Ronan R, Fairwell T, Brewer HB, Jr (1986) Human apolipoprotein A-I: posttranslational modification by fatty acid acylation. J Biol Chem 261:3911–3914

Hui DY, Brecht WJ, Hall EA, Friedman G, Immerarity TL, Mahley RW (1986) Isolation and characterization of the apolipoprotein E receptor from canine and human liver. J Biol Chem 261:4256–4267

Karanthanasis SK, Zannis VI, Breslow JL (1983) Isolation and characterization of the human apolipoprotein A-I gene. Proc Natl Acad Sci USA 80:6147–6151

Kay LL, Ronan R, Schaefer EJ, Brewer HB, Jr (1981) Tangier disease: a structural defect in apolipoprotein A-I (apoA-I$_{Tangier}$). Proc Natl Acad Sci USA 79:2485–2489

Knott TJ, Priestley LM, Urlea M, Scott J (1984) Isolation and characterization of a cDNA encoding the precursor for human apolipoprotein A-II. Biochem Biophys Res Commun 120:734–740

Lackner K, Law S, Brewer HB, Jr (1984) Human apolipoprotein A-II: complete nucleic acid sequence of preproapoA-II. FEBS Lett 175:159–164

Lackner KJ, Edge SB, Gregg RE, Hoeg JM, Brewer HB, Jr (1985) Isoforms of apolipoprotein A-II in human plasma and thoracic duct lymph. J Biol Chem 260:703–706

Law SW, Brewer HB, Jr (1984) Nucleotide sequence and the encoded amino acids of human apolipoprotein A-I mRNA. Proc Natl Acad Sci USA 81:66–70

Law SW, Brewer HB, Jr (1985) Tangier disease: the complete mRNA sequence encoding for preproapoA-I. J Biol Chem 260:12810–12814

Law SW, Brewer HB, Jr (1983) Nucleotide sequence and the encoded amino acids of human apolipoprotein A-I mRNA. Proc Natl Acad Sci USA 81:66–70

Law SW, Gray G, Brewer HB, Jr (1983) cDNA cloning of human apoA-I: amino acid sequence of preproapoA-I. Biochem Biophys Res Commun 112:257–264

Miller GJ, Miller NE (1975) Plasma-high density lipoprotein concentration and development of ischemic heart disease. Lancet 1:16–20

Monge JC, Law SW, Hoeg JM, Brewer HB, Jr (1985) Modulation of human apolipoprotein A-I mRNA by mevinolin and low density lipoproteins. Circulation 72:III-285

Ordovas JM, Schaefer EJ, Salem D, Ward RH, Glueck CJ, Vergani C, Wilson PWF, Karathanasis SK (1986) Apolipoprotein A-I gene polymorphism associated with premature coronary artery disease and familial hypoalphalipoproteinemia. N Engl J Med 314:671–677

Rhoads G, Gulbrandsen CL, Kagan A (1976) Serum lipoproteins and coronary heart disease in a population study of Hawaii-Japanese men. N Engl J Med 294:293–295

Roma P, Gregg RE, Zech LA, Glueck C, Vergani C, Bishop C, Brewer HB, Jr (1987) Metabolism of apolipoprotein A-I in hypoalphalipoproteinemic (hypoα) subjects with an apoA-I linked PstI restriction endonuclease fragment length polymorphism (RFLP). Am Fed Clin Res (in press)

Schaefer EJ, Blum CB, Levy RI, Jenkins LL, Alaupovic P, Foster DM, Brewer HB, Jr (1978) Metabolism of high density apolipoproteins in Tangier disease. N Engl J Med 299:905–910

Schmitz G, Assmann G, Robenek H, Brennausen B (1985) Tangier disease: a disorder of intracellular membrane traffic. Proc Natl Acad Sci USA 82:6305–6309

Sharpe CR, Sidoli A, Shelley CS, Lucero MA, Shoulders CC, Baralle FE (1984) Human apolipoprotein A-I, A-II, C-II, and C-III. cDNA sequences and mRNA abundance. Nucl Acid Res 12: 3917–3932

Shoulders CC, Kornblihtt AR, Munro BS, Baralle FE (1983) Gene structure of human apolipoprotein A-I. Nucl Acid Res 11:2827–2837

Third JLHC, Montag J, Flynn M, Friedel J, Laskarzewoski P, Glueck CJ (1984) Primary and familial hypoalphalipoproteinemia. Metabolism 33:136–146

Zannis VI, Karathanasis SK, Keutmann HT, Goldberger G, Breslow JL (1983) Intracellular and extracellular processing of human apolipoprotein A-I: secreted apolipoprotein A-I isoprotein 2 is a propeptide. Proc Natl Acad Sci USA 80:2574–2578

Zech LA, Schaefer EJ, Bronzert TJ, Aamodt RL, Brewer HB, Jr (1983) Metabolism of human apolipoproteins A-I and A-II: compartmental models. J Lipid Res 24:60–71

Roles of LCAT and Lipid Transfer Protein in HDL Metabolism

P. J. BARTER and G. J. HOPKINS [1]

1 HDL and Plasma Cholesterol Transport

The metabolism of HDL is central to plasma cholesterol transport. In the postabsorptive state cholesterol enters the plasma mainly as free cholesterol, either as a constituent of lipoproteins secreted by the liver or intestine or by diffusion from tissues down a concentration gradient maintained by the esterification of cholesterol within the plasma. This esterification is catalyzed by LCAT which interacts preferentially with HDL (Fielding and Fielding 1971). Most of the newly formed cholesteryl esters are initially incorporated into HDL (Rajaram and Barter 1985). The esterification also maintains a concentration gradient between the different plasma lipoprotein fractions, ensuring that free cholesterol in lipoproteins secreted from the liver and intestine is also channelled into HDL. Thus, regardless of its origin, a major proportion of the cholesterol entering plasma is converted into cholesteryl esters and incorporated into the core of HDL. Subsequent activity of the lipid transfer protein (LTP) then promotes a redistribution of cholesteryl esters from HDL to other plasma lipoprotein fractions. These transfers represent the major pathway by which cholesteryl esters become incorporated into the VLDL and LDL in human plasma (Rajaram and Barter 1985). Given that VLDL are catabolized to LDL in vivo, the end result of the transfer process is a delivery of cholesteryl esters to the LDL fraction and thus a regulated uptake by whatever tissues possess the appropriate receptors.

Plasma cholesteryl ester formation and transfer may be linked (Ihm et al. 1982) with evidence that the rate of transfer may influence the rate of formation. We have shown that transfers of cholesteryl esters from HDL to triglyceride-rich lipoproteins in vitro enhance the reactivity of HDL with LCAT (Hopkins and Barter 1984). We have also observed an enhanced reactivity with LCAT of the HDL isolated from subjects with hypertriglyceridaemia (Barter et al. 1985), an apparent consequence of the small particle size of HDL in such subjects.

---

1   Baker Medical Research Institute, Melbourne, Australia

Drugs Affecting Lipid Metabolism
Ed. by R. Paoletti et al.
© Springer-Verlag Berlin Heidelberg 1987

2 Roles of LCAT and LTP in Regulating HDL Particle Size

Several factors are known to be involved in the modulation of HDL particle size. Here, we consider the effects of LCAT and LTP. LCAT acts in vitro to increase the cholesteryl ester content and hence the particle size of HDL. In the absence of an additional source of free cholesterol, changes in particle size may be minimal during incubation in vitro; however, in the presence of LDL to provide free cholesterol for the reaction, activity of LCAT has been shown to expand even very small HDL_3 particles into particles of HDL_2 size (Rajaram and Barter 1986).

HDL particle size is also affected by activity of LTP. When human HDL are incubated in vitro with triglyceride-rich lipoproteins and a source of LTP there is an appearance of both smaller and larger HDL particles (Hopkins et al. 1985). The mechanism of formation of the small particles and their relationship to the lipid transfer process is uncertain. The formation of larger HDL, however, can be readily explained in terms of a partial replacement of HDL cholesteryl esters by triglyceride. However, this effect of lipid transfers in promoting HDL particle expansion in vitro contrasts markedly with the situation in vivo.

Human subjects with hypertriglyceridaemia, who also exhibit enhanced rates of lipid transfer between HDL and VLDL in vivo, have HDL particles that are smaller, not larger than normal (Chang et al. 1985). Conversely, rat plasma is almost totally deficient in· LTP but contains HDL particles that are considerably larger than those in human plasma (Ha et al. 1985). Furthermore, injection of human LTP into rats is associated in vivo with a reduction in HDL particle size (Ha et al. 1985). Thus, the observation in vitro that HDL particle size is increased by neutral lipid transfers is not consistent with a number of observations in vivo. In rabbits, however, we have found that the situation in vivo does mimic what takes place in vitro.

In rabbit plasma, a rich source of LTP, the HDL are triglyceride-rich and larger than those in human plasma. Rabbits, however, are deficient in activity of hepatic lipase and when rabbit plasma is mixed with rat hepatic lipase and incubated in vitro much of the HDL triglyceride is hydrolyzed and the particles are reduced to a size comparable to that of human HDL_3.

Consequently, we investigated the effects of hepatic lipase on human lipoproteins. There was little effect on the particle size of unmodified human HDL. After human HDL had become enriched with triglyceride as a result of neutral lipid transfers, however, activity of hepatic lipase had a dramatic effect, hydrolyzing triglyceride and promoting the appearance of very small particles comparable to the HDL found in subjects with hypertriglyceridaemia.

3 Activity of the HDL Conversion Protein

Changes in the size and density of HDL may occur in vitro under circumstances when there is no activity of LCAT, LTP or either hepatic or lipoprotein lipase. The existence of a previously unidentified protein has been established (Rye and Barter 1986). This protein, designated HDL conversion protein has now been isolated in pure form from human plasma. It promotes dramatic changes in the size of human HDL_3 during incubation in vitro. It does not, however, promote cholesterol esterification, lipid transfer or lipolysis. Addition of the conversion protein to a homogeneous population of human HDL_3 promotes the formation of a series of new populations of particles, some smaller and others larger than the parent lipoprotein. The physiological significance of this process is still quite unknown. It will be of interest to investigate its interaction with factors such as LCAT and LTP in the regulation of HDL metabolism.

References

Barter PJ, Hopkins GJ, Gorjatschko L (1985) Lipoprotein substrates for plasma cholesterol esterification – influence of particle size and composition of the high density lipoprotein subfraction 3. Atherosclerosis 58:97–107

Chang LBF, Hopkins GJ, Barter PJ (1985) Particle-size distribution of high-density lipoproteins as a function of plasma triglyceride concentration in human subjects. Atherosclerosis 56:61–70

Fielding CJ, Fielding PE (1971) Purification and substrate specificity of lecithin-cholesterol acyl transferase from human plasma. FEBS Lett 15:355–358

Ha YC, Chang LBF, Barter PJ (1985) Effects of injecting exogenous lipid transfer protein into rats. Biochim Biophys Acta 833:203–210

Hopkins GJ, Barter PJ (1984) Capacity of lipoproteins to act as substrates for lecithin: cholesterol acyltransferase. Enhancement by pre-incubation with an artificial triacylglycerol emulsion. Biochim Biophys Acta 794:31–40

Hopkins GJ, Chang LBF, Barter PJ (1985) Role of lipid transfers in the formation of a subpopulation of small high density lipoproteins. J Lipid Res 26:218–229

Ihm J, Ellsworth JL, Chataing B, Harmony JAK (1982) Plasma protein-facilitated coupled exchange of phosphatidylcholine and cholesteryl ester in the absence of cholesterol esterification. J Biol Chem 257:4818–4827

Rajaram OV, Barter PJ (1985) Reactivity of human lipoproteins with purified lecithin: cholesterol acyltransferase during incubations in vitro. Biochim Biophys Acta 835:41–49

Rajaram OV, Barter PJ (1986) Increases in the particle size of high-density lipoproteins induced by purified lecithin-cholesterol acyltransferase – effect of low-density lipoproteins. Biochim Biophys Acta 877:406–414

Rye K-A, Barter PJ (1986) Changes in the size and density of human high-density lipoproteins promoted by a plasma-conversion factor. Biochim Biophys Acta 875:429–438

The Profile of an HMG-CoA Reductase Inhibitor, CS-514 (SQ 31,000)

Y. Goto[1]

1 Introduction

Since more than 70% of the total input of body cholesterol is derived from de novo synthesis in humans (Dietschy and Wilson 1970), it is expected that the high level of serum cholesterol can be reduced in consequence of inhibition of cholesterol synthesis. The most desirable target for this inhibition is 3-hydroxy-3-methylglutaryl coenzyme A (HMG-CoA) reductase, the rate-limiting enzyme of cholesterogenesis. ML-236B (also called compactin) was discovered first by Sankyo Co. from the culture broth of a micro-organism as a potent competitive inhibitor of HMG-CoA reductase (Endo et al. 1976), and showed remarkable hypolipidemic effect on various experimental animals as well as humans (Kuroda et al. 1979; Tsujita et al. 1979; Yamamoto et al. 1980).

Among related compounds of ML-236B, CS-514 (SQ 31,000) was finally selected because of its strong potency and tissue selectivity. Figure 1 shows the structures of

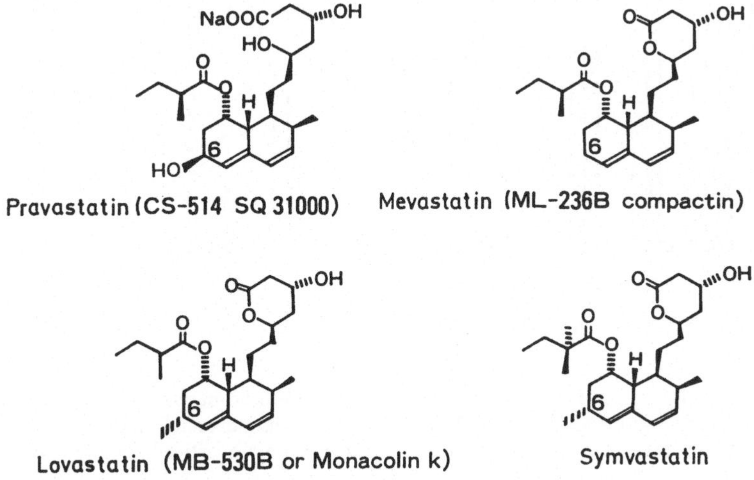

Pravastatin (CS-514 SQ 31000)   Mevastatin (ML-236B compactin)

Lovastatin (MB-530B or Monacolin k)   Symvastatin

Fig. 1. The structures of HMG-CoA reductase inhibitors

---

1 Tokai University School of Medicine, Bohseidai, Isehara, Kanagawa, 259-11, Japan

Drugs Affecting Lipid Metabolism
Ed. by R. Paoletti et al.
© Springer-Verlag Berlin Heidelberg 1987

ML-236B, CS-514 and other related compounds, Lovastatin (also called MB-530B or monacolin K) and synvinolin. This report briefly describes biochemical and pharmacological characteristics of CS-514 (Tsujita et al. 1986).

2 Tissue Selective Inhibition of Sterol Synthesis

In the cell-free enzyme system from rat liver and in freshly isolated rat hepatocytes, the concentrations required for 50% inhibition of sterol synthesis from acetate (I_{50}) were 0.8 ng ml^{-1} and 2.2 ng ml^{-1} respectively, corresponding to one-tenth and one-third to those of ML-236B. On the other hand, the inhibitory activity of CS-514 was weak in the cells from extrahepatic tissues, whereas that of ML-236B was strong as in rat hepatocytes. As shown in Fig. 2, orally administered CS-514 selectively inhibited the sterol synthesis in liver and intestine, the major sites of cholesterogenesis. In the case of ML-236B and Lovastatin, although the inhibition of sterol synthesis in liver and intestine was most potent, that in other organs was significant as well. Judging from the experiment of cell uptake using ^{14}C-CS-514 and ^{14}C-ML-236B sodium salt, the less potent inhibitory activity of CS-514 in extrahepatic tissues can be ascribed to lower uptake of the drug by those tissues.

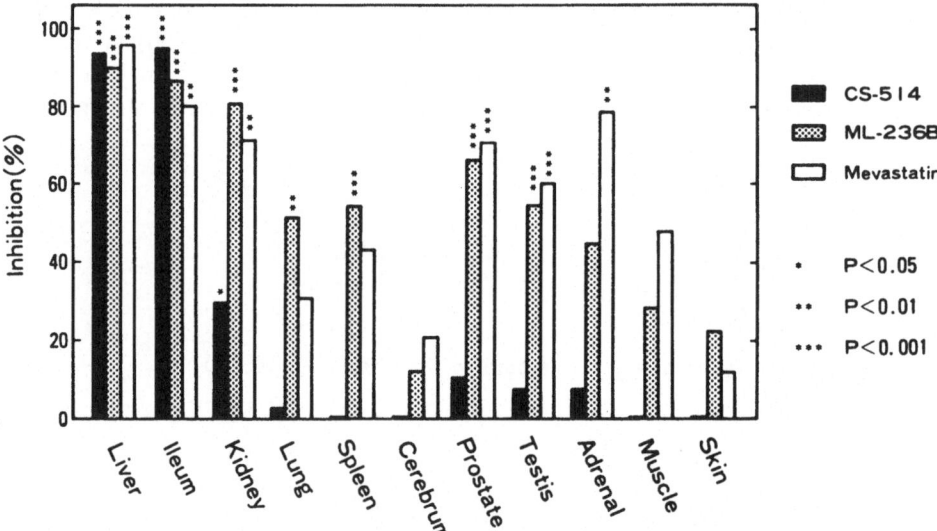

Fig. 2. Inhibitory activity of CS-514, ML-236B and Lovastatin (MB-530B or monacolin K) on sterol synthesis of various tissues in rats. CS-514, ML-236B or Lovastatin was administered orally at a dose of 25 mg kg^{-1} to male Wistar-Imamichi rats (n = 6). After 2 h, the rats were sacrificed and incorporation of the radioactivity from ^{14}C-acetate to sterols in slices was measured

3 Hypolipidemic Activity of CS-514

CS-514 decreased significantly the serum cholesterol levels in dogs, monkeys, rabbits and WHHL rabbits, an animal model of familial hypercholesterolemia in man, at doses of 0.625 mg to 50 mg kg^{-1} day^{-1} (Table 1). In contrast, CS-514 had no cholesterol-lowering effect in rats and mice except for the animals in which serum cholesterol levels were increased by Triton treatment. Among lipoprotein cholesterols, the atherogenic ones were preferentially reduced by the drug. The biliary neutral sterol levels were significantly reduced by 37% at a dose of 50 mg kg^{-1} day^{-1} for 5 weeks in dogs, resulting in a reduction of the lithogenic index by 43% ($P < 0.001$).

Table 1. Effect of CS-514 on serum lipids in various animal species. CS-514 was administered twice a day at the indicated doses, except for rats which received the drug once a day. The initial values of serum lipids in each animal group were obtained from the average value of at least three point assays

Dose (mg kg^{-1} per day)	Percent of initial value (mean ± SD)		
	Total cholesterol	Phospholipid	Triglyceride
A. *Beagle dog* (18 days, n = 6)			
Control	96 ± 5	96 ± 4	103 ± 10
0.625	88 ± 5[a]	88 ± 6[a]	83 ± 8[a]
1.25	82 ± 6[b]	64 ± 6[a]	89 ± 14
B. *Cynomolgus monkey* (18 days, n = 4)			
Control	96 ± 4	85 ± 8	110 ± 37
20	85 ± 6[c]	87 ± 9	96 ± 7
50	69 ± 11[a]	84 ± 11	94 ± 15
C. *Japanese white rabbit* (18 days, n = 6)			
Control	96 ± 9	96 ± 3	94 ± 35
6.25	78 ± 10[a]	84 ± 7[a]	79 ± 18
12.5	68 ± 11[a]	73 ± 7[a]	77 ± 33
D. *WHHL rabbit* (12 days, n = 4)			
Control	100 ± 20	93 ± 4	108 ± 13
12.5	82 ± 5[b]	88 ± 5	92 ± 11
50	72 ± 9[b]	84 ± 5[a]	107 ± 10
E. *Wistar-Imamichi rat* (14 days, n = 8)			
500	118[d]	101[d]	69[d]

[a] $p < 0.01$; [b] $p < 0.05$; [c] $p < 0.001$; [d] percent of control.

4 Prevention of Atherosclerosis in Coronary and of Xanthoma in WHHL Rabbits (Watanabe et al., in prep.)

CS-514 was administered orally to 2–3-month-old WHHL rabbits (n = 12) at a dose of 50 mg kg^{-1} day^{-1} for 6 months. Although the aortic atherosclerosis was not affected, the incidence of coronary atherosclerosis was markedly reduced from 43% (control group) to 19% (treated group, $P < 0.02$), and the development of lesions of coronary arteries was significantly prevented from 19.2% (control group) to 9.2% (treated group, $P < 0.05$). The occurrence of xanthoma in digital joints was also prevented.

5 Metabolic and Toxicological Characteristics of CS-514

A half-life of CS-514 in blood after oral administration to dogs was 2.5 h, suggesting that this drug is rapidly excreted from the body. CS-514 was mainly excreted to bile in rats and dogs, and the bioavailability was 45–65%.

The LD_{50} value of CS-514 orally administered to rats was more than 12,000 mg kg^{-1}, indicating that this drug is low toxic. There was no remarkable toxic effect on the 2-year chronic toxicity study using dogs, and on reproduction, teratogenicity and antigenicity.

From these results, CS-514 is expected to have a strong preventive effect on coronary heart disease in patients with extremely high blood cholesterol levels.

References

Dietschy JM, Wilson JD (1970) Regulation of cholesterol metabolism. N Engl J Med 282:1179–1183

Endo A, Kuroda M, Tsujita Y (1976) ML-236A, ML-236B and ML-236C, new inhibitors of cholesterogenesis produced by *Penicillium citrinum*. J Antibiot 29:1346–1348

Kuroda M, Tsujita Y, Tanzawa K, Endo A (1979) Hypolipidemic effects in monkeys of ML-236B, a competitive inhibitor of 3-hydroxy-3-methylglutaryl coenzyme A reductase. Lipids 14:585–589

Tsujita Y, Kuroda M, Tanzawa K, Kitano N, Endo A (1979) Hypolipidemic effects in dogs of ML-236B, a competitive inhibitor of 3-hydroxy-3-methylglutaryl coenzyme A reductase. Atherosclerosis 32:307–313

Tsujita Y, Kuroda M, Shimada Y, Tanzawa K, Arai M, Kaneko I, Tanaka M, Masuda H, Tarumi C, Watanabe Y, Fujii S (1986) CS-514, a competitive inhibitor of 3-hydroxy-3-methylglutaryl coenzyme A reductase: tissue-selective inhibition of sterol synthesis and hypolipidemic effect on various animal species. Biochim Biophys Acta 877:50–60

Yamamoto A, Sudo H, Endo A (1980) Therapeutic effects of ML-236B in primary hypercholesterolemia. Atherosclerosis 35:259–266

Induction of Hepatic Low-Density Lipoprotein Receptors in Heterozygous WHHL Rabbits Treated with CS-514 Alone or CS-514 in Combination with Cholestyramine

T. Kita[1], N. Kume[1], Y. Tsujita[2], T. Ito[3], Y. Watanabe[3], and C. Kawai[1]

1 Introduction

Epidemiologic and clinical data reveal the suprising fact that more than half of the people in industrialized societies have a level of circulating LDL that puts them at high risk for developing atherosclerosis. The more LDL in the blood, the more rapidly atherosclerosis develops (Goldstein and Brown 1982). However, we do not know yet the ideal level of cholesterol in the human blood. The plasma concentration levels of cholesterol now believed "normal" may well be too high in view of such a prevailing high risk of atherosclerotic lesions in industrialized societies. What determines the blood level of LDL? A specialized protein, called LDL receptor, plays an important role (Goldstein et al. 1983). The number of receptors on the surface of cells varies with the demand for cholesterol of the cells. When the need is low, the cells make fewer receptors and take up LDL at a reduced rate. This will protect cells against excess cholesterol, but at a high price: the reduction in the number of receptors decreases the rate at which LDL is removed from the circulation; the plasma level of LDL rises; and atherosclerosis is accelerated. The liver is a key organ in cholesterol and lipoprotein homeostasis, because of its large size and its high concentration of LDL receptors (Kita et al. 1982). It both synthesizes cholesterol (de novo synthesis) and acquires cholesterol from chylomicron remnants and other lipoproteins. Biliary excretion is the dominant pathway for net removal of cholesterol from the body. The bile acids are largely reabsorbed from the intestine, returned to the bloodstream, taken up by the liver and again secreted into the upper intestine.

Therefore, we thought inhibition of cholesterol synthesis (de novo) might force the liver to rely more on LDL uptake and thus stimulate greater production of receptors. To block cholesterol synthesis we took advantage of the discovery by Endo, of a remarkable natural inhibitor of HMG-CoA reductase, called compactin, and its derivative

1   3rd Division, Department of Internal Medicine, Faculty of Medicine, Kyoto University, 54 Kawaracho Shogoin, Sakyo-ku Kyoto 606, Japan
2   Fermentation Research Laboratories, Sankyo, Co. Ltd. 1-2-58, Hiromachi, Shinagawa-ku Tokyo 140, Japan
3   Institute for Experimental Animals, Kobe University 7-5-1 Kusunoki-cho, Chuo-ku Kobe 650, Japan

Drugs Affecting Lipid Metabolism
Ed. by R. Paoletti et al.
© Springer-Verlag Berlin Heidelberg 1987

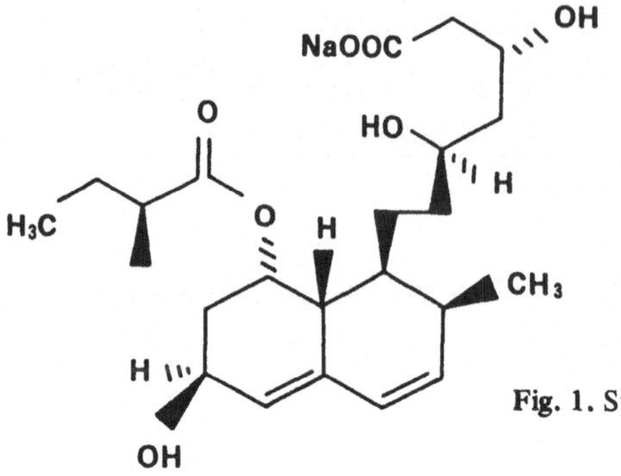

Fig. 1. Structure of CS-514

CS-514 (Fig. 1) (Tsujita et al. 1986). A side chain of the compactin molecule closely mimics the structure of the natural substrate of HMG-CoA reductase, and so it binds to the enzyme's active site and inhibits the activity. Furthermore, we reasoned that if the recycling of bile acids could be interrupted, the liver would be called on to convert more cholesterol into bile acids and this should lead the liver cells to make more LDL receptor. A class of drugs that interrupts the recycling of bile acids was already known, i.e. cholestyramine (Brown and Goldstein 1984).

2 How Does the Defect or Reduction of LDL Receptor Elevate Lipoproteins Containing Apoprotein B-100?

During the past several years, much has been learned concerning the biochemistry and the genetics of the abnormal LDL receptor in WHHL rabbit using a tissue culture system (Tanzawa et al. 1980; Kita et al. 1981). Furthermore, we are now learning about how the receptor defect affects body tissue. Especially the metabolic fate of lipoproteins (VLDL, IDL and LDL) containing apoprotein B-100. There is a marked delay in the clearance of VLDL, IDL and LDL in WHHL rabbit. The fraction catabolic rate for plasma LDL in WHHL rabbit is approximately one-third the FCR in normal rabbit (Bilheimer et al. 1982). The low catabolic rate of LDL in WHHL rabbit is insufficient to account fully for their 20-fold increase in plasma LDL levels. In addition to the reduced clearance, there must be a six-fold overproduction of LDL. How does deficiency of LDL receptors lead to excessive overproduction of LDL? According to the radiolabelled VLDL turnover study, in control rabbit, VLDL is converted to IDL at a normal rate and IDL is rapidly taken up by the liver via an LDL receptor. On the other hand, in the case of WHHL rabbit, VLDL is converted to IDL at a relatively normal rate, but most IDL fails to enter the liver for catabolism and are converted to LDL. As shown in Fig. 2, a decrease in hepatic LDL receptors can lead not only to impaired catabolism of LDL, but also to overproduction of LDL by the shunt mechanism (Kita et al. 1982; Goldstein et al. 1983). Therefore, if one can induce the hepatic LDL receptor, the level of plasma LDL becomes normalized.

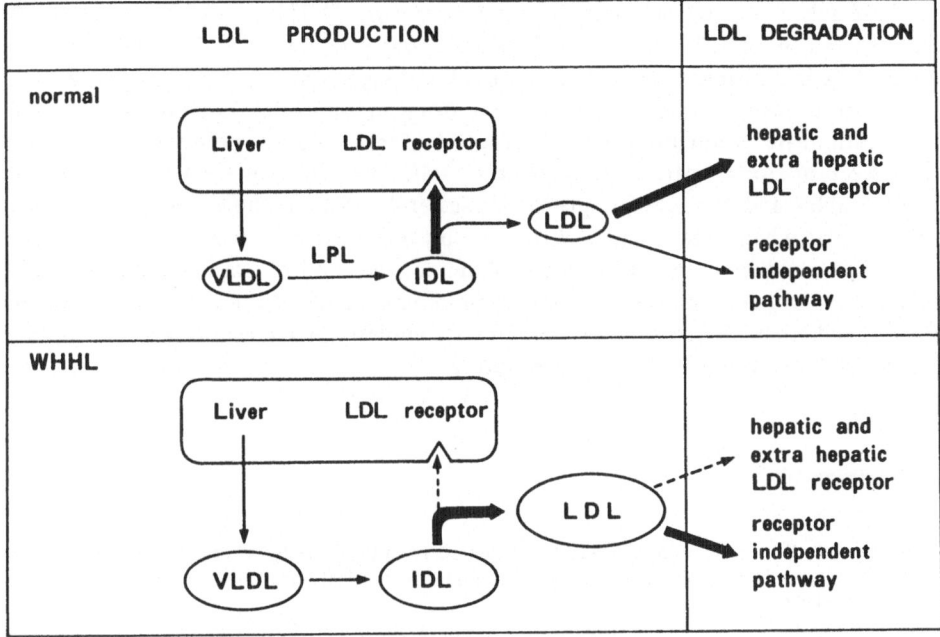

Fig. 2. Shunt pathway for the production of LDL in rabbit

3 How can the Synthesis of LDL Receptor be Induced?

If one can increase the hepatic demand for cholesterol, the liver can make more LDL receptor than before treatment. Previous studies in human beings with heterozygous familial hypercholesterolemia (FH) have raised the possibility that the single normal receptor gene in the liver can be made to produce an increased number of receptors if it is stimulated by agents that lower the hepatic content of cholesterol. Two mechanisms have been used for this stimulation: treatment with the inhibitor of cholesterol synthesis (compactin, CS-514) and with bile acid-binding resins. The studies in human

Table 1. Effect of CS-514 plus cholestyramine on LDL receptor protein in liver and on plasma cholesterol levels

Rabbit	Number of animals	LDL receptor activity ng mg^{-1} protein	Plasma cholesterol (mg dl^{-1})		
				Total	LDL
Heterozygote No drug	3	9.8 ± 1.7	Before After	118.7 ± 26.1 93.0 ± 21.1	44.9 ± 7.8 48.1 ± 15.9
CS-514	3	12.3 ± 2.2	Before After	82.7 ± 15.2 34.7 ± 6.3	26.2 ± 10.2 9.0 ± 3.9
CS-514 + Cholestyramine	3	16.6 ± 1.7	Before After	84.7 ± 17.0 15.7 ± 0.54	31.3 ± 13.2 3.1 ± 0.24

Mean ± SE.

FH have relied on measurements of the turnover in [^{125}I]-labelled LDL to establish that the fall in LDL levels is caused by increased production of LDL receptors. It should now be possible to use membrane-binding assays in liver of heterozygous WHHL rabbits to determine directly whether these drugs, or any other potential therapeutic agents, stimulate production of LDL receptors. Table 1 shows the activity of hepatic LDL receptor from control heterozygous WHHL rabbit on drug-treated heterozygous WHHL rabbit and the level of plasma cholesterol and LDL cholesterol. Data clearly showed that when heterozygous WHHL rabbit was treated with either CS-514 alone or CS-514 in combination with cholestyramine, the level of plasma cholesterol and LDL cholesterol decreased and the activity of hepatic LDL receptor in rabbits increased. We conclude that CS-514 alone or CS-514 with cholestyramine lead to a lower production of LDL by raising hepatic LDL receptors.

References

Bilheimer DW, Watanabe Y, Kita T (1982) Impaired receptor-mediated catabolism of low density lipoprotein in the WHHL rabbit, an animal model of familial hypercholesterolemia. Proc Natl Acad Sci USA 79:3305–3309

Brown MS, Goldstein JL (1984) How LDL receptors influence cholesterol and atherosclerosis. Sci Am 251:58–66

Goldstein JL, Brown MS (1982) The LDL receptor defect in familial hypercholesterolemia implications for pathogenesis and therapy. Med Clin N Am 66:335–362

Goldstein JL, Kita T, Brown MS (1983) Defective lipoprotein receptors and atherosclerosis. Lessons from an animal counterpart of familial hypercholesterolemia. N Engl J Med 309:288–296

Kita T, Brown MS, Watanabe Y, Goldstein JL (1981) Deficiency of low density lipoprotein receptors in liver and adrenal gland of the WHHL rabbit, an animal model of familial hypercholesterolemia. Proc Natl Acad Sci USA 78:2268–2272

Kita T, Brown MS, Bilheimer DW, Goldstein JL (1982) Delayed clearance of very low density and intermediate density lipoproteins with enhanced conversion to low density lipoprotein in WHHL rabbits. Proc Natl Acad Sci USA 79:5693–5697

Tanzawa K, Shimada Y, Kuroda M, Tsujita Y, Arai M, Watanabe Y (1980) WHHL-rabbit: a low density lipoprotein receptor-deficient animal model for familial hypercholesterolemia. FEBS Lett 118:81–84

Tsujita Y, Kuroda M, Shimada Y, Tanzawa K, Arai M, Kaneko I, Tanaka M, Masuda H, Tarumi C, Watanabe Y, Fujii S (1986) CS-514, a competitive inhibitor of 3-hydroxy-3-methylglutaryl coenzyme A reductase: tissue-selective inhibition of sterol synthesis and hypolipidemic effect on various animal species. Biochim Biophys Acta 877:50–60

The Clinical Pharmacology of SQ 31,000 (CS 514) in Healthy Subjects

H. Y. Pan [1], D. A. Willard [2], P. T. Funke [1], and D. N. McKinstry [1]

1 Introduction

Atherosclerosis, with the attendant consequences of coronary heart disease and stroke, is the major cause of death in the United States and most developed countries. Elevated serum cholesterol and, in particular, the low-density lipoprotein (LDL) fraction is a major risk factor in the development of arteriosclerotic disease in both animals and humans (Consensus Conference 1985, AHA Special Report 1984). Currently available lipid-lowering drugs are either poorly tolerated, cumbersome to ingest, or lack substantial effect on cholesterol or LDL. SQ 31,000: (3R,5R)-3,5-dihydroxy-7-[(1S,2S,6S, 8S, 8aR)-1,2,6,7,8,8a-hexahydro-6-hydroxy-2-methyl-8-[(S)-2-methyl-1-oxobutoxy]-1-naphthalenyl]-heptanoic acid, monosodium salt (Fig. 1), is a new HMG-CoA reductase inhibitor, structurally related to lovastatin and compactin. These agents are competitive inhibitors of HMG-CoA reductase, the rate-limiting enzyme in cholesterol biosynthesis. Both lovastatin and compactin have been reported to lower total cholesterol and LDL in healthy volunteers (Tobert et al. 1982) and in patients with familial hypercholesterolemia (Mabuchi et al. 1983; Illingworth and Sexton 1984). The cholesterol-lowering effects of these inhibitors are believed to be due to modest reductions in cellular pools of cholesterol, resulting in compensatory increases in the number of LDL receptors and enhanced clearance of cholesterol from the circulation (Bilheimer et al. 1983; Grundy and Bilheimer 1983). SQ 31,000 has been shown to be equipotent to lovastatin and compactin but is a more tissue-specific inhibitor of sterol synthesis (Tsujita et al. 1986).

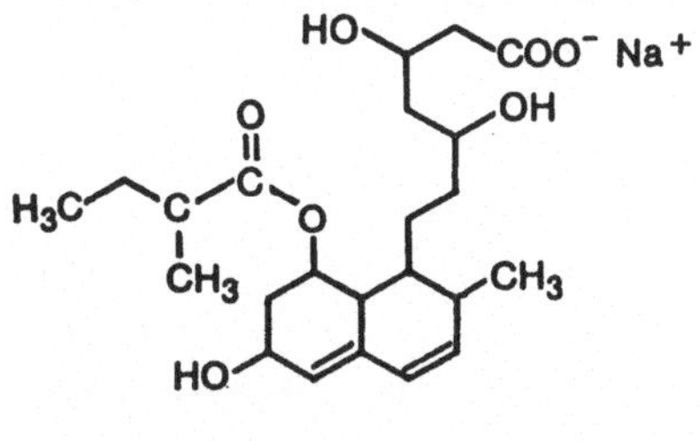

Fig. 1. SQ 31,000

1   The Squibb Institute for Medical Research, P.O. Box 4000, Princeton, NJ 08543, USA
2   The Medical Center at Princeton, Princeton, NJ 08540, USA

Drugs Affecting Lipid Metabolism
Ed. by R. Paoletti et al.
© Springer-Verlag Berlin Heidelberg 1987

In this study, the safety, tolerability, pharmacokinetic and pharmacodynamic effects of SQ 31,000 were investigated at three different doses in healthy normocholesterolemic subjects.

2 Methods

Subjects. 36 healthy male subjects between 18 and 39 years of age (mean, 27.8 ± 1.0) were enrolled. Informed consent was obtained and prestudy screening consisted of a complete physical examination, routine hematology, serum chemistry, urinalysis, and ECG. All subjects had serum cholesterol levels of 150 to 250 mg dl^{-1}. They were admitted to hospital 4 days before dosing and put on a diet comprised of 35 Cal kg^{-1} body wt., with 30% fat, 55% carbohydrates, and 15% protein. The daily cholesterol intake ranged from 275–325 mg.

Study Design. The 36 subjects were divided into three groups of 12. Placebo tablets were administered to four and SQ 31,000 tablets to eight subjects in each group in a double-blind, randomized fashion. Three dose levels were examined, 10 mg bid, 20 mg bid, and 40 mg bid. On day 1, a single dose of placebo, 10, 20, or 40 mg was administered followed by blood and urine sampling over 24 h for SQ 31,000 and RMS-416 (major metabolite) levels. Placebo or SQ 31,000 tablets were then administered twice daily for 10 days (days 2–11). On day 12, single doses were again administered followed by frequent blood and urine collections for steady-state pharmacokinetic analysis. Lipid and safety parameters were assessed at 0, 4, 8, and 12 days. Safety parameters included hematology, clotting function, urinalysis, serum chemistry, plasma and urinary steroids, and special parameters to examine possible effects on renal glomerular (β_2-microglobulin) or tubular (N-acetyl-β-glucosaminidase and alanine aminopeptidase) functions.

Analytical Procedures. Analysis of all samples for safety and lipid parameters was carried out in CDC (Center for Disease Control) certified laboratories. Plasma SQ 31,000 and RMS-416 levels were determined by a GC/MS method and urine levels were assayed by an HPLC procedure.

Clinical Evaluation. All subjects had complete physical examinations and ECG prior to dosing and on days 1, 2, 4, 8, and 12. These evaluations were also carried out on day 14 prior to discharge from the study.

3 Results

Tolerability and Safety. All 36 subjects completed the study. One subject receiving 10 mg bid developed a mild maculopapular skin rash involving the back and extremities on day 8. He was continued on therapy to day 12 with noticeable fading of the rash on day 10. No other symptoms or complaints were noted. Another subject in the 10-mg bid group developed a slight elevation of SGPT on day 12 (level = 78 IU l^{-1}, normal range = 10–55 IU l^{-1}). Two days after stopping treatment (day 14) his SGPT was 98 and a week later it returned to 19. All other liver enzymes remained within the normal range for this subject throughout the study. All other parameters in the remaining 35 subjects, including plasma and urinary steroids, and special renal function tests, did not show any significant changes compared to pretreatment values or the values in the placebo-treated group.

Pharmacokinetics. The initial and steady-state plasma pharmacokinetic parameters for SQ 31,000 and a hydroxylated metabolite, RMS-416, at the three dose levels are shown in Tables 1 and 2. Trough values (C_{min}) taken at day 4, 8, and 12 of the study showed no drug accumulation. Plasma protein-binding studies showed that SQ 31,000 was approximately 50 to 65% bound, while RMS-416 was 35 to 60% bound. There was no difference in binding after the initial dose and at steady state. Urinary recovery (SQ 31,000 + RMS-416) was approximately 15% of the single daily dose on day 1.

Pharmacodynamics. Percent changes in total cholesterol and LDL-cholesterol after SQ 31,000 treatment are illustrated in Figs. 2 and 3, respectively. Although the absolute HDL-cholesterol values showed a decrease at the end of treatment (–17% with 40 mg bid, p = NS, compared to –11.3% for the placebo group), the change in the percent of HDL-cholesterol in the total cholesterol content was significantly elevated in the 40-mg bid group (+17.5%, $p < 0.01$, compared to placebo). Apolipoprotein B also decreased significantly ($p < 0.001$) in the 40-mg bid group. Triglyceride and VLDL-cholesterol levels also showed decreases but, in general, were not statistically significant.

Table 1. Initial and steady state SQ 31,000 pharmacokinetics in healthy subjects

Dose		Initial[a]				Steady state[a]			
(mg)	N	C_{max}	T_{max}	AUC	$T^1/_2$	C_{max}	T_{max}	AUC	$T^1/_2$
10	8	9.1	1.0	22.7	1.5	8.8	1.3	21.8	1.5
20	8	26.5	1.0	59.9	1.5	19.0	1.1	43.7	1.5
40	8	45.8	1.3	100.6	1.4	40.8	1.3	91.6	1.7

[a] C_{max} = ng ml^{-1}; T_{max} = h; AUC = ng $ml^{-1} \times$ h; $T^1/_2$ = h.

Table 2. Initial and steady state RMS-416 pharmacokinetics in healthy subjects

Dose		Initial[a]				Steady state[a]			
(mg)	N	C_{max}	T_{max}	AUC	$T^1/_2$	C_{max}	T_{max}	AUC	$T^1/_2$
10	8	11.1	0.9	19.2	0.9	12.9	1.2	24.4	0.8
20	8	27.5	1.1	50.0	1.0	40.2	1.2	78.2	1.2
40	8	56.3	1.3	103.0	0.9	71.3	1.3	148.4	1.3

[a] C_{max} = ng ml^{-1}; T_{max} = h; AUC = ng $ml^{-1} \times$ h; $T^1/_2$ = h.

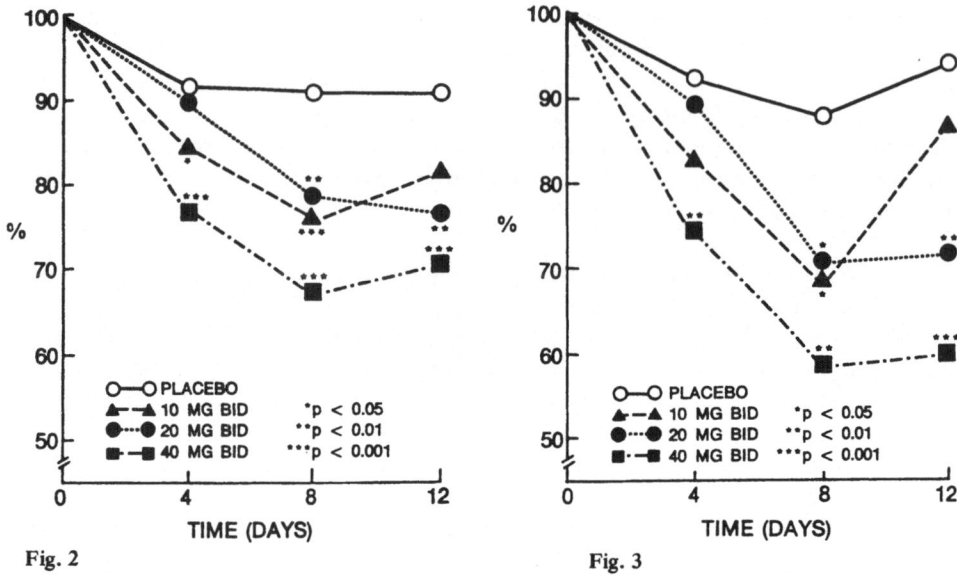

Fig. 2

Fig. 3

Fig. 2. Percent change in total cholesterol after SQ 31,000 treatment

Fig. 3. Percent change in LDL-cholesterol after SQ 31,000 treatment

4 Discussion and Conclusion

The safety, pharmacokinetics, and pharmacodynamics of SQ 31,000 were examined in 36 healthy male subjects (24 given active drug) over a 12-day period under well-controlled conditions. Other than the case of mild rash and another with isolated transient elevation of SGPT, the drug was well tolerated by all subjects. Time of peak plasma levels (T_{max}) for both SQ 31,000 and RMS-416 was approximately 1 h. The plasma elimination half-life ($t^1/_2$ β) for SQ 31,000 was slightly longer than that for RMS-416, 1.5 versus 1.0 h. Areas under the plasma concentration-time curves (AUC) for both SQ 31,000 and RMS-416 after initial and steady-state dosing demonstrated dose proportionality. Peak plasma levels (C_{max}) also showed linear dose relationships. Steady-state pharmacokinetic parameters indicated no drug accumulation on these bid regimes. The slightly lower SQ 31,000 AUC values (p = NS vs initial dose) at steady state suggest induction of liver enzymatic biotransformation. The higher metabolite (RMS-416) AUC values at steady state also suggest the possibility of enzyme induction. These data must be corroborated and further studies are required to investigate possible induction of metabolism. SQ 31,000 was a very effective cholesterol and LDL-cholesterol lowering agent even in these subjects with normal cholesterol levels. At 80 mg day^{-1}, total cholesterol and LDL-cholesterol decreased by about 30 and 40%, respectively, after 10 days of bid treatment. These decreases are greater than those generally achieved with other classes of hypocholesterolemic agents. This study demonstrated that SQ 31,000 is a safe and potent lipid-lowering agent with desirable pharmacokinetic properties.

References

AHA Special Report (1984) Recommendations for treatment of hyperlipidemia in adults. Ad hoc committee to design a dietary treatment of hyperlipoproteinemia. Circulation 69:1067A

Bilheimer DW, Grundy SM, Brown MS, Goldstein JL (1983) Mevinolin and colestipol stimulate receptor-mediated clearance of low-density lipoprotein from plasma in familial hypercholesterolemia heterozygotes. Proc Natl Acad Sci USA 80:4124

Consensus Conference (1985) Lowering blood cholesterol to prevent heart disease. Nat Inst Health, Bethesda, MD. J Am Med Assoc 253:2080

Grundy SM, Bilheimer DW (1984) Inhibition of 3-hydroxy-3-methylglytaryl-CoA reductase by mevinolin in familial hypercholesterolemia heterozygotes: effect on cholesterol balance. Proc Natl Acad Sci USA 81:2538

Illingworth DR, Sexton GJ (1984) Hypocholesterolemic effects of mevinolin in patients with heterozygous familial hypercholesterolemia. J Clin Invest 74:1972

Mabuchi H, Sakai T, Sakai Y et al. (1983) Reduction of serum cholesterol in heterozygous patients with familial hypercholesterolemia. Additive effects of compactin and cholestyramine. N Engl J Med 308:609

Tobert JA, Bell GD, Birtwell J et al. (1982) Cholesterol-lowering effect of mevinolin, an inhibitor of 3-hydroxy-3-methylglutaryl-coenzyme A reductase, in healthy volunteers. J Clin Invest 69: 913

Tsujita Y, Kuroda M, Shimada Y et al. (1986) CS-514, a competitive inhibitor of 3-hydroxy-3-methylglutaryl coenzyme A reductase: tissue-selective inhibition of sterol synthesis and hypolipidemic effect on various animal species. Biochim Biophys Acta 877:50

Long-Term Effects of CS-514 on Serum Lipoprotein Lipid and Apolipoprotein Levels in Patients with Familial Hypercholesterolemia

H. Mabuchi, N. Kamon, H. Fujita, I. Michishita, M. Takeda, K. Kajinami, H. Itoh, T. Wakasugi, and R. Takeda[1]

1 Introduction

Familial hypercholesterolemia is an autosomal dominant disorder characterized by elevated levels of low-density lipoprotein (LDL) cholesterol, tendon xanthomas, and premature coronary artery disease (CAD) (1). The disorder results from a complete or partial defect of the LDL receptor (1) that normally controls the degradation of LDL. Reduction of LDL-cholesterol levels has proved to be effective in decreasing the incidence of CAD (2, 3). Compactin (4) and mevinolin (5), competitive inhibitors of 3-hydroxy-3-methylglutaryl-coenzyme A (HMG-CoA) reductase, are effective in lowering cholesterol levels of patients with familial hypercholesterolemia. These drugs lower plasma LDL by decreasing the rate of LDL production and stimulating the production of LDL receptor in the liver (6) (Fig. 1). LDL receptor has a specific binding activity

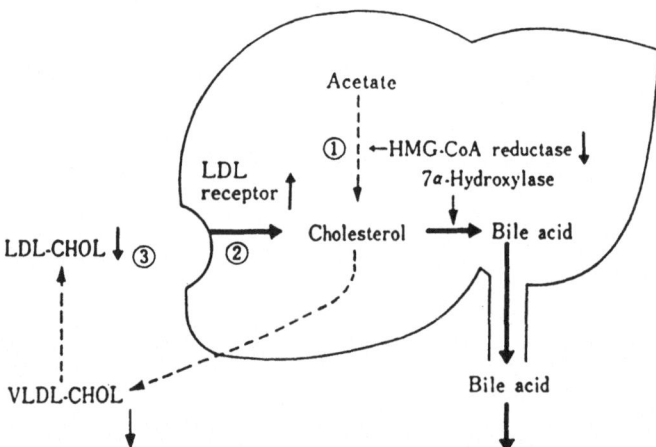

Fig. 1. Postulated mechanisms of actions of compactin and its analogs on LDL-cholesterol metabolism. *1* Compactin and its analogs inhibit cholesterol biosynthesis. *2* Hepatic cholesterol is supplied by increasing receptor-mediated LDL-cholesterol uptake. *3* Consequently, the plasma LDL-cholesterol falls

1   The Second Department of Internal Medicine, Kanazawa University School of Medicine, Kanazawa, Japan

Drugs Affecting Lipid Metabolism
Ed. by R. Paoletti et al.
© Springer-Verlag Berlin Heidelberg 1987

Compactin   CS-514

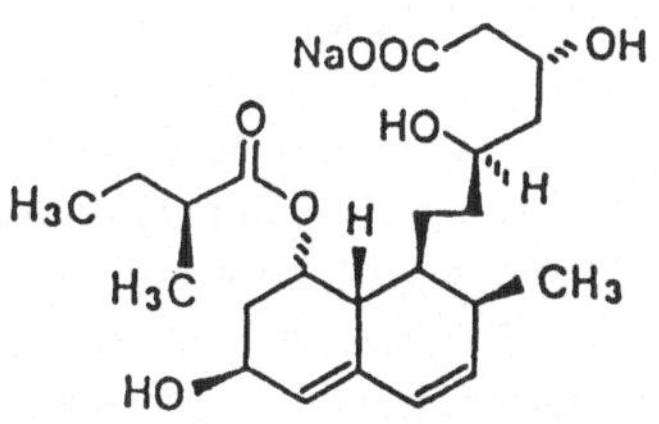

Structure		
Molecular formula	$C_{23}H_{34}O_5$	$C_{23}H_{35}O_7Na$
Molecular weight	390.52	446.52

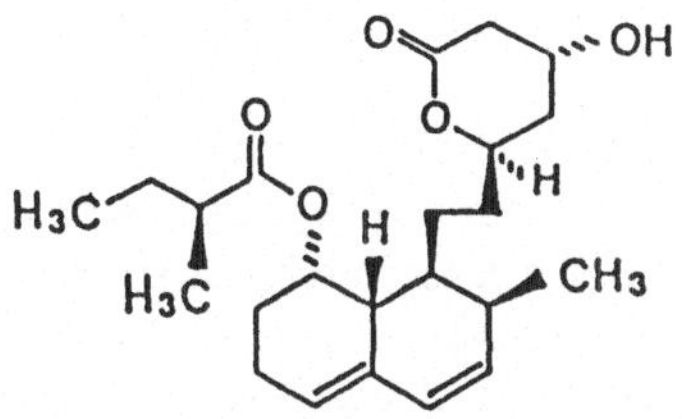

Fig. 2. Chemical structures of compactin and CS-514

for apolipoproteins B and E (7). The apolipoprotein B is present in very low-density lipoprotein (VLDL), intermediate density lipoprotein (IDL), and LDL (8), which are highly associated with CAD (9, 10). Apolipoproteins A-I and A-II are primary apolipoproteins in high-density lipoprotein (HDL) (8), and apolipoprotein A-I in patients with · CAD has been demonstrated to be decreased (11).

CS-514 is a new derivative of compactin (Fig. 2). We studied the long-term effects of CS-514 on serum lipoproteins and apolipoproteins in the treatment of heterozygous patients with familial hypercholesterolemia. The short-term results of CS-514 were reported elsewhere (12).

2 Materials and Methods

Nineteen heterozygous patients with familial hypercholesterolemia were chosen for this study. The diagnosis of familial hypercholesterolemia was based on the following two criteria: the presence of primary hypercholesterolemia with tendon xanthomas and the presence of primary hypercholesterolemia with or without tendon xanthomas in a first-degree relative of a patient with familial hypercholesterolemia. Achilles tendon xanthomas were diagnosed by radiographic measurement of the thickness of the tendons (13). The clinical and laboratory data are summarized in Table 1.

Table 1. Clinical data in 19 heterozygous patients with familial hypercholesterolemia

	Age (yr)	Height (cm)	Weight (kg)	Cholesterol (mg dl^{-1})	Triglyceride (mg dl^{-1})	Phospholipid (mg dl^{-1})	Achilles tendon thickness (mm)
Range	19–69	142–179	37–73	268–456	79–403	261–373	9.0–26.0
Mean ± SEM	50±3	159±3	58±3	353±13	165±22	295±8	13.4±1.1

All patients were given instructions for a diet low in cholesterol and saturated fat. After at least 4 weeks of stabilization, drug therapy with CS-514 at a dose of 5 mg twice daily was started. After 6 weeks, dosage of CS-514 was increased to 10 mg twice daily.

Combined drug therapy with CS-514, cholestyramine, and probucol was studied in an other six heterozygous patients with familial hypercholesterolemia. Initially, drug therapy with cholestyramine (4 g three times daily) was started. After several months of cholestyramine therapy, additional therapy with probucol (500 mg twice daily) was started. Finally, CS-514 was added to the cholestyramine-probucol regimen.

Methods of lipoprotein fractionations and lipid determinations were shown in our previous paper (4). All apolipoproteins were determined by the single radial immuno-diffusion method (14) (Daiichi Chemical Co Ltd, Tokyo) with an agarose gel plate containing 3% goat antiserum against purified human apolipoproteins A-I, A-II, B, C-II, C-III, and E. Statistical calculations were performed with Student's t-test.

3 Results

Table 2 shows the concentrations of serum cholesterol, triglyceride, and phospholipids in 19 patients with familial hypercholesterolemia, by sequential therapy with a diet low in cholesterol and saturated fat, the diet plus 5 mg CS-514 twice daily, and the diet plus 10 mg CS-514 twice daily. The individual serum cholesterol responses are shown in Fig. 3. Mean serum cholesterol levels ($\pm$SEM) fell by 17%, from 342 ± 16 mg dl^{-1} to 284 ± 12 mg dl^{-1} during therapy with 5 mg CS-514 twice daily ($p < 0.001$). The higher dosage of CS-514 produced further reduction in serum cholesterol levels by 10% to 255 ± 12 mg dl^{-1} ($p < 0.01$). Mean serum phospholipid levels fell from 295 ± 10 mg dl^{-1} during the diet alone to 256 ± 11 mg dl^{-1} ($p < 0.01$). Table 2 and Fig. 4 show the cholesterol levels in lipoprotein fractions during sequential treatment. When compared with the diet alone, 5 mg CS-514 twice daily reduced LDL-cholesterol concentrations by 25% from 287 ± 20 mg dl^{-1} to 216 ± 14 mg dl^{-1} ($p < 0.001$). When CS-514 was used in doses of 10 mg twice daily, LDL-cholesterol levels further decreased to 210 ± 17 mg dl^{-1} ($p < 0.05$). The total reduction from the values during the diet alone was 27% ($p < 0.001$). HDL-cholesterol levels increased significantly from 44 ± 4 mg dl^{-1} with diet alone to 49 ± 4 mg dl^{-1} with 10 mg CS-514 twice daily ($p < 0.01$). VLDL and IDL cholesterol levels showed no significant changes.

Serum apolipoprotein A-I levels increased significantly from 119 ± 21 mg dl^{-1} with diet alone to 130 ± 5 mg dl^{-1} with 5 mg CS-514 twice daily and increased to 157 ± 8 mg dl^{-1} with 10 mg of CS-514 twice daily ($p < 0.001$) (Table 3, Fig. 5). Serum apolipoprotein A-II increased slightly from 30 ± 1 mg dl^{-1} to 37 ± 2 mg dl^{-1} after the treatment with 10 mg CS-514 twice daily ($p < 0.01$). Serum apolipoprotein B levels decreased significantly from 221 ± 12 mg dl^{-1} with diet alone to 185 ± 10 mg dl^{-1} with 5 mg CS-514 twice daily ($p < 0.001$), and fell further to 164 ± 10 mg dl^{-1} with 10 mg CS-514 twice daily ($p < 0.001$). Serum apolipoprotein C-II and C-III levels remained unchanged during the treatment period. Serum apolipoprotein E levels changed significantly from 6.4 ± 0.4 mg dl^{-1} to 5.2 ± 0.4 mg dl^{-1} ($p < 0.01$) with 10 mg CS-514 twice daily.

Table 2. Serum and HDL cholesterol, triglyceride, and phospholipid levels during the treatment with CS-514[a]

	Diet alone	CS-514 (10 mg day⁻¹)		CS-514 (20 mg day⁻¹)				
		Week 2	Week 4	Week 8	Week 12	Week 24	Week 36	Week 48
Serum cholesterol	342 ± 16	292 ± 14	284 ± 12	263 ± 13	260 ± 11	268 ± 12	269 ± 14	255 ± 12
HDL-cholesterol	44 ± 4	47 ± 3	49 ± 4	50 ± 3	50 ± 4	50 ± 4	50 ± 4	49 ± 4
Serum triglyceride	137 ± 17	137 ± 14	151 ± 17	124 ± 15	124 ± 13	134 ± 14	117 ± 14	117 ± 15
Serum phospholipid	295 ± 10	173 ± 9	272 ± 8	260 ± 10	257 ± 8	267 ± 9	268 ± 11	256 ± 11

[a] All values are mg dl⁻¹ and are given as mean ± SEM.

Table 3. Serum apolipoprotein levels during the treatment with CS-514[a]

Apolipoprotein	Diet alone	CS-514 (10 mg day⁻¹)		CS-514 (20 mg day⁻¹)				
		Week 2	Week 4	Week 8	Week 12	Week 24	Week 36	Week 48
A-I	119 ± 5	123 ± 5	130 ± 5	134 ± 5	132 ± 5	145 ± 5	150 ± 6	157 ± 8
A-II	30 ± 1	21 ± 1	32 ± 1	33 ± 1	34 ± 1	37 ± 2	39 ± 2	37 ± 2
B	226 ± 12	185 ± 9	185 ± 10	172 ± 11	169 ± 8	185 ± 11	168 ± 9	164 ± 10
C-II	4.9 ± 0.4	4.6 ± 0.4	4.7 ± 0.4	4.3 ± 0.4	4.2 ± 0.3	4.6 ± 0.4	4.5 ± 0.5	4.1 ± 0.4
C-III	10.9 ± 0.6	10.5 ± 0.6	10.9 ± 0.8	10.5 ± 0.9	10.6 ± 0.6	10.9 ± 0.8	9.5 ± 0.7	9.4 ± 0.6
E	6.2 ± 0.4	5.2 ± 0.2	5.4 ± 0.2	5.2 ± 0.4	5.0 ± 0.2	5.7 ± 0.3	5.4 ± 0.4	5.2 ± 0.4

[a] All values are mg dl⁻¹ and are given as mean ± SEM.

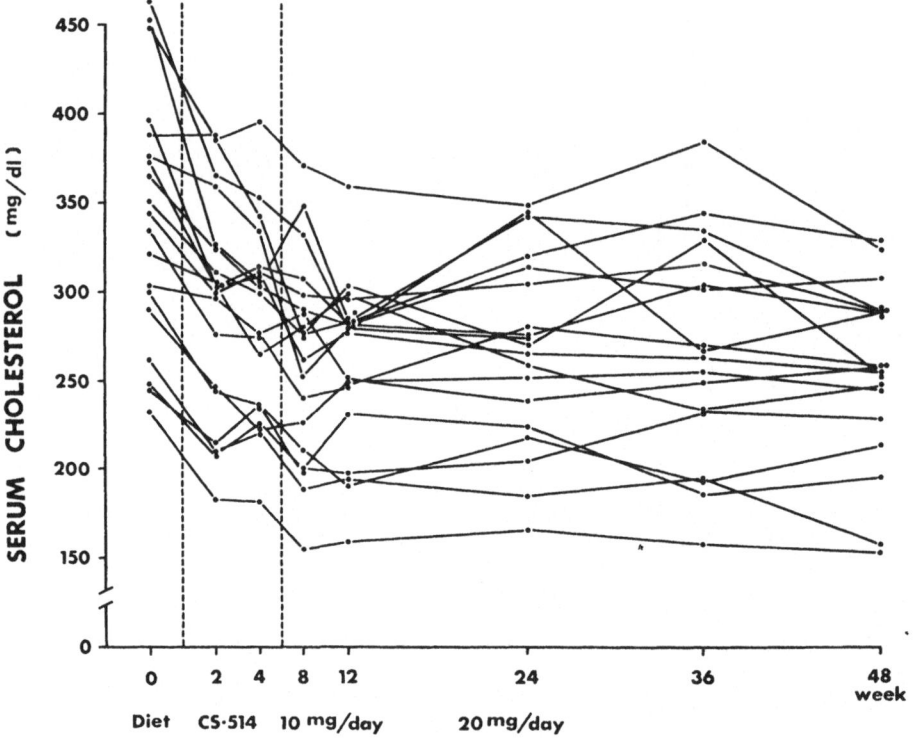

Fig. 3. Serum cholesterol levels (mg dl⁻¹) in 19 heterozygous patients with familial hypercholesterolemia

With the combination therapy with cholestyramine, probucol, and CS-514, serum cholesterol decreased by 40% from 370 ± 27 mg dl^{-1} to 224 ± 18 mg dl^{-1} ($p < 0.001$), and LDL-cholesterol levels decreased by 48% from 274 ± 19 mg dl^{-1} to 143 ± 13 mg dl^{-1} ($p < 0.001$) (Fig. 6). However, HDL-cholesterol levels decreased insignificantly.

No changes of xanthomas were observed during the treatment period. No side effects, such as gastrointestinal, hematologic, neurologic, or other biochemical abnormalities, were observed in any of the patients.

4 Discussion

Compactin and its derivatives are novel competitive inhibitors of HMG-CoA reductase (15) and have been found to be effective in lowering serum cholesterol in heterozygous patients with familial hypercholesterolemia (4, 5). A new derivative (CS-514) of compactin was produced by microbial transformation of compactin. Both CS-541 and compactin have a portion that resembles the HMG moiety of HMG-CoA, a substrate of HMG-CoA reductase (15), and they bind to the enzyme's active site and inhibit the enzyme activity. Thus, the mechanism of action of these compounds is competitive

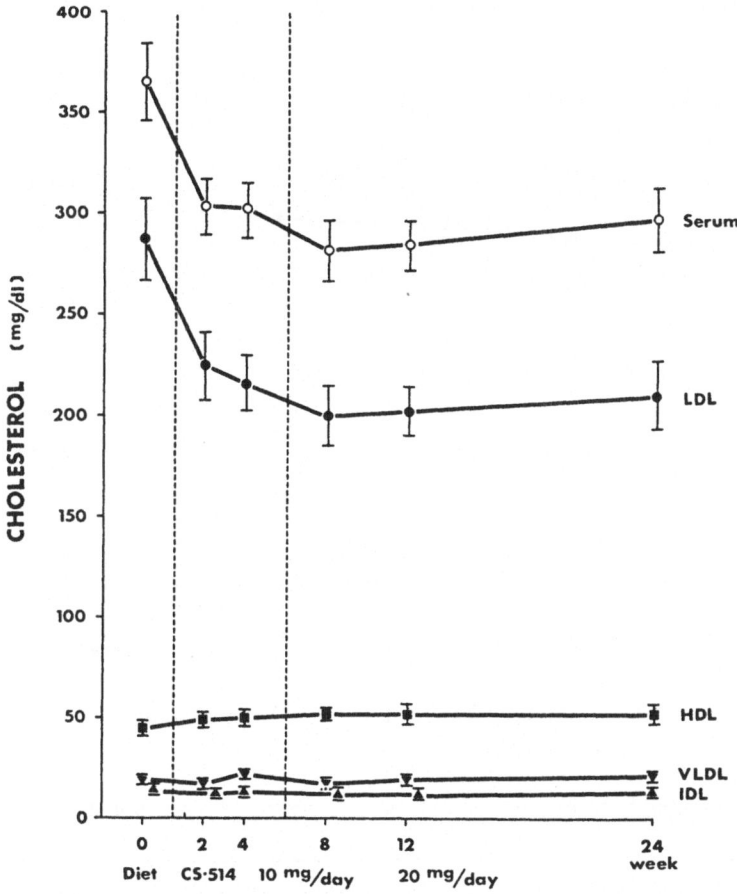

Fig. 4. Effects of CS-514 on serum, LDL, HDL, VLDL, and IDL cholesterol levels

with regard to HMG-CoA reductase. Compactin and its derivatives are known to in-crease LDL receptor activities in the liver (6), and the LDL receptor binds both apo-lipoproteins B and E of the lipoproteins (7). Thus, lipoproteins containing apolipopro-tein B and E (VLDL, IDL, and LDL) should be selectively decreased with CS-514. In this study, apolipoprotein B and E levels decreased significantly, while apolipopro-tein A-I and A-II levels increased.

Higher apolipoprotein B levels were reported in subjects with CAD compared to those without CAD (16). On the other hand, many researchers have demonstrated a decrease in apolipoprotein A-I in subjects with CAD (17). Thus, CS-514 might be promising in correcting abnormal apolipoprotein levels in familial hypercholesterol-emia.

When the relative potencies of compactin, mevinolin, and CS-514 were compared, the effects of 20 mg day^{-1} CS-514 is equivalent to those of 40 mg day^{-1} mevinolin and 15–60 mg day^{-1} compactin (Table 4) (4, 5, 18, 19). Combination therapy with several drugs sometimes normalizes the LDL-cholesterol levels (20, 21), and the three-

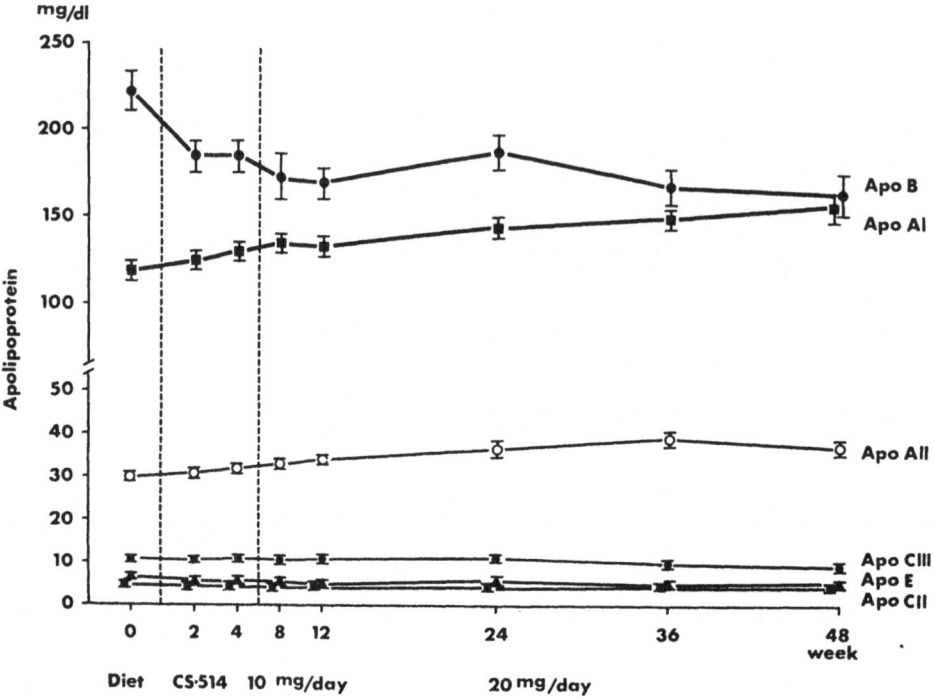

Fig. 5. Effects of CS-514 on serum apolipoprotein A-I, A-II, B, C-II, C-III, and E levels

Table 4. Effects of compactin, mevinolin and CS-514 on LDL-cholesterol levels in hypercholesterolemic patients

Authors	Subjects (n)	Drug	Dose (mg day⁻¹	LDL-cholesterol[a]		
				Before treatment	After treatment	Percentage changes
Mabuchi et al.	7	Compactin	30–60	299 ± 24	211 ± 29	−29
Yamaguchi et al.	11	Compactin	15–60	195 ± 49	139 ± 47	−29
Bilheimer et al.	6	Mevinolin	40	262 ± 43	191 ± 33	−27
Illingworth et al.	13	Mevinolin	40	369 ± 43	240 ± 50	−35
Mabuchi et al.	19	CS-514	20	287 ± 84	210 ± 61	−27

[a] Mean ± SD, mg dl⁻¹.

drug therapy with cholestyramine, probucol, and CS-514 reduced LDL-cholesterol levels by 48%, as in the other combined-drug therapies (Table 5).

Thus, CS-514 reduces atherogenic lipoproteins and apolipoproteins B and increases HDL and apolipoprotein A-I and A-II, and appears to be a useful drug for heterozygous familial hypercholesterolemia.

Acknowledgment. We are indebted to the Sankyo Company, Tokyo, for providing the CS-514.

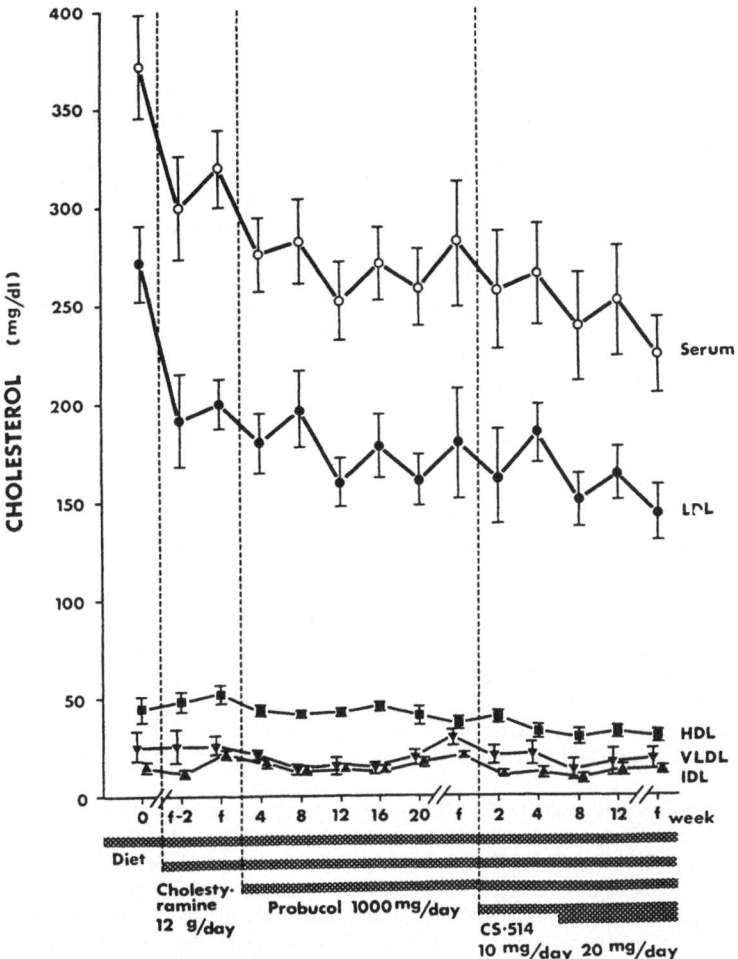

Fig. 6. Effects of combined drug therapy with cholestyramine, probucol, and CS-514 on serum and lipoprotein cholesterol levels in six patients with familial hypercholesterolemia

Table 5. Effects of combined drug therapy in familial hypercholesterolemia

Authors	Sub-jects (n)	Diet alone	LDL-cholesterol levels (mg dl⁻¹)[a]			Percentage reduction
			Initial drug therapy		Additional drug therapy	
Mabuchi et al.	10	263 ± 39	Cholestyramine	190 ± 28	Compactin	125 ± 31	53
Illingworth	10	409 ± 26	Mevinolin	254 ± 15	Colestipol	188 ± 12	54
Mabuchi et al.	6	274 ± 42	Cholestyramine	201 ± 30	Probucol + CS-514	174 ± 57 143 ± 33	48

[a] Mean ± SD.

References

1. Goldstein JL, Brown MS (1983) Familial hypercholesterolemia. In: Stanbury JB, Wyngaarden JB, Fredrickson DS, Goldstein JL, Brown MS (eds) The metabolic basis of inherited disease, 5th edn. McGraw-Hill, New York, pp 672–712
2. Lipid Research Clinics Program (1984) The lipid research clinics coronary primary prevention trial results. I. Reduction in incidence of coronary heart disease. JAMA 251:351–364
3. Idem. II. The relationship of reduction in incidence of coronary heart disease to cholesterol lowering. JAMA 25:365–374
4. Mabuchi H, Haba T, Tatami R et al. (1981) Effects on an inhibitor of 3-hydroxy-3-methylglutaryl coenzyme A reductase on serum lipoproteins and ubiquinone-10 levels in patients with familial hypercholesterolemia. N Engl J Med 305:478–482
5. Illingworth DR, Sexton GJ (1984) Hypocholesterolemic effects of mevinolin in patients with heterozygous familial hypercholesterolemia. J Clin Invest 74:1972–1978
6. Kovanen PT, Bilheimer DW, Goldstein JL, Jaramillo JJ, Brown MS (1971) Regulatory role for hepatic low density lipoprotein receptors in vivo in the dog. Proc Natl Acad Sci USA 78: 1194–1198
7. Brown MS, Kovanen PT, Goldstein JL (1981) Regulation of plasma cholesterol by lipoprotein receptors. Science 212:628–635
8. Breslow JL (1985) Human apolipoprotein. Molecular biology and genetic variation. Annu Rev Biochem 54:699–727
9. Mahley RW (1982) Atherogenic hyperlipoproteinemia. The cellular and molecular biology of plasma lipoproteins altered by dietary fat and cholesterol. Med Clin N Am 66:375–402
10. Tatami R, Mabuchi M, Ueda K et al. (1981) Intermediate-density lipoprotein and cholesterol-rich very low density lipoprotein in angiographically determined coronary artery disease. Circulation 64:1174–1184
11. Maciejko JJ, Homes DR, Kottke BA, Zinsmeister AR, Dinh DM, Mao JT (1983) Apolipoprotein A-I as a marker of angiographically assessed coronary-artery disease. N Engl J Med 309: 385–389
12. Mabuchi H, Kamon N, Fujita H et al. (1987) Effects of CS-514 on serum lipoprotein lipid and apolipoprotein levels in patients with familial hypercholesterolemia. Metabolism 36:475–479
13. Mabuchi H, Ito S, Haba T et al. (1977) Discrimination of familial hypercholesterolemia and secondary hypercholesterolemia by Achilles' tendon thickness. Atherosclerosis 28:61–68
14. Goto Y, Akanuma Y, Harano Y et al. (1986) Determination of serum apolipoproteins, A-I, A-II, B, C-II, C-III and E in normolipidemic healthy Japanese subjects; determined by SRID method. J Clin Biochem Nutrit 1:73–88
15. Endo A (1981) Biological and pharmacological activity of inhibitors of 3-hydroxy-3-methylglutaryl coenzyme A reductase. Trends Biochem Sci 6:10–12
16. Sniderman AD, Wolfson C, Teng B et al. (1982) Association of hyperapobetalipoproteinemia with endogenous hypertriglyceridemia and atherosclerosis. Ann Int Med 97:833–839
17. Brunzell JD, Sniderman AD, Albers JJ, Kwiterovich PO, Jr (1984) Apoproteins B and A-I and coronary artery disease in humans. Arteriosclerosis 4:79–83
18. Bilheimer DW, Grundy SM, Brown MS, Goldstein JL (1983) Mevinolin and colestipol stimulate receptor-mediated clearance of low density lipoprotein from plasma in familial hypercholesterolemia heterozygotes. Proc Natl Acad Sci USA 80:4124–4128
19. Yamaguchi K, Nakamura N, Uzawa H (1984) Blocking 3-hydroxy-3-methylglutaryl coenzyme A reductase with ML236B enhances the plasma cortisol response to adrenocorticotropin in patients with hypercholesterolemia. J Clin Endocrinol Metab 58:786–789
20. Mabuchi H, Sakai T, Sakai Y et al. (1983) Reduction of serum cholesterol in heterozygous patients with familial hypercholesterolemia. Additive effects of compactin and cholestyramine. N Engl J Med 308:609–613
21. Illingworth DR (1984) Mevinolin plus colestipol in therapy for severe heterozygous familial hypercholesterolemia. Ann Int Med 101:598–604

Intensive Drug Treatment
for Familial Hypercholesterolemia

A. Yamamoto, S. Yokoyama, and T. Yamamura [1]

1 Introduction

To prevent the development of atherosclerotic diseases, it is essential to reduce serum cholesterol to a level near or below 220 mg dl^{-1}. The discovery of compactin (ML-236 B) and its derivatives (Endo et al. 1976) made it possible to treat hypercholesterolemia by suppressing the key enzyme for cholesterol synthesis (Yamamoto et al. 1980). In this study we made a trial to evaluate the cholesterol-lowering effect of CS-514, 3β-hydroxy-compactin (Tsujita et al. 1986), in patients with familial hypercholesterolemia (FH) and to establish a potent combination therapy which enables us not only to reduce low-density lipoprotein (LDL) to an optimum level, but also to efficiently achieve the regression of xanthomas and atherosclerotic vascular lesions.

2 Subjects and Methods

Twenty patients with heterozygous FH (7 males and 13 females aged between 33 and 70 years) are the subjects of this study. CS-514 was given at a dose of 5 to 40 mg day^{-1} (mostly bid); cholestyramine 4g per each meal (twice or three times a day); and probucol 0.75 to 1.5 g day^{-1} (bid or tid). These drugs were used in either single or combined regimen. Four patients with ordinary hyperlipidemia (type IIa, non-FH) were used as a control group. Blood was obtained after overnight fasting and the serum lipids were measured by enzymatic assay. Average values of serum lipids at each stage of the treatment were obtained following several measurements at 2 or 4 week intervals.

3 Results

3.1 Effect of Varying Doses of CS-514 on Serum Cholesterol Levels

In ten FH patients, the change in LDL-cholesterol concentration was followed by varying the dose of CS-514. In almost all cases the dose-response curve appeared in

---

1 Department of Etiology and Pathophysiology, National Cardiovascular Center Research Institute, 5-7-1, Fujishiro-dai, Suita-shi, Osaka-fu 565, Japan

a saturable manner, diminishing the slope with increasing the dose of the drug and converging to a level around 220 mg dl^{-1}. In contrast, the dose-response curve went straight down to the level near or below 150 mg dl^{-1} in non-FH patients (Fig. 1). At a relatively low dosage, the effect of CS-514 was very close to or slightly (up to 50%) larger than the effect of compactin as already reported elsewhere (Yamamoto et al. 1986a).

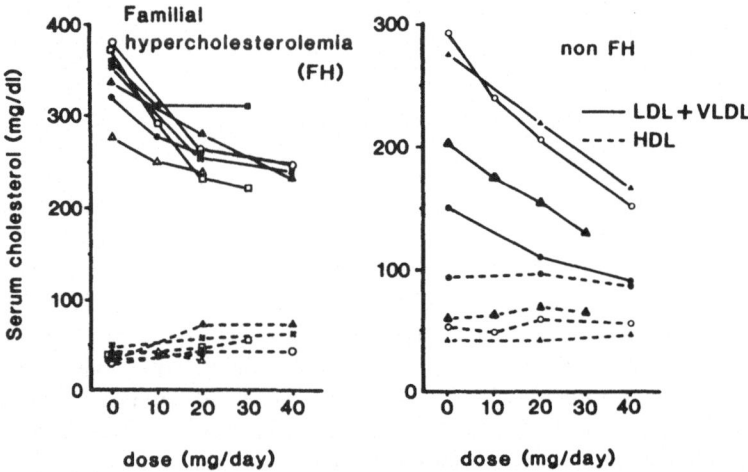

Fig. 1. Effect of CS-514 on serum LDL- and HDL-cholesterol levels in hypercholesterolemia; comparison of the dose-response curve between FH and non-FH patients

An escape phenomenon was seen in some patients. In two patients, who are cousins and the parents of a homozygous child with a receptor-negative type, a paradoxical rebound of serum choleterol (return to the pretreatment level) was observed by increasing the dose from 10 to 30 mg day^{-1} either in single regimen or in combined use with cholestyramine and probucol (Fig. 2a). In another patient, plasma cholesterol tended to increase during the course of treatment at a fixed dose (Fig. 2b). In the early phase, HDL-cholesterol mainly contributed to such an increase. But, later on, it went down to the pretreatment level, when the LDL-cholesterol level was recovered to a certain extent. Such a phenomenon was similar to the shift of HDL- and LDL-cholesterol following the treatment of FH patients with plasmapheresis (Parker et al. 1985), suggesting that the increase in cholesterol synthesis is first reflected on the HDL-cholesterol level and later on LDL.

3.2 Combination Drug Therapy Using CS-514 and Cholestyramine

Eight patients were given CS-514 (at a dose of 10 mg day^{-1}) and cholestyramine in both single and combined regimens. Use of cholestyramine alone resulted in 23% reduction in LDL-cholesterol and CS-514 alone 14% reduction. Combination of these drugs resulted in 33% reduction in LDL-cholesterol showing the additive effect (expected value: 34%) as already reported elsewhere (Yamamoto et al. 1986a). HDL-

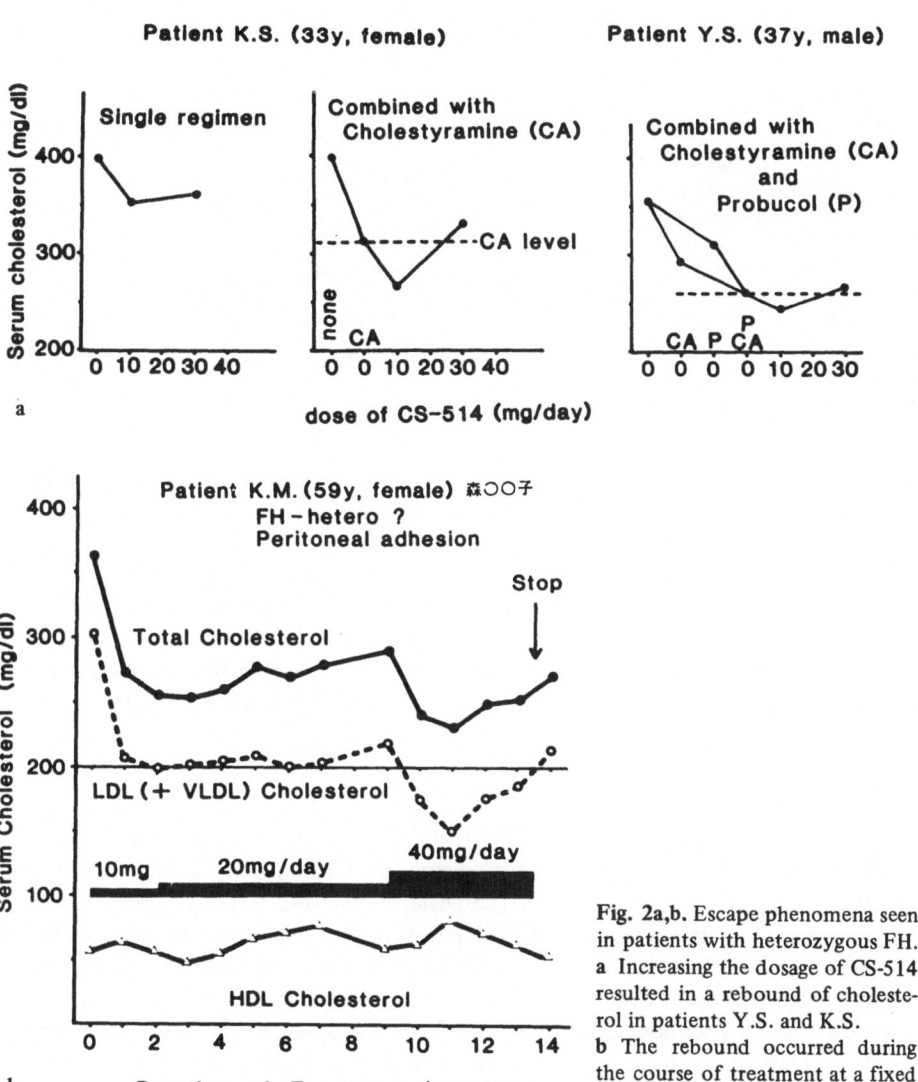

Fig. 2a,b. Escape phenomena seen in patients with heterozygous FH. a Increasing the dosage of CS-514 resulted in a rebound of cholesterol in patients Y.S. and K.S. b The rebound occurred during the course of treatment at a fixed dose in patients K.M.

cholesterol increased by using each one of these drugs and the increase by combination therapy also showed an additive effect.

3.3 Intensive Drug Treatment Using Three Drugs in Combination

A combination drug therapy with CS-514, cholestyramine, and probucol was tested in eight patients. All of them tolerated well such an intensive treatment without any serious side effects and their serum cholesterol levels fell down into the optimum range; to near or below 200 mg dl^{-1} (Fig. 3).

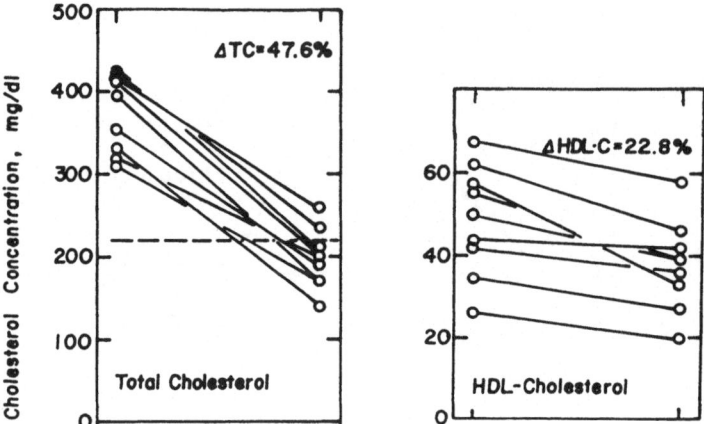

Fig. 3. Effects of an intensive drug treatment using three durgs in combination, CS-514, cholesty-ramine, and probucol, on serum total and HDL-cholesterol in patients with heterozygous FH

In five of these patients probucol was given first, and then, cholestyramine and CS-514 was successively added. HDL-cholesterol once decreased by probucol showed a good recovery in all patients by the addition of cholestyramine and further recovery was·ob-served in three patients by the addition of CS-514 (Yamamoto et al. 1986b).

4 Discussion

The clearance of plasma cholesterol was mainly performed by LDL receptors in the liver and in peripheral tissues and the cholesterol-lowering effect of cholestyramine and compactin (or its analoges) mostly depends upon the increase in the number of receptors. Probably because of the restriction in number and function of the recep-tors, the effect of these drugs on FH patients was limited compared with the effect on non-FH patients without a deficiency in LDL receptors. We have already reported that it is necessary to double the dose of compaction in patients with heterozygous FH to achieve the same extent of the reduction in serum cholesterol as in the cases of combined hyperlipidemia (Yamamoto et al. 1984). The escape phenomenon observed in three patients in this study gave us additional evidence that an induction of 3-hy-droxy-3-methylglutaryl coenzyme A reductase took place by lowering cholesterol. Such a situation forced us to use an intensive combination therapy by adding another kind of drug with a different mode of action. Probucol seems to be valuable in this sense, because it prevents oxidative modification of LDL (Steinberg D 1986) and also prevents macrophages from foamy change and causes a marked regression of xan-thomatous lesions (Yamamoto et al. 1986b,c).

References

Endo A, Kuroda M, Tanzawa K (1976) Competitive inhibition of 3-hydroxy-3-methylglutaryl coenzyme A reductase by ML-236A and ML-236B fungal metabolites having hypocholesterolemic activity. FEBS Lett 72:323–326

Parker T, Gordon BR, Saal SD, Brun AL, Ahrens EH, Jr (1985) Plasma high density lipoprotein is increased in man when low density lipoprotein is lowered by LDL-pheresis. Proc Natl Acad Sci USA 83:777–781

Steinberg D (1986) Studies on the mechanism of action of probucol. Am J Cardiol 57:16H–21H

Tsujita Y, Kuroda M, Shimada Y, Tanzawa K, Arai M, Kaneko M, Tanaka M, Masuda H, Tarumi C, Watanabe Y, Fujii S (1986) CS-514, a competitive inhibitor of 3-hydroxy-3-methylglutaryl coenzyme A reductase: tissue-selective inhibition of sterol synthesis and hypolipidemic effect on various animal species. Biochim Biophys Acta 877:50–60

Yamamoto A, Sudo H, Endo A (1980) Therapeutic effect of ML-236B in primary hypercholesterolemia. Atherosclerosis 35:259–266

Yamamoto A, Yamamura T, Yokoyama S, Sudo H, Matsuzawa Y (1984) Effect of combined drug regimen, cholestyramine and compactin, on heterozygous familial hypercholesterolemia. Int J Clin Pharm Ther Toxicol 22:493–497

Yamamtoo A, Yokoyama S, Yamamur T (1986a) Combined drug treatment and plasmapheresis for familial hypercholesterolemia. In: Fears R (ed) Pharmacological control of hyperlipidemia. Prous, Barcelona, pp 333–342

Yamamoto A, Matsuzawa Y, Yokoyama S, Funahashi T, Yamamura T, Kishino B (1986b) Effects of probucol on xynthomata regression in familial hypercholesterolemia. Am J Cardiol 57:29H–. 35H

Yamamoto A, Taka-ichi S, Hara H, Nishikawa O, Yokoyama S, Yamamura T, Yamaguchi T (1986c) Probucol prevents lipid storage in macrophages. Atherosclerosis 62:209–217

Effect of CS-514 on Hypercholesterolemic Patients

N. NAKAYA [1] and Y. GOTO [2]

The efficacy and adverse effects of CS-514 have been studied in the phase II trial conducted in Japan.

A total of 446 hypercholesterolemic patients, whose cholesterol levels were more than 220 mg dl⁻¹, were given CS-514 in 53 institutes. However, the medication procedure was out of protocol in 97 cases and 349 cases were adopted for efficacy analysis. They consist of 157 heterozygous familial hypercholesterolemic (FH) patients and 192 nonfamilial hypercholesterolemic (nonFH) patients.

CS-514 was given at a daily dose of 10 mg (5 mg b.i.d.) for 12 weeks in most patients, but the dose was increased to 20 mg day⁻¹ (10 mg b.i.d.) in 32 nonFH patients and 57 FH patients.

Average total serum cholesterol decreased by 16.6% (284.8 ± 3.3 mg dl⁻¹ to 235.6 ± 2.8 mg dl⁻¹) ($p < 0.001$) in nonFH patients and by 17.5% (355.5 ± 5.7 mg dl⁻¹ to 292.4 ± 5.6 mg dl⁻¹) ($p < 0.001$) in FH patients (Fig. 1).

When patients were classified into three groups, according to the initial value, the higher the initial value was, the larger was the rate of reduction. Serum cholesterol decreased by 19.1% in the highest group (300 mg dl⁻¹ ≤ TC); by 15.5% in the middle group (250 mg dl⁻¹ ≤ TC < 300 mg dl⁻¹); and by 12.4% in the lowest group (220 mg dl⁻¹ ≤ TC < 250 mg dl⁻¹).

LDL-cholesterol was estimated according to the formula of Friedewald. It decreased by 23.2% (195.0 ± 3.2 mg dl⁻¹ to 145.7 ± 2.8 mg dl⁻¹) ($p < 0.001$) in nonFH patients and by 23.6% (279.7 ± 6.5 mg dl⁻¹ to 214.3 ± 6.1 mg dl⁻¹) ($p < 0.001$) in FH patients.

HDL-cholesterol increased by 8.7% (51.3 ± 1.0 mg dl⁻¹ to 54.6 ± 1.1 mg dl⁻¹) ($p < 0.001$) in nonFH patients and by 12.8% (46.2 ± 1.2 mg dl⁻¹ to 52.0 ± 1.3 mg dl⁻¹) ($p < 0.001$) in FH patients (Fig. 2).

Serum triglyceride was reduced from 193.3 ± 11.0 mg dl⁻¹ to 170.8 ± 8.1 mg dl⁻¹ ($p < 0.01$) in nonFH patients and from 150.0 ± 6.7 mg dl⁻¹ to 136.8 ± 6.8 mg dl⁻¹ ($p < 0.05$) in FH patients (Fig. 3). When patients were classified into three groups, according to the initial value, the higher the initial value was, the larger was the rate of reduction. It decreased by 20.9% (442.8 ± 34.3 mg dl⁻¹ to 329.0 ± 25.3 mg dl⁻¹) ($p <$

1 Tokai University School of Medicine Tokyo Hospital, Yoyogi 1-2-5, Shibuya-ku, Tokyo, 151 Japan
2 Tokai University School of Medicine, Bohseidai, Isehara Kanagawa, 259-11 Japan

Drugs Affecting Lipid Metabolism
Ed. by R. Paoletti et al.
© Springer-Verlag Berlin Heidelberg 1987

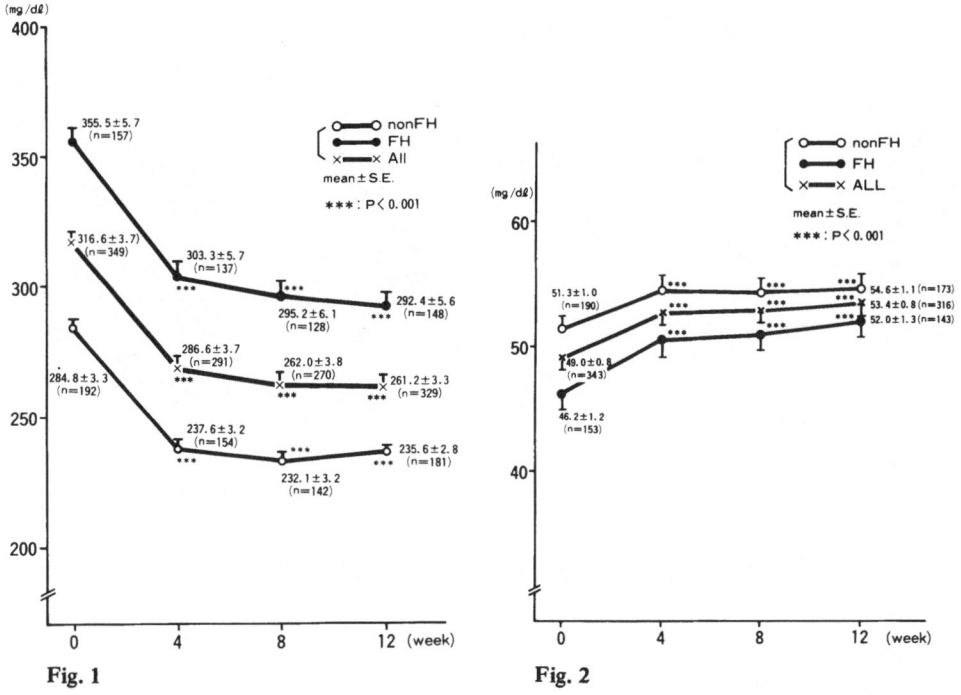

Fig. 1. Total serum cholesterol change in all, FH and non FH patients

Fig. 2. HDL-cholesterol change in all, FH and non FH patients

0.001) in the highest group (300 mg dl^{-1} ≤ TG) and by 8.8% (197.3 ± 3.6 mg dl^{-1} to 178.3 ± 6.4 mg dl^{-1}) (p < 0.01) in the middle group (150 mg dl^{-1}< TG ≤ 300 mg dl^{-1}), whereas it increased by 4.6% (104.0 ± 1.9 mg dl^{-1} to 105.4 ± 3.0 mg dl^{-1}) in the lowest group (TG < 150 mg dl^{-1}).

The atherogenic index, which was calculated from the formula (A.I.= TC-(HDL-C) /HDL-C), improved markedly in both nonFH and FH patients. It decreased from 5.02 to 3.77 in nonFH patients and from 7.78 to 5.25 in FH patients.

Apolipoproteins were measured by a single radial immunodiffusion method. Apo A-I and apo A-II increased significantly whereas apo-B, apo C-II and apo-E decreased significantly. Apo C-III also decreased, but the reduction did not attain statistical significance (Fig. 4). Apo A-I increased by 15.7% (141.2 ± 2.4 mg dl^{-1} to 161.9 ± 2.8 mg dl^{-1}) (p < 0.001) in nonFH patients and by 12.1% (132.9 ± 3.5 mg dl^{-1} to 145.9 ± 3.4 mg dl^{-1}) (p < 0.001) in FH patients. Apo A-II increased by 16.0% and 13.1% in non-FH and FH patients, respectively. Apo-B decreased by 14.3% (166.8 ± 3.7 mg dl^{-1} to 144.4 ± 3.8 mg dl^{-1}) (p < 0.001) in nonFH patients and by 17.5% (214.7 ± 6.4 mg dl^{-1} to 175.1 ± 5.3 mg dl^{-1}) (p < 0.001) in FH patients. Apo C-II and apo C-III decreased by 7.5% (p < 0.05) and 4.2%, respectively, in nonFH patients and by 7.9% (p < 0.01) and 1.4%, respectively, in FH patients. Apo-E decreased by 7.44% (6.17 ± 0.43 mg dl^{-1} to 5.72 ± 0.33 mg dl^{-1}) (p < 0.01) in nonFH patients and by 8.24% (6.34 ± 0.22 mg dl^{-1} to 5.69 ± 0.24 mg dl^{-1}) (p < 0.01) in FH patients.

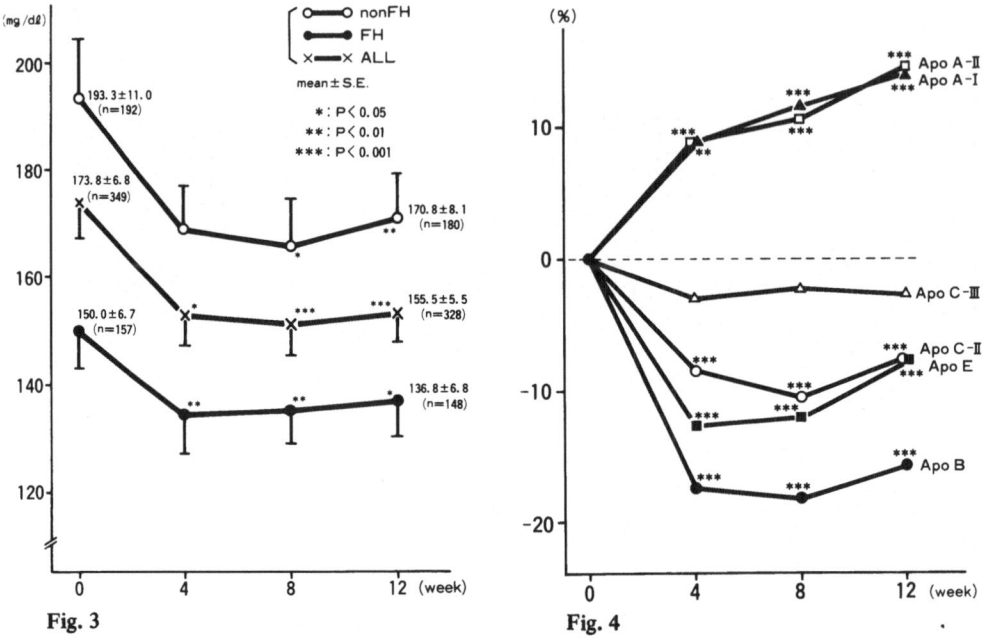

Fig. 3. Serum triglyceride change in all, FH and non FH patients

Fig. 4. Percentage change in apolipoproteins

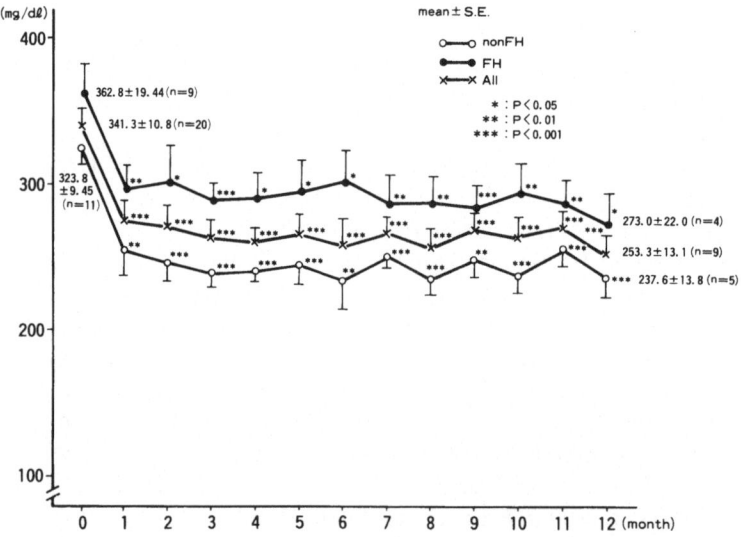

Fig. 5. Total serum cholesterol change in 1-year follow-up study

Adverse effects were examined in 446 patients. Abnormal increase in GOT (7 cases), GPT (7), γ-GTP (5), alkaline phosphatase (2), CPK (4), and eosinophil (1) were observed and gastrointestinal symptoms (7), skin rash (3), general fatigue (1), dull headache (1), stiffness of finger (1), lumbago (1), and facial edema (1) were reported.

In the long-term follow-up study, 20 cases were being followed for more than 6 months, including 9 cases for over 1 year, in our hospital. They consist of 9 FH patients and 11 nonFH patients. Total serum cholesterol attained almost maximum reduction after 1 month and this level was maintained for 1 year (Fig. 5). LDL-cholesterol and apo-B showed the same trend as total cholesterol. Serum triglyceride, apo C-II apo C-III, and apo-E levels decreased more slowly and had more fluctuation, however, the reduced levels were also maintained for 1 year. On the other hand, HDL-cholesterol, apo A-I, and apo A-II increased gradually and the increased levels were maintained for 1 year. No adverse effect was observed.

It is concluded that CS-514 is a very useful drug in treatment of hypercholesterolemic patients including heterozygous familial hypercholesterolemia.

Hyperlipoproteinemia and Therapy: Results of a Multicenter Trial with Bezafibrate, a New Lipid-Lowering Drug

D. Sommariva [1] and A. Branchi [2]

1 Introduction

For several years evidence has been accumulating to show a strong association of elevated serum lipid levels with atherosclerotic vascular disease. Recent studies have demonstrated that lowering serum cholesterol reduces the incidence of heart attacks (LRC-CPPT 1984) and can prevent or delay the rate of progression of atherosclerotic lesions (Levy et al. 1984; Arntzenius et al. 1985). Thus, the treatment of hyperlipidemic states, together with the correction of other atherosclerotic risk factors, seems to be advisable.

The basic therapeutic approach to the hyperlipoproteinemic patient is diet. When diet, as frequently occurs in severe hyperlipoproteinemia, fails to normalize lipoprotein pattern, drug therapy is suggested. Many effective hypolipidemic drugs are commercially available; among these, bezafibrate is one of the most powerful and tolerated. The drug, 2-[4-[2-(4-chlorobenzamido)-ethyl]-phenoxy]-2-methylpropionic acid, a clofibrate analogue, has been demonstrated in different experimental models to have multiple effects on lipid metabolism. The inhibition of peripheral lipolysis and of hepatic fatty acid synthesis and the increase in hepatic fatty acid β-oxidation (Kohlmeier and Schlierf 1982) may play critical roles in reducing hepatic triglyceride production and hence in decreasing the secretion of VLDL into the blood.

Moreover, catabolism of triglyceride-rich lipoproteins is enhanced through the activation of lipoprotein lipase (Klose et al. 1979) and this may account also for the increase of the HDL-cholesterol level.

Studies in rats demonstrated that bezafibrate suppresses the activity of hepatic 3-hydroxy-3-methylglutaryl coenzyme A reductase, the rate-limiting enzyme in cholesterolgenesis (Berndt et al. 1978). The inhibition of sterol biosynthesis is followed by an increased uptake and degradation of LDL owing to the increase of the expression of the high affinity receptors for LDL (Stewart et al. 1982).

The effects of bezafibrate on lipid metabolism can explain, at least in part, the activity of the drug on the lipoprotein pattern of hyperlipidemic patients. These mainly consist in a fall of apo-B and of total and LDL-cholesterol in type IIa and in type IIb patients and in a decrease of serum triglycerides and VLDL lipids in type IV and IIb

---

1 II Department of Medicine, L. Sacco Hospital, Milan, Italy
2 Institute of Medical Pathology, University of Milan, Milan, Italy

Drugs Affecting Lipid Metabolism
Ed. by R. Paoletti et al.
© Springer Verlag Berlin Heidelberg 1987

patients. Apo A-I and HDL-cholesterol rise particularly in patients with hypertriglyceridemia (Weisweiler and Schwandt 1980; Sommariva et al. 1985). The changes in lipoprotein pattern are related to the pretreatment lipoprotein levels, i.e. the greater the lipoprotein abnormality, the greater the hypolipidemic effect (Sommariva et al. 1985). This makes the drug of particular interest in severe hyperlipoproteinemic states.

2 The Multicenter Trial

In 1983 a trial with bezafibrate started in 86 Italian Centers to verify in a large number of hyperlipoproteinemic patients the effectiveness, tolerance and safety of the drug. A total of 1,132 hyperlipoproteinemic patients were selected, 1,085 of them completed the study. Their ages ranged from 18 to 85 years with a mean of 55 ± 0.3 (SEM), 539 were males and 542 females (in 4 cases sex was not reported).

Sixty percent (660) of them was diabetic (385 on oral hypoglycemic agents and 90 on insulin) and 40% (425) non-diabetic.

3 Results

All the patients have been treated with bezafibrate 200 mg t.i.d. for at least 3 months. Patients who normalized after 3 months continued the treatment with the drug 200 mg b.i.d., the other patients remained at full dosage.

After 12 weeks of bezafibrate 200 mg t.i.d., serum cholesterol fell by 23% in type IIa, by 20% in type IIb and by 21% in type IV non-diabetic patients. The changes in the serum cholesterol level in diabetic patients were similar: -24% in type IIa, -20% in type IIb and -14% in type IV (Table 1).

Serum triglycerides decreased on the average by 17% in type IIa, by 38% in type IIb and by 62% in type IV non-diabetic patients. In diabetic patients the changes were of -24%, -41% and -56% respectively (Table 2). In both diabetic and non-diabetic patients a highly significant correlation was found between basal serum lipid levels and their changes during the treatment. The highest correlation coefficient was the one between serum cholesterol and its change in type IV non-diabetics (r=0.97) and the lowest one between the serum triglyceride level and its diminution in type IIa non-diabetic patients (r=0.20).

In patients who continued the treatment with bezafibrate 200 mg t.i.d. (203 diabetics and 94 non-diabetics) the serum level of cholesterol and triglycerides recorded at the end of the fourth month were slightly but significantly lower than the ones observed at the third month of therapy (in diabetics, cholesterol decreased from 253.4 ± 3.9 to 231.7 ± 3.2 mg dl^{-1} and triglycerides from 251.2 ± 12.6 to 208.4 ± 9.6; in non-diabetics, cholesterol decreased from 263.3 ± 6.0 to 243.5 ± 5.1 and triglycerides from 233.9 ± 17.3 to 193.7 ± 12.7). The same occurred in patients who continued the treatment with 200 mg bezafibrate b.i.d. In fact, in diabetic patients (265) serum cholesterol decreased on the average from 242.1 ± 3.2 mg dl^{-1} at the third month control

Table 1. Changes in serum cholesterol (x ± SEM) during the treatment with bezafibrate 200 mg t.i.d.

Patients	No.	Basal (mg dl^{-1})	12 weeks (mg dl^{-1})	Probability
Non-diabetic	425			
Type IIa	120	344.9 ± 4.3	265.6 ± 4.2	<0.001
Type IIb	207	294.6 ± 3.9	235.4 ± 3.4	<0.001
Type IV	98	277.6 ± 6.8	218.6 ± 5.6	<0.02
Diabetic	660			
Type IIa	116	344.5 ± 6.7	261.5 ± 4.9	<0.001
Type IIb	336	303.5 ± 3.7	243.2 ± 2.7	<0.001
Type IV	208	274.8 ± 6.3	236.1 ± 4.0	<0.001

Table 2. Changes in serum triglycerides (x ± SEM) during the treatment with bezafibrate 200 mg t.i.d.

Patients	No.	Basal (mg dl^{-1})	12 weeks (mg dl^{-1})	Probability
Non-diabetic	425			
Type IIa	120	160.1 ± 4.1	133.5 ± 4.5	<0.001
Type IIb	207	315.7 ± 6.4	197.1 ± 5.3	<0.001
Type IV	98	782.9 ± 72.3	301.4 ± 32.4	<0.001
Diabetic	660			
Type IIa	116	167.9 ± 4.7	128.4 ± 3.9	<0.001
Type IIb	336	325.3 ± 4.9	192.3 ± 4.2	<0.001
Type IV	208	749.2 ± 33.9	332.3 ± 15.8	<0.001

to 225.6 ± 2.9 mg dl^{-1} at the fourth month and serum triglycerides from 199.2 ± 7.2 to 165.9 ± 4.2 mg dl^{-1}. In non-diabetic patients (220) serum cholesterol decreased from 229.9 ± 3.2 to 213.7 ± 2.9 and triglycerides from 185.7 ± 6.1 to 150.1 ± 4.0 mg dl^{-1}. Two hundred and eighty-eight (68%) non-diabetic and 369 (56%) diabetic patients achieved the normalization of serum lipids (cholesterol < 250 and triglycerides < 200 mg dl^{-1}) during bezafibrate therapy.

During the study, 47 patients dropped out, 5 because of side effects (1 abdominal pain and nausea, 2 heartburn, 1 abdominal discomfort and 1 epigastric pain) and 42 for causes unrelated to the assumption of the drug. In two diabetic patients a slight raise of blood urea and serum creatinine occurred. In 15 patients there was an increase in serum creatine phosphokinase, in 9 of aspartate aminotransferase and in 4 of alanine aminotransferase. One hundred and eighty-seven patients complained of subjective side effects, mainly gastrointestinal (Table 3).

Table 3. Prevalence and types of reported untoward side effects

Type	No.	(%)	Type	No.	(%)
None	898	(82.8)	Dyspepsia	9	(0.8)
Nausea	61	(5.8)	Abdominal discomfort	6	(0.6)
Epigastric pain	41	(3.9)	Asthenia	5	(0.5)
Headache	18	(1.7)	Flatulence	2	(0.2)
Diarrhea	15	(1.4)	Dryness of mouth	2	(0.2)
Vomiting	14	(1.3)	Impotence	2	(0.2)
Heartburn	13	(1.2)	Constipation	2	(0.2)
Dizziness	11	(1.1)	Myalgia	1	(0.1)
Itching	10	(1.0)	Exanthema	1	(0.1)

4 Discussion

Although multicenter trials with drugs may be biased in different ways, results are often of interest mainly because of the large sample size that can be studied in short time.

Of course, a short-term enlarged study does not completely respond to all questions about the long-term effectiveness, tolerance and safety of the therapy, however, it permits one to acquire valuable experience.

Results of this study, one of the largest case series treated with bezafibrate, confirm previous observations about the hypolipidemic effectiveness of the drug in different types of hyperlipoproteinemia (Weisweiler and Schwandt 1980; Sommariva et al. 1985; Olsson et al. 1985). On the whole, bezafibrate was well tolerated and in only five patients withdrawal of the drug was necessary because of side effects. Safety laboratory parameters showed minor changes in 30 patients.

Bezafibrate may be considered then a safe and effective hypolipidemic drug suitable for the treatment of the commonest types of hyperlipoproteinemia.

List of Participants to the Multicenter Trial:

Adda G., Albani A., Allochis G., Altomonte L., Ame' C., Angileri G., Baffelli E., Baggio E., Bargero G., Barbera R., Bellogini G.C., Bellomaria C., Bentley R., Bertello P.D., Bettale G., Biagioli R., Bindelli T., Biondo S., Bonfiglioli D., Bossi A., Branchi A., Buratti G.P., Burrafato S., Caimi S., Capra Marzani M., Capretti L., Carta Q., Cascone A., Caselle M.T., Casertano F., Castellazzi R., Caviezel F., Cazzalini C., Cecchini E., Cinquegrani M., Colucci B., Conconi G., Comaschi M., Corgiat L., Costa C., Dani F., D'Antonio R., Daprati A, De La Pierre V., Diana A., Di Piazza A., Erle G., Fazias M., Fatati G., Ferrari Bravo A., Ferrari M., Ferrari V., Ferraris G.M., Foglini P., Fontana S., Franchino A.M., Frediani R., Furlani M., Fusi M.G., Gamba P.L., Gandini R., Ghilardi G., Giovannelli A., Gnesotto M., Gosso P., Grandi E., Guardamagna C., Incontro C., Inzoli S., Lambardo A., Lazanio T., Leonardi R., Lorenti I., Lunetta M., Lupinacci G., Lutterotti A., Macri G., Maestripieri P.L., Maghenzani G., Mairino M.P., Malvicino F., Mangano N., Maraffi F.,

Margiotta A., Martina V., Mazzi C., Mian P., Michelotto U., Montalenti P., Montemezzani M., Montini M., Moronesi F., Mughini L., Motta L., Nicodano A., Nicrosini F., Noacco C., Nosari I., Oliva G., Pace G.P., Pagani G., Paleari F., Papa A., Parodi F.A., Pastean H., Pennici F., Perdomini A.G., Pina P., Pino G., Posca M., Ragonese F., Ravagnani E., Rebagliati M., Rivolta M.R., Sacco P., Saitta A., Savagnone E., Savastano A., Scavino S., Scornavacca G., Scotti L., Sommariva D., Speroni G., Spina M.R., Squadrito S., Strata A., Sudomo L., Taglioretti D., Tenconi M.T., Tironi S., Tirrito M., Tosi M., Travaglini A., Troili F., Uccella R., Vario S., Vecchiet L., Velussi M., Virgili F., Vitacolonna E., Vitolo E., Zampini A.

References

Arntzenius AC, Kromhout D, Barth JD, Reiber JHC, Bruschke AVG, Buis B, Gent CM Van, Kempen-Voogd N, Strikwerda S, Velde EA Van der (1985) Diet, lipoproteins, and the progression of coronary atherosclerosis. The Leiden intervention trial. New Engl J Med 312:805–811

Berndt J, Ganmert R, Still J (1978) Mode of action of the lipid lowering agents, clofibrate and BM 15075, on cholesterol biosynthesis in rat liver. Atherosclerosis 30:147–152

Klose G, Behrendt J, Vollmar J, Greten H (1979) Effect of bezafibrate on the activity of lipoprotein lipase and hepatic triglyceride hydrolase in healthy volunteers. In: Greten H, Lang PD, Schettler G (eds) Lipoproteins and coronary heart disease. Witzstrock, New York Baden-Baden Cologne, pp 182–184

Kohlmeier M, Schlierf G (1982) Mode of action of bezafibrate: a hypothesis. In: Crepaldi G, Greten H, Schettler G, Baggio G (eds) Lipoprotein metabolism and therapy of lipid disorders. Excerpta Medica, Amsterdam Oxford Princeton, pp 93–96

Levy RI, Brenside JF, Epstein SE, Kelsey SF, Passaniani ER, Richardson JM, Loh IK, Stone NJ, Aldrich RF, Battaglini JW, Moriarty DJ, Fisher ML, Friedman L, Friedewald W, Detre KM (1984) The influence of changes in lipid values induced by cholestyramine and diet on progression of coronary artery disease: results of the NHLBI type II coronary intervention study. Circulation 69:325–337

Lipid Research Clinics Program (1984) The lipid research clinics coronary primary prevention trial results: I. Reduction in incidence of coronary heart disease. JAMA 251:351–364

Olsson AG, Lang PD, Vollmar J (1985) Effect of bezafibrate during 4.5 years of treatment of hyperlipoproteinaemia. Atherosclerosis 55:195–203

Sommariva D, Tirrito M, Bonfiglioli D, Pogliaghi I, Branchi A, Ottomano C, Bellintani L (1985) Changes in serum lipoprotein pattern following bezafibrate. Differential effects in type IIa and in type IIb hyperlipoproteinemic patients. Pharmacol Res Comm 17:1181–1191

Steward JM, Packard CJ, Lorimer AR, Boag DE, Shepherd J (1982) Effects of bezafibrate on receptor mediated and receptor independent low density lipoprotein catabolism in type II hyperlipoproteinaemic subjects. Atherosclerosis 44:355–365

Weisweiler P, Schwandt P (1980) Lipoprotein lipids and apolipoproteins after a six month bezafibrate treatment. Artery 6:402–406

Lipoprotein Changes Induced by Bezafibrate – 200 mg t. i. d. – and by Bezafibrate in a Slow-Release Preparation – 400 mg Once a Day – in Patients with Primary Hyperlipoproteinaemia

S. Bertolini, S. Valice, N. Elicio, A. Daga, S. Cuzzolaro, G. Montagna, G. Pistocchi, and R. Balestreri [1]

Abbreviations: apo= apoprotein; BZ= Bezafibrate conventional form; BZ-R= Bezafibrate sustained-release form; FCHL= Familial combined hyperlipidaemia; FHTG= Familial hypertriglyceridaemia; HDLc= High density lipoprotein cholesterol; IDL= Intermediate density lipoproteins; LDLc= Low density lipoprotein cholesterol; Tc= Total cholesterol; TG= Triglycerides; VLDLc= Very low density lipoprotein cholesterol; VLDLtg= Very low density lipoprotein triglycerides.

1 Introduction

Bezafibrate (2,4,2-(4-chlorobenzamido)-ethyl-phenoxy-2-methylpropionic acid) has been shown to be an effective lipid-lowering drug in different types of hyperlipoproteinaemia, when administered at a dose of 200 mg three times a day (Gavish et al. 1986). Recently, a sustained-release form of Bezafibrate (400 mg tablet) has been introduced in order to improve patient's compliance, due to the single daily dose which has to be taken. This sustained-release preparation, in comparison to the conventional form at the same dosage, is characterized by a similar relative biological availability (Ledermann and Kaufmann 1981), by a lower plasma peak concentration (34%), by a longer mean retention time (6.0 vs 2.4 h) and elimination half-life (2.6 vs 1.6 h) (Von Möllendorff 1984).

Here, we first give the results obtained during an open, short-term study in different types of hyperlipoproteinaemic patients assuming the sustained-release form, and then the results of a single blind medium-term comparative study in type IIb and IV subjects assigned at random to treatment with Bezafibrate 200 mg t.i.d., or with Bezafibrate 400 mg u.i.d. in the sustained-release form.

2 Patients and Methods

2.1 Short-Term Study

Nineteen hyperlipoproteinaemic patients (8 males and 11 females; ranging in age from 35 to 66 years, average age 53.5 ± 8.9 years; ranging in body mass index from 20.2 to

1 Atherosclerosis Prevention Center, Department of Internal Medicine, University of Genoa, Viale Benedetto XV n. 6, 16132 Genoa, Italy

Drugs Affecting Lipid Metabolism
Ed. by R. Paoletti et al.
© Springer-Verlag Berlin Heidelberg 1987

Table 1. Plasma lipids (mmol l^{-1}) and apoproteins (mg dl^{-1}) in hyperlipoproteinaemic patients (type: 2 IIa, 9 IIb, 1 III, 2 IV, 5 V) before and after 2 months of treatment with Bezafibrate R 400 mg u.i.d. (mean ± 1 SD)

	Type IIa + IIb			Type III			Type IV + V		
	Before	After	Δ%	Before	After	Δ%	Before	After	Δ%
b.wt. (kg)	65.5 ± 13.9	65.2 ± 14.0		68.0	67.0		71.3 ± 13.0	71.2 ± 12.9	
Tc	7.44 ± 1.07	6.09 ± 1.01^c	-18	12.7	5.63	-56	6.42 ± 1.52	6.47 ± 2.03	
VLDLc	0.83 ± 0.46	0.44 ± 0.42^b	-47	8.07	1.87	-77	3.14 ± 1.84	1.26 ± 0.38^a	-60
LDLc	5.41 ± 1.16	4.26 ± 1.13^c	-21	3.72	2.24	-40	2.46 ± 0.68	4.09 ± 1.68^a	+66
HDLc	1.19 ± 0.27	1.39 ± 0.30^b	+16	0.91	1.52	+67	0.82 ± 0.17	1.12 ± 0.19^b	+36
TG	2.78 ± 1.33	1.63 ± 0.90^c	-41	6.84	1.47	-78	8.32 ± 4.15	3.33 ± 1.16^a	-60
VLDL TG	2.21 ± 1.19	1.30 ± 0.88^c	-41	5.69	1.23	-78	7.54 ± 3.86	2.88 ± 1.07^a	-62
apo AI	129 ± 19	157 ± 30^c	+21	110	176	+60	128 ± 22	149 ± 20^b	+16
apo AII	38 ± 6	51 ± 6^c	+35	31	58	+86	36 ± 5	49 ± 4^c	+34
AI/HDLc	2.85 ± 0.44	2.94 ± 0.43		3.10	2.99		4.06 ± 0.44	3.44 ± 0.27^a	-15
AI/AII	3.44 ± 0.53	3.03 ± 0.41^b	-12	3.51	3.02	-14	3.52 ± 0.30	3.05 ±0.25^c	-13
apo CII	7.2 ± 1.8	5.7 ± 1.7^c	-21	13.0	8.60	-34	11.0 ± 3.2	8.3 ± 1.9	-25
apo CIII	15.0 ±4.8	10.5 ± 3.2^c	-30	29.5	17.8	-40	28.8 ± 10.4	18.4 ± 6.7^b	-36
CII/CIII	0.49 ± 0.09	0.55 ± 0.08^a	+12	0.44	0.48	+9	0.39 ± 0.06	0.47 ± 0.08^c	+20
apo-B	176 ± 17	152 ± 20^c	-13	152	98	-35	130 ± 15	161 ± 33^a	+23
apo-E	6.6 ± 2.3	5.4 ± 1.6^b	-18	28.5	16.8	-41	11.6 ± 5.5	6.4 ± 1.2^a	-45

$^a p$ <0.05. $^b p$ <0.01. $^c p$ <0.001: significance of differences with respect to the pretreatment value (Student's t-test for paired data).

31.2 kg/h^2, average BMI 25.3 ± 3.2) were typed according to WHO criteria (Beaumont et al. 1970): two type IIa, nine type IIb, one type III, two type IV, five type V (Table 1). All patients had primary hyperlipidaemia; familial combined hyperlipidaemia (FCHL) was documented in six of them and familial hypertriglyceridaemia (FHTG) in four. Five patients (two IIa, two IIb, one V) suffered from ischaemic heart disease (previous myocardial infarction) and one (III) from peripheral arterial disease.

The patients were put on "prudent" isocaloric diet (50% carbohydrate, 30% fat, 20% protein, cholesterol ⩽ 300 mg day^{-1}) for 1 month in order to stabilize their body weight and plasma lipids. Alcoholic beverages were proscribed. Then the patients gave their informed consent for the study, and received after the evening meal one placebo tablet for 1 month and afterwards one tablet of Bezafibrate 400 mg in a sustained-release form (BZ-R) for a further 2 months. The planned diet was followed during the whole period of the study.

Blood samples were taken after a 12–14 h overnight fast using Venoject Terumo containing EDTA/Na2 (final conc. 1.5 mg ml^{-1}); samples were collected at the first

Table 2. Effect of Bezafibrate 200 mg t.i.d. (BZ) and Bezafibrate R 400 mg u.i.d. (BZ-R) on plasma lipids (mmol l^{-1}) and apoproteins (mg dl^{-1}) in hyperlipoproteinaemic patients type IIb (n = 21 vs 21) (mean ± 1 SD)

Months		Placebo (1st) 0	Treatment (Bezafibrate) 2	4	6	Mean % change[d]	Placebo (2nd) 7
Tc	BZ	7.78 ± 0.75	6.07 ± 0.80[c]	6.12 ± 0.71[c]	6.07 ± 0.68[c]	- 21	7.33 ± 0.55[b]
	BZ-R	7.98 ± 1.09	6.29 ± 1.01[c]	6.09 ± 0.87[c]	5.99 ± 0.66[c]	- 22	7.51 ± 0.75[b]
TG	BZ	2.76 ± 0.45	1.61 ± 0.58[c]	1.55 ± 0.36[c]	1.49 ± 0.33[c]	- 43	2.55 ± 0.66[a]
	BZ-R	2.98 ± 0.89	1.83 ± 0.50[c]	1.65 ± 0.53[c]	1.68 ± 0.50[c]	- 41	2.69 ± 0.66[a]
LDLc	BZ	5.48 ± 0.77	3.94 ± 0.83[c]	4.01 ± 0.63[c]	3.99 ± 0.70[c]	- 26	4.89 ± 0.88[b]
	BZ-R	5.63 ± 1.00	4.19 ± 1.00[c]	4.02 ± 0.83[c]	3.92 ± 0.74[c]	- 27	5.19 ± 0.86[b]
HDLc	BZ	1.27 ± 0.26	1.49 ± 0.33[c]	1.52 ± 0.29[c]	1.50 ± 0.30[c]	+18	1.35 ± 0.25[b]
	BZ-R	1.24 ± 0.31	1.43 ± 0.37[c]	1.46 ± 0.38[c]	1.45 ± 0.36[c]	+16	1.29 ± 0.31[b]
HDL2c	BZ	0.34 ± 0.14	0.47 ± 0.18[c]	0.47 ± 0.17[c]	0.46 ± 0.15[c]	+35	0.39 ± 0.15[a]
	BZ-R	0.33 ± 0.17	0.41 ± 0.15[c]	0.44 ± 0.17[c]	0.43 ± 0.16[c]	+29	0.36 ± 0.16
HDL3c	BZ	0.92 ± 0.15	1.02 ± 0.17[c]	1.05 ± 0.16[c]	1.04 ± 0.17[c]	+12	0.96 ± 0.12[a]
	BZ-R	0.91 ± 0.18	1.02 ± 0.25[c]	1.02 ± 0.22[c]	1.01 ± 0.21[c]	+11	0.94 ± 0.17 ·
apo AI	BZ	135 ± 21	149 ± 24[c]	153 ± 21[c]	152 ± 19[c]	+11	141 ± 19[b]
	BZ-R	132 ± 21	150 ± 23[c]	151 ± 20[c]	150 ± 22[c]	+12	139 ± 19[c]
apo-B	BZ	165 ± 25	134 ± 22[c]	131 ± 20[c]	133 ± 22[c]	- 18	163 ± 28
	BZ-R	162 ± 22	132 ± 20[c]	126 ± 17[c]	127 ± 16[c]	- 19	153 ± 18[b]

[a] $p < 0.05$. [b] $p < 0.01$. [c] $p < 0.001$: significance of differences with respect to value of the first placebo period (Student's t-test for paired data). [d] Mean % changes from 1st placebo during the whole period of treatment.

visit, at the start and at the end of the placebo period, and at the end of the first and second month of Bezafibrate treatment.

VLDL was separated by the tube-slicing technique after being centrifugated at plasma density (40,000 rpm, for 18 h, at 10°C) in a Beckman L8-70M ultracentrifuge with 50.3 Ti rotor. Aliquots of the VLDL and of the lipoprotein fraction of density > 1.006 g ml^{-1} were taken for cholesterol and triglycerides determination. In the infranate, HDL was separated from LDL by precipitation of the latter using heparin and manganese chloride (final conc. 184 UI ml^{-1}, 92 mmol^{-1} respectively) (Lipid Research Clinics Program: Manual of Laboratory Operations 1982). The concentration of triglycerides was determined using the GPO-PAP method (Boehringer Mannheim GmbH) in whole plasma, in the top fraction (VLDL), and in the bottom fraction before precipitation (LDL + HDL). The concentration of cholesterol was determined using the CHOD-PAP method with Tris-buffer (Monotest Cholesterol High Performance, Boehr-

Table 3. Effect of Bezafibrate 200 mg t.i.d. (BZ) and Bezafibrate R 400 mg u.i.d. (BZ-R) on plasma lipids (mmol l^{-1}) and apoproteins (mg dl^{-1}) in hyperlipoproteinaemic patients type IV (n= 14 vs 14) (mean ± 1 SD)

Months		Placebo (1st) 0	Treatment (Bezafibrate) 2	4	6	Mean % change[d]	Placebo (2nd) 7
Tc	BZ	6.20 ± 1.09	5.50 ± 0.87[c]	5.27 ± 0.74[b]	5.20 ± 0.61[b]	- 14	5.84 ± 0.80[a]
	BZ-R	6.63 ± 1.28	5.45 ± 0.79[c]	5.39 ± 0.66[c]	5.20 ± 0.63[c]	- 19	6.45 ± 1.24
TG	BZ	5.08 ± 2.26	2.20 ± 1.05[c]	2.17 ± 1.07[c]	2.24 ± 1.17[c]	- 57	4.38 ± 2.42[a]
	BZ-R	5.96 ± 2.90	2.31 ± 0.89[c]	2.14 ± 0.82[c]	2.33 ± 0.76[c]	- 60	5.45 ± 2.94
HDLc	BZ	0.95 ± 0.19	1.19 ± 0.16[c]	1.16 ± 0.19[c]	1.18 ± 0.16[c]	+23	1.03 ± 0.15[a]
	BZ-R	0.90 ± 0.15	1.12 ± 0.21[c]	1.16 ± 0.20[c]	1.11 ± 0.18[c]	+24	0.95 ± 0.17
HDL2c	BZ	0.20 ± 0.09	0.27 ± 0.12[c]	0.25 ± 0.10[a]	0.27 ± 0.09[c]	+34	0.22 ± 0.08
	BZ-R	0.19 ± 0.07	0.28 ± 0.08[c]	0.29 ± 0.08[c]	0.25 ± 0.08[b]	+46	0.21 ± 0.07
HDL3c	BZ	0.75 ± 0.12	0.93 ± 0.10[c]	0.91 ± 0.14[c]	0.91 ± 0.11[c]	+21	0.80 ± 0.08
	BZ-R	0.72 ± 0.11	0.83 ± 0.16[c]	0.87 ± 0.14[c]	0.87 ± 0.13[c]	+18	0.74 ± 0.12
apo AI	BZ	112 ± 10	126 ± 12[c]	129 ± 11[c]	131 ± 11[c]	+13	116 ± 11
	BZ-R	107 ± 8	120 ± 15[c]	126 ± 12[c]	123 ± 9[c]	+14	113 ± 9[b]
apo-B	BZ	123 ± 21	121 ± 21	109 ± 23[a]	110 ± 15[a]	- 7	120 ± 21
	BZ-R	127 ± 26	123 ± 29	117 ± 28[a]	114 ± 25[b]	- 7	126 ± 19

[a] p <0.05. [b] p <0.01. [c] p <0.001: significance of differences with respect to value of the first placebo period (Student's t-test for paired data). [d] Mean % changes from 1st placebo during the whole period of treatment.

inger Mannheim GmbH) in whole serum, in the top fraction (VLDL), in the bottom fraction before precipitation (LDL + HDL), and in the supernatant after precipitation (HDL). LDL-cholesterol concentration was obtained indirectly by subtracting the HDL-cholesterol level from the cholesterol concentration of the bottom fraction after centrifugation at d= 1.006. The recoveries of triglycerides and cholesterol in the isolated density classes were within 100 ± 8% in all samples. The concentration of apolipoproteins AI, AII, B, CII, CIII and E was measured in whole plasma by the RID method using Daiichi plates (Daiichi Pure Chemicals Co, LTD, Tokyo). To dissociate the lipids from apoproteins before the assay, the plasma samples were treated with 3% Tween 20 (apo CII and CIII determinations) or with 3% Triton·X100 (apo-E determination). The interassay precisions for cholesterol, triglycerides and apoproteins, expressed as coefficients of variation (%), were 3.3, 4.2 and 2.5–3.5 respectively. The statistical significance of differences between treatment values and placebo values was checked by Student's t-test for paired data; linear regression analysis was used to determine the coefficients of correlation between lipid and apoprotein values before and after treatment.

Table 4. Body weight, plasma glucose and uric acid levels (mg dl^{-1}) during treatment with Bezafibrate 200 mg t.i.d. (BZ) and Bezafibrate R 400 mg u.i.d. (BZ-R) in hyperlipoproteinaemic patients type IIb and IV (n= 35 vs 35) (mean ± 1 SD)

Months		Placebo (1st) 0	Treatment (Bezafibrate) 2	4	6	Mean % change[d]	Placebo (2nd) 7
B.wt. (kg)	BZ	67.2 ± 9.8	66.9 ± 10.1	67.3 ± 10.2	67.6 ± 10.1		67.5 ± 10.1
	BZ-R	68.7 ± 11.9	68.5 ± 12.0	68.9 ± 12.1	69.0 ± 11.9		69.0 ± 11.8
Glucose	BZ	95.5 ± 10.1	90.6 ± 10.1^c	91.3 ± 11.0^b	90.9 ± 10.8^c	- 4	93.7 ± 10.5
	BZ-R	99.1 ± 12.1	96.9 ± 13.7	95.5 ± 12.0^b	93.8 ± 9.9^c	- 4	98.6 ± 12.6
Uric acid	BZ	5.15 ± 1.44	4.98 ± 1.23	4.84 ± 1.06	4.85 ± 0.99^a	- 4	5.07 ± 1.44
	BZ-R	5.29 ± 1.73	4.76 ± 1.36^b	5.13 ± 1.57	4.83 ± 1.39^a	- 7	5.25 ± 1.44

$^a p$ <0.05. $^b p$ <0.01. $^c p$ <0.001: significance of differences with respect to value of the first placebo period (Student's t-test for paired data). ^{d}Mean % changes from 1st placebo during the whole period of treatment.

2.2 Comparative Medium-Term Study

Seventy subjects with primary hyperlipoproteinaemia, after a 2-month period of stabilization on a prudent diet, were allocated at random in two groups. All patients received a placebo for 1 month and then started Bezafibrate treatment; drug therapy lasted 6 months and was followed by a final placebo period of a further 4 weeks. Group BZ (21 type IIb, 14 type IV; 21 males, 14 females; 51.8 ± 9.1 years, from 29 to 63; 24.5 ± 2.4 body mass index, from 20.4 to 30.1; 9 with documented familial hyperlipidaemia, 7 FCHL and 2 FHTG; 9 with ischaemic heart disease, 3 with peripheral arterial disease, 1 with transitory ischemic attack) and group BZ-R (21 type IIb, 14 type IV; 23 males, 12 females; 51.8 ± 9.7 years, from 34 to 66; 24.6 ± 2.3 body mass index, from 19.9 to 29.4; 8 with documented familial hyperlipidaemia, 6 FCHL and 2 FHTG; 8 with ischaemic heart disease, 5 with peripheral arterial disease) were treated with Bezafibrate (200 mg t.i.d.) and Bezafibrate in a sustained-release form (400 mg once a day after the evining meal) respectively (Table 4). Fourteen patients who dropped out (side effects mainly gastro-intestinal discomfort = 9; treatment ineffective = 3; motiveless = 2) were not included in these groups. The clinical and biochemical evaluation of the patients were done during the diet period alone, at the end of the first placebo period, during treatment (every month for 6 months) and at the end of the second placebo period. Plasma samples were collected at each visit to determine cholesterol (Tc), LDL-cholesterol (LDLc), HDL-cholesterol (HDLc) and its fractions (HDL2c, HDL3c, triglycerides (TG), apoproteins AI and B (apo AI, apo-B), glucose and uric acid. Cholesterol and triglycerides in whole plasma, cholesterol in HDL and HDL3 fractions, glucose and uric acid were measured with the enzymatic method (Monotest Cholesterol High Performance, CHOD-PAP method, Boehringer Mannheim; triglycerides, GPO-

PAP method, Boehringer Mannheim; Glucinet, GOD-POD method, Sclavo; Urica Color, uricase-POD method, Boehringer Mannheim). HDL-cholesterol was determined in the supernatant after precipitation of the apo B-containing lipoproteins (VLDL, LDL) by heparin and manganese chloride (final conc. 184 UI ml^{-1}, 92 mmol^{-1} respectively); HDL fraction from lipaemic samples was ultrafiltered as suggested by Warnick and Albers (1978). HDL2 and HDL3 cholesterol subfractions were obtained according to Gidez et al. (1982). LDL-cholesterol concentrations in patients of type IIb were calculated as follows: LDLc = Tc − HDLc − (0.166 x TG) (Wilson et al. 1985). Apolipoproteins AI and B were measured in whole plasma by rocket immunoelectrophoresis (apo AI: 1% agarose REO 15 Behringwerke, 4% Dextran T10 Pharmacia and 0.8% sheep anti-apo AI Boehringer Mannheim Lot. 10543-02 in 0.05 M barbital buffer, pH 8.6; 2 V cm^{-1} for 14 h at 15°C; Apo B: 1% agarose REO 15 Behringwerke and rabbit anti-apo B Behringwerke Lot. 104914 in 0.05 M barbital buffer, pH 8.6; 2 V cm^{-1} for 16 h at 15°C). All lipid assessments were done within 4 h of blood sampling; apolipoproteins were assayed within 1 month in plasma aliquots kept at −30°C until analysis. Quality assurance of the test systems resulted in the following: the day-to-day coefficients of variation were 3.3% for Tc, 4.2% for TG, 3.0% for apo AI, 4.8% for apo-B, 2.4% for glucose and 4.0% for uric acid; the within-day coefficients of variation were 2.2% for HDLc and 3.2% for HDL3c.

The statistical significance of differences between treatment values and the first placebo values was checked for each group and each type by Student's t-test for paired data. Variance and covariance analyses (two-factor experiment with repeated measures on one factor), with and without inclusion of values of the second placebo period, were used to evaluate differences between the effects of Bezafibrate 200 mg t.i.d. and Bezafibrate sustained-release 400 mg u.i.d. (Winer 1971).

3 Results and Comments

3.1 Short-Term Study

Before treatment, taking into account the whole group (n=19), we found the following highly significant ($p < 0.001$) correlations: VLDLc vs apo-E (r=0.918), VLDLtg vs apo CII (r=0.906), VLDLtg vs apo CIII (r=0.851), LDL vs apo-B (r=909), VLDLtg vs LDLc (r=0.711), TG vs HDLc (r=−0.734). On the other hand, we found a weakly significant ($p < 0.05$) correlation between HDLc and apo AI (r=0.574) or apo AII (r=0.538). The effects of Bezafibrate treatment in the different types of hyperlipoproteinaemia are shown in Table 1. Interesting points to note are the strong lipid-lowering effect in the patient with type III, and the increase of LDLc and apo-B in type IV + V (mainly in type V patients). Considering all the type IIa, IIb, IV and V patients together, and excluding the type III patient, the percent changes of LDLc after 2 months of treatment appear to be related to the initial plasma concentrations both of VLDL triglycerides (r=0.775, $p < 0.001$) and of LDL-cholesterol (r= −0.683, $p < 0.01$); in other words, the effect of Bezafibrate on LDL-cholesterol seems to depend on the pretreatment concentrations of VLDL triglycerides and LDL-cholesterol. Furthermore, in the whole group (n=19), the percent changes of HDL-cholesterol were positively related to initial VLDL triglycerides plasma concentrations (r=0.638, $p < 0.01$). After 2 months of Bezafibrate, the absolute changes of some variables (mg dl^{-1}) were correlated between themselves with a high level of significance ($p < 0.001$): VLDLc vs apo-E (r=0.889); VLDLtg vs apo CII (r=0.807), vs apo CIII (r=0.827), vs apo E (r=0.813); LDLc vs apo-B (r=0.931). In contrast, HDL-cholesterol changes did not show any significant correlation with either triglycerides or apoproteins AI and AII changes. Moreover, the ratio between apo AI and apo AII concentrations decreased, suggesting a change of the HDL composition during treatment.

The mechanisms involved in the lipoproteins changes induced by Bezafibrate have been partly elucidated. Bezafibrate increases both lipoproteinlipase and hepatic lipase activity, enhancing triglyceride removal and the VLDL→IDL→LDL conversion process (Gavish et al. 1986). In grossly hypertriglyceridaemic subjects, the LDL-cholesterol level rises because of the increase of LDL production from VLDL; a reduction in fractional LDL-apo B clearance seems also to contribute to the increase of LDL-cholesterol and apo-B in this kind of patient (Shepherd et al. 1984). In hypercholesterolaemic individuals with normal or moderately increased levels of triglycerides, LDL-cholesterol falls as Bezafibrate inhibits hepatic hydroxymethylglutaryl CoA reductase and cholesterol synthesis, leading to an accelerated LDL degradation via the LDL (B, E) receptors (Stewart et al. 1982). This receptor mechanism might be also operative in type III patients, in that it would remove some VLDL remnant particles (apo-E decrease) from circulation, but Bezafibrate does not seem to be able to change VLDL˙ lipid composition [in our patient VLDLc/VLDLtg ratio (mg dl^{-1}) was 0.62 during the placebo period and 0.66 after 2 months of Bezafibrate therapy].

3.2 Comparative Medium-Term Study

The results are shown in Tables 2 to 4. Variance and covariance analyses, the latter was performed in order to correct any difference between the groups at baseline level, did not show any significant difference between the effects of bezafibrate 200 mg did not show any significant difference between the effects of Bezafibrate 200 mg t.i.d. and Bezafibrate in sustained-release form 400 mg u.i.d.

The sustained-release preparation, even if at a lower daily dose (400 mg), appears to be as effective as the conventional preparation (600 mg). This finding may be explained by the fact that a single daily administration usually improves patient's compliance; moreover, the evening consumption might increase the effect of the drug owing to the activity of hydroxymethylglutaryl CoA reductase peaks at night (Hamprecht et al. 1969).

In our opinion, the sustained-release form seems to be subjectively better tolerated than the conventional one; in addition, with neither preparation did we find any significant or clinically relevant changes in the biological safety parameters (erythrocyte and leukocyte counts, haemoglobin, SGOT, SGPT, CPK, BUN and creatinine did not change; bilirubin, GGT and Alk. phosphatase showed a trend to decrease).

Suggestions have been made recently (Gavish et al. 1986) for the mechanisms which induce HDL-cholesterol increases during Bezafibrate therapy. Both the enhanced catabolism of VLDL, coupled with the transfer of VLDL surface components to HDL, and the decreased exchange of cholesteryl ester for triglycerides between HDL and VLDL (when the VLDL plasma pool decreases) are likely to contribute to an HDL-cholesterol increase. These mechanisms should above all increase the HDL2 subfraction, but Bezafibrate has also been reported to have a stimulatory effect on hepatic lipase; this enzyme, promoting conversion of HDL2 into HDL3 density range particles, may also lead to an increase of the HDL3 cholesterol.

References

Beaumont JL, Carlson LA, Cooper GR, Fejfar Z, Fredrickson DS, Strasser T (1970) Classification of hyperlipidemias and hyperlipoproteinemias. WHO Bull 43:891–908

Gavish D, Oschry Y, Fainaru M, Eisenberg S (1986) Change in very low-, low-, and high-density lipoproteins during lipid lowering (Bezafibrate) therapy: studies in type IIa and type IIb hyperlipoproteinemia. Eur J Clin Invest 16:61–68

Gidez LI, Miller GJ, Burstein M, Slagle S, Eder HA (1982) Separation and quantitation of subclasses of human plasma high density lipoproteins by a simple precipitation procedure. J Lipid Res 23:1206–1223

Hamprecht B, Nüssler C, Lynen F (1969) Rhythmic changes of hydroxymethyl-glutaryl coenzyme A reductase activity in livers of fed and fasted rats. FEBS Lett 4:117–121

Ledermann H, Kaufmann B (1981) Comparative pharmacokinetics of 400 mg Bezafibrate after a single oral administration of a new slow-release preparation and the currently available commercial form. J Int Med Res 9:516–520

Lipid Research Clinics Program (1982) Manual of laboratory operations. Lipid and lipoprotein analysis. US Dep Health Human Serv. NIH, 2nd edn. Bethesda, pp 63–77

Möllendorf E von (1984) Pharmacokinetics of Bezafibrate after administration of Bezalip l.a. in comparison with Bezalip. In: Boehringer Mannheim's internal report 18/4/1984 (report n. 16–01)

Shepherd J, Packard CJ, Stewart JM, Atmeh RF, Clark RS, Boag DE, Carr K, Lorimer AR, Ballantyne D, Morgan HG, Lawrie TDL (1984) Apolipoprotein A and B (S_f 100–400) metabolism during Bezafibrate therapy in hypertriglyceridemic subjects. J Clin Invest 74:2164–2177

Stewart JM, Packard CJ, Lorimer AR, Boag DE, Shepherd J (1982) Effects of Bezafibrate on receptor-mediated and receptor-independent low density lipoprotein catabolism in type II hyperlipoproteinaemic subjects. Atherosclerosis 44:335–355

Warnick GR, Albers JJ (1978) Heparin-Mn^{2+} quantitation of high–density-lipoprotein cholesterol: an ultrafiltration procedure for lipemic samples. Clin Chem 24:900–904

Wilson PWF, Zech LA, Gregg RE, Schaefer EJ, Hoeg JM, Sprecher DL, Brewer HB jr (1985) Estimation of VLDL cholesterol in hyperlipidemia. Clin Chim Acta 151:285–291

Winer BJ (1971) Statistical principles in experimental design. Mc Graw-Hill, New York

Modifications of Apoprotein, Lipoprotein Parameters and HDL$_2$ and HDL$_3$ Subfractions During Treatment with Bezafibrate Retard

A. Ventura, E. Mannarino, G. Ciuffetti, D. Siepi, and G. Lupattelli[1]

1 Introduction

Bezafibrate is today one of the most commonly used hypolipidemic drugs in the treatment of type IIA, IIB and IV hyperlipidemias. This depends on the specific activity of this drug. Bezafibrate in fact has a selective hypolipidemic activity according to the hyperlipemic phenotype: in type IIA it mainly reduces TC (total cholesterol) levels, while in types IIB and IV it decreases especially TG (triglycerides) levels. Bezafibrate is also able to increase HDL-C (Fellin et al. 1981; Mannarino et al. 1982) and to modify apoprotein patterns: apo A$_1$ generally increases (Weisweiler et al. 1980) in hypertriglyceridemic patients. Recently, a slow preparation of Bezafibrate has been prepared to allow the administration of a single daily dose (400 mg) (Ledermann and Kaufmann 1981). The aim of our study was to investigate the effects of Bezafibrate Retard (BfR) administration on both lipoproteins and apoproteins in three groups of patients affected by primary hyperlipidemias.

2 Materials and Methods

Sixty outpatients affected by primary hyperlipidemias were admitted to our study: 20 type IIA (10 males, 10 females, average age 47 ± 2 years), 20 type IIB (12 males, 8 females, average age 54 ± 2 years) and 20 type IV (14 males, 8 females, average age 43 ± 3 years). All hypolipidemic drugs were suspended at least 4 weeks before the trial began. No patient was overweight. The single-blind study lasting 12 weeks was divided into three phases: (1) 6 weeks of an isocaloric diet (less than 30% of total calories as lipids and less than 300 mg cholesterol per day; the ratio polyunsaturated/saturated fatty acids about 2:1); (2) 8 weeks of the same diet with the random administration of BfR (400 mg in a single dose after the evening meal) to 10 patients of each phenotype and placebo to the remaining 10; (3) 4 weeks of isocaloric diet. The determination of TC (method of Abell et al. 1952), TG (method of Wahlefeld

---

1 Institute of 2 Clinica Medica Generale e Terapia Medica, University of Perugia, Policlinico Monteluce, 06100 Perugia, Italy

Drugs Affecting Lipid Metabolism
Ed. by R. Paoletti et al.
© Springer Verlag Berlin Heidelberg 1987

1974), apo A_I, A_{II} and B (method of Mancini et al. 1975), HDL, HDL_2, HDL_3-C (method of Gidez et al. 1982) and LDL-C (calculated by Friedwald formula, 1972) was performed at the sixth and the fourth week before treatment, at the beginning of the treatment, at the end of the second, fourth, sixth and eighth week of this study and finally at the end of the second and fourth week after the drug/placebo suspension. The general clinical evaluation and the control of all the "safety parameters" were performed at the beginning of the first phase and at the end of each successive phase. Student's t-test was used for statistical analysis of paired results.

Table 1. Effects of BfR on type IIA hyperlipidemia

Weeks	0	2	4	6	8
BfR (400 mg/die)					
TC (mg%)	301 ± 15	289 ± 12[a]	281 ± 9[a]	273 ± 11[b]	263 ± 15[b]
TG (mg%)	107 ± 7	101 ± 5	87 ± 2	84 ± 3	80 ± 1
LDL-C (mg%)	243 ± 5	230 ± 7	224 ± 10[a]	216 ± 11[b]	208 ± 15[b]
HDL-C (mg%)	37 ± 2	38 ± 2	39 ± 2[a]	39 ± 2[a]	39 ± 2[a]
HDL_2-C (mg%)	15 ± 2	16 ± 1	16 ± 2	17 ± 2[b]	17 ± 2[b]
HDL_3-C (mg%)	22 ± 1	22 ± 1	23 ± 1	22 ± 1	22 ± 1
Apo AI (mg%)	130 ± 3	132 ± 2	134 ± 2	136 ± 4[a]	139 ± 7[b]
Apo AII (mg%)	54 ± 2	55 ± 4	55 ± 4	55 ± 4	55 ± 3
Apo-B (mg%)	146 ± 2	144 ± 2	143 ± 2	141 ± 3	138 ± 4[a]

vs basal values: [a]$p < 0.05$; [b]$p < 0.02$.

Table 2. Effects of BfR on type IIB hyperlipidemia

Weeks	0	2	4	6	8
BfR (400 mg/die)					
TC (mg%)	285 ± 5	286 ± 7	266 ± 8	258 ± 14	262 ± 9
TG (mg%)	350 ± 32	318 ± 33[a]	266 ± 21[c]	283 ± 30[b]	256 ± 25[b]
LDL-C (mg%)	177 ± 12	166 ± 9	175 ± 9	162 ± 11	172 ± 8
HDL-C (mg%)	37 ± 2	38 ± 2	38 ± 2	39 ± 2	39 ± 2
HDL_2-C (mg%)	14 ± 1	15 ± 1	15 ± 1	16 ± 1[b]	16 ± 1[b]
HDL_3-C (mg%)	25 ± 1	23 ± 1	23 ± 2	23 ± 2	23 ± 2
Apo AI (mg%)	142 ± 8	144 ± 6	147 ± 7	147 ± 6	150 ± 7[b]
Apo AII (mg%)	75 ± 6	76 ± 7	76 ± 5	77 ± 6	78 ± 6
Apo-B (mg%)	128 ± 3	124 ± 3	125 ± 3	122 ± 5	116 ± 6

vs basal values: [a]$p < 0.05$; [b]$p < 0.02$; [c]$p < 0.01$.

Table 3. Effects of BfR on type IV hyperlipidemia

Weeks	0	2	4	6	8
BfR (400 mg/die)					
TC (mg%)	227 ± 3	229 ± 3	230 ± 3	224 ± 4	220 ± 5
TG (mg%)	375 ± 12	274 ± 13	234 ± 16^a	245 ± 11^c	234 ± 11^c
LDL-C (mg%)	116 ± 4	137 ± 7	138 ± 5	135 ± 5	135 ± 5
HDL-C (mg%)	36 ± 2	37 ± 2	39 ± 2^a	39 ± 2^a	39 ± 2^a
HDL_2-C (mg%)	11 ± 1	12 ± 1	13 ± 1^a	14 ± 1^b	14 ± 2^b
HDL_3-C (mg%)	25 ± 1	25 ± 1	26 ± 2	26 ± 2	26 ± 2
Apo AI (mg%)	128 ± 4	128 ± 4	128 ± 4	130 ± 4	133 ± 5
Apo AII (mg%)	53 ± 3	54 ± 4	55 ± 4	54 ± 4	53 ± 4
Apo-B (mg%)	93 ± 7	97 ± 5	100 ± 4	101 ± 3	101 ± 3^b

vs basal values: $^a p < 0.05$; $^b p < 0.02$; $^c p < 0.01$.

3 Results

The results of this study are shown in Tables 1 to 3.

Placebo. No important variations in the mean values of lipids, lipoproteins and apo-proteins were observed in any hyperlipidemic group.

BfR. After 8 weeks' treatment the following results were observed: in type IIA the drug significantly lowered TC (−13%), LDL-C (−14%) and apo B (−5%) mean values, while it raised HDL-C (+5%), HDL_2-C (+13%) and apo A_I (+7%) mean values (Table 1); in type IIB we observed a significant decrease in TG (−27%) mean values and an increase in HDL_2-C (+14%) and apo A_I (+6%) mean values (Table 2); in type IV, BfR administration resulted in a significant decrease in TG (−38%) mean values and in an increase in HDL-C (+8%), HDL_2-C (+27%) and apo B (+9%) mean values (Table 3).

4 Conclusions

Our results clearly show that BfR, because of its efficacy and tolerability, is to be considered a "true" hypolipidemic agent. Therefore, it can be recommended in types IIA, IIB and IV even if its prevalent hypotriglyceridemic effect makes it a first choice for type IV.

Acknowledgements. The authors would like to thank Mrs. G.A. Boyd Mancinelli B.A. (hons) for her help in the translation of this paper.

References

Abell LL, Levy BB, Kendal FE (1952) A simplified method for the estimation of total cholesterol in serum and demonstration of its specificity. J Biol Chem 195:357–363

Fellin R, Martini S, Crepaldi G, Senin U, Mannarino E, Avellone G, Notarbartolo A, Capurso A, D'Agostino C, Montaguti U, Celin D, Descovich GC, Mantovani E (1981) Multicenter trial with Bezafibrate in primary hyperlipidemias. Curr Ther Res 29:657–665

Gidez LI, Miller GJ, Burstein M, Slage S, Eder HA (1982) Separation and quantification of subclasses of human plasma high density lipoprotein by a simple precipitation procedure. J Lipid Res 23:1206–1212

Ledermann H, Kaufmann B (1981) Comparative pharmacokinetics of 400 mg Bezafibrate after a single oral administration of a new slow-release preparation and the currently available commercial form. J Int Med Res 9:516–520

Mancini G, Carbonara AO, Heremans IF (1975) Immunochemical quantitation of antigens by single radial immunodiffusion. Immunochmistry 2:235–242

Mannarino E, Senin U, Fioroni G, Ventura S (1982) Effects of Bezafibrate treatment on lipids and lipoproteins. In: Crepaldi G, Greten H, Schettler G, Baggio G (eds) Lipoproteins metabolism and therapy of lipids discorders. Int Symp 16–17 April 1982, Florence. Excerpta Medica, Amsterdam Oxford Princeton, pp 113–119

Wahlefeld AW (1974) Determination after enzymatic hydrolysis. In: Bergmeyer HU (ed) Methods of enzymatic analysis, 2nd Engl edn. Chemie, Weinheim; and Academic Press, London, New York, p 1831

Weisweiler P, Schwandt P (1980) Lipoprotein lipids and apolipoproteins after six months treatment with Bezafibrate. Artery 6:402–408

Effect of Bezafibrate Retard on Plasma Lipoproteins in Hypertriglyceridemic Patients With and Without Diabetes Mellitus

G. Riccardi, G. Saldalamacchia, S. Genovese, L. Patti, G. Marotta, A. Postiglione, A. Rivellese, B. Capaldo, and M. Mancini [1]

1 Introduction

Bezafibrate (2-[4-(2-(4-chlorobenzamido)-ethyl-phenoxy]2-methyl propionic acid) is a new hypolipidemic drug which is more effective than its analogue clofibrate in reducing plasma lipid levels. Fully adsorbed in the gut it binds by 95% to plasma proteins, has a mean half-life of 2 h and is excreted through urine within 24–48 h (Abshagen et al. 1979). Bezafibrate increases hepatic and extra-hepatic lipoprotein lipase activity, thus enhancing the plasma removal of triglyceride-rich lipoproteins. The increase of lipoprotein lipase activity takes place at the muscle tissue level but not in the adipose tissue.

In addition to the effects on very low density lipoproteins it has been shown that Bezafibrate increases low density lipoprotein (LDL) receptor-dependent catabolism (Klose et al. 1980; Packard et al. 1982; Vessby et al. 1982).

In rat liver Bezafibrate increases B-oxidation of fatty acids in peroxisomes, thus reducing the substrate for the synthesis of VLDL. Moreover, in isolated rat hepatocytes it inhibits the HMG-CoA reductase activity (Berndt et al. 1980; Lazarow 1980).

This chapter evaluates the effects of Bezafibrate on plasma lipoprotein levels in hypertriglyceridemic patients with and without diabetes. A new preparation has been utilized which is administered once a day, in the evening, at the dose of 400 mg, instead of the traditional 200 mg t.i.d.

2 Patients and Methods

Sixteen patients of both sexes and in the age range 30 to 65 years were investigated. Their relative body weight was 130%. Eight patients had hyperlipoproteinemia type IIB and 8 hyperlipoproteinemia type IV (WHO 1970). Six of 16 patients were affected by non-insulin-dependent diabetes mellitus (NIDDM). None had liver or kidney disease. A lipid-lowering normocaloric diet was followed during the study. Diabetic

---

1 Institute of Internal Medicine and Metabolic Disease, Second Medical School, University of Naples, Italy

patients who were on oral hypoglycemic drugs (n=6) maintained the antidiabetic therapy throughout the experiment.

The study was performed according to a double-blind crossover design: patients underwent, in a random order, a period of placebo therapy and another period in which they received a single daily dose of Bezafibrate Retard (400 mg) administered in the evening. Each period lasted 2 months.

Total plasma and lipoprotein concentration of cholesterol and triglyceride was measured at monthly intervals by enzymatic methods. Lipoprotein separation was performed by preparative ultracentrifugation at density 1,006 g ml^{-1}. High density lipoprotein (HDL) cholesterol was determined after polyanion precipitation (Hatch and Lees 1968).

Fasting blood glucose was estimated each month by an enzymatic method (God Perid Biochemia).

For plasma lipids and lipoproteins the average of the two determinations for each treatment period was calculated. Statistical analysis was performed according to Snedecor (Snedecor and Cochran 1967). Logarithmic transformation was employed when appropriate (total and VLDL triglyceride). Values in the text are given as mean ± standard deviation.

3 Results

No significant change in body weight was observed during the study. Table 1 shows the effects of Bezafibrate Retard upon plasma lipid and lipoprotein concentrations in all 16 patients under study: a marked significant reduction of plasma triglyceride (−43%) and cholesterol (−16%) levels was observed. Both cholesterol and triglyceride were decreased in VLDL, while no significant variation was observed in LDL.

Tables 2 and 3 show the marked reduction of plasma and VLDL cholesterol and triglyceride levels in type II and type IV patients. In type IIB patients Bezafibrate induced, in addition, a significant increase of high density lipoprotein (HDL) cholesterol levels after 2 months of treatment (Table 4).

The possible interference of diabetes mellitus on the effect of Bezafibrate Retard upon plasma lipoprotein fractions was evaluated by two-way analysis of variance. No difference was observed between patients with and without diabetes in relation to their response to the hypolipidemic treatment (Table 5). By this type of statistical approach Bezafibrate was shown to reduce significantly plasma and VLDL triglyceride as well as VLDL cholesterol levels irrespective of the presence of diabetes.

In normoglycemic patients fasting blood glucose was not changed after Bezafibrate (70 ± 16 mg dl^{-1}) in comparison with placebo (80 ± 12 mg dl^{-1}). The same is true for patients with NIDDM who did not modify their fasting blood glucose concentration after placebo (158 ± 60 mg dl^{-1}) and after the active treatment (141 ± 67 mg dl^{-1}).

Bezafibrate was well tolerated by all patients. During the placebo period five patients reforted headache and vertigo, two had increased hunger, one patient had abdominal discomford and two reforted diarrhea. During Bezafibrate treatment headache was never present. One patient was affected by abdominal pain, two pa-

Table 1. Total plasma lipids and lipoprotein fraction at the end of the wash-out period and during placebo and Bezafibrate R treatment (all patients, n = 16); mg dl^{-1}, mean ± SD)

	Wash-out	Placebo	Bezafibrate[a]
Total triglyceride	485 ± 399	509 ± 480	279 ± 151***oo
VLDL triglyceride	380 ± 70	395 ± 362	199 ± 126***oo
Total cholesterol	266 ± 83	256 ± 81	226 ± 56**o
VLDL cholesterol	95 ± 90	67 ± 33	48 ± 31***oo
LDL cholesterol	140 ± 54	137 ± 58	142 ± 51
HDL cholesterol	31 ± 10	39 ± 9	40 ± 10

[a]Student's paired t-test: vs wash-out   vs placebo
                                 * $p < 0.05$ o $p < 0.05$
                              ** $p < 0.02$ oo $p < 0.001$
                             *** $p < 0.001$

Table 2. Total serum and lipoprotein triglyceride and cholesterol at the end of the wash-out period and during placebo and Bezafibrate R treatment (type IIB hyperlipidemic patients, n=8; mg dl^{-1}; mean ± SD)

	Wash-out	Placebo	Bezafibrate[a]
Total triglyceride	375 ± 151	406 ± 140	232 ± 81**oo
VLDL triglyceride	255 ± 199	313 ± 137	156 ± 70*o
Total cholesterol	289 ± 77	278 ± 67	248 ± 64
VLDL cholesterol	83 ± 74	72 ± 31	39 ± 19***o
LDL cholesterol	174 ± 53	169 ± 52	168 ± 55
HDL cholesterol	32 ± 11	38 ± 10	41 ± 9

[a]Student's paired t-test: vs wash-out   vs placebo
                                 * $p < 0.05$ o $p < 0.01$
                              ** $p < 0.02$ oo $p < 0.002$
                             *** $p < 0.001$

tients had diarrhea and two reforted increased hunger. Safety blood parameters (serum transaminase, serum creatinine and uric acid levels, RBC and WBC) were unchanged in all patients during the placebo and the active treatment period.

4 Discussion

This study shows that Bezafibrate Retard (400 mg once a day) is effective in reducing plasma cholesterol and triglyceride levels. The parallel decrease of VLDL triglyceride

Table 3. Total serum and lipoprotein triglyceride and cholesterol at the end of the wash-out period and during placebo and Bezafibrate R treatment (type IV hyperlipidemic patients n=8; mg dl^{-1}; mean ± SD)

	Wash-out	Placebo	Bezafibrate[a]
Total triglyceride	595 ± 538	612 ± 670	326 ± 192*oo
VLDL triglyceride	504 ± 494	477 ± 496	247 ± 192**ooo
Total cholesterol	243 ± 88	234 ± 91	204 ± 39
VDL cholesterol	107 ± 108	63 ± 38	58 ± 40**ooo
LDL cholesterol	106 ± 29	101 ± 40	112 ± 24
HDL cholesterol	30 ± 10	40 ± 9	38 ± 11°

[a]Student's paired t-test: vs wash-out   vs placebo
                          * $p < 0.02$ o $p < 0.05$
                          ** $p < 0.01$ oo $p < 0.02$
                                               ooo $p < 0.001$

Table 4. Variations in HDL-cholesterol concentration during the study (mg dl^{-1}; mean ± SD)

	Wash-out	Placebo		Bezafibrate[a]	
		I Month	II Month	I Month	II Month
Type II B (n=8)	32 ± 11	41 ± 12	34 ± 10	36 ± 12	46 ± 9*oo
Type IV (n=8)	30 ± 1	38 ± 9	41 ± 12	37 ± 8	40 ± 16

[a] * Student's paired t-test vs wash-out period $p < 0.05$.
oo Student's paired t-test vs placebo II month $p < 0.01$.

Table 5. Bezafibrate R lipid response in hypertriglyceridemic patients with and without diabetes, (mg dl^{-1}; mean ± SD)

	Non-Diabetic patients (n=10)		Diabetic patients (n=6)		Analysis of variance (p-value)		
	Placebo	Bezafibrate	Placebo	Bezafibrate	Between treatments	Between groups	Interaction
Triglyceride							
Total	423 ± 173	259 ± 137	654 ± 772	312 ± 180	0.025	N.S.	N.S.
VLDL	331 ± 162	196 ± 121	502 ± 569	203 ± 146	0.01	N.S.	N.S.
Cholesterol							
Total	268 ± 67	235 ± 64	246 ± 106	211 ± 41	N.S.	N.S.	N.S.
VLDL	75 ± 38	47 ± 27	53 ± 12	49 ± 40	0.05	N.S.	N.S.
LDL	149 ± 66	157 ± 62	112 ± 24	121 ± 13	N.S.	N.S.	N.S.
HDL	38 ± 11	39 ± 9	39 ± 7	41 ± 11	N.S.	N.S.	N.S.

and cholesterol supports the hypothesis that Bezafibrate affects the lipoprotein turnover more than VLDL relative lipid composition. These observations are in agreement with other studies showing that the drug activates lipoprotein lipase activity, thus enhancing the catabolism of triglyceride-rich lipoproteins (Klose et al. 1980; Vessby et al. 1982).

An increased lipoprotein lipase activity after Bezafibrate is also in line with the finding of increased plasma HDL levels during the administration of the active drug (Tall et al. 1978).

A consequence of the enhanced VLDL catabolism is the increased production of LDL through the metabolic cascade VLDL-IDL-LDL. Therefore, in theory, treatment with Bezafibrate should result in a higher LDL plasma concentration. However, this is not the case. In our study, as well as in other studies on hypertriglyceridemic patients, Bezafibrate does not induce any significant change in LDL (Fellin et al., 1982). This is probably the consequence of the increased LDL receptor-mediated catabolism induced by the drug (Packard et al. 1982).

In our study Bezafibrate Retard has a metabolic efficacy similar to that expressed by the traditional preparation (200 mg t.i.d.). Olsson et al. (1977) showed a plasma triglyceride reduction by 39% in patients with type IV hyperlipoproteinemia after 3 months of treatment with the usual Bezafibrate preparation. Similar results were obtained by other authors (Kaffarnik et al., 1978). In these two studies a 6% and 20% plasma cholesterol reduction was obtained after 6 and 12 weeks of treatment with Bezafibrate 200 mg t.i.d. in patients with type II B hyperlipoproteinemia. All these figures are very similar to the ones observed in our study with Bezafibrate Retard.

Bezafibrate Retard has, beyond doubt, several advantages in comparison to the traditional preparation: once a day administration in the evening, a lower dose, excellent tolerability and better patient compliance (Ledermann et al. 1984). Moreover, Bezafibrate Retard once a day has a similar hypolipidemic effect as compared to the usual 200 mg tablet t.i.d. despite the fact that plasma concentrations of the drug are lower. Therefore, no correlation seems possible between plasma levels of Bezafibrate and hypolipidemic efficacy (Oster et al. 1980). It is possible that during the night the activity of the drug upon lipoprotein metabolism is optimized and therefore the retard preparation administered in the evening is more effective. Enzymes influencing lipid metabolism are, in fact, known to have a circadian rhythm (Steimeier et al. 1980).

The similar low prevalence of side effects during the active treatment and the placebo period should be underlined. Moreover, Bezafibrate Retard has no adverse effect upon blood glucose concentrations in patients with and without diabetes, being in both groups particularly effective in lowering plasma lipid concentrations.

In conclusion, Bezafibrate Retard is able to lower plasma cholesterol and triglyceride concentrations in hypertriglyceridemic patients. This effect is mediated by the reduction of plasma VLDL lipoprotein concentration. Moreover, in patients with type II B hyperlipoproteinemia Bezafibrate Retard increases HDL levels. These effects are similar in patients with and without diabetes. Bearing this in mind, we believe that Bezafibrate Retard can be considered as a first choice drug for the treatment of patients with isolated (type IV) or combined (type II B) hypertriglyceridemia both with and without diabetes mellitus.

References

Abshagen U, Bablok W, Koch K, Lang PD, Schmidt HAE, Senn M, Stork H (1979) Disposition, pharmacokinetics of Bezafibrate in man. Eur J Clin Pharm 16:31–34

Beaumont JL, Carlson LA, Cooper GR, Fejfer T, Fredrickson DS, Strasser T (1970) Classifications of hyperlipidemias and hyperlipoproteinemias. Bull WHO 43:891–915

Berndt JR, Gaumert R, Still J (1980) Inhibition by Bezafibrate of hydroxymethylglutaryl CoA reductase, the enzyme regulating cholesterol biosynthesis in the rat liver. In: Greten H, Lang PD, Schettler G (eds) Lipoprotein and coronary heart disease. Witzstrock, New York Baden-Baden Cologne, pp 83–86

Fellin R, Martin J, Crepaldi G, Senin U, Mannarino E, Avallone G, Notarbartolo A, Capurso A, D'Agostino C, Mandaputi V, Celin D, Descovich GC, Mantovani E (1982) Effects of Bezafibrate on lipid and lipoprotein concentration in primary hyperlipidemias. In: Crepaldi G, Greten H, Schettler G, Paggio G (eds) Lipoprotein metabolism and therapy of lipid disorders. Excerpta Medica, Amsterdam Oxford Princeton, p 153

Hatch FT, Lees RW (1968) Practical methods for plasma lipoprotein analysis. Adv Lipid Res 6:1–68

Kaffarnik H, Scheider J, Schubart H, Muhlfellner O, Muhlfellner G, Haunsmann L, Zofel P (1978) Long-term results with Bezafibrate a new derivate of clofibrate. In: Carlson LA (ed) Int Conf Atherosclerosis. Raven, New York, p 129

Klose G, Bahrendt J, Vollmar J, Greten H (1980) Effect of Bezafibrate on the activity of lipoprotein lipase and hepatic triglyceride hydrolase in healthy volunteers. In: Greten H, Lang PD, Schettler G (eds) Lipoproteins and coronary heart disease. Witzstrock, New York Baden-Baden Cologne, pp 182–184

Lazarow PB (1980) Elevation of peroxisomal B-oxidation by Bezafibrate. In: Greten H, Lang PD, Schettler G (eds) Lipoprotein and coronary heart disease. Witzstrock, New York Baden-Baden, Cologne, pp 96–100

Ledermann H, Kaufmann B (1981) Comparative pharmacokinetics of 400 mg Bezafibrate after a single oral administration of a new slow release preparation and the currently available commercial form. J Int Med Res 9:516–520

Olsson AG, Rossner S, Walddius G, Carlson LA, Lang PD (1977) Effect of BM 15075 on lipoprotein concentrations in different types of hyperlipoproteinemia. Atherosclerosis 27:279–283

Oster P, Schlierf G, Lang PD, Andreas J, Marhlbeyer W, Schellenberg B, Valmar J (1980) Effect of high doses of Bezafibrate on lipids and lipoproteins in patients with various types of hyperlipoproteins. In: Grethen H, Lang PD, Schettler G (eds) Lipoproteins and coronary heart disease. Witzstrock, New York Baden-Baden Cologne, pp 116–118

Packard CJ, Stewart JM, Loimer AR, Morgan HG, Lawric TDV, Shepherd J (1982) Drug-induced modulation of the low density lipoprotein receptor pathways in the treatment of hypercholesterolemia. In: Crepaldi G, Greten H, Schettler G, Baggio G (eds) Lipoprotein metabolism and therapy of lipid disorders. Excerpta Medica, Amsterdam Oxford Princeton, pp 107–111

Snedecor GW, Cochran WG (1967) Statistical methods. State Univ Press, Iowa

Steimeier K, Stork H, Lenz H, Lenschner G, Liede V (1980) Pharmacology and mode of action of Bezafibrate in animals In: Greten H, Lang PD, Schettler G (eds) Lipoproteins and coronary heart disease. Witzstrock, New York Baden-Baden Cologne, p 76

Tall AR, Small DM (1978) Current concepts: plasma high density lipoproteins. New Engl J Med 299:1232–1236

Vessby B, Lithell H, Gustafson S, Ledermann H (1982) Increase in lipoprotein lipase activity in skeletal muscle following Bezafibrate. In: Crepaldi G, Greten H, Schettler G, Paggio G (eds) Lipoprotein metabolism and therapy of lipid disorders. Excerpta Medica, Amsterdam Oxford Princeton, p 101

Long-Term Experience of a Single Daily Dose of Bezafibrate Retard 400 in Hyperlipoproteinaemia of Types IIa, IIb, and IV

W. Schwartzkopff [1]

The once daily dosing of lipid-lowering agents of the clofibrate series is gaining increasing importance, because with multimorbidity, especially in elderly patients, the number of drugs prescribed increases and consequently compliance falls. We have therefore investigated in 70 patients with hyperlipoproteinaemias whether a single daily dose of 400 mg Bezafibrate Retard led during 42 weeks' long-term therapy to a clinically relevant reduction of cholesterol., triglycerides, LDL-CH and apoprotein B, as well as a constant rise in HDL-CH and of the anti-atherogenic apoproteins A-I and A-II. The question in which we were particularly interested werde:

1. What percentage of the patients on a single daily dose of 400 mg Bezafibrate responded with a reduction of cholesterol in HLP type IIa/IIb of more than 10% and in HLP types IIb and IV with a triglyceride reduction of more than 20%?
2. Is the responder rate to Bezafibrate retard the same in men and women and is it influenced by the HLP type?
3. Does the single daily dose of Bezafibrate 400 have an equally good effect on blood lipids and apoproteins as a dose of 200 mg Bezafibrate three times daily which was used in earlier investigations?
4. Does the lowering of the lipids correlate positively with the Bezafibrate concentration and negatively with the reduction in alkaline phosphatase?
5. Does Bezafibrate 400 affect the function of the liver and kidney and does it have any effect on the purine and carbohydrate metabolism?
6. What subjective side effects are to be expected during therapy with Bezafibrate 400?

Study Structure. Seventy patients with HLP types IIa, IIb and IV were treated. Sex: m 46, f 24; age: $\bar{x}$ = 63,9 years weight: $\bar{x}$ = 71,3 kg, height: $\bar{x}$ = 169.3 cm with HLP type IIa (n = 29, m 16; f 13) IIb (n = 18, m 9; f 9) and IV (n = 23, m 21; f 2).

After a 6-week placebo and diet phase, Bezafibrate 400 was prescribed as a single dose in the evening for a period of 42 weeks. Control examinations and blood collection were carried out in the fasting state at intervals of 6 weeks. Triglycerides, cholesterol, LDL- and HDL-cholesterol and the apoproteins A-I, A-II and B were determined.

---

1 Fett- u. Stoffwechselambulanz, Abt. Innere Medizin u. Poliklinik Klinikum Rudolf Virchow (Standort Charlottenburg), Freie Universität Berlin, Soorstr. 83, 1000 Berlin 19

Drugs Affecting Lipid Metabolism
Ed. by R. Paoletti et al.
© Springer-Verlag Berlin Heidelberg 1987

Compliance was determined by the measurement of Bezafibrate concentration in the serum and by measurement of alkaline phosphatase. The effect of Bezafibrate 400 on fasting and postprandial blood sugar, as well as on renal glucose excretion, was investigated in a further 47 patients with type I and type II diabetes and secondary HLP types IV/V.

Results

1. Checking the responder rate. There were considerable differences with respect to the target of more than 10% CH reduction in the HLP type IIa, IIb and IV. Only about 10 to 30% of the patients with HLP type IV responded during the course of the 42 weeks treatment with a clinically relevant lowering of cholesterol. With HLP type IIa there was about 60% response and for HLP type IIb about 76%.

A more than 20% lowering of triglycerides was found in 56% of the patients with HLP type IIa, 62% with HLP type IIb. It could be seen from the course of the responder rates for cholesterol and triglycerides that the responder rate fell for all three types of HLP with increasing duration of treatment.

2. How high is the responder rate in men and women? In men with HLP type IIa the responder rate with respect to a clinically relevant lowering of CH was 33%, but in women with type IIa 80%. The responder rate for triglycerides in men was 38% and in women about 80%. With type IIb the responder rates for cholesterol and triglycerides did not differ between the sexes. 78% of the men and 75% of the women exhibited a reduction of CH of more than 10%. The rate for triglycerides was 100% in each case. There were no comparative figures for HLP type IV available as the number of women investigated was too low. The slight effect of bezafibrate retard on the cholesterol (+ 1%) in the HLP type IV is associated with the fact that at an LDL-CH level of <150 mg dl^{-1} LDL-CH and apoprotein B increase as a result of the breakdown of VLDL to LDL. The responder rate is best with HLP type IIb. Men with type IIa either responded less well to the treatment with Bezafibrate 400 than women or were less reliable in taking the drug.

3. Does Bezafibrate 400 reduce the lipids and lipoproteins as well as normal Bezafibrate? The effect of Bezafibrate 400 on the lipids and lipoproteins is in our experience from various long-term studies less than that of normal Bezafibrate. The cholesterol was reduced in all types together with Bezafibrate 400 by about 9%, LDH-CH by 10% and triglycerides by 25% (Table 1). With 200 mg Bezafibrate t.i.d., the cholesterol fell by 14%, the LDL-CH by 12% and triglycerides by 39%. If one exludes the non-responders from the analysis, however, Bezafibrate retard has approximately as good an action as normal Bezafibrate. It is striking that the Lp(a) fell in both sexes by about 15 to 20%. A reduction of Lp(a) is not generally found during short-term therapy. The Lp(a) before treatment was increased to more than 30 mg dl^{-1} in about 18% of the patients. About half of the patients with raised Lp(a) had a history of coronary heart disease or peripheral vascular disorders. The values of Lp(a) above 30 mg dl^{-1} also fell slowly during the course of treatment with Bezafibrate 400 by more than 40%.

Table 1. Effect of Bezafibrate Retard 400 during 42 weeks

	Change (%)			
	Type IIa (n = 29)	Type IIb (n = 18)	Type IV (n = 23)	All patients (n = 70)
Cholesterol	− 12	− 18	+ 1	− 9
LDL-cholesterol	− 15	− 17	+ 4	− 10
Triglycerides	− 12	− 42	− 27	− 25
HDL-cholesterol	+ 11	+ 23	+ 25	+ 19
Apoprotein A-I	+ 19	+ 3	+ 17	+ 14
Apoprotein A-II	+ 50	+ 36	+ 40	+ 42
Apoprotein B	− 5	− 22	− 1	− 8

4. Behaviour of the serum concentrations of Bezafibrate and alkaline phosphatase in response to Bezafibrate 400. The difference in the reduction of triglycerides and cholesterol in men and women with HLP type IIa caused us to compare the Bezafibrate concentrations and the alkaline phosphatase levels in responders and non-responders.

In the responders we found a mean Bezafibrate concentration of $4.4 \pm 0.4 \, \mu g \, ml^{-1}$. In the non-responders there was only a level of $2.8 \pm 0.2 \, \mu g \, ml^{-1}$ in the serum. Alkaline phosphatase fell by $23 \, U \, l^{-1}$ in the responders, but in the non-responders by only $12 \, U \, l^{-1}$. These different levels in the Bezafibrate concentrations of responders and non-responders could be explained with the single dose of 400 mg Bezafibrate Retard, on the one hand, by poorer compliance and, on the other, by a higher body weight, which with the constant daily bezafibrate dose (400 mg) automatically leads to a lower Bezafibrate concentration in the serum. The responders were in fact 6.5 kg or 9.5% lighter than the non-responders.

The measurement of Bezafibrate and alkaline phosphatase justifies the following conclusion. A clinically relevant success of therapy on cholesterol and triglycerides with Bezafibrate 400 is to be expected if the Bezafibrate concentrations in the serum are over $4 \, \mu g \, ml^{-1}$ or the alkaline phosphatase falls by more than $20 \, U \, l^{-1}$.

5. Behaviour of the function of liver and kidneys as well as of the purine and carbohydrate metabolism. Bezafibrate 400 did not have any negative effects on liver function. The transaminases sGPT, sGOT and γ-GT remained constant. In rare cases there may be a rise in CPK, which may be associated with myalgia.

In the presence of normal renal function, there is a moderately significant rise in serum creatinine ($0.1 \, mg \, dl^{-1}$). At creatinine levels above $1.5 \, mg \, dl^{-1}$, Bezafibrate normal should be used only at reduced daily doses and Bezafibrate 400 not at all. With respect to the purine metabolism it can be said that the uric acid rises significantly but does not reach the pathological range. Haemoglobin (−4%), erythrocytes (−3%) and leukocytes (−6%) were reduced, but there was a rise in the number of platelets (+11%).

The body weight remained constant during Bezafibrate 400 therapy. Both normal Bezafibrate and Bezafibrate 400 lower the fasting blood sugar. In the 70 non-diabetics the fasting blood sugar fell significantly by 4 mg dl^{-1}. In 47 types II or type I diabetics with secondary HLP, who were treated in addition to Bezafibrate 400 with diet, glibenclamide or insulin, there was a significant reduction of the fasting blood sugar only in the type II diabetics treated with glibenclamide during the 42-week Bezafibrate 400 therapy.

While the postprandial blood sugar (−9%) and renal glucose excretion in the urine (−1.3 g l^{-1}) of tablet- and insulin-dependent diabetics were affected, the reductions found were not significant.

6. Subjective side effects. During therapy with normal Bezafibrate or Bezafibrate 400, side effects may occur in individual cases. Myositis-like symptoms with a rise in CPK have been observed, especially if no dosage adjustment was undertaken in the presence of impaired renal function. Dicoumarol effects are potentiated by Bezafibrate. More frequent checks of the thromboplastin time with possible reduction of the dicoumarol dose are necessary.

Further side effects include intestinal symptoms, nausea, vomiting, allergic exanthema and occasionally loss of hair. Such symptoms were observed in seven patients and four dropped out of the study because of these symptoms.

In summary, it can be said that the HLP types IIa, IIb and IV can be successfully treated with a single dose of Bezafibrate 400 if compliance is good. The best effects were found in the HLP type IIb. Once daily administration of Bezafibrate 400 had, however, less effect on blood lipids than 200 mg Bezafibrate t.i.d. as the result of inadequate compliance, because omission of one tablet of Bezafibrate retard means loss of the entire daily dose.

In HLP type IIa, the responder rates on once daily dosage differed between men and women. In men it was only 38%, in women, in contrast, 80%. We set a reduction of 10% cholesterol and 20% TG as measures for clinical effects. Normal Bezafibrate and Bezafibrate 400 are intended for the treatment of hypertriglyceridaemia of types IIb, III, IV and V that are refractory to diet. In familial HCL Bezafibrate normal or Bezafibrate retard should be combined with lipid-lowering agents with other mechanisms of action. Lowering of triglycerides and reduction of fasting blood sugar and of renal glucose excretion can be induced in diabetics with secondary HLP of types IIb/IV with normal Bezafibrate or Bezafibrate 400.

The mechanism of the Bezafibrate action is activation of the lipoprotein lipase, reduction of the endogenous VLDL synthesis and an increased LDL catabolism via receptor-dependent routes as the result of an inhibition of HMG-CoA reductase. This would explain the cholesterol reduction in type IIa. The side-effect rate is low, but can lead to premature stoppage of the therapy in individual cases. Alkaline phosphatase and, in special hospitals, the measurement of Bezafibrate concentration in the serum, can be used as measures for monitoring compliance.

Effects of Bezafibrate on Lipoprotein Metabolism in Cell Culture

S. Eisenberg [1]

1 Introduction

Bezafibrate (BZ) is an efficient lipid-lowering drug that is effective in several types of hyperlipidemia (1). BZ reduces high plasma triglyceride (TG) levels in patients with hypertriglyceridemia (HTG) type IV (2), high plasma LDL in patients with familial and non-familial hypercholesterolemia (HCH) type IIA (3), and both triglyceride and cholesterol in combined hyperlipidemia (CHL) type IIB (3). BZ is also effective in dyslipoproteinemia type III and some patients with type V hyperlipoproteinemia (4). Another important action of BZ is elevation of low HDL levels, especially in patients with high plasma TG, either type IV or IIB (1, 3). While the effects of BZ on plasma lipids and lipoprotein-lipid levels have been documented in many studies (1), it is not common knowledge that the hypolipidemic therapy also normalizes abnormal structure, composition and metabolism of the various plasma lipoproteins. The purpose of the present text is to summarize our investigations on the effects of BZ administration on the lipoprotein system in various forms of hyperlipidemia. These data are presented according to the lipoprotein family affected.

2 Very Low Density Lipoprotein

Bezafibrate effectively reduces high VLDL levels in HTG and CHL states (2,3) and normal levels in HCH (3). This action reflects the BZ-induced increase of plasma and tissue lipoprotein lipase (LPL) activity. Very significant and high correlations are found between the increments of post-heparin plasma LPL and decrements of TG levels (3). Metabolic studies in humans indeed demonstrate a three-fold increase of VLDL-apo B fractional catabolic rate (from 7.0 to 22.9 pools day^{-1}), a decrease of residence time (from 3.4 to 1.0 h) and a slight albeit insignificant decrease in VLDL-apo B synthetic rate (from 11.6 to 9.6 mg kg^{-1} day^{-1} (5). This effect not only reduces the plasma VLDL mass, but also induces a very significant change in several

1 Lipid Research Laboratory, Department of Medicine B, Hadassah University Hospital, Jerusalem, Israel

Drugs Affecting Lipid Metabolism
Ed. by R. Paoletti et al.
© Springer-Verlag Berlin Heidelberg 1987

other metabolic pathways. In our study (2), several abnormalities are found in HTG-VLDL, namely enrichment of the lipoprotein with free and esterified cholesterol, but subnormal content of protein and triglycerides. We consider the main abnormality to be the enrichment of the VLDL with CE molecules. This abnormality, present predominantly in the large-sized and less dense particles (6), reflects the long circulating time of VLDL in HTG states and excessive CE transfer to VLDL from both LDL and HDL (2,6). Such particles, because of their expanded CE content, cannot form normal LDL (6), and therefore must be cleared from the plasma as CE-rich "remnant" particles. BZ treatment corrects the abnormal composition of HTG-VLDL (2), and presumably the defective metabolism of the lipoprotein (5,6).

3 Low Density Lipoprotein

The effects of BZ on the LDL system in patients with different forms of HLP are complex. In HTG, BZ corrects abnormal structure and composition of the LDL and causes a substantial increase of LDL-cholesterol (LDL-C) levels (2). An opposite effects is found in HCH, 15–30% decrease of LDL-cholesterol level, while in CHL the effect is variable: in some patients the LDL-cholesterol level remains unchanged but in others, it either decreases or increases (3). This perplexing effect of BZ on LDL can be understood only when it is realized that the LDL system is different in the different patients. HTG-LDL is denser and smaller than N-LDL, and the HTG-lipoprotein contains subnormal amounts of free and esterified cholesterol but elevated amounts of triglycerides and protein (2). Some of these abnormalities can also be detected in CHL but not in HCH patients (3). In HTG and CHL states, all six abnormalities tend to revert towards normal when BZ therapy is instituted (2, 3). We regard the abnormal structure and composition of HTG-LDL to reflect excessive transfer of TG and CE molecules between an expanded pool of VLDL and a normal, or even decreased pool of LDL. This excessive lipid transfer reaction causes HTG-LDL to be a small, dense, cholesterol poor lipoprotein. On theoretical considerations alone, the HTG particle should be an inefficient regulator of cellular metabolic processes that depend on entry of cholesterol to the cells, e.g. regulation of LDL receptor activity and of cellular cholesterol synthesis. Indeed, such abnormal behaviour of HTG-LDL is found in cell culture experiments and is shown to reflect the amount of cholesterol in the particles (estimated by CE to protein ratio (7). Not surprisingly, BZ therapy corrects these metabolic abnormalities (7). In HCH, in contrast, when LDL structure and composition is normal, BZ exerts its known effect on LDL metabolism: increased LDL receptor activity, increased LDL catabolism and decreased LDL levels. In our studies (2, 3), the change of LDL-cholesterol levels in patients with HTG, CHL and HCH were strongly related to the initial plasma TG and LDL-C levels. We believe that at least three different mechanisms that regulate LDL levels are affected by Bezafibrate. These are: rates of VLDL to LDL conversions, rates of LDL degradation and altered LDL composition due to the activity of plasma lipid transfer proteins. BZ affects all three directly or indirectly. In HTG states, BZ causes increased LDL formation from VLDL (by decreasing VLDL-CE content), normalizes the subnormal cholesterol content of HTG-LDL and decreases cellular

LDL receptor protein synthesis. All three effects would cause an increase of the low LDL-C and LDL mass concentration. In HCH, when VLDL to LDL conversion is normal or even elevated and LDL composition is unalatered, the main effect of BZ would be increased LDL degradation and reduced LDL concentration. The effects of BZ in CHL should be variable, depending on the degree of abnormalities of the LDL system and the effects of the drug on each of the three processes. That indeed has been found (3).

4 High Density Lipoproteins

The effects of BZ on HDL are less clear. Similar to LDL, HTG-HDL is smaller and denser than N-HDL and the lipoprotein is poor in cholesterol but enriched with triglycerides and protein (2). These abnormalities presumably reflect excessive lipid transfer activity in HTG states (2). Although less marked, similar abnormalities are found in CHL but not HCH patients (3). BZ causes an increased HDL-C concentration in all patients without a marked effect on apo A-I (2, 3). Interestingly, the absolute and percentage increase of HDL-C is the highest in patients with the highest pretreatment plasma TG levels (i.e. HTG) and lowest in those with normal initial TG levels (i.e. HCH). The effect in CHL patients is intermediate. Two mechanisms have been proposed to explain this phenomenon (3). The first is decreased CE and TG transfer between HDL and VLDL when the expanded plasma pool of triglyceride-rich lipoproteins is reduced. The second mechanism relates the BZ-induced increase of HDL to the increase of LPL activity. Both mechanisms are expected to affect predominantly HDL-C concentrations with only a modest effect, or no effect on apo A-I (8). Whether HTG-HDL behaves abnormally in metabolic cycles and whether BZ normalizes such abnormalities (if they exist), has not been determined.

5 Bezafibrate, Lipoprotein Metabolism and Atherogenesis

The considerations detailed above demonstrate the complexity of the plasma lipoprotein system. Single metabolic defects appear to cause multiple abnormalities of the structure, composition and metabolism of all plasma lipoproteins. The example shown in Fig. 1 is based on our observations in HTG states. Ideally, cholesterol should leave the plasma with lipoproteins that interact with specific cellular receptors (e.g. LDL receptors) or with lipoproteins that are involved with reverse cholesterol transport (e.g. HDL). Both LDL and HDL channel cholesterol into re-utilization metabolic cycles that are, therefore, anti-atherogenic. In the presence of metabolic defects, however, some or much cholesterol leaves the plasma through other pathways including those that direct the cholesterol to atheroma-forming cells. Correction of metabolic defects by bezafibrate (and other lipid-lowering drugs) causes normalization of the plasma fat transport system and is expected to be associated with a decreased tendency to develop atherosclerotic diseases.

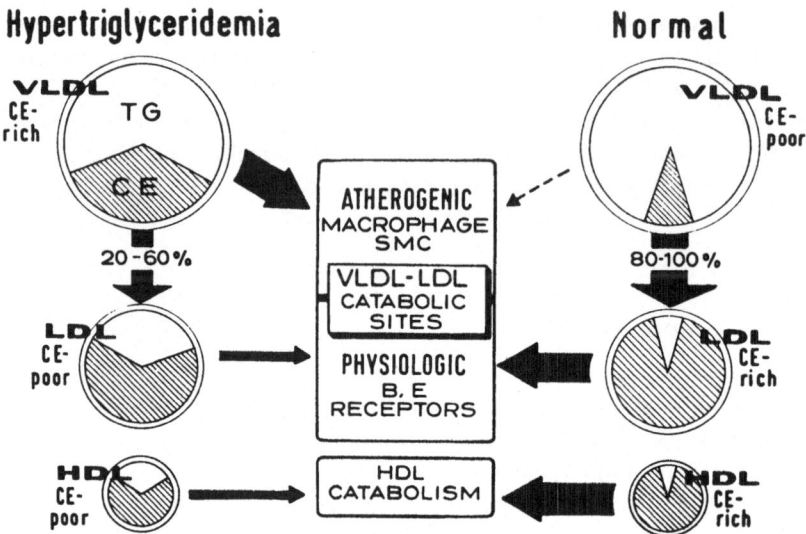

Fig. 1. Lipoprotein metabolism and cholesterol distribution in normo- and hypertriglyceridemic state

References

1. Eisenberg S, Gavish D, Kleinman Y (1986) Bezafibrate. In: Levy RI, Shepherd J, Packard CJ, Miller JE (eds) Pharmacological control of hyperlipidaemia. Prous, Barcelona, p 145
2. Eisenberg S, Gavish D, Oschry Y, Fainaru M, Deckelbaum RJ (1984) Abnormalities in very low, low, and high density lipoproteins in hypertriglyceridemia. Reversal toward normal with beza-fibrate treatment. J Clin Invest 74:470−482
3. Gavish D, Oschry Y, Fainaru M, Eisenberg S (1986a) Changes in very low-, low-, and high-density lipoproteins during lipid lowering (bezafibrate) therapy: studies in type IIa and type IIb hyperlipoproteinaemia. Eur J Clin Invest 16:61−68
4. Gavish D, Eisenberg S, Berry EM, Kleinman Y, Witztum E, Norman J, Leitersdorf E (1987) Bulimia − an underlying behavioral disorder in hyperlipidemic subjects. Arch Int Med 147:705−708
5. Shepherd J, Packard CJ, Stewart JM, Atmeh RF, Clark RS, Boag DE, Carr K, Lorimer AR, Ballantyne D, Morgan HG, Lawrie TDV (1984) Apolipoprotein A and B (S_f 100−400) metabolism during Bezafibrate therapy in hypertriglyceridemic subjects. J Clin Invest 74:2164−2177
6. Oschry Y, Olivecrona T, Deckelbaum RJ, Eisenberg S (1985) Is hypertriglyceridemic very low density lipoprotein a precursor of normal low density lipoprotein. J Lipid Res 26:158−167
7. Kleinman Y, Eisenberg S, Oschry Y, Gavish D, Stein O, Stein Y (1985) Defective metabolism of hypertriglyceridemic low density lipoprotein in cultured human skin fibroblasts. Normalization with Bezafibrate therapy. J Clin Invest 75:1796−1803
8. Eisenberg S (1984) High density lipoprotein metabolism. J Lipid Res 25:1017−1058

Efficacy and Tolerance of Fenofibrate: A Ten-Year Experience

G. F. BLANE[1]

1 Introduction

My objective is to paint with a very broad brush to give you some general idea of the clinical effects of this compound.

Fenofibrate (Lipanthyl)[2] is structurally related to clofibrate (Fig. 1). However, replacement of the chlorine atom with a parachlorobenzoyl element changes the geometry and appears to have conferred on fenofibrate considerably greater potency and a greater specificity (Boucherle 1980; Chazan 1984).

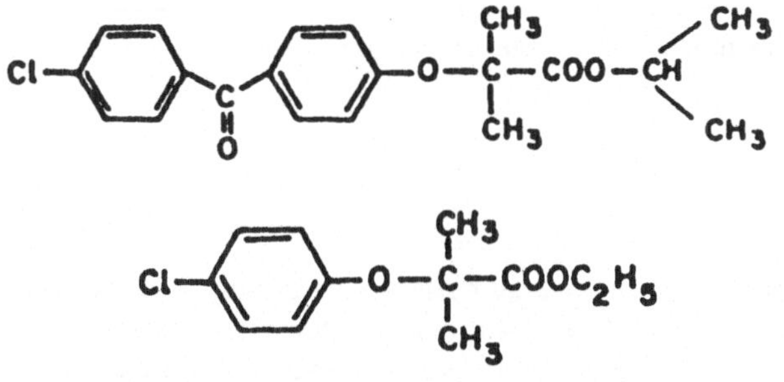

Fig. 1. Structure of fenofibrate and clofibrate

2 Clinical Studies

Fenofibrate was first introduced into clinical practice in France in 1975 and has since been widely used in Europe in all of the hyperlipidemias, with the exception of type 1 patients. It has been estimated from capsule sales that the global clinical experience with fenofibrate to date corresponds to some 6 million patient-years of treatment.

---

1 Laboratoires Fournier, Centre de Recherches de Daix, 50 Rue de Dijon, Daix, 21121 Fontaine-les-Dijon, France
2 Lipanthyl Rx: manufactured by Laboratoires Fournier S.A., Dijon, France

Drugs Affecting Lipid Metabolism
Ed. by R. Paoletti et al.
© Springer-Verlag Berlin Heidelberg 1987

Some 80 clinical trials have been conducted using fenofibrate. The distribution of these trials by country is shown in Table 1 from which is seen that more than 3.600 patients have been involved and that the number of patient-years of fenofibrate exposure in trials is close to six thousand.

Table 1. Clinical trials with fenofibrate[a]

Country	First trial	Number of trials	Number of patients	Number of patient-years
France	1973	22	1538	4734
West Germany	1976	15	782	636
Belgium	1976	6	165	73
Italy	1979	14	315	94
Switzerland	1976	4	75	19
Poland	1978	5	137	43
Sweden	1977	2	60	104
U.S.A.	1981	1	226	107
Others	1977	12	320	117
Total		81	3618	5927

[a]Mostly open with preliminary 1−2 month diet-only period to establish baseline.

2.1 Efficacy

In order to obtain a clear perspective on the overall pattern of response to fenofibrate we combined and reanalysed the evidence from 14 European short-term studies of up to 6 months duration where the dose was most often the recommended one capsule of 100 mg with each of the three principal daily meals (Blane et al. 1986).

For the 807 type IIa patients in these 14 trials the mean total cholesterol reduction after 1 month of treatment was 24% and this value changed negligibly over subsequent visits up to 6 months. A similar pattern was seen for the 688 type IIb patients.

Corresponding total triglyceride means in types IIb and IV patients after a 6-month treatment period represented percentile reductions of 38 and 57% from the respective baselines.

For the longer-term analysis of the same basic plasma lipid parameters data were combined from six studies of 6 months to 6 and more years duration, for all of which good raw data were available 701 type IIa, 599 type IIb, 15 type III and 236 type IV patients were included and the results are shown by type in Fig. 2 for cholesterol and Fig. 3 for triglycerides. The mean lipid reductions seen at 6 months were maintained virtually unchanged over the 6-year observation period.

For triglycerides, as for cholesterol, one sees the greatest percentile reductions where baseline levels are highest.

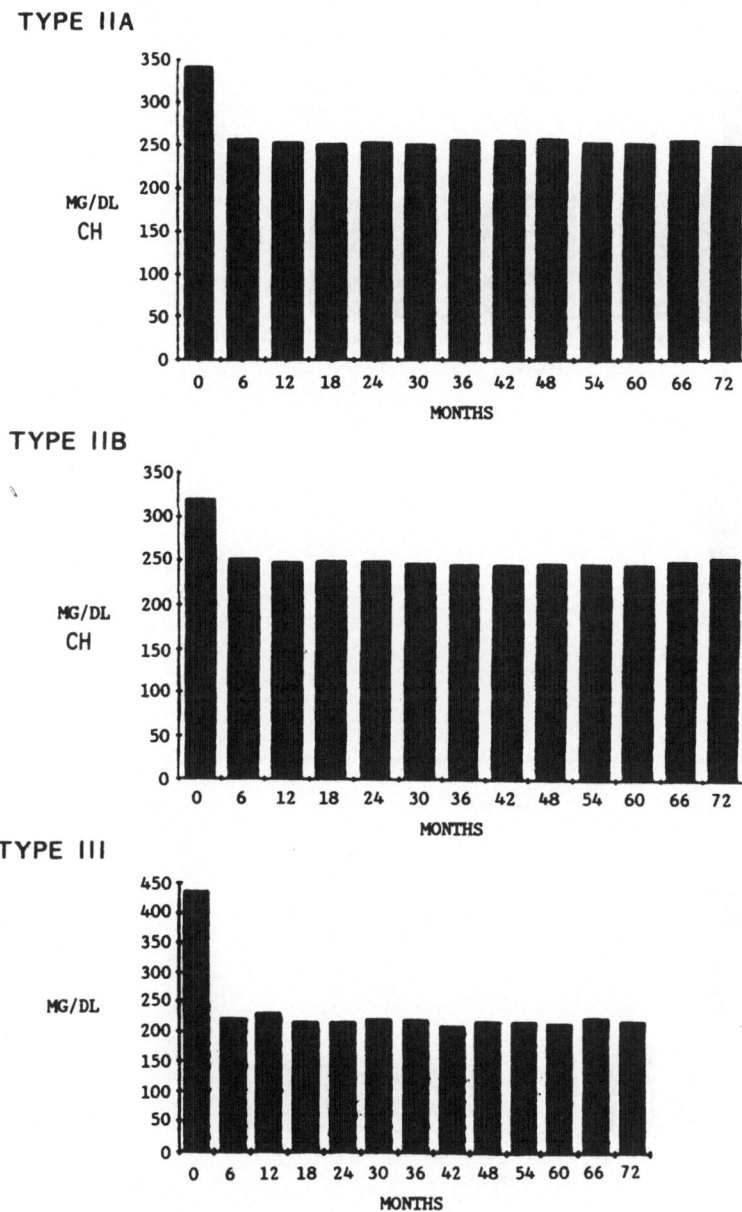

Fig. 2. Mean total plasma cholesterol over 6 years

It is of interest to compare these results from shorter mostly open European studies with those obtained in the recently concluded double-blind placebo-controlled study conducted in 11 lipid clinics in the United States (Table 2) (Brown et al. 1986). For the 92 type IIa patients the mean reduction in total cholesterol after 6 months of fenofibrate treatment was 17.5% and that for total triglycerides 37.9%.

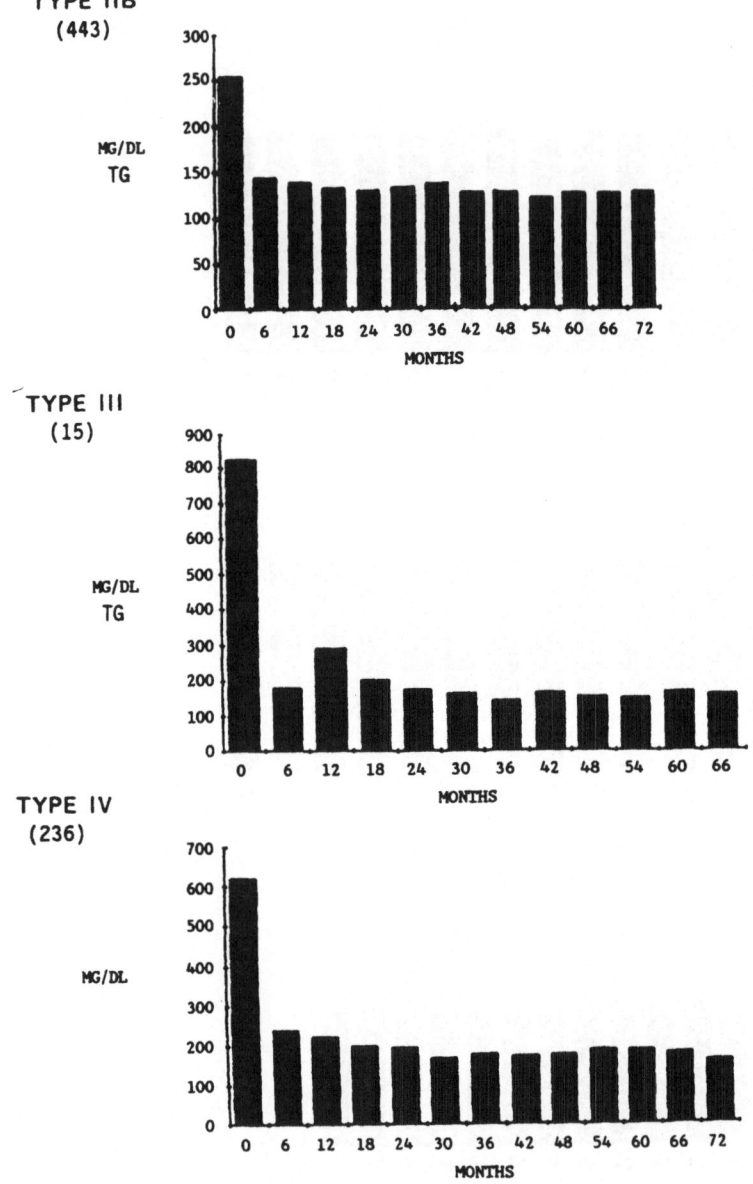

Fig. 3. Mean plasma triglycerides over 6 years

The corresponding figures for the 24 type IIb patients were 15.8 and 44.6%.

Thus, the lipid changes seen in this placebo-controlled US trial, during which patients were rigorously monitored for compliance to both drug treatment and diet, fit into the range observed during previous studies and tend to validate the mass of existing open European data.

Table 2. Percentage change at end point from baseline values: 24 week double-blind period of US study

	Type IIA		Type IIB	
	Fenofibrate N = 92	Placebo N = 88	Fenofibrate N = 24	Placebo N = 22
Total cholesterol	− 17.5	− 0.4	− 15.8	+ 4.6
LDL-cholesterol	− 20.3	+ 0.4	− 6.1	− 0.5
HDL-cholesterol	+ 11.1	− 1.2	+ 15.3	− 3.5
Total triglycerides	− 37.9	− 4.2	− 44.6	+ 22.3
LDL/HDL-cholesterol	− 27.1	− 1.9	− 13.3	0.0
VLDL-cholesterol	− 38.4	− 2.5	− 52.7	+ 8.4

Table 3. Summary table of changes in plasma lipid parameters following treatment with fenofibrate

	Trend	Type IIa (%)	Type IIb (%)	Type III (%)	Type IV (%)
Cholesterol	↓↓	16−30%	10−30%	38−51%	−
Triglycerides	↓↓↓	−	30−67%	48−85%	35−60%
VLDL-cholesterol	↓↓	26−61%	30−70%	>20%	60%
LDL-cholesterol	↓↓ or ↗	17−29%	2−29%		−10 to + 40%[a]
HDL-cholesterol	= or ↗	0−40%	0−26%		0−23%
Apo AI	= or ↗		3−38%		
Apo B	↓↓		10−37%		

[a]Very dependent on baseline level which is often low in type IV.

Seven European and two North American groups have included useful lipoprotein assays in their fenofibrate studies and six have studied Apo AI and Apo B. The range of changes seen in these and the simple lipid parameters are summarized in Table 3.

Three brief comments:
1. HDL levels are consistently increased where the baseline is low.
2. LDL levels are consistently reduced except where the baseline is already normal, e.g. some IIb and many type IV patients.
3. Apoprotein changes reflect those seen in the relevant simple lipids.

As a parenthesis, fenofibrate consistently reduces towards normal the elevated plasma uric acid levels seen commonly in hyperlipidemic patients.

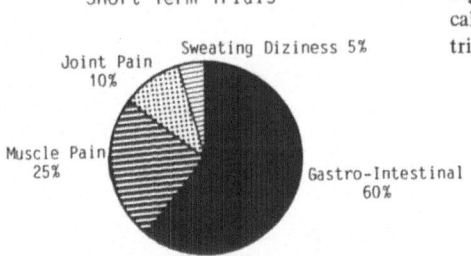

Short Term Trials *

* Circle represents 6,3% of 1838 patients

Fig. 4. Nature and relative frequency of clinically observed unwanted effects in clinical trials

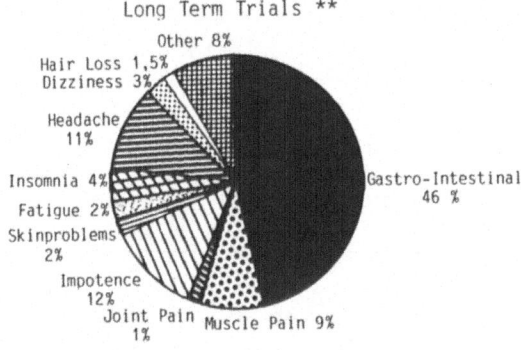

Long Term Trials **

** Circle represents 11,3% of 1571 trial patients

2.2 Unwanted Effects

2.2.1 Clinical Unwanted Effects

The overall incidence of unwanted clinical effects in European open clinical studies was 6.3% in short-term trials and 11.3% in trials of greater than 6 months duration (Fig. 4).

Approximately half of all problems fell into the gastro-intestinal category, were apparently mild, and varied from constipation to diarrhea. Skin problems were typically classified as pruritus, urticaria or erythema and were occasionally bothersome, but cleared up rapidly when treatment was stopped.

Post-marketing surveillance data accumulated since 1975 reflect the list of unwanted effects seen in trials but with a much lower incidence.

2.2.2 Biological Unwanted Effects

Sporadic and short-lived increases in transaminases have been reported by a number of authors. These seem to subside without discontinuing fenofibrate treatment in most cases. There appears to be some slight tendency towards increase in plasma urea and creatinine values during long-term trials but there is no evidence of any associated renal dysfunction or pathology. Blood glucose levels are not significantly affected by fenofibrate although there is a tendency towards reduction: in diabetic patients previous anti-diabetic medication did not need adjustment. Other small changes in biochemical values appear also to be without significance.

There is no evidence of any association between fenofibrate and gallstones in normal use or trials. However, five investigations of bile during administration of fenofibrate have shown small upturns in indices of bile lithogenicity, which are associated with increased cholesterol clearance and which seem unlikely to be of clinical significance.

3 Conclusion

Given that fenofibrate appears to be an effective drug for the regulation of perturbed plasma lipids, it is appropriate here to consider what clinical benefit might be predicted.

The LRC/CPPT cholestyramine study published in 1984 proved finally, that lowering LDL-cholesterol levels in asymptomatic men can diminish the incidence of coronary heart disease morbidity and mortality (Lipid Research Clinics Program 1984). No epidemiological-scale study has been made with fenofibrate. Howwever, the long-term total cholesterol and LDL-cholesterol reductions with this drug are considerably greater than those that were seen with cholestyramine in the LRC/CPPT trial, and fenofibrate also increases HDL-cholesterol while cholestyramine has little effect (Table 4).

The prediction would be that if fenofibrate is, as appears to be the case, considered to be relatively free from long-term major adverse effects, it should be more clinically active than the resin at the level of cardiovascular morbidity and mortality.

Table 4. Lipid changes after 1 year of treatment with fenofibrate compared with values from LRC/CPPT study

Plasma lipid parameters	Changes from baseline vis a vis placebo (%)	
	Fenofibrate US trial[a]	Cholestyramine LRC/CPTT trial[b]
Total cholesterol	− 20	− 12
LDL cholesterol	− 22	− 16
HDL-cholesterol	+ 9	+ 3
Total triglycerides	− 38	− 4
HDL-chol/total-chol	+ 36	+ 21

[a] 85 Male and female patients receiving fenofibrate throughout both double-blind (24 weeks) and to the end of the open (24 weeks) period of the study.

[b] 1900 Men at start of trial in cholestyramine-treated group.

References

Blane GF, Bogaievsky Y, Bonnefous F (1986) Fenofibrate: influence on circulating lipids and side-effects in medium and long-term clinical use. In: Fears R (ed) Pharmacological control of hyperlipidaemia. Prous, Barcelona, pp 187–216

Boucherle A (1980) Le fénofibrate: présentation chimique – différences avec le clofibrate. Nouv Press Med 9:3721–3723

Brown WV, Dujovne CA, Farquhar JW, Feldman EB, Grundy SM, Knopp RH, Lasser NL, Mellies MJ, Palmer RH, Samuel P, Schonfeld G, Superko HR (1986) Effects of fenofibrate on plasma lipids. Double-blind multicenter study in patient with type II A or II B hyperlipidemia. Arterosclerosis 6:670–678

Chazan JB (1984) New Developments concerning the mode of action of Fenofibrate. In: Carlson LA, Olsson AG (eds) Treatment of hyperlipoproteinemia. 41st Meet Eur Atherosclerosis Group, Stockholm (Sweden), June 2–3, 1984. Raven, New York, pp 171–174

Lipid Research Clinics Program (1984) The lipid research clinics coronary primary prevention trial results: I. Reduction in incidence of coronary heart disease. JAMA 251:351–364

Mechanism of Action of Fenofibrate: New Data

M. Pascal, H. Cao Danh, and C. Legendre [1]

1 Introduction

Fenofibrate [2-[4-(4-chlorobenzoyl)-phenoxy]-2-methyl-propanoic acid isopropyl ester] is a hypolipidemic drug used in the prevention of cardiovascular diseases (Blane, Bonnefous and Bogaievsky, 1986). In man and animals this compound, administered by the oral route, is readily and entirely transformed into fenofibric acid (LF 153) which in turn can be partially metabolized in a reduced derivative, LF 433. The potent cholesterol-lowering effect has been demonstrated with fenofibrate (Gurrieri et al. 1976). Liver cholesterol hemeostasis plays a central role in the control of circulating cholesterol levels and in the development of atherosclerosis. The key interrelated events are liver uptake of cholesterol by the LDL receptors, the endogenous biosynthesis of cholesterol, its transformation and elimination as bile acids and its storage in the form of cholesteryl ester by the action of acyl coenzyme A cholesterol acyltransferase (ACAT). Inhibition of cholesterol synthesis and esterification have been proven successful in the control of hypercholesterolemia and its consequences (Alberts et al. 1980; De Vries et al. 1986). Therefore, investigations were carried out on these two metabolic pathways using fenofibrate.

2 Liver Cholesterol Biosynthesis

Previous work (Warner-Lambert, internal report) indicated that $[^{14}C]$-acetate incorporation in liver cholesterol was inhibited in fenofibrate-treated rats (100 mg kg^{-1} day^{-1} during 1 week, oral route). However, $[^{14}C]$-mevalonate incorporation was not impaired in the same animals, thus suggesting inhibition of cholesterol biosynthesis at a step preceding mevalonate formation.

One reasonable target for fenofibrate action could be hydroxymethylglutaric coenzyme A reductase (HMG CoA reductase) which converts HMG CoA to mevalonate and has been shown to be the rate-limiting step in this biosynthetic pathway. Using the assay described by Alberts et al. (1980), with the liver enzyme solubilized by a freeze-

---

1 Laboratoires Fournier, Centre de Recherches de Daix, 21121 Fontaine-les-Dijon, France

Drugs Affecting Lipid Metabolism
Ed. by R. Paoletti et al.
© Springer-Verlag Berlin Heidelberg 1987

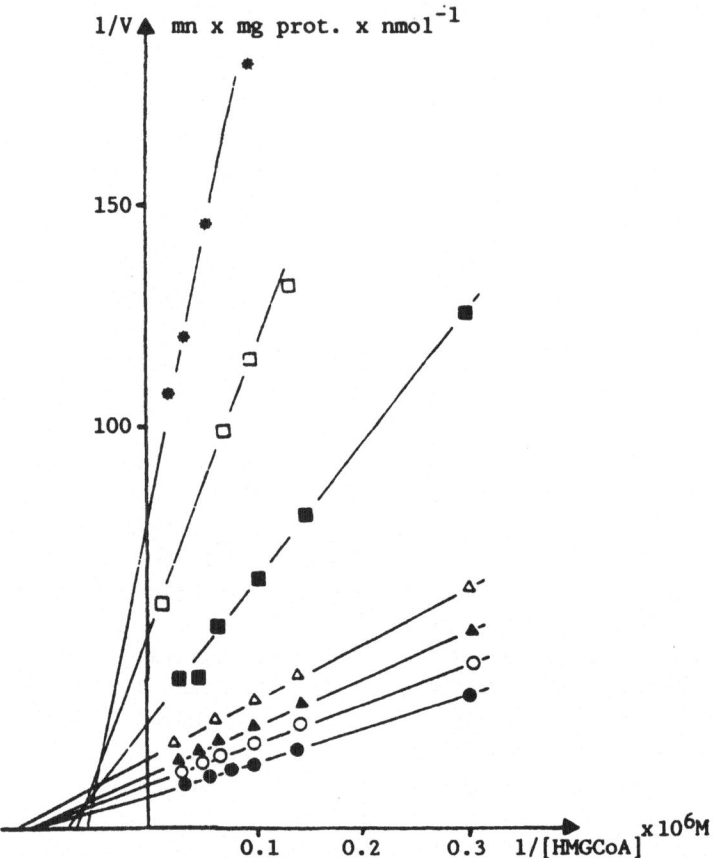

Fig. 1. Lineavez-Burk plot of solubilized rat liver HMG CoA reductase inhibition by fenofibric acid (LF 153): * 5 10^{-3} M; □ 3.5 10^{-3} M; ■ 2.5 10^{-3} M; △ 1.25 10^{-3}; ▲ 10^{-3} M; ○ 0.5 10^{-5} M; ● 0

thawing procedure, in vitro inhibition could be demonstrated for LF 153 but not for LF 433. A Lineavez-Burk plot (Fig. 1) showed that LF 153 has an inhibitory pattern close to that of a non-competitive type inhibitor. Calculated K_I (1.2 10^{-3} M) was far above currently observed circulating levels, suggesting that such an inhibition does not take place in vivo. Similar results were obtained with the microsomal enzyme.

However, liver microsomal HMG CoA reductase activity was clearly decreased in fenofibrate-treated rats: when a daily oral administration (100 mg kg^{-1}) was used, depressed activity was observed after 2 days and was maximum after 7 days (Fig. 2). When rats were treated with LF 433 (50 mg kg^{-1}) an equivalent maximum inhibition was obtained, however, the onset was earlier. Dose-effect curves, after 7 days of treatment, (Fig. 3) indicated a good correlation between the hypocholesterolemic activity and the ex vivo HMG CoA reductase inhibition for both compounds. Meanwhile, on a dose basis, LF 433 appears more potent than fenofibrate.

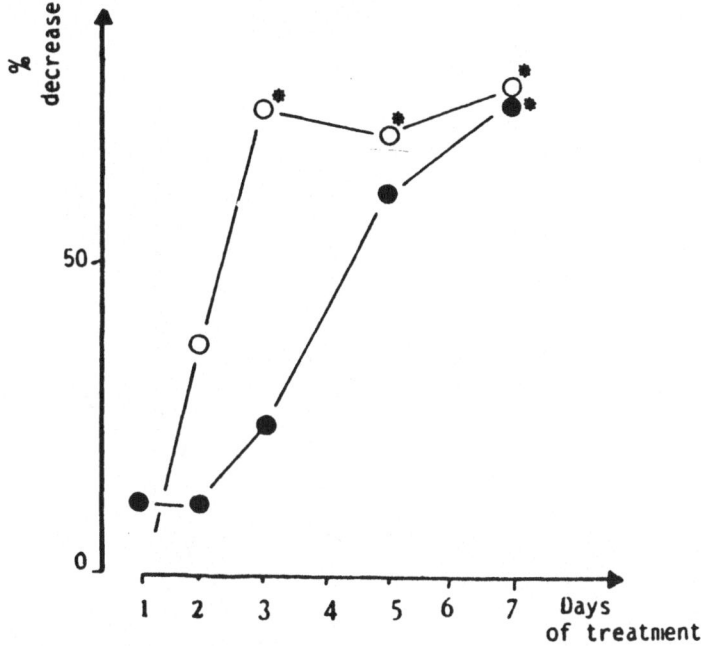

Fig. 2. Liver microsomal HMG CoA reductase activity decrease in fenofibrate (●) and LF 433 (○) treated rats at daily oral doses of 100 mg kg^{-1} and 50 mg kg^{-1} respectively. Six to eight rats were used for each assay. Results are expressed as % decrease compared to control rats: Asterisk= $p <$ 0.05 (Mann-Whitney test)

3 Cholesterol Esterification by ACAT

Using the in vitro assay described by Severson and Fletcher (1981), LF 153 and 433 were found to be inhibitors of liver microsomal ACAT, with an IC$_{50}$ of 0.46 and 0.36 mM respectively. Compared to HMG CoA reductase, inhibitory doses were lower but still superior to the commonly observed circulating levels.

When rats were treated daily with fenofibrate (100 mg kg^{-1}) or LF433 (50 mg kg^{-1}) by oral administration, a time-dependent decrease in ACAT activity was observed up to 7 days (Fig. 4).

Dose-effect curves (Fig. 5) indicated that LF 433 is more potent than fenofibrate. However, even for doses up to 200 mg kg^{-1} inhibition does not exceed 50% which is lower than the 75% inhibition obtained with HMG CoA reductase. As ACAT activity appears to be regulated by the free cholesterol level in microsomes (Hashimoto and Daton, 1977), free cholesterol and esterified cholesterol levels in microsomes were evaluated according to Watson (1960) after diethylether ethanol (1/1) extraction. As shown in Fig. 5, a slight decrease in free cholesterol was observed with a dose pattern similar to that of ACAT inhibition. The same effect was observed for esterified cholesterol (data not shown).

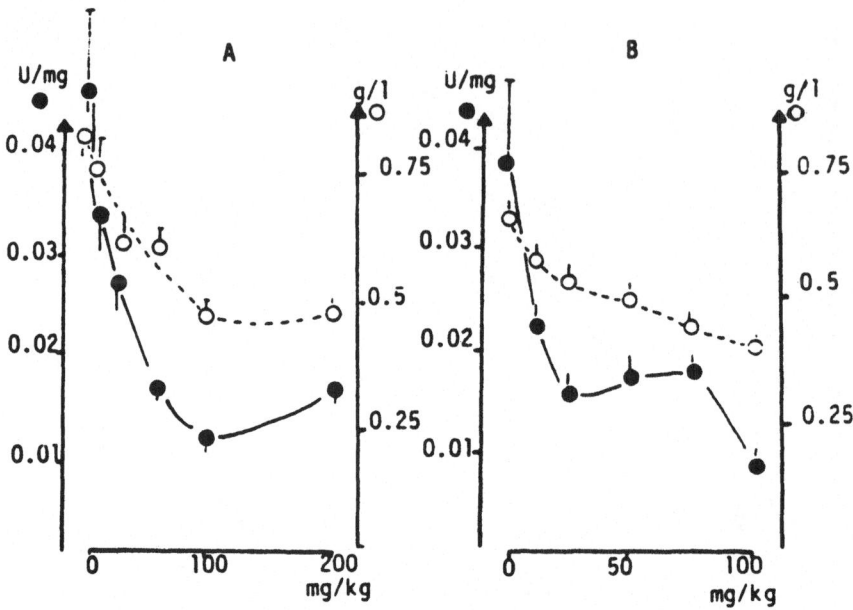

Fig. 3A, B. Dose-effect relationship of fenofibrate (*A*) and LF 433 (*B*) ex vivo inhibition of rat liver microsomal HMG CoA reductase activity and serum cholesterol-lowering effect: ● HMG CoA reductase activity nmol min^{-1} mg^{-1} proteins (U mg^{-1}); ○ serum cholesterol level g l^{-1}. Results are expressed as mean ± mean SD (n= 8)

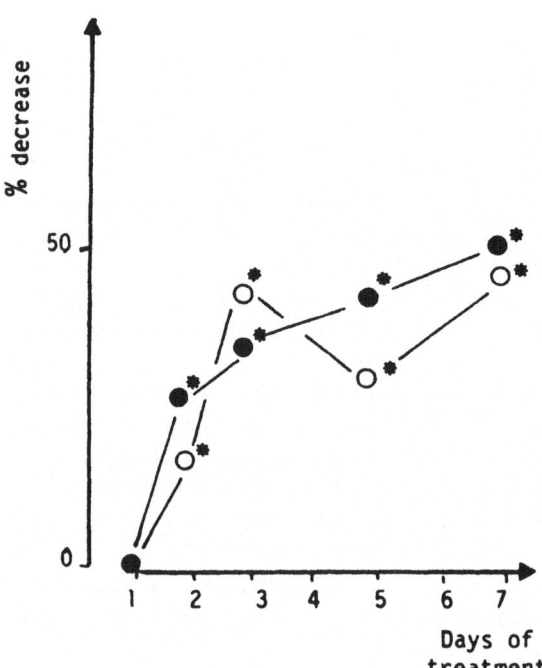

Fig. 4. Liver microsomal ACAT activity decrease in fenofibrate (●) and LF 433 (○) treated rate at a daily oral dose of 100 mg kg^{-1} and 50 mg kg^{-1} respectively. Six to eight rats were used for each assay. Results are expressed as % decrease compared to control rats: Asterisk= $p < 0.05$ (Mann-Whitney test)

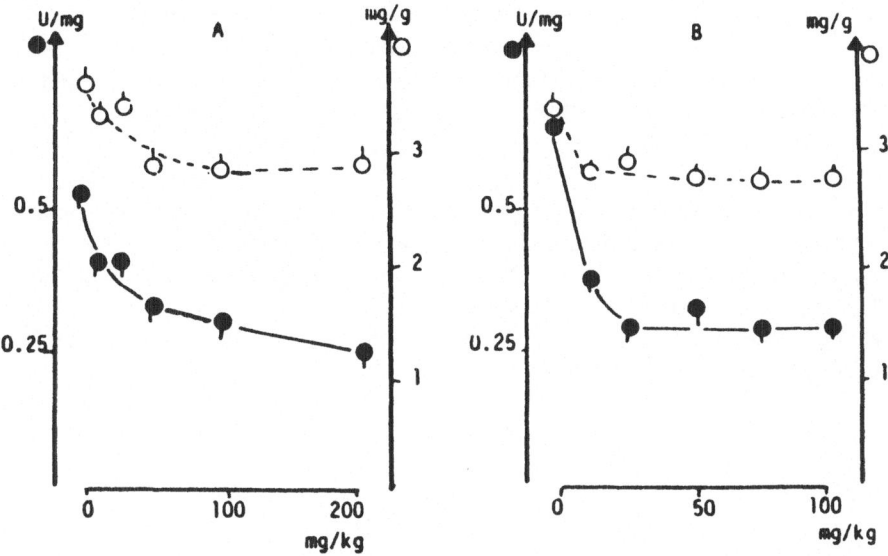

Fig. 5 A,B. Dose-effect relationship of fenofibrate (**A**) and LF 433 (**B**) ex vivo inhibition of rat liver microsomal ACAT and liver microsome-free cholesterol level: ● ACAT activity, nmol min^{-1} mg^{-1} proteins (U mg^{-1}); ○ free cholesterol level, mg g^{-1} microsomes. Results are expressed as mean˙ ± SD (n= 8)

4 Discussion

Our results show that fenofibrate treatment of rats induces a decrease of both HMG CoA reductase and ACAT activities. The former result is in accordance with the observation of Schneider et al. (1985) that fenofibrate lowers leukocyte HMG CoA reductase activity in type IIa and IIb patients. Such an effect can be at least partially mediated by the reduced metabolite, LF 433, which appears more potent. Consistent with this hypothesis, is the fact that in rats the reduction of LF 153 to LF 433 is a particularly important metabolic pathway compared to humans (Caldwell et al. 1986).

A direct inhibition of the two enzymatic activities is not likely since the in vitro inhibitory concentrations are of a higher magnitude than the corresponding circulating drug levels. Concerning HMG CoA reductase activity, different mechanisms of regulation have been described. These include: regulation of enzyme synthesis (Clarke et al. 1983), of enzyme degradation (Gil et al. 1985) and of the activation state of the enzyme by a phosphorylation-dephosphorylation process (Kennelly and Rodwell 1985). Additionally, a thiol-dependent allosteric modulation of enzymatic activity has also been described (Roitelman and Shechter 1984). Investigations of the effect of fenofibrate on such mechanisms are in progress. The observed ACAT ex vivo inhibition may be the consequence of decreased microsomal-free cholesterol level. However, it remains to be shown whether the 20% diminution observed is sufficient to explain the 50% inhibition of the enzymatic activity.

Inhibition of HMG CoA reductase and ACAT coupled to the already described increased output of cholesterol in the form of bile acids in rat liver (Bentzen et al. 1978) should lead to increased uptake of LDL cholesterol by the receptor pathway and increase elimination of cholesterol. In fact, enhanced uptake of LDL-cholesterol by a receptor-mediated pathway has been demonstrated in type IV patients (Shepherd et al. 1985).

The effects of fenofibrate on liver cholesterol homeostasis are conceivably related to its hypocholesterolemic activity as judged by the correlation between the latter and enzyme inhibitions. The expectation of an anti-atherosclerotic property for the compound is further supported by the fact that ACAT inhibition should prevent the cholesteryl ester accumulation in arterial tissues which is one of the characteristics features of the atherosclerotic plaques.

References

Alberts AW, Chen J, Hunt V, Huff J, Hoffman C, Rothrock J, Lopez M, Joshua H, Harris E, Patchett A, Monoghan R, Currie S, Stapley E, Albers-Shonbert G, Hensens O, Hirshfield J, Hoogsteen K, Liesch J, Springer J (1980) Mevilonin: a highly potent competitive inhibitor of hydroxymethyl glutaryl-coenzyme A reductase and a cholesterol-lowering agent. Proc Natl Acad Sci USA 77(7):3957–3961

Bentzen C, Tourne C, Wulfert E (1978) Effect of procetofen on hepatic uptake and biliary secretion by the isolated perfused liver. In: Kritchevsky D, Paoletti R, Holmes WL (eds) Advances in experimental medicine and biology, vol 109. Plenum, New York, p 368

Blane GF, Bogaievsky Y, Bonnefous F (1986) Fenofibrate: influence on circulating lipids and side-effects in medium and long-term clinical use. In: Sears R (ed) Pharmacological control of hyperlipidaemia. Prous, Barcelona, pp 187–216

Caldwell J, Strolin Benedetti M, Weil A (1986) Comparative metabolism of fenofibrate in rats and human volunteers. Br Pharmacol Soc Meet, Bath (England), April 9–11, 1986, Abstr C69

Clarke CF, Edwards PA, Lan SF, Tanaka RD, Fogelman AM (1983) Regulation of 3-hydroxy-3-methylglutaryl-coenzyme A reductase mRNA levels in rat liver. Proc Natl Sci USA 80:3304–3308

De Vries VG, Schaffer SA, Largis EE, Dutia MO, Wang CH, Bloom JD, Katocs AS (1986) Potential antiatherosclerotic agents 5. An acyl-CoA: cholesterol O-acyltransferase inhibitor with hypocholesterolemic acitvity. J Med Chem 29:1131–1133

Gil G, Faust JR, Chin DJ, Goldstein JL, Brown MS (1985) Membrane bound domain of HMG CoA reductase is required for sterol-enhanced degradation of the enzyme. Cell 41:249–258

Gurrieri J, Le Lous M, Renson FJ, Tourne C, Voegelin H, Majoie B, Wulfert E (1976) Antilipidemic drugs, part 2:Experimental study of a new potent hypolipidemic drug, isopropyl -[4'-(p-chlorobenzoyl) 2-phenoxy-2-methyl] propionate (LF 178). Arzneimitt Forsch 28:3–20

Hashimoto S, Daton S (1977) Studies of the mechanism of augmented synthesis of cholesterol ester in atherosclerotic rabbit aortic microsomes. Atherosclerosis 28:447–452

Kennelly PJ, Rodwell VW (1985) Regulation of 3-hydroxy-3-methylglutaryl coenzyme A reductase by reversible phosphorylation-dephosphorylation. J Lipid Res 26:903–914

Roitelman J, Shechter I (1984) Regulation of rat liver 3-hydroxy-3-methylglutaryl coenzyme A reductase: evidence for thiol-dependent allosteric modulation of enzyme activity. J Biol Chem 259(2):870–877

Schneider A, Stange EF, Ditschuneit HH, Ditschuneit H (1985) Fenofibrate treatment inhibits HMG-CoA reductase activity in mononuclear cells from hyperlipoproteinemic patients. Atherosclerosis 56:257–262

Severson DL, Fletcher T (1981) Studies on the activity of Acyl-CoA cholesterol-o-acyl transferase and acyl cholesterol ester synthetase in rat aortas. Biochim Biophys Acta 664:475–486

Shepherd J, Caslake MJ, Lorimer R, Vallance BO, Packard CJ (1985) Fenofibrate reduces low density lipoprotein catabolism in hypertriglyceridemic subjects. Arteriosclerosis, 5(2):162–168

Watson D (1960) A simple method for the determination of serum cholesterol. Clin Chim Acta 5: 637–643

Species Differences in the Metabolic Disposition of Fenofibrate

A. WEIL [1], J. CALDWELL [1], M. STROLIN-BENEDETTI [2], and P. DOSTERT [2]

1 Introduction

Fenofibrate (isopropyl 4-(p-chlorobenzoyl)phenoxyisobutyrate: Lipanthyl) is a hypo-lipidaemic agent structurally related to clofibrate. In view of current concern over the safety-in-use of this class of drug, there is a need for information about their metabolic disposition, in terms of rates and routes of metabolism and excretion and various factors which might influence these, such as species differences, chronic administration and dose dependency. As an adjunct to both the toxicological evaluation and clinical pharmacological investigation of fenofibrate, we now present the results of a study of its fate in rat, dog and man, and some potential influencing factors.

2 Methods

[*carboxy*-^{14}C]-Fenofibrate was administered in gelatine capsules to male Wistar rats, male beagle dogs and eight healthy human volunteers (4 M, 4 F) at a dose of 5 mg kg^{-1}, and their urine and faeces collected daily for 7 days. The elimination of [^{14}C] was determined as described by Sangster et al. (1983). In addition, the fate of [^{14}C]-fenofibrate was studied in bile-duct cannulated rats.

Urinary, biliary and faecal metabolites were characterized by the general methods of Caldwell and Hutt (1986), by selective solvent and solid-phase extractions, TLC and HPLC before and after treatment with β-glucuronidase or mild alkali, and their identities confirmed by MS and NMR. The enantiomeric composition of the chiral major metabolite was determined by HPLC separation of its diastereoisomeric naphthylethyl-amides, according to Hutt et al. (1986), and by chiral HPLC on a Cyclobond I (β-cyclo-dextrin) column. The ester glucuronide of fenofibric acid was isolated from human urine by TLC and the influence of preincubation pH upon its lability to β-glucuronid-ase investigated according to Sinclair and Caldwell (1982).

1 Department of Pharmacology, St. Mary's Hospital Medical School, London W2 1PG, England
2 Laboratoires Fournier, Centre de Recherche, 21121 Fontaine-les-Dijon, France

Drugs Affecting Lipid Metabolism
Ed. by R. Paoletti et al.
© Springer-Verlag Berlin Heidelberg 1987

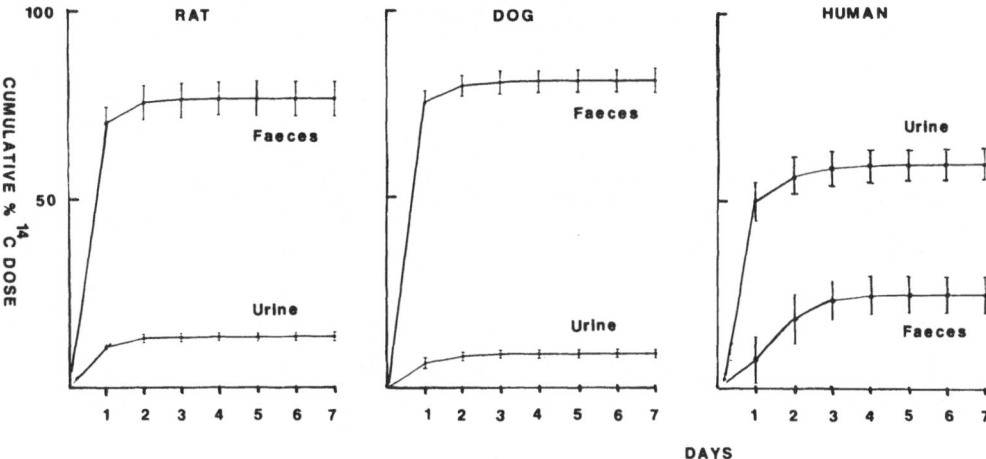

Fig. 1. Species comparison of rate and route of excretion of $[^{14}C]$ after administration of $[^{14}C]$-fenofibrate. Dose= 5 mg kg^{-1} p.o. in gelatine capsules. Data points are means ± SD, n= 5 for animals, n= 8 for human

3 Results

3.1 Excretion Pattern of $[^{14}C]$

The excretion balances of $[^{14}C]$ obtained in the three species are presented in Fig. 1. The patterns in rat and dog are very similar, with the bulk of the excreted material being found in the faeces and only ca. 10% in the urine. The majority of the excretion occurred in the first 24 h. The low recovery of $[^{14}C]$ in the urine suggests poor absorption of fenofibrate from the gut, but studies in bile-duct cannulated rats indicate the extent of absorption to be higher and that fenofibrate metabolites undergo a limited enterohepatic circulation. After dosing fenofibrate in capsules, cannulated rats excreted 4% of dose in the urine, 14% in the faeces and 22% in the bile in 48 h, while in those given the bile of fenofibrate-treated rats intraduodenally, 3% was in the urine, 3% in the faeces and 28% in the bile. In humans the excretion pattern is very different, with 60% being recovered in the urine and at least 25% in the faeces in 7 days. Again, the majority of the excretion occurred in the first 24 h after dosing.

3.2 Urinary Metabolites

Fenofibrate undergoes three metabolic reactions, namely ester hydrolysis, carbonyl reduction and ester glucuronidation. In all three species, ester hydrolysis was complete, with no intact fenofibrate recovered in the urine. There occurred considerable species differences in the relative extents of carbonyl reduction and ester glucuronidation of the fenofibric acid so liberated, and in the stereochemistry of the reduction. In rats and dogs, carbonyl reduction was the major pathway with a marked product stereoselectivity for the (-) isomer: indeed, in dogs, ester glucuronidation could not be detec-

Table 1. Species comparison of urinary metabolites of fenofibrate[a]

| | | % [14]C dose as that compound in 0–24 h urine of | | |
		Rat	Dog	Man
Fenofibrate		n.d.	n.d.	n.d.
Fenofibric acid	free	1.5 (1.0–2.0)	0.3 (0–0.8)	7.8 ± 2.2
	glucuronide	0.9 (0.6–1.1)	n.d.	32.9 ± 2.4
"Reduced	free	5.4 (4.6–6.1)	4.1 (1.8–5.1)	3.7 ± 0.6
fenofibric acid"	glucuronide	0.3 (0.2–0.4)	n.d.	3.1 ± 0.4
Unknown metabolite(s)		2.7 (2.3–3.3)	2.1 (1.2–4.8)	0.1 ± 0.1
Enantiomeric composition (−/+) of "reduced fenofibric acid"		96/4	95/5	52/48

[a]Dose= 5 mg kg^{-1} p.o. in gelatine capsules. Figures are means, with ranges in parentheses for animals (n=5) and ± SD for human subjects (n=8); n.d.= not detected.

ted. In man, however, ester glucuronidation very greatly predominated, and the minor reduction pathway exhibited little or no stereoselectivity. These data are presented in Table 1.

3.3 Biliary Metabolites in the Rat

The biliary metabolites comprised fenofibric acid, "reduced fenofibric acid" and their ester glucuronides, in approximately equal proportions (ca. 5% of dose each).

3.4 Faecal Metabolites

TLC and HPLC analysis of methanolic extracts of all three species (containing > 90% of faecal [^{14}C]) showed that the major (> 70% of faecal [^{14}C]) metabolite present was unchanged fenofibrate, accompanied by smaller amounts of fenofibric acid together with trace quantities of "reduced fenofibric acid".

3.5 Stability of Fenofibryl Glucuronide

To provide an indication of the reactivity of fenofibryl glucuronide, its pH-dependent intramolecular rearrangement to β-glucuronidase-resistant isomers has been investigated. Rearrangement was not observed at pHs from 5.2–7.4, and at pH 8.6, 35% was converted to enzyme-resistant conjugates. The half-life for this reaction is ca. 40 h.

4 Discussion

This study has revealed significant species differences in the excretion pattern and metabolism of fenofibrate. In rat and dog, excretion is principally faecal as the unchanged drug, in part a consequence of the poor absorption of fenofibrate, although there does occur some biliary excretion and an enterohepatic circulation. This is to be expected on the basis of the molecular weights of the biliary metabolites (318 and greater). The urinary metabolites arise by ester cleavage and carbonyl reduction, and glucuronidation is low or absent. In humans, by contrast, the bulk of the drug is excreted in the urine as fenofibryl glucuronide.

The metabolic pathways of fenofibrate are those to be expected from its structure, but the species differences observed would not be expected on the basis of analogy with related structures, e.g. clofibrate (Emudianughe et al. 1983). Unlike aromatic and arylacetic acids, the structure-metabolism relationships for aryloxyacetic acids cannot be explained simply at the present time (Van der Waterbeemd et al. 1986). The difference in the metabolism and disposition of fenofibrate in humans in comparison with rats and dogs is very marked, and a number of other instances of extensive glucuronidation as a feature of human metabolism exist (Caldwell 1985). The extreme stability of the ester glucuronide of fenofibric acid, in comparison with certain other ester glucuronides (see Van Breemen and Fenselau 1985), suggests that this metabolite has. little toxicological significance in this case.

Acknowledgement. Supported by a grant from Pole de Toxicologie Cellulaire de Bourgogne.

References

Caldwell J (1985) Glucuronic acid conjugation in the context of the metabolic conjugation of xenobiotics. In: Matern S, Bock KW, Gerok W (eds) Advance in Glucuronide Conjugation. MTP, Lancaster, UK, pp 7–20

Caldwell J, Hutt AJ (1986) Methodology for the isolation and characterization of conjugates of xenobiotic carboxylic acids. In: Bridges JW, Chasseaud LF (eds) Progress in Drug Metabolism, vol 9. Taylor and Francis, London, pp 11–51

Emudianughe TS, Caldwell J, Sinclair KA, Smith RL (1983) Species differences in the metabolic conjugation of clofibric acid and clofibrate in laboratory animals and man. Drug Metab Dispos 11:97–102

Hutt AJ, Fournel S, Caldwell J (1986) The application of a radial compression column to the high performance liquid chromatographic separation of the enantiomers of some 2-arylpropionic acids as their diastereoisomeric R-(+)-1-[naphthen-1-yl]ethylamides. J Chrom 378:409–418

Sangster SA, Caldwell J, Hutt AJ, Smith RL (1983) The metabolism of *p*-propylanisole in the rat and mouse and its variation with dose. Fd Chem Toxicol 21:263–271

Sinclair KA, Caldwell J (1982) The formation of β-glucuronidase resistant glucuronides by the intramolecular rearrangement of glucuronic acid conjugates. Biochem Pharmacol 31:953–957

Van Breemen R, Fenselau C (1985) Acylation of albumin by 1-O-acyl glucuronides. Drug Metab Dispos 13:318–320

Van der Waterbeemd H, Testa B, Caldwell J (1986) The influence of conformational factors on the metabolic conjugation of aryloxyacetates. J Pharm Pharmacol 38:14–18

Pharmacokinetics of Lipanthyl in the Elderly

J. P. Guichard [1], M. Strolin-Benedetti [1], G. Houin [2], and
J. L. Albarede [3]

1 Introduction

Up to about 10 years ago, dyslipidaemia discovered in patients aged more than 65 years was not treated, the associated cardiovascular risk being considered weak in comparison with other risk factors. More recently, because of the increase of the duration of life, these dyslipidaemias in elderly patients are increasingly being treated with lipid-lowering drugs such as fenofibrate.

Furthermore, patients who began a normolipidaemic treatment in younger life, continue this treatment beyond 65 years of age.

Thus, in order to avoid potential risks which could be due to pharmacokinetic and metabolism changes of aging and to adapt the treatment, it is advisable to study the pharmacokinetics of normolipidaemic drugs in the elderly. This is the aim of the present study. The results presented here are partial results from the on-going study.

2 Method

Five hospitalized patients (two men and three women) aged 77—87 years were admitted to the study. All patients had normal renal and hepatic functions for their age except patient No. 4 whose renal function was impaired as assessed by a high serum creatinine level of 170 μmol l^{-1}. Each subject received a single oral dose of 100 mg fenofibrate given as one capsule of Lipanthyl. The capsule was administered during a standardized meal consisting of: one egg, 50 g bread, 50 g ham, 25 g cheese, 20 g butter, 100 ml orange juice and 100 ml water, following an overnight fast. No other food was allowed during the 4 h following dosing.

Venous blood samples were drawn into heparinized tubes containing NaF (50 mM) before drug administration and at 0.25, 0.5, 1, 2, 3, 4, 5, 6, 7, 8, 12, 14, 24, 36, 48, 72, 96 and 120 h after dosing. Plasma was immediately separated by centrifugation and

1 Laboratoires Fournier, Centre de Recherches, 50, rue de Dijon, Daix, 21121 Fontaine-les-Dijon, France
2 Unité de Pharmacocinétique Clinique, CHU Purpan, 31059 Toulouse, France
3 Centre de Gériatrie, CHU Purpan, 31059 Toulouse, France

Drugs Affecting Lipid Metabolism
Ed. by R. Paoletti et al.
© Springer-Verlag Berlin Heidelberg 1987

kept frozen at −20°C until analysis. Urine was collected in several fractions for 120 h after drug administration. A sample of each fraction was kept frozen at −20°C until assayed.

After oral administration of $[^{14}C]$-fenofibrate to young adults during a standardized meal, no unchanged compound was detectable in plasma, the radioacitivty consisting almost entirely of fenofibric acid (Strolin Benedetti et al. 1986). Fenofibric acid, the active form of fenofibrate, was therefore assayed in plasma and urine by a specific HPLC method based on separation by reverse-phase chromatography with UV detection.

Plasma and urine samples were extracted by acetonitrile under acidic conditions. A specially synthesized analogue was used as internal standard.

The recovery of fenofibric acid was 75% from 500 ng ml^{-1} to 5000 ng ml^{-1}. The limit of determination was 50 ng ml^{-1}. At this limit the precision was 16.1%. The precision was inferior to 10% from 100 ng ml^{-1} to 10 000 ng ml^{-1}.

The pharmacokinetic analysis of the resulting plasma fenofibric acid concentration versus time data was carried out using the interactive computer program PHARM (Gomeni 1984). The following pharmacokinetic parameters were determined:

1. Observed time of peak plasma concentration (tmax).
2. Observed peak plasma concentration (Cmax).
3. Half-life of the terminal elimination phase ($t_{1/2}$) calculated from the relationship · $t_{1/2} = 0.693/\lambda z$, where λz is the slope of the last exponential phase.
4. Area under the plasma concentration-time curve extrapolated to infinity (AUC) calculated by the trapezoidal rule form 0 to the time of the last assayed sample and extrapolated to infinity (Gibaldi and Perrier 1982).
5. Apparent total body clearance Cl/f calculated as Dose/AUC.
6. Apparent volume of distribution Vz/f calculated as $D/\lambda z$. AUC.
7. Total amount of fenofibric acid (free + conjugate) excreted in urine in infinite time [Ae(∞)].
8. Renal clearance Cl_R calculated as: Ae (∞)/AUC.

Results are expressed as the mean ± SD. Individual values of kinetic parameters obtained in the elderly were compared to the values obtained in 12 healthy men aged 20 to 27 years (mean = 23.3 ± 2.2 years) who received a single dose of 100 mg fenofibrate during the same standardized meal. Statistical comparisons were by U-test of Mann-Whitney for tmax and by analysis of variance for other parameters.

3 Results

The plasma concentration-time curves for fenofibric acid in the elderly and in young adults are shown in Fig. 1. Kinetic values for each patient are listed in Table 1. The mean values of these parameters and the corresponding values obtained in young healthy adults are presented in Table 2.

The peak concentration occurred in the elderly at 4.4 ± 1.6 h. The time of the peak concentration was not different from that seen in adults: 4.8 ± 0.8 h. The peak concentration (Cmax) in the elderly was 3.0 ± 2.0 mg l^{-1}. The Cmax obtained in adults was

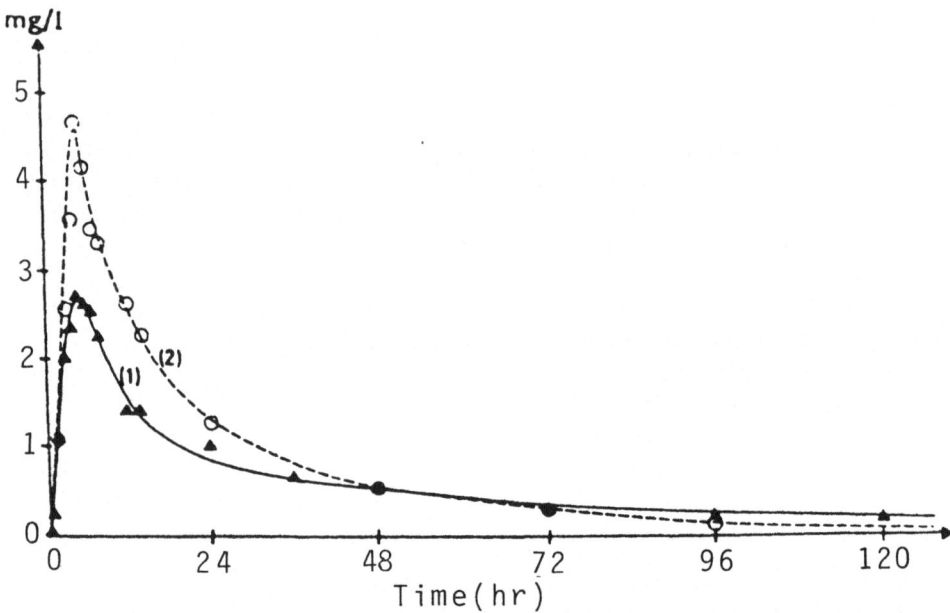

Fig. 1. Mean plasma concentration time profile of fenofibric acid after a single oral administration of 100 mg fenofibrate in the elderly (*1*) and in young healthy adults (*2*)

Table 1. Kinetic parameters of fenofibric acid in 5 elderly subjects receiving a single oral dose of 100 mg fenofibrate

Subject	Sex	Age (yr)	Weight (kg)	tmax (h)	Cmax (mg l^{-1})	t$^{1/2}$ (h)	AUC (mg l^{-1} h)	Cl/f (l h^{-1})	Vz/f (l)	Ae (∞) (mg)	Cl$_{R}$ (l h^{-1})
1	M	77	52	3	1.3	22.5	38.9	2.27	73.8	10.6	0.27
2	F	84	55	7	2.6	33.3	66.4	1.33	64.0	24.6	0.37
3	M	84	70	4	1.8	46.0	105.8	0.84	55.4	5.1	0.05
4	F	87	45	3	2.6	41.9	112.6	0.78	47.5	2.6	0.02
5	F	84	50	5	6.5	14.9	92.2	0.96	20.6	21.2	0.23
M		83	54.4	4.4	3.0	31.7	83.2	1.23	52.3	12.8	0.19
SD		3.7	9.4	1.6	2.0	13.0	30.4	0.62	20.2	9.7	0.15

significantly higher with a value of 4.8 ± 0.9 mg l^{-1}. The area under the curve was slightly but not significantly reduced in the elderly with a value of 83.2 ± 30.4 mg l^{-1} h and 97.6 ± 54.0 mg l^{-1} h in adults.

The elimination half-life was 31.7 ± 13.0 h in the elderly. The half-life measured in adults was significantly shorter with a value of 20.8 ± 5.9 h.

The apparent total body clearance in the elderly was 1.23 ± 0.62 l h^{-1}. This value is slightly but not significantly higher than the apparent total body clearance measured in adults which was 1.09 ± 0.44 l h^{-1}.

Table 2. Kinetic parameters of fenofibric acid in 12 healthy young adults and in 5 elderly patients receiving a single dose of 100 mg fenofibrate during a standardized meal

Parameters		Elderly	Adult
tmax	(h)	4.4 ± 1.6	4.8 ± 0.8
Cmax	(mg l^{-1})	3.0 ± 2.0*	4.8 ± 0.9
$t^{1/2}$	(h)	31.7 ± 13.0*	20.8 ± 5.9
AUC	(mg l^{-1} h)	83.2 ± 30.4	97.6 ± 54.0
Cl/f	($l\,h^{-1}$)	1.23 ± 0.62	1.09 ± 0.44
Vz/f	(l)	52.3 ± 20.2*	30.3 ± 8.5
Ae(∞)	(mg)	12.8 ± 9.7*	27.6 ± 12.5
Cl_R	($l\,h^{-1}$)	0.19 ± 0.15	0.36 ± 0.23

*$p < 0.05$.

The apparent volume of distribution was significantly increased in the elderly with a value of 52.3 ± 20.2 l compared to the value of 30.3 ± 8.5 l obtained in adults.

The total urinary excretion was significantly reduced in the elderly: 12.8 ± 9.7 mg of fenofibric acid free and conjugate was excreted in urine instead of 27.6 ± 12.6 mg in adults after the same dose.

The renal clearance of total fenofibric acid was 0.19 ± 0.15 l h^{-1} in the elderly. This value was less than the value of 0.36 ± 0.23 l h^{-1} measured in young adults but the difference was not statistically significant because of the large interindividual variations.

4 Discussion

After oral administration of 100 mg fenofibrate to the elderly, the main pharmacokinetic modifications compared to the young adult were an increase of the elimination half-life of 52%, a slight increase of 13% of the apparent total body clearance, an increase of 73% of the apparent volume of distribution and a reduction of 47% in the total renal clearance.

These findings show that the increase of the elimination half-life in the elderly is due to the increase of the volume of distribution and not to a decrease in the total clearance, although the renal clearance is reduced. This may be explained by a lower plasma albumin concentration in elderly patients, this reduction reaching 20–25% in patients over 60 years of age compared to those under 40 years (Greenblatt 1979). Indeed, fenofibric acid is more than 99% bound to serum albumin (Urien et al. 1980). A decline in plasma albumin concentration may therefore reduce the degree of binding of fenofibric acid. Consequently, the volume of distribution of fenofibric acid would increase. The total body clearance of fenofibric acid is low: 1.09 ± 0.44 l h^{-1} in the young adult. An increase of the unbound fraction of fenofibric acid will therefore tend to provoke an increase of its total body clearance.

From a clinical point of view the observed pharmacokinetic modifications in the elderly are not important and do not justify a change in the dosage regimen. The total body clearance being slightly increased, the mean steady-state concentration of the total fenofibric acid will be slightly lower in the elderly than in younger adults. However, the mean steady-state concentration of the unbound drug, which is considered as the active fraction, will remain unchanged or at most marginally increased (Greenblatt et al. 1982).

References

Gibaldi M, Perrier D (1982) Pharmacokinetics, 2nd edn. Dekker, New York Basel, p 445

Gomeni R (1984) Pharm: an interactive graphic program for individual and population pharmacokinetics parameters estimation. Comput Biol Med 14:25–34

Greenblatt DJ (1979) Reduced serum albumin concentration in the elderly: a report from the Boston collaborative drug surveillance program. J Am Geriatr Soc 27:20–22

Greenblatt DJ, Sellers EM, Shader RI (1982) Drug disposition in old age. N Engl J Med 306:1081–1088

Strolin Benedetti M, Guichard JP, Vidal R, Donath A (1986) Kinetics and metabolic fate of ^{14}C fenofibrate in human plasma. IIIrd World Conf Clinical pharmacology and therapeutics, Stockholm, 27 July–1 August 1986

Urien S, Albengres E, Zini R, d'Athis P, Tillement JP (1980) Serum binding and interactions of chlorophenoxyisobutyric acid, itanoxone and fenofibric acid according to their different HSA binding sites. In: Fumagalli R, et al. (eds) Drug affecting lipid metabolism. Elsevier, Amsterdam New York

Effect of Fenofibrate on Lipoprotein Particles Predictive of Coronary Atherosclerosis

J. C. Fruchart[1], C. Fievet[1], J. M. Bard[1], S. Marcovina[2], P. Drouin[3], P. Douste-Blazy[4], I. Luyeye[1], N. Slimane[1], and P. Puchois[1]

1 Introduction

Lipoprotein density classes and electrophoretic bands have been accepted as the fundamental physicochemical and metabolic entities of the lipid transport system (Gofman et al. 1954; Hatch and Lees 1968). This conceptual view has been used for the classification of the various dyslipoproteinemic states (Fredrickson et al. 1967). Recently, the discovery of several apolipoproteins, widely distributed throughout the density spectrum, with important structural and metabolic functions led to their use as specific markers for the classification of the lipoprotein species (Alaupovic et al. 1972; Alaupovic 1982). According to this view, the plasma lipoprotein system consists of a mixture of particles which may contain only one apolipoprotein (simple lipoprotein particles) or more than one apolipoprotein (complex lipoprotein particles). Changes in molecular structure can lead to variable expression of apolipoprotein epitopes. Using carefully selected monoclonal antibodies and newly developed immunological procedures we are now able to distinguish between atherogenic and non-atherogenic apo B-containing lipoprotein particles (Fievet et al. 1985; Fruchart et al. 1985c). These methods have been applied to a double-blind randomized study using fenofibrate in 37 type II_A hypercholesterolemic patients.

2 Material and Methods

2.1 Study Design

Thirty-seven patients of both sexes (22 men and 15 females) between 18 and 73 years of age, suffering from primary hypercholesterolemia were recruited into a double-blind, randomized placebo-controlled study carried out over 5 months:

1   SERLIA, Institut Pasteur, 1 rue du Pr. Calmette, 59019 Lille Cédex, France
2   Laboratori Di Ricerca, Istituto S. Raffaele, Via Olgettina 60, 20132 Milano, Italy
3   Service de Médecine Générale G, Hôpital Jeanne d'Arc, 54201 Toul Cédex, France
4   Laboratoire de Biochimie, Hôpital Purpan, 31052 Toulouse Cédex, France

Drugs Affecting Lipid Metabolism
Ed. by R. Paoletti et al.
© Springer-Verlag Berlin Heidelberg 1987

1. Drugs which might affect lipid metabolism were withdrawn and for 2 months patients followed a prudent diet and received placebo only.
2. The diet was continued for a further 3 months but, at the same time, 20 patients received 300 mg day^{-1} fenofibrate and 17 patients received placebo. Drug and placebo were presented and were taken under identical conditions. Plasma lipoprotein particles were determined after 1 month (T_{-1}) and 2 months (T_0) of the first placebo period, and after 1 and 3 months of the double-blind period (T_1, T_3).

2.2. Quantification of Lipoprotein Particles

Apolipoprotein B-containing particles were quantified as previously described (Fruchart et al. 1985b). Lipoprotein particles recognized by six well-characterized anti-apo B monoclonal antibodies $(BL_3, BL_5, BL_7, 2A, 9A, 6B)$ were determined using an immunoenzymometric assay. BL_3, BL_5 and BL_7 have been used in a sandwich system (Fievet et al. 1985) and 2A, 9A, 6B in a competitive system with coated LDL. Lipoprotein particles containing apolipoproteins B and C-III and apolipoproteins B and E were quantified using a two-site differential immunoenzymometric assay (Fruchart et al. 1985a; Puchois et al. 1985).

Anti-apolipoprotein C-III or anti-apolipoprotein E antibodies were used as solid-phase antibody. Apolipoprotein B present in the retained particles was evaluated using peroxidase-labelled anti-apolipoprotein B antibodies.

3 Results and Discussion

In a preliminary study of apo B-containing lipoproteins, as recognized by our six monoclonal antibodies, in 24 patients suffering from familial hypercholesterolemia and 24 healthy volunteers matched for age and sex, we showed increases in all the apo B-containing lipoproteins in the patients. Nevertheless, particles recognized by BL_3, 9A and 6B seem to contribute to a better discrimination between patients and controls (Fig. 1). Moreover, we have shown in a separate study that particles recognized by BL_3 antibody $(LpBL_3)$ are increased in patients with angiographically proven coronary artery disease (Fievet et al. 1985; Fruchart et al. 1985c). Preliminary results indicate that lipoproteins recognized by 2A, 9A and 6B might also be significantly increased in these patients with coronary artery disease.

From the above, it appears that patients with familial hypercholesterolemia have a significant increase of some atherogenic apo B-containing lipoproteins recognized by these specific monoclonal antibodies used for quantification.

In the present study fenofibrate induced a variety of effects on lipoprotein particles recognized by the six monoclonal antibodies used (Fig. 2). After 1 month of treatment (T_1), BL_3, BL_7, 2A, 9A and 6B are significantly lower in the treated than in the placebo group; the observed differences are respectively, 33.2, 28.5, 36.8, 29 and 23.6%. After 3 months of treatment (T_3), these differences become 32.4, 49.7, 33.7, 23 and 21.6%. BL_5 is lowered by 24.3 and 12.6% at T_1 and T_3 respectively, but these

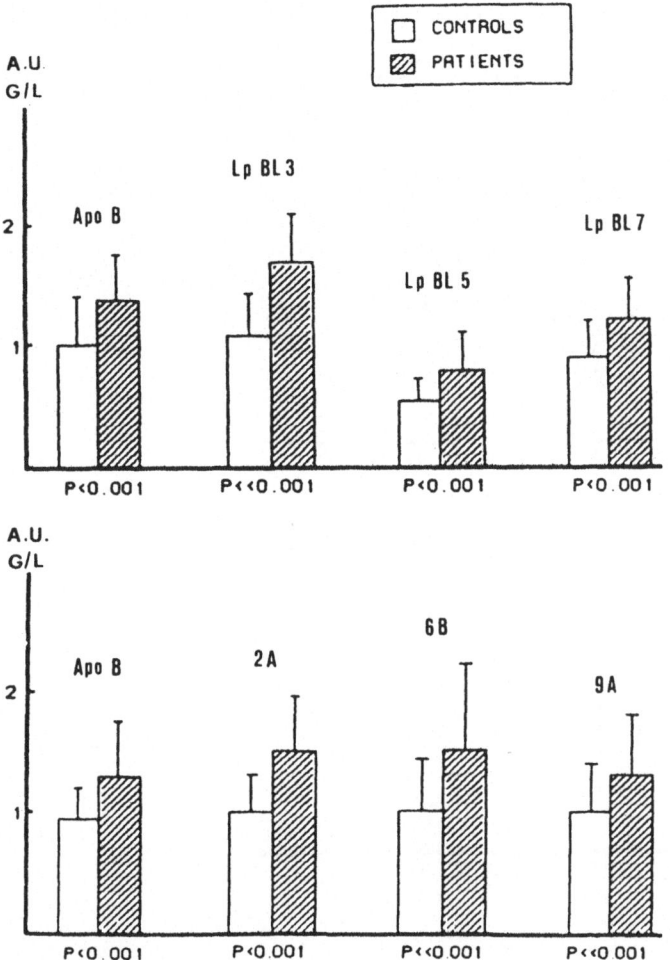

Fig. 1. Apolipoprotein B and apo B-containing lipoproteins in controls (n=24) ☐ and patients with familial hypercholesterolemia (type IIa) (n=24) ▨

differences are not significant. These results confirm LpB immunoheterogeneity and indicate that fenofibrate substantially reduces the proportion of atherogenic particles.

Characterization of the $LpBL_3$ particles showed that they contain mainly apolipoprotein B but some of them contain also apolipoprotein C-III and E (data not published). The decrease of LpE:B (36.9% at T_1 and 32.3% at T_3) and LpC-III:B (35.6% at T_1 and 39.2% at T_3) induced by fenofibrate were very close to those observed for $LpBL_3$ and LpB2A (Fig. 3). This suggests that fenofibrate could affect this type of complex lipoprotein particle to the same extent as the others. Concerning the effect of fenofibrate on LpE:B and LpC-III:B, three hypotheses can be proposed. First, a decrease of apo B synthesis by fenofibrate could lead to less association with apo E and apo C-III. Second, the impact of fenofibrate on lipoprotein particles containing

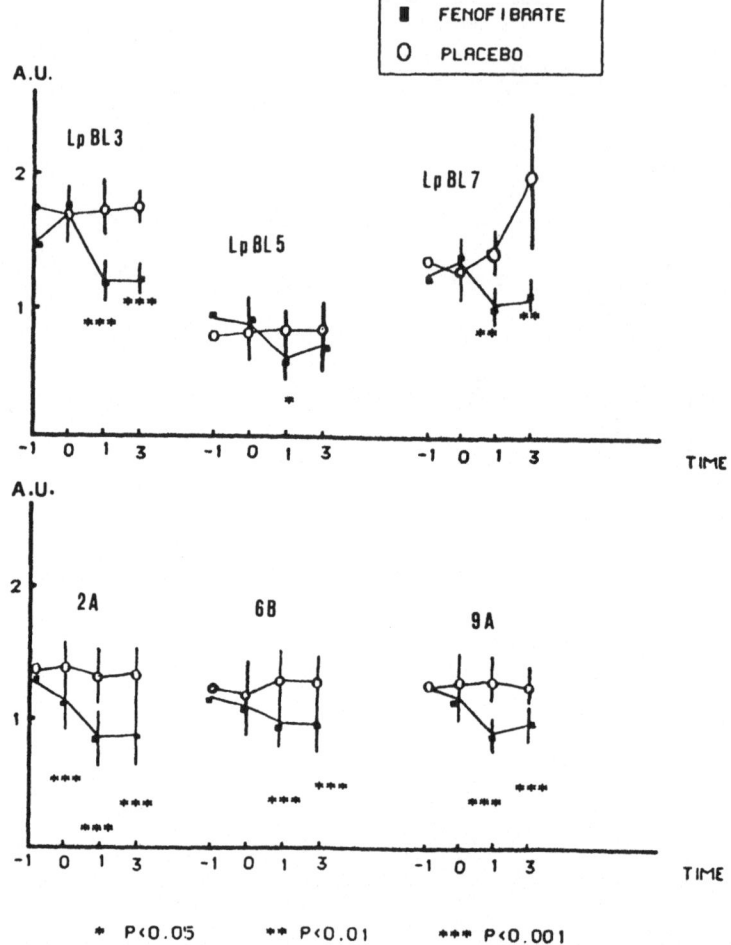

Fig. 2. Effect of fenofibrate on lipoprotein particles recognized by anti-apo B monoclonal antibodies of 37 patients with familial hypercholesterolemia (type IIa)

E and B could be explained by a fenofibrate induction of the B, E receptor-specific pathway for LpE:B catabolism. Third, apo B-containing lipoprotein particles could be exposed to a higher rate of lipolysis.

4 Conclusion

Our results in hyperlipidemic patients indicate that atherogenic apo B-containing lipoproteins, as recognized by different monoclonal antibodies, are reduced by fenofibrate treatment.

Further investigations of the effects of this drug on other apo B-containing lipoproteins could lead to a better knowledge of its mechanism of action at the molecular level.

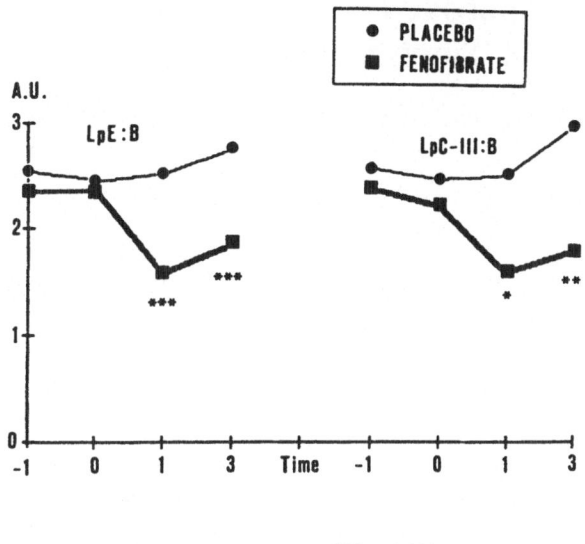

Fig. 3. Effect of fenofibrate on LpE:B and LpC-III:B lipoprotein particles of 37 patients with familial hypercholesterolemia (type IIa)

* p < 0.05   ** p < 0.01   *** p < 0.001

References

Alaupovic P (1982) The role of apolipoproteins in lipid transport processes. Ric Clin Lab 12:3–21

Alaupovic P, Lee DM, McConathy WJ (1972) Studies of the composition and structure of plasma lipoproteins. Distribution of lipoprotein families in major density classes of normal human plasma lipoproteins. Biochim Biophys Acta 260:689–707

Fievet C, Demarquilly C, Luyeye I, Fievet P, Moschetto Y, Fruchart JC (1985) Utilisation d'anticorps polyclonaux et monoclonaux pour le dépisage de l.athérosclérose' Intérèt de nouveaux marqueurs. Ann Biol Clin 43:500–504

Fredrickson DS, Levy RI, Lees RS (1967) Fate transport in lipoproteins – an intergrated approach to mechanisms and disorders. New Engl J. Med 276:32–77, 94–103, 148–156, 215–226, 273–281

Fruchart JC, Kandoussi A, Parsy D, Koren E, Puchois P (1985a) Measurement of lipoprotein particles defined by their apolipoprotein composition using an enzyme linked differential immunosorbent assay. J Clin Chem Clin Biochem 23(9):619

Fruchart JC, Luyeye I, Parra H, Slimane N, Fievet C (1985b) Les lipoprotéines athérogènes et leur détection immunologique. Bull Acad Natl Med 169(6):719–722

Fruchart JC, Parra H, Kandoussi A, Fievet C, Lablanche JM, Bertrand M (1985c) Monoclonal antibody mapping of lipoprotein particles in the prediction of coronary atherosclerosis. Circulation 72:92

Gofman JW, Delalla O, Glazier F, Nichols AV (1954) The serum lipoprotein transport system in health, metabolic disorders, atherosclerosis and coronary heart disease. Plasma 2:413–484

Hatch FT, Lees RS (1968) Practical methods for plasma lipoprotein analysis. Adv Lipid Res 6:1–68

Puchois P, Kandoussi A, Parsy D, Koren E, Fruchart JC (1985) Measurement of lipoprotein immunosorbent assay. Ann Biol Clin 43:679

Tentative Biochemical Approach to Fenofibrate Efficacy in Tinnitus Aurium

M. Strolin-Benedetti, E. Chesne, P. Dostert, and H. Cao Danh [1]

1 Introduction

Reduced blood flow in the inner ear may be one cause of tinnitus and similar disorders of hearing. In this case, inner ear tissue could be in a chronic hypoxic state, that is the oxygen supply to the tissue is insufficient, as postulated by Frèche and Tronche (1983).

Improvements of hearing and superficial face vascularization have been observed when fenofibrate was administered to hyperlipidemic patients with tinnitus aurium and other symptoms related to inner ear disorders (Bogaievsky et al., this volume).

Accordingly, we have studied the capacity of this compound to modify the levels of the endogenous regulators of oxygen affinity to hemoglobin or to change a biochemical parameter linked to erythrocyte deformability. 2,3-Diphosphoglycerate (2,3-DPG) is the main endogenous regulator of oxygen affinity to hemoglobin. According to the model of Perutz (1970), 2,3-DPG binds to the central cavity of partially deoxygenated hemoglobin and thus facilitates the unloading of oxygen. Similar effects have been observed with other endogenous organic phosphate compounds: among these, only ATP is present in a concentration sufficient to influence the oxygenation state of hemoglobin (Benesch and Benesch 1967). However, from the quantitative standpoint, 2,3-DPG is the most important. An increase of 2,3-DPG or ATP concentration in the red cell may contribute to the maintenance of an adequate oxygen supply to the tissue. Furthermore, ATP has a major role in the maintenance of red cell deformability and contributes to blood flow in microvessels (Weed et al. 1969). Such effects have already been described for drugs used in the treatment of inner ear disorders. For example, pentoxifylline was shown to increase 2,3-DPG levels among patients with an acute and severe arteriopathy (Le Devehat et al. 1981). Pentoxifylline is also able to increase red cell ATP content in vitro and in vivo, as shown by Vigneron and Stoltz (1977). Therefore, we have investigated whether fenofibrate or fenofibric acid, the active principle of fenofibrate, is able to modify the concentration of 2,3-DPG or ATP in vivo.

---

1   Laboratoires Fournier, Centre de Recherches de Daix, 21121 Fontaine les Dijon, France

Drugs Affecting Lipid Metabolism
Ed. by R. Paoletti et al.
© Springer-Verlag Berlin Heidelberg 1987

2 Methods

Experiments were carried out with healty human volunteers and with mice (Oncins France, strain 1, 25–30 g) under normoxic conditions. In mice, hypoxic conditions were also studied. 2,3-DPG concentration of blood was measured by the enzymatic method of Michal (1974); ATP content was evaluated by the enzymatic method of Jaworek (1974). Hemoglobin levels were measured by the method of Drabkin (1948). Unless otherwise stated, Scheffe's parametric test was used for statistical analysis.

3 Results

No modification of the hemoglobin levels was observed during the experiments. As shown in Table 1, fenofibrate when administered as a single or repeated oral doses did not modify the levels of 2,3-DPG in human volunteers ($p > 0.05$). Furthermore, no modification of blood 2,3-DPG and ATP levels was observed in normoxic mice treated either by single or repeated oral administrations of fenofibrate or intravenous administration of fenofibric acid (Table 2, $p > 0.05$). After 1 h of hypoxia, the blood levels of 2,3-DPG were increased (about 16%, $p \leqslant 0.025$) and remained constant up to 2 h. When such hypoxic mice were treated with fenofibrate or fenofibric acid, no significant modification of the 2,3-DPG increase was observed (Table 3, $p > 0.05$).

Furthermore, 2 h after the end of hypoxia, the mean levels of 2,3-DPG observed in the treated groups of animals were comparable to those of the control groups and that of animals under normoxic conditions, indicating that fenofibrate did not prevent the decrease of 2,3-DPG levels when mice returned to a normoxic state. Unlike the levels of 2,3-DPG, the levels of ATP were significantly decreased after 1 h of hypoxia (about 26%, Mann and Whitney's non-parametric test, $p \leqslant 0.05$). When mice were treated

Table 1. Effects of treatments with fenofibrate on blood concentrations of 2,3–DPG in healthy volunteers[a]

Treatment	(2,3-DPG) (mmol l^{-1} blood)	
	Before treatment	After treatment
Fenofibrate single oral dose (300 mg)	2.136 ± 0.068	1.948 ± 0.064
Fenofibrate Repeated oral doses (300 mg) once daily over 4 days	2.176 ± 0.046	2.180 ± 0.053
Fenofibrate Repeated oral doses (300 mg) once daily over 8 days	2.192 ± 0.062	2.057 ± 0.040

[a]The concentrations of DPG are evaluated 5 h after the last administration of fenofibrate. Results are mean ± SEM (n=6 to 13).

Table 2. Effects of treatments with fenofibrate or fenofibric acid on the concentrations of 2,3-DPG and ATP in blood of normoxic mice[a]

Treatment	Time of sacrifice After treatment (h)	(2,3-DPG) (mmol l^{-1} blood)		(ATP) (mmol l^{-1} blood)	
		Treated	Control	Treated	Control
Fenofibrate Single dose (200 mg/kg)	2	4.262 ± 0.093	4.191 ± 0.126	0.999 ± 0.037	1.059 ± 0.037
	4	4.252 ± 0.075	4.174 ± 0.188	1.083 ± 0.034	1.023 ± 0.035
Fenofibrate Repeated oral doses (200 mg kg^{-1}) twice daily over 3.5 days	2	4.224 ± 0.104	4.251 ± 0.114	1.041 ± 0.029	0.965 ± 0.030
	4	3.972 ± 0.101	3.937 ± 0.082	1.001 ± 0.028	1.008 ± 0.030
Fenofibric acid Single I.V. dose (120 mg kg^{-1})	1	4.218 ± 0.114	4.168 ± 0.061	0.067 ± 0.036	0.997 ± 0.018
	2	4.297 ± 0.083	4.279 ± 0.066	1.053 ± 0.048	0.951 ± 0.031
	4	4.178 ± 0.109	4.249 ± 0.097	0.988 ± 0.023	0.968 ± 0.022

[a]Results are mean ± SEM (n = 5 to 8). Control animals received methylcellulose or physiological serum.

Table 3. Effects of treatments with fenofibrate or fenofibric acid on hypoxia-induced variation of blood concentrations of 2,3-DPG and ATP in mice[a]

Treatment	Time of sacrifice After hypoxia Period (h)	(2,3-DPG) (mmol l^{-1} blood)		(ATP) (mmol l^{-1} blood)	
		Treated	Control	Treated	Control
Fenofibrate Single oral dose (200 mg kg^{-1})	0	4.747 ± 0.069	4.815 ± 0.086	0.759 ± 0.055	0.664 ± 0.051
	2	4.440 ± 0.210	4.442 ± 0.207	0.739 ± 0.047	0.802 ± 0.054
Fenofibrate Repeated oral doses (200 mg kg^{-1}) twice daily over 3.5 days	0	4.728 ± 0.038	4.910 ± 0.150	0.712 ± 0.046	0.631 ± 0.043
	2	4.222 ± 0.150	4.141 ± 0.167	0.752 ± 0.047	0.707 ± 0.035
Fenofibric acid Single i.v. dose (120 mg kg^{-1})	0	4.752 ± 0.071	4.780 ± 0.131	1.009 ± 0.020 ***	0.688 ± 0.039
	2	4.187 ± 0.042	4.207 ± 0.122	0.966 ± 0.036 ***	0.724 ± 0.025

[a]Mice were exposed to gas mixture containing 10% vol. O_2 (flow rate 38 l h^{-1}) and 90% vol. N_2 (flow rate 347 l h^{-1}) during 1 h, at room temperature and atmospheric pressure. Fenofibrate was administered 2 h before hypoxia; fenofibric acid was administered just before hypoxia. Results are mean ± SEM (n = 5 to 8). Statistical analysis: Scheffe's parametric test; ***$p < 0.001$.

with fenofibrate, no significant modification of the amplitude of this decrease was noted ($p > 0.05$). Two hours after the end of hypoxia, no difference of ATP levels was observed between treated and control animals. In contrast, the injection of fenofibric acid, under our experimental conditions, prevented the decrease of ATP concentrations induced by 1 h of hypoxia. The mean level of ATP observed in treated animals was significantly higher than the mean level observed in control animals. Furthermore, 2 h after the end of hypoxia when ATP levels tend to return to normal values, higher ATP concentrations were observed in treated mice.

4 Discussion

It has been reported that fibrates such as bezafibrate or clofibrate lower the oxygen affinity to hemoglobin (Perutz and Poyart 1983). These drugs seem to act synergistically rather than competing with the natural effector 2,3-DPG, as X-ray crystallographic analysis has shown that these fibrates bind to sites between the two α-chains of the hemoglobin molecule, which are far removed from the binding site of 2,3-DPG. Structural similarity between fenofibrate and bezafibrate or clofibrate suggests that also fenofibrate should be able to promote the dissociation of oxygen from hemoglobin and. that the binding sites of this compound would be the same as those of the other fibrates studied. However, these in vitro effects required concentrations of fibrates in large excess of circulating drug levels (Faed and McQueen 1974; Ledermann and Kaufmann 1981) and may not be relevant in vivo.

In order to further assess whether fenofibrate has any effect on oxygen dissociation, we measured 2,3-DPG and ATP levels after treatment in normoxic and hypoxic mice. Since hypoxia leads to increased 2,3-DPG and decreased ATP levels, these conditions may be more relevant when studying drug effects.

When we administered fenofibrate to mice by the oral route at dose levels known to modify plasma lipid levels in rodents (Gurrieri et al. 1976), no modification of the two endogenous modulators was observed. However, after injection of fenofibric acid at a dose leading to high circulating drug levels (about 3 mM compared to about 0.03 mM after oral treatment), ATP was increased in hypoxic mice. This concentration is in the range of the doses used in vitro with fibrates in order to demonstrate oxygen dissociation and may, at least in part, explain the observed in vitro effects.

Finally then, our results do not favour the hypothesis that fenofibrate can increase oxygen dissociation from hemoglobin in vivo. Furthermore, as ATP has been described to have a key role in erythrocyte membrane deformability, ensuring the flow of red cells in the finest capillaries (Weed et al. 1969), an in vivo action of fenofibrate on this rheological parameter is not likely. However, we do not rule out the possibility that the clinically observed beneficial effect of fenofibrate on inner ear disorders is related to decreased blood viscosity. Blood viscosity is dependent on a variety of factors, i.e. plasma viscosity, erythrocyte aggregation, hematocrit and erythrocyte fluidity (Müller, 1981). The first two of these parameters are controlled by high molecular weight plasma proteins, in particular fibrinogen (Lowe et al. 1982; Maeda et al. 1984). Indeed, results from ongoing work show decreased fibrinogen levels in hyperfibrinogenic spontaneously hypertensive rats after fenofibrate treatment (J. Millet, pers. commun.). Therefore, investigations are continuing on these parameters related particularly to viscosity.

References

Benesch R, Benesch RE (1967) The effect of organic phosphates from the human erythrocyte on the allosteric properties of hemoglobin. Biochim Biophys Res Commun 26:162–167

Drabkin DL (1948) The standardization of hemoglobin measurement. Am J Med Sci 215:110–111

Faed EM, McQueen EG (1974) Serum clofibrate concentrations in patients on continuous medication. Pharmacology 12:144–151

Frèche H, Tronche R (1983) Les pièges de la pathologie ischémique en ORL. In: Le malade ischémique. Gaz Med, Fr 90(30):22–24

Gurrieri J, Le Lous M, Renson FJ, Tourne C, Voegelin H, Majoie B, Wulfert E (1976) Antilipidemic drugs, Part 2: Experimental study of a new potent hypolipidemic drug, isopropyl - [4'(p chlorobenzoyl) 2 phenoxy−2−methyl] propionate (LF 178). Arzneimitt Forsch 28:3–20

Jaworek O, Grüber W, Bergmeyer HU (1974) Adenosine -5'-triphosphate. Determination with 3-phosphoglycerate kinase. In: Bergmeyer HU (ed) Methods of enzymatic analysis, 2nd edn. Chemie International, Deerfield Beach, pp 2097–2101

Ledermann H, Kaufmann B (1981) Comparative pharmacokinetics of 400 mg bezafibrate after a single oral administration of a new slow-release preparation and the currently available commercial form. J Int Med Res 9:516–520

Le Devehat C, Lemoine A, Cirette B, Ramet M (1981) Pharmacological influences of pentoxifylline on red cell filterability and 2,3-diphosphoglycerate. Scand J Clin Lab Invest 41(156):301–303

Lowe GDO, Stromberg P, Forbes CD, McArdle BM, Lorimer AR, Prentice CRM (1982) Increased blood viscosity and fibrinolytic inhibitor in type II hyper-lipoproteinaemia. Lancet 1:472–475

Maeda N, Imaizumi K, Sekiya M, Shiga T (1984) Rheological characteristics of desialylated erythrocytes in relation to fibrinogen-induced aggregation. Biochim Biophys Acta 776:151–158

Michal G (1974) 2,3-diphosphoglycerate. In: Bergmeyer HU (ed) Methods of enzymatic analysis, 2nd edn. Chemie International, Deerfield Beach, pp 1433–1438

Müller R (1981) Hemorheology and peripheral vascular diseases: a new therapeutic approach. J Med 12(4):209–235

Perutz MF (1970) Stereochemistry of cooperative effects in hemoglobin Bohr effect and combination with organic phosphates. Nature (London) 228(5273):734–739

Perutz MF, Poyart C (1983) Bezafibrate lowers oxygen affinity of haemoglobin. Lancet 2:881–882

Vigneron C, Stoltz JF (1977) Modifications physico-chimique et rhéologique de l'hématie sous l'influence de la pentoxifylline. Med Acta 4:200–204

Weed RI, La Celle PL, Merrill W (1969) Metabolic dependence of red cell deformability. J Clin Invest 48:795–809

Clinical Effect of Lipanthyl on Tinnitus Aurium and Other Symptoms Related to Inner Ear Disorders

Y. Bogaievsky [1], F. Torossian [2], G. Jost [3], and F. Bonnefous [1]

1 Introduction

Some neuro-sensorial troubles of the inner ear have been shown to be possibly related to metabolic disorders, among them hyperlipidemia (Spencer 1973; Martin and Oudot 1977). Except Lidocaine intravenously, no drug has shown real effectiveness in a controlled double-blind placebo trial to treat tinnitus aurium. In a previous pilot study (Torossian and Laredo 1986), fenofibrate, Lipanthyl, a well-known hypolipidemic compound (Blane et al. 1986), has already demonstrated an activity against not only hyperlipidemia but also against some neuro-sensorial troubles of the inner ear. The aim of the present trial was to evaluate more precisely the effectiveness of fenofibrate on some cochleo-vestibular problems, notably tinnitus and hypacusia, in hyperlipidemic patients.

2 Patients and Methods

Forty-nine hyperlipidemic adult (plasma cholesterol $\geq$ 260 mg dl^{-1} or 6.7 mmol l^{-1} and/ or triglycerides $\geq$ 150 mg dl^{-1} or 1.7 mmol l^{-1}) patients were included in a double-blind placebo controlled trial. All presented hearing disorders: hypacusia, lack of intelligibility, tinnitus, vertigo, lack of balance or other related symptoms. These patients were not being treated by an hypolipidemic drug but were following a diet. Fenofibrate, 200 mg b.i.d. or placebo was allocated randomly to the patients for a 6-month period of treatment.

Lipid levels and other routine safety assays were performed at entry and at the third and sixth month of treatment. The clinical characteristics of the symptoms were evaluated at the same times with a four-step scale (0 to 3). Because of a lack of uniformity in the distribution of the symptoms, an index was calculated for each patient: sum of the number of characteristics x score of the symptom for the six symptoms.

1 Laboratoires Fournier, Centre de Daix, 50 rue de Dijon, Daix, 21121 Fontaine les Dijon, France
2 Clinique du Château, 34 quai de Tounis, 31000 Toulouse, France
3 53 rue de Prony, 75017 Paris, France

Drugs Affecting Lipid Metabolism
Ed. by R. Paoletti et al.
© Springer-Verlag Berlin Heidelberg 1987

Before and at the end of the trial, auditory function was determined by subjective audiological tests which consisted of tonal audiometry by air and bone conduction and vocal audiometry. At the same moment a facial thermogram was obtained on a polaroid film using an Aga Thermovision 680 camera coupled with a Tigrane colorizer for visual interpretation; the pictures were also treated by a computer program to calculate the isothermic area. Three face areas were assessed: the Wood triangle (WT), temporal area (TA) and internal oculo-orbital angle (IOOA). The comparative assessment of two thermograms of the same patient was made at the same level of thermic sensitivity, obtained when the external oculo-orbital angle (EOOA) area appeared.

The comparison of both groups was tested by a Student's t-test when the variance homogeneity was checked by a Bartlett's test. Time evolution and comparison between both treatments were tested by a two-way variance analysis for the numerical data and by χ^2 tests for the distribution data.

3 Results

3.1 Homogeneity of the Two Groups

The main characteristics of the patients are described in Table 1. For nearly all of the measured items, the two groups are comparable including history of the diseases, associated diseases, drug combination and such safety values as uric acid, blood glucose, fibrinogen and platelets. The small differences observed on the type of hyperlipidemia distribution, the levels of VLDL-, LDL- and HDL-cholesterol and the weight and blood pressure are not sufficient to exclude the comparison.

3.2 Effects on Serum Lipid Concentration

Fenofibrate decreased significantly plasma levels of cholesterol, triglycerides, VLDL- and LDL-cholesterol by 16, 23,5, 39 and 18% respectively, while placebo was without effect. Fenofibrate increased HDL-cholesterol by 11% again with no activity in the placebo group. Apo A increased and Apo B decreased slightly with fenofibrate.

3.3 Clinical Improvement (Table 2)

The number of patients presenting the clinical symptoms changes with the symptom considered. Tinnitus which was present in 21 patients in each group, was improved in 81% of the patients in the fenofibrate group compared to 19% in the placebo group. For 3 patients out of 21, tinnitus disappeared completely with fenofibrate. Lack of intelligibility improved in 61% of the patients in the fenofibrate group compared to 5% in the placebo group and a third of the fenofibrate group patients improved their hypacusia and none in the placebo group. The other symptoms are not sufficiently present to make a good comparison. Nevertheless, fenofibrate appeared to improve them compared to placebo. The clinical index varied from 7.6 ± 3.2 to 3.87 ± 2.6 in the fenofi-

Table 1. Characteristics of the patients at the onset of the study

Items	Fenofibrate group	Placebo group	Statistical comparison	
Number of patients	23	26		
Sex	14 F – 9 M	9 F – 17 M	χ^2	NS
Age (years) (mean ± SD)	45.6 ± 13.2	50.7 ± 7.8	t	NS
Type of hyperlipidemia IIa	9	17		
IIb	14	9	χ^2	p=0.05
Cholesterol (m mol l^{-1})				
(mean ± SD)	7.32 ± 0.65	7.40 ± 0.50	t	NS
Triglycerides (m mol l^{-1})				
(mean ± SD)	1.80 ± 0.80	1.60 ± 0.90	t	NS
Symptoms (No. of patients)				
- Hypacusia	18	22		
- Lack of intelligibility	18	21		
- Tinnitus	21	21	χ^2	NS
- Vertigo	4	5		
- Lack of balance	8	11		
- Others	10	9		
Index (mean ± SD)	7.6 ± 3.2	6.85 ± 2.9	t	NS
Audiometric tests				
Hearing loss				
Right ear (dB)(mean ± SD)	29.13 ± 13.6	34.00 ± 14.40	t	NS
Left ear (dB) (mean ± SD)	28.04 ± 14.70	31.40 ± 16.60	t	NS
Speech reception threshold				
Right ear (dB)(mean ± SD)	34.60 ± 17.00	32.30 ± 14.50	t	NS
Left ear (dB) (mean ± SD)	32.80 ± 15.6	33.60 ± 17.10	t	NS

brate group (49% improvement). It varied from 6.85 ± 2.9 to 6.58 ± 2.9 in the placebo group (4% improvement).

3.4 Results of the Audiometric Tests

After 6 months of treatment with fenofibrate the average of hearing loss decreased by 6.6 and 5.5% for the right and left ear respectively, while there was no change in the placebo group. In the same way, the speech reception threshold decreased by 19.4 and 8.5% in the right and left ear respectively, in the fenofibrate group compared to an increase of 9.3 and 25% in the placebo group. Table 2 shows that 35 to 43% of the patients in the fenofibrate group improved their hearing loss and 48 to 61% improved their speech reception threshold compared to 8–12% and 0–8% respectively, in the placebo group.

Y. Bogaievsky et al.

Table 2. Comparison of the efficacy of the two treatments

		Group	Number of patients involved	Disap- pearance	Improve- ment	No change	Worse	Statistical comparison x^2
Clinical symptoms								
Tinnitus		F	21	3	14	4	0	$p<0.01$
		P	21	0	4	16	1	
Hypacusia		F	18	2	4	12	0	$p<0.001$
		P	22	0	0	22	0	
Lack of intelligibility		F	18	2	9	7	0	$p<0.001$
		P	21	0	1	20	0	
Vertigo		F	4	3	1	0	0	$p<0.01$
		P	5	0	0	4	1	
Lack of balance		F	8	4	1	3	0	$p<0.05$
		P	11	0	2	9	0	
Others		F	10	4	4	2	0	$p<0.01$
		P	9	1	0	8	0	
Hearing loss	Right	F	23		10[a]	11	2[b]	$p<0.05$
	ear	P	25		2	20	3	
	Left	F	23		8	14	1	$p<0.05$
	ear	P	25		3	16	6	
Speech reception threshold	Right	F	23	9[c]	5[d]	8	1[b]	$p<0.001$
	ear	P	25	1	1	12	11	
	Left	F	23	4	7	9	3	$p<0.001$
	ear	P	25	0	0	14	11	

[a] Improvement is a gain $\geqslant 5$ dB.

[b] Loss of $\geqslant 5$ dB.

[c] Improvement $\geqslant 10$ dB.

[d] Improvement $\geqslant 1$ dB and < 10 dB.

3.5 Thermographic Improvement [in the Patients with Tinnitus (Fig. 1)]

The facial thermograms have been analyzed by two methods, considering only the six previously defined areas. A visual assessment was translated in score from: - worse, = equal, + to + + + improvement. Using this technic, 17 patients were assessed in the fenofibrate group and 16 in the placebo group. The results showed a significant improvement in the treatment group compared to the placebo group. In the fenofibrate group, the thermograms were improved in 15 patients and unchanged in 2 patients. The thermograms were improved in 3 patients, unchanged in 7 and worsened in 6 patients of the placebo group.

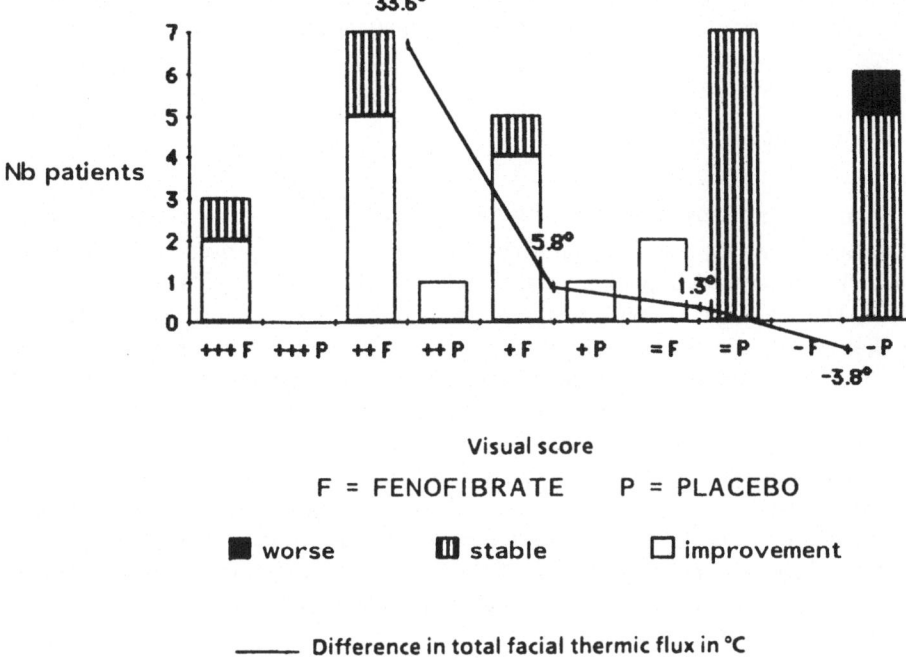

Fig. 1. Tinnitus improvement and thermograms assessment

A quantitative measurement was also performed. Thermographic photographs were digitalized and the thermic flux for each of the 60 possible levels was measured. The results were classified in ten different classes and the surfaces of each thermic flux class measured by computer in PIXELS. An average thermic flux was calculated for each area. To evaluate the improvement, we have taken the difference of the total facial thermic flux before and after the treatment. If we exclude the (+++) visual class because of the low number of patients (4) there is a good correlation between the computer and the visual assessments. The improvement starts with an increase of the thermic flux $\geq 5.8°$C and a negative change in this thermic flux means a poor result. The total facial thermic flux calculated varied from $-112°$C to $+92°$C. The results show a link between tinnitus improvement and facial vascularization.

4 Safety Data

Two side effects have been reported in the treated group. A gastric pain after 4 months was treated successfully with an antiacid drug over 10 days. Fenofibrate was withdrawn. The second side effect was a skin rash with pruritus which appeared after 15 days of fenofibrate and spontaneously disappeared after 10 days without any change of treatment. No changes in the blood pressure, heart rate and weight have been noticed in any group during the trial. No significant changes have been shown in the biological assays: uric acid, glucose, fibrinogen and platelets in any group.

5 Discussion

The clinical symptoms of inner ear disorders are poorly modified by conventional drug treatment, especially tinnitus for which only lidocaine has proved its efficacy in a double-blind controlled placebo trial (Duckert and Rees 1983; Martin and Colman 1980). The present trial confirms the previous one (Torossian and Laredo 1986). No correlation between the clinical improvement of the tinnitus and the lipid parameters has been demonstrated. Since the structurally related clofibrate and bezafibrate are known to shift the oxygen equilibrium curve of hemoglobin to the right (Perutz and Poyart 1983), fenofibrate can probably do the same, at the same high in vitro concentrations. A study of 2,3-DPG and ATP levels in fenofibrate-treated mice and volunteers has confirmed that this oxygenation hypothesis is unlikely in vivo (Strolin Benedetti et al., this Vol.). On the other hand, the improvement of the facial thermogram leads us to conclude that an improvement of the microcirculation due to a so far unidentified rheologic effect of fenofibrate, possibly a decrease of blood viscosity as observed by Arntz et al. (1985), may have occurred.

6 Conclusion

In a double-blind placebo controlled trial, fenofibrate has been shown to have an effect on clinical and audiometric measurements of inner ear disorder. This action does not seem to be correlated with the hypolipidemic properties of the drug. To explain the thermographically assessed improvement, it is suggested that fenofibrate has an effect on some rheologic parameters, rather than on the oxygen dissociation curve.

Acknowledgement. We are grateful to Pr Mignot (Hôpital Saint Jacques, Besançon) for his help in the computer assessment of the facial thermograms.

References

Arntz HR, Heitz J, Schäfer JH, Oeff M, Zingler G (1985) Influence of fenofibrate on pathological blood rheology in type II hyperlipoproteinemia. Abstract 457 of the 7th Int Symp Atherosclerosis, Melbourne (Australia), Oct 1985, Proc Poster Commun PJ Nestel

Blane GF, Bogaievsky Y, Bonnefous F (1986) Fenofibrate: influence on circulating lipids and side-effects in medium and long-term clinical use. In: Sears R (ed) Pharmacological control of hyperlipidaemia. Prous, Barcelona, pp 187−216

Duckert LG, Rees TS (1983) Treatment of tinnitus with intravenous lidocaine: a double-blind randomized trial. Otolaryngol Head Neck Surg 91:550−555

Martin FW, Colman BH (1980) Tinnitus: a double-blind cross-over controlled trial to evaluate the use of lignocaine. Clin Otolaryngol 5:3–11

Martin H, Oudot J (1977) Dysmétabolisme lipidique et surdites d'oreille interne. J Fr Oto-Rhino-Laryng 26:725–728

Perutz MF, Poyart C (1983) Bezafibrate lowers oxygen affinity of haemoglobin. Lancet 2:881–882

Spencer JT (1973) Hyperlipoproteinemias in the etiology of inner ear disease. Laryngoscope 83:639–678

Torossian F, Laredo C (1986) Cochleovestibular disorders and hyperlipidemia. A controlled trial with fenofibrate. Monogr Atherosclerosis 14:222–229

Effect of Gemfibrozil Administration on Biliary Lipid Secretion: A Crossover Study with Clofibrate

L. Barbara, G. Mazzella, N. Villanova, P. Simoni, M. Ronchi, A. Roda, E. Roda, and F. Bazzoli [1]

1 Introduction

It is generally recognised that several drugs successfully used to reduce serum cholesterol and triglycerides also interfere with biliary lipid metabolism (Von Bergmann and Leiss 1984).

Clofibrate treatment, in particular, seems to increase the risk of cholesterol gallstone formation by increasing saturation of bile with cholesterol (Cooper et al 1975, Bateson et al. 1978). It has been hypothesised that supersaturation of bile during clofibrate treatment is related to an enhanced secretion of cholesterol in bile and to a decreased synthesis of bile acids. However, to date, biliary lipid secretion studies during clofibrate treatment have not been performed.

Gemfibrozil, another fibric acid derivative, has been recently introduced for treatment of hyperlipidemia (Kankola et al. 1981). This agent, in fact, similarly to clofibrate and to other fibric acid derivatives, is effective in lowering both serum cholesterol and triglycerides and in increasing HDL (O'Connor 1982). On the other hand, the knowledge of its effects on biliary lipids is still limited and preliminary results are conflicting. In a comparative study with clofibrate, Gemfibrozil administration induced a slight and not significant increase of biliary cholesterol saturation, which in turn was not significantly different with that of clofibrate (Hall et al. 1981).

In another study it was shown that Gemfibrozil definitely increased cholesterol saturation and that this effect was mainly related to increased biliary output of cholesterol and to a decreased output of bile acids (Leiss et al. 1985). However, both studies were conducted on normal volunteers and not on patients with hyperlipidemia.

The aim of this study was, therefore, to evaluate the effects of Gemfibrozil administration on bile cholesterol saturation and hepatic secretion rates of cholesterol, bile acids and phospholipids, in normal subjects and in patients with hyperlipidemia and to compare them with that of clofibrate.

---

1 Istituto di Clinica Medica III e Gastroenterologia, Universita di Bologna, Policlinico S. Orsola, Via Massarenti, 9, Bologna, Italy

Drugs Affecting Lipid Metabolism
Ed. by R. Paoletti et al.
© Springer Verlag Berlin Heidelberg 1987

2 Materials and Methods

Subjects. Ten patients with hyperlipidemia (total cholesterol > 250 mg dl^{-1}; triglycerides > 170 mg dl^{-1}; 3 males and 7 females; age range 26–59) and 10 control subjects (3 males; 7 females; age range 24–69) participated in the study.

Experimental Design. Each subject received, in a single-blind crossover design, Gemfibrozil (1.2 g day^{-1}) and clofibrate (2.0 g day^{-1}) for 6 weeks with a 4-week "wash-out" period between each preparation. The sequence Gemfibrozil-clofibrate or clofibrate-Gemfibrozil was assigned according to a random table. Serum cholesterol, triglycerides, serum HDL-cholesterol, biliary cholesterol saturation index, biliary lipid secretion rates and total bile acid pool were evaluated prior to treatment and subsequently after each treatment period.

Analytical Methods. Total cholesterol and triglycerides were measured on a device for automated chemical analysis. Bile lipid composition and biliary lipid secretion rates were measured by an intestinal perfusion technique according to the methods of Grundy and Metzger (Grundy and Metzger 1972). A triple lumen tube was positioned with two proximal outlets adjacent to the ampulla of Vater and a third outlet 10 cm distally. A liquid formula, containing 43% of calories as fat, 15% as protein and 42% as carbohydrates, was infused at a rate of 2.6 ml min^{-1}; the caloric infusion rate was 3.4 cal min^{-1} and PEG was adopted as dilution marker. This type of formula and infusion rates were chosen in order to obtain a steady and valid gallbladder contraction throughout the study.

In preliminary experiments, in fact, we could not obtain a satisfactory gallbladder contraction, as documented by ultrasound monitoring, using either an amino acid infusion or a liquid formula with a lower fat content. After 4 h for stabilization of hepatic bile secretion, hourly samples were obtained for 6 h. Hourly outputs of cholesterol, bile acids and phospholipids were calculated according to the equations of Grundy and Metzger.

Bile acid pool size was evaluated by the Duane method modified by Grundy for intestinal perfusion studies.

3 Results

Total serum cholesterol and triglycerides levels significantly decreased after Gemfibrozil and similarly after clofibrate, in both groups of subjects. HDL-cholesterol significantly increased after Gemfibrozil, but not after clofibrate treatment. As expected clofibrate administration induced (Table 1) a significant increase of the cholesterol saturation index which was observed in both groups of patients.

In contrast, Gemfibrozil raised the saturation index in normal subjects, but not in patients with hyperlipidemia where a slight, although significant, decrease was observed. Hepatic secretion rates of cholesterol were significantly increased after clofibrate administration in normal subjects and in patients with hyperlipidemia, while after

Table 1. Biliary lipids[a]

	Biliary lipid outputs (mmol h^{-1}; mean ± SD)			Pool size (mmol; mean±SD)	S.I. (mean ± SEM)
	Cholesterol	Bile Acids	Phospholipids		
Control subjects					
Basal	0.16 ± 0.04	2.59 ± 0.6	0.30 ± 0.1	7.3 ± 2.8	1.34 ± 0.11
Gemfibrozil	0.166 ± 0.04	2.18 ± 0.27(a)	0.26 ± 0.09	5.49 ± 1.77	1.46 ± 0.13(b)
Clofibrate	0.18 ± 0.03 (c)	2.5 ± 0.43	0.35 ± 0.08	6.23 ± 1.8	1.50 ± 0.1(d)
Hyperlipidemics patients					
Basal	0.32 ± 0.078	2.96 ± 0.95	0.44 ± 0.12	6.78 ± 2.94	1.93 ± 0.2
Gemfibrozil	0.25 ± 0.11(e)	2.11 ± 1.08(f)	0.40 ± 0.24	4.19 ± 1.39(g)	1.82 ± 0.1(h)
Clofibrate	0.43 ± 0.075 (i)	3.16 ± 0.98	0.56 ± 0.21	5.05 ± 0.96	2.20 ± 0.2 (k)

[a]Letters in parentheses have the following denotation: (a) $P<0.02$ vs basal and clofibrate; (b) $P< 0.02$ vs basal; (c) $P< 0.02$ vs basal and Gemfibrozil; (d) $P< 0.01$ vs basal; (e) $P< 0.02$ vs basal; (f) $P<0.005$ vs basal and clofibrate; (g) $P< 0.001$ vs basal; (h) $P< 0.01$ vs basal; (i) $P< 0.001$ vs basal and Gemfibrozil; (k) $P< 0.01$ vs basal and Gemfibrozil.

Gemfibrozil administration and only in hyperlipidemic patients, biliary cholesterol output was significant reduced.

Hepatic secretion rates of bile acids were similarly decreased by Gemfibrozil administration in normal and hyperlipidemic subjects. Clofibrate did not significantly affect bile acid output. Gemfibrozil and clofibrate administration did not significantly affect phospholipid output. Similarly, a lack of significant changes was observed for the total bile acid pool; however, in this case a trend towards a smaller bile acid pool after treatment was present in all groups.

4 Discussion

Our study confirms that Gemfibrozil, similarly to clofibrate, is effective in lowering serum lipids and indicates that the two agents were administered for an adequate period to obtain their therapeutical effects. The effects induced on bile lipid metabolism after a 6-week treatment period should therefore be representative of their definitive effect; however, since we do not know their precise mechanism of action and their influence on the overall cholesterol metabolism, we cannot state that our observations on bile lipids are the ultimate results also in patients requiring chronic treatment for months or years.

As expected, clofibrate induced an increase of cholesterol saturation in bile; this effect was mainly realted to an increased hepatic secretion of cholesterol, while bile acid and phospholipid outputs were almost unchanged. The fact that clofibrate administration increases hepatic cholesterol secretion and does not significantly influence bile acid secretion was not shown before; however, our findings are in agreement with previous studies showing that clofibrate enhances fecal excretion of neutral steroids and only slightly decreases fecal excretion of bile acids (Kesäniemi and Grundy 1984).

The effect of Gemfibrozil administration in normal subjects was similar to that of clofibrate and our results are in agreement with those reported by Leiss and Von Bergmann (Leiss et al. 1985).

In contrast, in patients with hyperlipidemia, Gemfibrozil induced a slight, but significant, decrease of the biliary cholesterol saturation index; this effect was observed, together with a significant decrease of biliary cholesterol output, and despite a decreased bile acid output.

It has been previously suggested that cholesterol synthesis is decreased during Gemfibrozil administration (Leiss et al. 1985) and our results, showing that bile acid secretion rates and bile acid pool size are reduced, also support this hypothesis.

On the other hand, our findings of reduced hepatic secretion rates of cholesterol, together with the observation by Kesäniemi (Kesäniemi and Grundy 1986) of an increased fecal excretion of neutral sterols, suggest that also cholesterol intestinal absorption should be reduced during Gemfibrozil administration.

Although the effects of Gemfibrozil on overall cholesterol metabolism is still poorly understood, our study indicate that in contrast to clofibrate, it induces a fall of biliary cholesterol output, which despite a decrease in bile acid output, determines a decrease of the biliary saturation index. The reason for the different effect observed between normal subjects and patients with hyperlipidemia is not apparent; however, it should not be surprising that a drug affecting cholesterol metabolism could act in normal subjects differently than in patients with disturbances of the cholesterol metabolism itself.

In conclusion, our results indicate that Gemfibrozil administration, in contrast to clofibrate does not increase biliary cholesterol saturation and suggest that it should not increase the risk of gallstone formation in patients with hyperlipidemia.

References

Bateson MC, Maclean D, Ross PE (1978) Clofibrate therapy and gallstone induction. Dig Dis 23: 623–628

Cooper T, Geizerova M, Oliver MF (1975) Clofibrate and gallstones. Lancet I:1083

Grundy SM, Metzger AL (1972) A physiological method for estimation of hepatic secretion of biliary lipids in man. Gastroenterology 62:1200–1217

Hall MJ, Nelson LM, Russel RI (1981) Gemfibrozil, the effect on biliary cholesterol saturation of a new lipid-lowering agent and its comparison with clofibrate. Atherosclerosis 39:511–516

Kankola S, Manninen V, Halkonen M (1981) Gemfibrozil in the treatment of dyslipidemias in middle-aged male survivors of myocardial infarction. Acta Med Scand 209:69–73

Kesäniemi YA, Grundy SM (1984) Influence of Gemfibrozil on metabolism of cholesterol and plasma triglycerides in man. JAMA 251:2241–2246

Leiss O, von Bergmann K, et al. (1985) Effect of Gemfibrozil on biliary lipid metabolism in normolipidemic subjects. Metabolism 34:74–82

O'Connor RE (1982) Gemfibrozil: results of clinical studies in the United States. In: Ricci G, Paoletti R, Pocchiari F (eds) Therapeutic selectivity and risk benefit assessment of hypolipidemic drugs. Raven, New York, pp 59–61

Von Bergmann K, Leiss O (1984) Effect of short-term treatment with bezafibrate and fenofibrate on biliary lipid metabolism in patients with hyperlipidemia. Eur J Clin Invest 14:150–154

Plasma Lipids, Lipoproteins, and Apoproteins During Gemfibrozil Treatment in Primary Hyperlipidemias

E. Manzato, S. Zambon, R. Marin, G. Baggio, and G. Crepaldi[1]

1 Introduction

Several authors have reported that Gemfibrozil decreases the plasma concentration of triglycerides (TG) and (to a lesser extent) cholesterol (CT) and raises HDL in hyperlipoproteinemias (Olsson et al. 1976; Manninen 1983; Pickering 1983; Samuel 1984).

In hypertriglyceridemic patients Gemfibrozil decreases the production of VLDL-TG and enhances its clearance (Kesäniemi and Grundy 1984) and normalizes LDL metabolism by reducing LDL production and catabolism (Vega and Grundy 1985).

Recent studies have reported that in type V patients, Gemfibrozil: increases the synthetic rates of apo A-I and A-II with the appearance in plasma of smaller (and heavier) HDL particles; reduces TG levels producing smaller VLDL with a reduced apo C-III/C-II ratio; and increases lipoprotein lipase activity (Saku et al. 1985). Moreover, Gemfibrozil prevents the accumulation of abnormal lipoproteins in CT-fed rats normalizing apo E, A-I, and B levels (Krause and Newton 1985).

Few data are available in the literature about the effects of Gemfibrozil on HDL subclasses and plasma apoproteins in patients with primary hyperlipoproteinemias. For this reason the aim of our study was the analysis of plasma lipoproteins (in particular HDL subfractions) and apoproteins in these patients during Gemfibrozil treatment.

2 Patients and Methods

Fourteen patients (all males) with primary hyperlipoproteinemia (6 type IIa, 3 type IIb, 5 type IV) were studied. Two months before and during the study period the patients followed a standard hypolipidemic isocaloric diet (20% protein, 45% carbohydrates, 35% fat) containing less than 500 mg CT day^{-1} with a P/S ratio of 1.8.

Patients were treated for 4 months with Gemfibrozil (LIPOZID, Pierrel, Italy) 600 mg b.i.d. Clinical examination, lipoprotein and apoprotein analysis, and safety parameter determination (including CBC, glucose, liver and kidney function tests) were performed before and after 2 and 4 months of drug treatment.

---

1 Department of Internal Medicine, University of Padua, Via Giustiniani, 2, 35128 Padua, Italy

Drugs Affecting Lipid Metabolism
Ed. by R. Paoletti et al.
© Springer-Verlag Berlin Heidelberg 1987

Lipids and lipoproteins were determined as described (Manzato et al. 1986). HDL subfractions were separated by polyanion precipitation (Martini et al. 1984). Apoproteins were determined by radial immunodiffusion. Statistical comparisons were carried out by Student's two-tailed t-test for paired data.

3 Results and Discussion

The mean levels of plasma lipids, lipoproteins and apoproteins before and after 4 months of treatment are reported in Tables 1 and 2.

The mean percent variations of plasma lipids, lipoproteins and apoproteins after 4 months of Gemfibrozil treatment are reported in Table 3.

No side effects and no modifications in the safety parameters were observed during Gemfibrozil treatment. During the same period the body weight remained stable.

In our study Gemfibrozil was effective in reducing serum TG in all types of hyperlipoproteinemia. These results were obtained without a significant variation in CT/TG ratio in VLDL suggesting that the VLDL lipid composition was not particularly modified.

Table 1. Plasma lipid and lipoprotein mean ($\pm$SD) levels (mg dl^{-1})

	IIa (n = 6)		IIb (n = 3)		IV (n = 5)	
	0	+4	0	+4	0	+4
CT-WS	348 ± 61	274 ± 64^a	311 ± 27	262 ± 57	197 ± 19	206 ± 30
VLDL	12 ± 7	5 ± 4^a	43 ± 18	17 ± 14^a	53 ± 13	23 ± 14^a
LDL	276 ± 64	202 ± 66^a	219 ± 21	188 ± 46	106 ± 23	136 ± 26
HDL$_2$	28 ± 7	36 ± 9^a	24 ± 2	27 ± 1^b	16 ± 5	23 ± 4^b
HDL$_3$	27 ± 3	30 ± 4^a	24 ± 2	25 ± 2	18 ± 3	23 ± 4^c
TG-WS	117 ± 28	67 ± 33^a	266 ± 57	146 ± 83^a	348 ± 88	170 ± 69^a

^{a} $p < 0.01$; ^{b} $p < 0.05$; ^{c} $p < 0.001$.

Table 2. Plasma apoprotein mean ($\pm$SD) levels (mg dl^{-1})

APO	Type IIa		Type IIb		Type IV	
	0	+4	0	+4	0	+4
A-I	148 ± 13	155 ± 25	147 ± 12	151 ± 20	123 ± 16	138 ± 16^b
A-II	38 ± 5	38 ± 6	39 ± 3	33 ± 1	32 ± 5	34 ± 5
B	207 ± 36	171 ± 40^a	178 ± 13	177 ± 43	116 ± 21	140 ± 22^a
C-II	4.5 ± 1.2	3.4 ± 0.7	7.2 ± 1.8	4.8 ± 1.4	6.3 ± 1.6	4.7 ± 2.0
C-III	11.0 ± 1.8	8.3 ± 2.1	15.3 ± 1.5	10.3 ± 2.5^a	16.8 ± 3.5	10.6 ± 1.1^a
E	5.8 ± 1.6	5.1 ± 1.4	6.8 ± 1.9	4.9 ± 1.0	6.2 ± 1.8	4.7 ± 1.3

^{a} $p < 0.01$; ^{b} $p < 0.05$.

Table 3. Plasma lipid, lipoprotein and apoprotein percent variations (0 vs +4)

Type	IIa	IIb	IV	Type	IIa	IIb	IV
CT-WS	−21	−17	+ 5	A-I	+ 5	+ 3	+13
VLDL	−56	−66	−50	A-II	0	−16	+ 8
LDL	−27	−14	+31	B	−18	0	+22
HDL_2	+32	+13	+53	C-II	−13	−33	−25
HDL_3	+12	+ 4	+29	C-III	−23	−33	−35
TG-WS	−43	−48	−48	E	−10	−27	−20

As already reported (Olsson et al. 1976; Pickering 1983; Kesäniemi and Grundy 1984; Vega and Grundy 1985), Gemfibrozil normalized the LDL levels in hypercholesterolemic and hypertriglyceridemic patients. The LDL increase observed in hypertriglyceridemia does not seem to be related to the drug treatment since it has been reported also during dietary treatment (Manzato et al. 1986).

The increase of HDL subclasses during Gemfibrozil treatment is particularly interesting since it was observed in all types of hyperlipoproteinemia.

The apoprotein modifications might be related to the Gemfibrozil effects on VLDL (apo C-II, C-III, E) and LDL (apo B) metabolism.

The overall effect of this drug on plasma lipids, lipoproteins and apoproteins seems to be useful to improve the atherogenic risk in hyperlipoproteinemic patients.

References

Kesäniemi YA, Grundy SM (1984) Influence of Gemfibrozil and clofibrate on metabolism of cholesterol and plasma triglycerides in man. JAMA 251:2241–2246

Krause BR, Newton RS (1985) Apolipoprotein changes associated with the plasma lipid-regulating activity of Gemfibrozil in cholesterol-fed rats. J Lipid Res 26:940–949

Manninen V (1983) Clinical results with Gemfibrozil and background to the Helsinki heart study. Am J Cardiol 52:35B–38B

Manzato E, Marin R, Gasparotto A, Baggio G, Martini S, Gabelli C, Crepaldi G (1986) Lipoprotein modification during dietary treatment in patients with primary type V hyperlipoproteinemia. Eur J Clin Invest 16:149–156

Martini S, Baggio G, Baroni L, Baldo Enzi G, Fellin R, Baiocchi MR, Crepaldi G (1984) Evaluation of HDL_2 and HDL_3 cholesterol by a precipitation procedure in a normal population and in different hyperlipidemic phenotypes. Clin Chim Acta 137:291–298

Olsson AG, Rössner S, Walldius G, Carlson LA (1976) Effect of Gemfibrozil on lipoprotein concentrations in different types of hyperlipoproteinemia. Proc R Soc Med 69:28–31

Pickering JE (1983) Clinical results with Gemfibrozil. Am J Cardiol 52:39B–40B

Saku K, Gartside PS, Hynd BA, Kashyap ML (1985) Mechanism of action of Gemfibrozil on lipoprotein metabolism. J Clin Invest 75:1702–1712

Samuel P (1984) Efficacy of Gemfibrozil as a lipid regulator in a patient population in the United States. Vasc Med 2:8–15

Vega GL, Grundy SM (1985) Gemfibrozil therapy in primary hypertriglyceridemia associated with coronary heart disease. JAMA 253:2398–2403

Effects of Gemfibrozil on Lipoproteins in Patients with Dyslipoproteinemia

G. A. Giudici[1], F. Pagani[1], C. Selvini[2], P. Stefanoni[2], C. di Santo[2], and C. Vergani[2]

1 Introduction

Gemfibrozil is a non-halogenated derivative of fibric acid. It was synthesized in 1968 and its first administration for clinical purposes took place in 1971 (Marks 1982). Its hypolipidemic effect is primarily evident in decreasing plasma levels of triglyceride-rich lipoproteins in hypertriglyceridemic patients who do not respond to diet. Moreover, Gemfibrozil treatment is followed by an increase in high-density lipoproteins (HDL) (Brown and Goldstein 1985).

The drug has always been considered to be "hypolipidemic", but quite recently it has also begun to be considered a "lipid-regulating" agent, because of its ability to modify lipoprotein composition (Krause and Newton 1985).

We designed a double-blind study to compare the effects of Gemfibrozil with those of a placebo in 40 patients with type IIa, IIb and IV hyperlipoproteinemia (HLP).

2 Patients

Ten patients with type IIa, 10 with type IIb and 20 with type IV HLP, according to WHO classification (Beaumont et al. 1970), were admitted to the study.

Using a double-blind design, patients were randomly allocated to either placebo or Gemfibrozil (1,200 mg/die) groups. Tables 1 and 2 summarize some patient characteristics.

Treatment was preceded by a 15-day wash-out period. Throughout the study the patients were maintained on their usual, prudent diet. Type IIa and IIb patients received a diet providing 48% of the total calories as carbohydrates, 32% as lipids and 20% as proteins, the daily intake of cholesterol being less than 300 mg with a P/S ratio of 1.2.

The diet for type IV patients provided 40% of the total calories as carbohydrates, 35% as lipids and 25% as proteins with a restricted alcohol intake.

Blood samples were drawn after an overnight fast on days 0–45 and 90.

---

1 Fondazione Rivetti, Laboratory of Biochemestry and Molecular Biology, Viale Monte Nero, 32, 20135 Milano, Italy
2 Institute of Internal Medicine, 9, Via Pace, 20122 Milano, Italy

Drugs Affecting Lipid Metabolism
Ed. by R. Paoletti et al.
© Springer-Verlag Berlin Heidelberg 1987

Table 1. Patient characteristics

	HLP phenotype	Gemfibrozil	Placebo
Number of patients	IIa	5	5
	IIb	5	5
	IV	10	10
Age (years)[a]	IIa	51.8 ± 11.2	46.0 ± 12.7
	IIb	50.8 ± 6.8	40.8 ± 17.2
	IV	45.8 ± 7.2	40.5 ± 6.4
Weight (kg)[a]	IIa	64.0 ± 11.5	63.4 ± 5.3
	IIb	64.0 ± 6.3	73.6 ± 11.7
	IV	67.9 ± 11.2	73.2 ± 12.5

Table 2. Total lipid and lipoprotein lipid base values

HLP type groups		Plasma cholesterol mg dl^{-1} [a]				Plasma triglyceride mg dl^{-1} [a]
		Total	VLDL	LDL	HDL	
IIa	Gemfibrozil	308.5	26.3	236.7	48.3	131.8
		± 25.6	± 6.9	± 36.3	±13.6	± 35.6
	Placebo	349.3	31.4	252.7	49.1	133.5
		± 83.1	±12.1	± 62.7	± 5.2	± 37.9
IIb	Gemfibrozil	292.4	52.0	187.4	47.7	313.1
		± 26.1	±18.0	± 18.9	±17.2	± 91.2
	Placebo	296.3	58.0	198.1	45.0	337.7
		± 39.4	±17.3	± 34.2	± 8.2	±134.8
IV	Gemfibrozil	231.7	52.6	133.3	39.8	352.4
		± 41.8	±14.2	± 37.3	± 6.0	±132.4
	Placebo	219.7	39.5	126.7	38.4	303.5
		± 49.3	±23.7	± 31.3	± 8.0	± 96.4
Controls		176	17	113	46	87
		± 27	± 7	± 27	±11	± 37

[a] Mean ± SD.
No significant differences were found between Gemfibrozil and placebo groups.

3 Materials and Methods

Cholesterol (C) and triglycerides (TG) were determined by enzymatic methods (Boehringer, Mannheim, FRG); HDL-cholesterol (HDL-C) was measured in the supernatant after precipitation of the apoprotein-B (Apo-B)-containing lipoproteins according to the Lopes-Virella method (Lopes-Virella et al. 1977). Very low density lipoprotein cholesterol (VLDL-C) was determined in the supernatant after ultracentrifugation at d. 1.006 g ml^{-1}. Low density lipoprotein cholesterol (LDL-C) was calculated as the difference between total cholesterol (TC) and the sum of HDL-C and VLDL-C.

Table 3. Apoprotein base values in type IV HLP patients and in controls

Groups	Apo-AI	Apo-AII	Apo-B	Apo-CII	Apo-CIII	Apo-E
			mg dl⁻¹ [a]			
Gemfibrozil	121.0	32.9	119.2	6.7	19.1	6.3
	± 44.5	±14.4	± 34.9	±2.3	± 6.2	±2.4
Placebo	119.4	33.6	124.0	5.7	14.0	4.9
	± 12.2	± 6.3	± 46.7	±1.5	± 6.3	±1.4
Controls	138	38	86.5	2.5	6.4	3.3
	± 20	± 6.0	± 17.2	±1.1	± 1.8	±0.9

[a] Mean ± SD.

In Type IV patients, the levels of Apo-AI, Apo-AII, Apo-B, Apo-CII, Apo-CIII and Apo-E were determined using the radial immunodiffusion (RID) technique (Daiichi Pure Chemical Co., Ltd., Tokyo, Japan).

Table 3 summarizes the apoprotein base values. Normal values obtained from 40 subjects (24 males and 16 females) in the age range of 22 to 46 years, are also reported.

Serum glutamic oxalacetic transaminase (SGOT), serum glutamic pyruvic transaminase (SGPT), plasma glucose, serum uric acid and hemogram were determined by standard laboratory techniques. Statistical evaluation was performed by using the paired t-test for comparison of the means.

4 Results

Table 4 shows our results. After 45 and 90 days of treatment, a significant decrease in the TG and VLDL-C levels was observed in the subjects given Gemfibrozil when compared to those given the placebo (Fig. 1). TC and LDL-C showed a significant decrease in the Gemfibrozil group when compared to placebo in type IIb patients.

A small, but not significant decrease was observed in type IIa, whereas the levels remained almost unchanged in type IV. An increase of HDL-C levels was observed in type IIb as well as in type IV patients treated with Gemfibrozil (Fig. 2). No significant changes were observed in the Apo-AI and Apo-AII levels in type IV.

In these subjects Apo-CII levels did not show any changes, while Apo-CIII and Apo-E levels significantly decrease in the Gemfibrozil group with an increase of the Apo-CII/Apo-CIII ratio (Fig. 3).

5 Discussion

At the daily dose of 1,200 mg for a period of 90 days, Gemfibrozil appears free of side effects. Only one dropout was observed in the Gemfibrozil group due to GI distress. There were also two dropouts in the placebo group.

Table 4. Effect of Gemfibrozil (1,200 mg/die) and placebo on lipid and lipoprotein parameters

| | % Change from base value observed on day: | | | |
| | 45 | | 90 | |
	G	P	G	P
HLP type IIa				
Total chol.	− 7.4	+ 2.4	− 8.0	+11.8
VLDL chol.	−12.3	−11.9	−36.0[a]	−15.6
LDL chol.	−10.5	+ 4.3	− 7.4	+10.1
HDL chol.	+ 6.8	+ 1.5	+ 4.3	− 2.1
Total triglycerides	−15.9	−15.4	−46.3[a]	−19.3
HLP type IIb				
Total chol.	−21.8[a]	− 6.2	−18.7[a]	− 1.6
VLDL chol.	−34.4[a]	−14.6	−31.8[a]	−14.6
LDL chol.	−14.0[a]	− 1.2	−12.8[a]	− 0.6
HDL chol.	+12.6[a]	+ 2.9	+16.0[a]	+ 3.8
Total triglycerides	−47.7[a]	−20.9	−44.8[a]	−23.5
HLP type IV				
Total chol.	+ 0.1	− 4.4	− 5.8	− 3.8
VLDL chol.	−33.4[a]	+18.7	−39.2[a]	− 5.3
LDL chol.	+ 2.0	+ 2.3	+ 4.0	+ 3.1
HDL chol.	+15.7[a]	− 7.6	+13.6[a]	− 2.5
Total triglycerides	−42.8[b]	+21.6	−49.7[b]	− 8.7
Apo-AI	− 2.9	+ 4.6	+ 4.5	+ 3.7
Apo-AII	− 3.6	+ 5.5	− 2.1	− 3.9
Apo-B	− 7.9	− 4.5	− 8.2	+ 1.7
Apo-CII	− 5.6	− 0.3	+ 4.1	− 3.5
Apo-CIII	−26.9[b]	− 3.1	−21.1[b]	+ 7.2
Apo-E	−33.6[b]	+ 9.8	−28.3[b]	+15.0
Apo-CII/CIII	+29.8[b]	− 3.4	+31.2[b]	−10.8

[a] $p < 0.05$, [b] $p < 0.01$; Gemfibrozil (G) versus placebo (P).

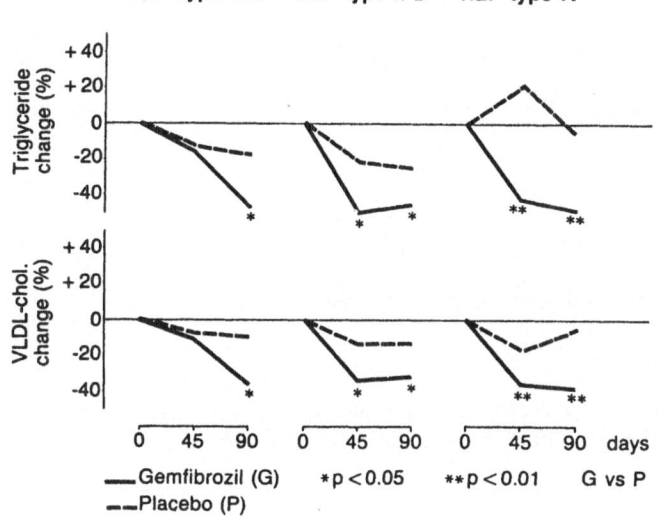

Fig. 1. Percentage changes of triglyceride and VLDL-cholesterol plasma levels in patients with type IIa, IIb, IV hyperlipoproteinemia. Comparison of Gemfibrozil (1,200 mg/die) with placebo

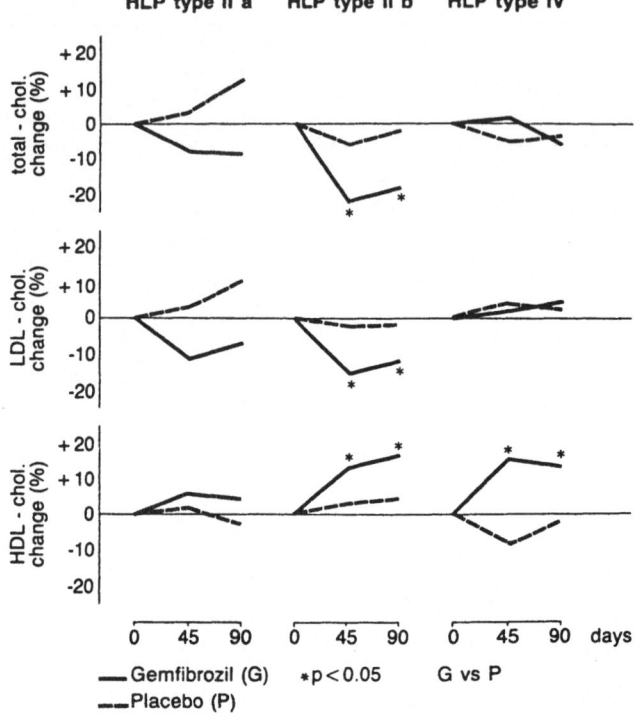

Fig. 2. Percentage changes of total cholesterol, LDL-cholesterol, HDL-cholesterol plasma levels in patients with type IIa, IIb, IV hyperlipoproteinemia. Comparison of Gemfibrozil (1,200 mg/die) with placebo

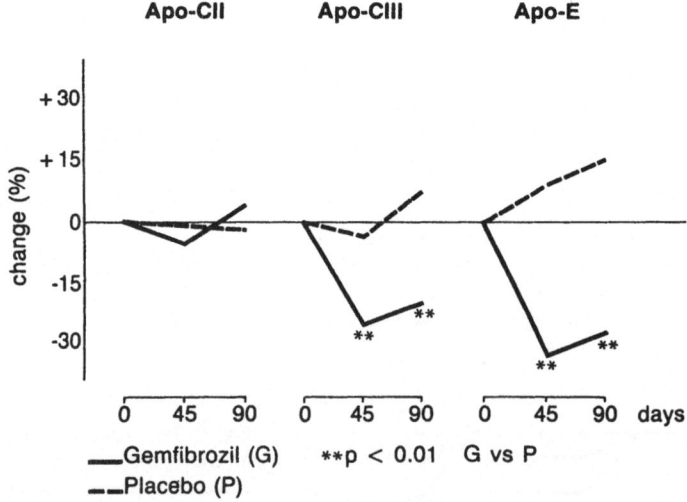

Fig. 3. Percentage changes of Apo-CII, Apo-CIII and Apo-E plasma levels in patients with type IV hyperlipoproteinemia. Comparison of Gemfibrozil (1,200 mg/die) with placebo

SGOT, SGPT, plasma glucose, serum uric acid and the hemogram did not show any significant variation during the Gemfibrozil treatment period.

In the Gemfibrozil group plasma lipid changes are characterized by a marked reduction of TG and VLDL-C. Apo-CIII and Apo-E, major VLDL components, which are increased in hypertriglyceridemic patients (Blum et al. 1980; Curry et al. 1980), tend to return towards normal values. The decrease of TG and VLDL-C is in agreement with other reports in the literature (Kaukola et al. 1981; Lewis 1982; Pickering 1983; Samuel 1983, 1984; Kashyap 1984; Virtamo et al. 1984; Saku et al. 1985; Vega and Grundy 1985).

Gemfibrozil decreases VLDL and LDL synthesis (Kissebach et al. 1976; Kesäniemi and Grundy 1984).

Our results agree with those of Luley et al. (1986), who observed a reduction of 33 and 44% respectively, of Apo-CIII and Apo-E levels in 20 hypertriglyceridemic patients given 900 mg/die of Gemfibrozil for 6 weeks. These results support the hypothesis of an inhibiting effect of the drug on the hepatic synthesis of VLDL. The reduction of Apo-CIII, which undergoes sialylation in the Golgi apparatus, suggests an effect at this level.

The isoelectric-focussing analysis of the C-peptides from the VLDL of one type IV patient, carried out before and after 90 days of Gemfibrozil treatment, did not show any changes in isoform distribution in spite of a decrease of the Apo-CIII levels as shown by RID (Fig. 4).

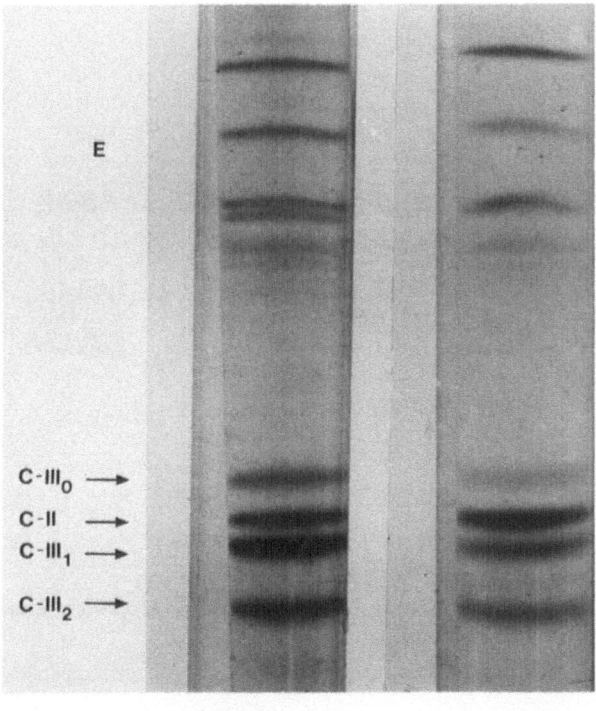

E

$C\text{-}III_0 \longrightarrow$

$C\text{-}II \longrightarrow$

$C\text{-}III_1 \longrightarrow$

$C\text{-}III_2 \longrightarrow$

B   A

Fig. 4A,B. Isoelectric-focussing of VLDL peptides of one patient with type IV hyperlipoproteinemia before (B) and after (A) 90 days Gemfibrozil (1,200 mg/die) treatment

Table 5. Primary hypoalphalipoproteinemia[a]

Patients		BMI	HDL-C	Apo-AI	Apo-AII	Apo-B	Apo-CII	Apo-CIII	Apo-E
					mg/dl				
1	Before	24.8	29	106	22.6	58	3	8	4
	After	24.6	37	119	27.1	68	2.8	7.2	3.7
	% Change	− 0.9	+28	+ 18	+20	+ 7	−7	−10	− 7.5
2	Before	23.3	30	108	27.2	87	2.8	6.7	3
	After	23.2	36	120	31.5	82	3	6.2	2.7
	% Change	− 0.4	+20	+ 11	+17	− 6	+7	− 7	−10
3	Before	22.4	28	99	26	70	2.4	4.8	2.6
	After	22.8	34	111	31	65	2.6	5.2	2.8
	% Change	+ 1.8	+21	+ 12	+19	− 7	+8	+ 8	+ 8

[a] Body mass index (BMI) and lipoprotein parameters before and after 90 days of Gemfibrozil treatment (1,200 mg/die) (preliminary results).

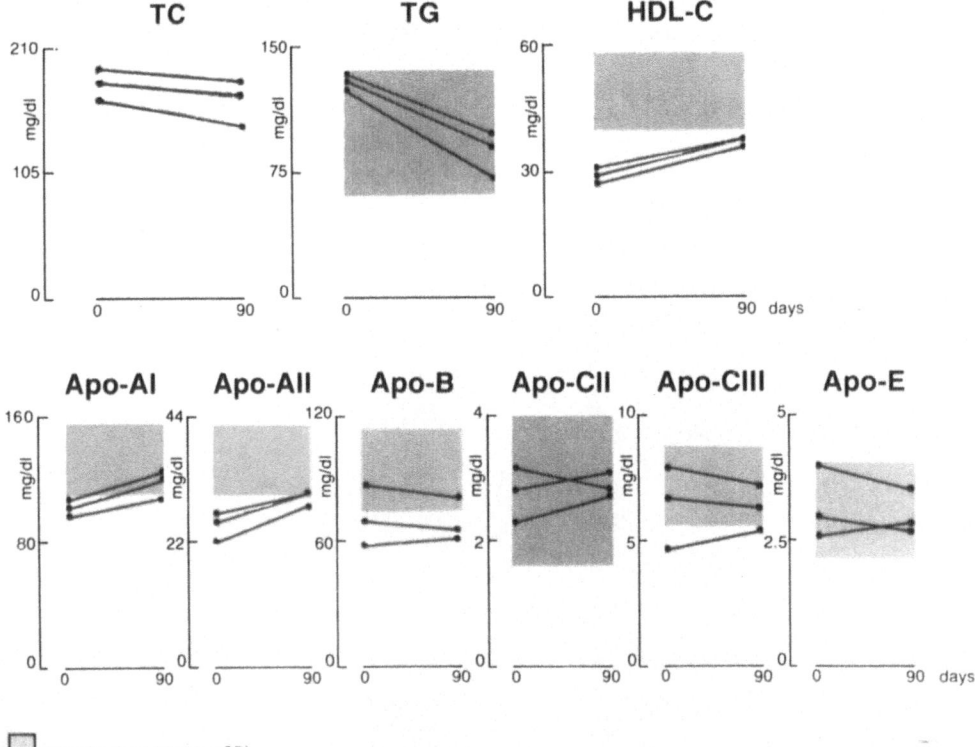

normal values (mean ± SD)

Fig. 5. Effects of Gemfibrozil (1200 mg/die) on total lipid, HDL-cholesterol and apoprotein levels in three patients with primary hypoalphalipoproteinemia

Apo-CII is an activator of lipoproteinlipase (LPL) (Havel et al. 1970; La Rosa et al. 1970), while under certain experimental conditions, Apo-CIII is capable of inhibiting the enzyme activity (Brown and Baginsky 1972; Krauss et al. 1973).

The increased Apo-CII/Apo-CIII ratio, also reported by Saku et al. (1985), is suggestive of an activation of LPL. In fact, increased LPL activity during Gemfibrozil treatment has been described by Saku et al. (1985) and by Nikkilä et al. (1976). According to Nikkilä et al. (1976) hepatic lipoproteinlipase (HLPL) activity was also increased.

It is well known that in most patients with hypertriglyceridemia the decrease of TG is accompanied by an increase of HDL (Schaefer et al. 1978; Witztum et al. 1980). This is paralleled by a transfer of surface components (free cholesterol, phospholipids, Apo-C and Apo-E) from VLDL to HDL_3, which in turn are transformed into HDL_2 (Mahley et al. 1978).

In the present study, following Gemfibrozil treatment, we have observed a moderate increase of HDL-C in type IIb and IV patients. Glueck (1983) observed an increase in HDL-C in type IIa, type IIb and type IV patients of 25, 20 and 17% respectively. In type IV patients Saku et al. (1985) observed a 27% increase of Apo-AI and a 34% increase of Apo-AII, while an increase of Apo-AII was reported by Kaukola et al. (1981).

In type IIa and IIb patients admitted to our study, TC and LDL-C show a trend towards lowering after Gemfibrozil treatment. A slight decrease of TC and LDL-C was reported by Lewis in type IIa and IV patients (1982).

In summary, Gemfibrozil appears to act mainly by regulating the apoprotein composition of VLDL. It is known that Apo-C and Apo-E have an important role in TG-rich lipoprotein metabolism (Curry et al. 1980; Blum et al. 1980). A lipid-regulating effect of the drug has been observed also in cholesterol-fed rats (Krause and Newton 1985).

We have also treated three patients with primary hypoalphalipoproteinemia with Gemfibrozil (1200 mg/die), a syndrome described by Vergani and Bettale (1981).

After 90 days of treatment an increase of HDL-C, Apo-AI and Apo-AII levels was observed (Table 5; Fig. 5). A 20 to 25% increase of HDL was also reported by Glueck (1983) in patients with primary hypoalphalipoproteinemia. In these patients, Gemfibrozil may act by interfering either with the synthesis or with the catabolism of HDL. Preliminary studies on the kinetics of antologous HDL in one subject with primary hypoalphalipoproteinemia indicate that the plasma residence time is decreased (B. Brewer, C. Vergani).

References

Beaumont JL, Carlson LA, Cooper GR et al. (1970) Classification of hyperlipidemias and hyperlipoproteinemias. Bull WHO 43:891

Blum CB, Aron L, Sciacca R (1980) Radioimmunoassay studies of human apolipoprotein. Eur J Clin Invest 66:1240

Brown MS, Goldstein JL (1985) Drugs used in the treatment of hyperlipoproteinemias. In: Goodman, Gilman (eds) The pharmacological basis of therapeutics, 7th edn. MacMillan, New York, pp 827–845

Brown WV, Baginsky ML (1972) Inhibition of lipoprotein lipase by an apoprotein of human very low density lipoprotein. Biochem Biophys Res Commun 46:375

Curry HD, Mc Conathy WJ, Fesmire JD et al. (1980) Quantitative determination of human apolipoprotein C-III by electro-immunoassay. Biochim Biophys Acta 617:503

Glueck CJ (1983) Influence of Gemfibrozil on high density lipoproteins. Am J Cardiol 52:31B

Havel RJ, Shore VG, Shore B et al. (1970) Role of specific glycopeptides of human serum lipoproteins in the activation of lipoprotein lipase. Circ Res 27:595

Kashyap ML (1984) The effects of Gemfibrozil on plasma lipids and lipoproteins in man. Vasc Med 2:16

Kaukola S, Manninen V, Malkonen M et al. (1981) Gemfibrozil in the treatment of dyslipidaemias in middle-aged male survivors of myocardial infarction. Acta Med Scand 209:69

Kesäniemi YA, Grundy SM (1984) Influence of Gemfibrozil and clofibrate on metabolism of cholesterol and plasma triglycerides in man. JAMA 251:2241

Kissebach AH, Alfarasi S, Adams PW et al. (1976) Transport kinetics of plasma free fatty acid, very low density lipoprotein triglycerides and apoprotein in patients with endogenous hypertriglyceridemia. Effect of 2,2 dimethyl, 5 (2,5 xylyloxy) valeric acid therapy. Atherosclerosis 24:199

Krause BR, Newton RS (1985) Apolipoprotein changes associated with the plasma lipid-regulating activity of Gemfibrozil in cholesterol-fed rats. J Lipid Res 26:940

Krauss RM, Hebert PN, Levy RI (1973) Further observation on the activation and inhibition of lipoprotein lipase by apolipoproteins. Circ Res 33:403

La Rosa JC, Levy RI, Hebert R (1970) A specific apoprotein activator for lipoprotein lipase. Biochem Biophys Res Commun 47:57

Lewis JE (1982) Long-term use of Gemfibrozil (LOPID) in treatment of dyslipidemia. Angiology 33:603

Lopes-Virella HF, Stone P, Ellis S et al. (1977) Cholesterol determination in high density lipoprotein separated by three different methods. Clin Chem 23:882

Luley C, Schwartzkopff W, Wang CS et al. (1986) Gemfibrozil treatment of hypertriglyceridemic patients alters the apolipoprotein composition in VLDL and HDL and activates lipolysis. 6th Int Washington Spring Symp, Cardiovascular disease '86, May, pp 20–23, Abstr 149

Mahley RW, Innerarity T, Bersat TP et al. (1978) Alterations in human high-density lipoproteins, with or without increased plasma cholesterol, induced by high diets in cholesterol. Lancet 2:807

Marks J (ed) (1982) Dyslipoproteinemia. Aspects of Gemfibrozil therapy. Res Clin Forums (Kent, England) 4:1–8

Nikkilä EA, Ylikahri R, Huttunen JK (1976) Gemfibrozil: effect on serum lipids, lipoproteins, postheparin plasma lipase activities and glucose tolerance in primary hypertriglyceridemia. Proc R Soc Med 69 (Suppl 2):58

Pickering JE (1983) Clinical results with Gemfibrozil. Am J Cardiol 52:39B

Saku K, Gartside PS, Hynd BA et al. (1985) Mechanism of action of Gemfibrozil on lipoprotein metabolism. J Clin Invest 75:1702

Samuel P (1983) Effects of Gemfibrozil on serum lipids. Am J Med 74:23

Samuel P (1984) Efficacy of Gemfibrozil as a lipid regulator in patient population in United States. Vasc Med 2:8

Schaefer EJ, Levy RI, Anderson DW et al. (1978) Plasma triglycerides in regulation of HDL cholesterol levels. Lancet 2:391

Vega JL, Grundy SM (1985) Gemfibrozil therapy in primary hypertriglyceridemia associated with coronary heart disease. JAMA 253:2398

Vergani C, Bettale G (1981) Familial hypoalphalipoproteinemia. Clin Chim Acta 114:45

Virtamo J, Manninen V, Malkonen M (1984) A placebo controlled, rising dose, double-blind trial with Gemfibrozil in dieting patients with primary hyperlipoproteinemia. Vasc Med 2:22

Witztum JL, Dillingham MA, Giese W et al. (1980) Normalization of triglycerides in type IV hyperlipoproteinemia fails to correct low levels of high-density-lipoprotein cholesterol. N Engl J Med 14:907

HDL Deficiency, Atherosclerosis, and Stimulation of HDL Synthesis: Role of Gemfibrozil

M. L. KASHYAP[1] and K. SAKU[1,2]

1 Background

The accretion of cholesterol in tissues (e.g., the arterial wall) is influenced by (a) the deposition of atherogenic lipoproteins, e.g., low density lipoproteins (LDL) and (b) removal by reverse cholesterol transport, a process mediated by HDL. The process of atherosclerosis may thus be retarded or reversed by (a) lowering plasma concentrations of atherogenic lipoproteins and/or (b) stimulating reverse cholesterol transport. During the last few decades clinical as well as basic experimental evidence has been obtained· to indicate that decreasing plasma levels of atherogenic lipoproteins, especially low density lipoproteins, is unequivocally associated with decreased risk for coronary heart disease. The results of the Coronary Primary Prevention Trial (1, 2) attest to the value of lowering LDL cholesterol in the prevention of coronary disease in man. More recent studies have also shown that lowering atherogenic lipoproteins (while raising HDL) can also result in lack of progression of coronary lesions (3).

There is very little information about removal of cholesterol that has already deposited in the arterial wall. In order to approach the problem of reversal of coronary lesions, we believe it is most important to understand the process of reverse cholesterol transport. In this process, excess cholesterol in various tissues, including the arterial wall, are removed from the body. The physiological events that are responsible for the removal of cholesterol excess are poorly understood at present. The following is an overview of the current concept of reverse cholesterol transport. In this process, shown schematically in Fig. 1, tissue cholesterol is taken up initially by high density lipoprotein (HDL). Only free (unesterified) cholesterol from tissue is taken up by HDL, where it forms a component of the outer amphipathic layer of HDL. In this layer are also various apolipoproteins (apo) of HDL which are important in cholesterol and triglyceride transport. In the context of cholesterol efflux or removal, apoprotein apo AI and apo-E are important. It is possible that other minor apoproteins, e.g., apo AIV and possibly apo AII may also be important in the uptake or further processing of the unesterified cholesterol in HDL. An enzyme, lecithin cholesterol acyl transferase (LCAT),

1 University of California at Irvine and Veterans Administration Medical Center, Long Beach, CA 90822, USA
2 Present address: School of Medicine, Fukuoka University, 45-1, 7-chome Nanakuma, Jonan-ku, Fukuoka 814-01, Japan

Drugs Affecting Lipid Metabolism
Ed. by R. Paoletti et al.
© Springer-Verlag Berlin Heidelberg 1987

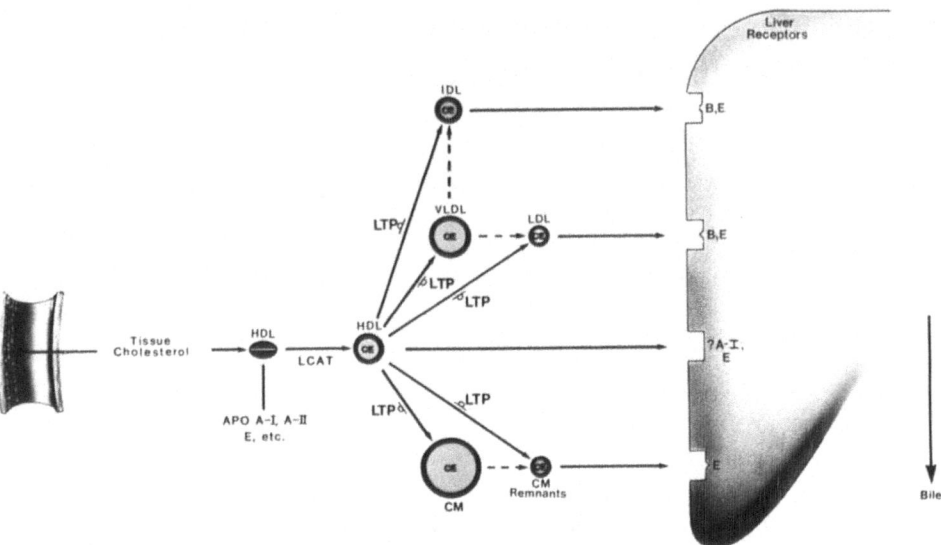

Fig. 1. Reverse cholesterol transport concept

is activated by apo AI and possibly apo CI in HDL. Activated LCAT results in the esterification of cholesterol which is hydrophobic and therefore is transferred to the interior core of HDL; thus, apo AI on the HDL surface may take up more unesterified cholesterol from tissues. HDL-cholesterol is removed by its transfer to the liver via HDL and excreted in bile. Recent studies have shown that the esterified cholesterol in HDL can be transferred to other lipoproteins, viz., very low density lipoproteins (VLDL), LDL, chylomicrons (CM), and chylomicron remnants (CM remnants) by lipid transfer proteins (LTPs) that effect this exchange. By this mechanism, shown in Fig. 1, the capacity of HDL for cholesterol removal is amplified and made more efficient. LDL and chyclomicron remnants are taken up by specific receptors on the liver which endocytose these lipoprotein particles by the apo-B, E receptor for LDL and apo-E receptor for chylomicron remnants.

On the basis of this scheme, any agent (diet, exercise, drugs) that stimulates this process will also enhance cholesterol removal. Thus, it is not surprising that low levels of apo AI are associated with severe atherosclerosis and accelerated accretion of cholesterol in certain tissues, e.g., coronary arteries. Epidemiologic studies as well as clinical studies attest to the central concept of the role of apo AI as a risk factor for atherosclerosis and marker for angiographically demonstrable coronary artery disease.

As pointed out above, agents that stimulate apo AI synthesis will be theoretically expected to enhance the process of reverse cholesterol transport. Likewise, agents that depress apo AI synthesis would be expected to result in cholesterol accretion and increased progression of arterial lesions. Agents that raise the plasma level of apo AI and HDL by depressing the catabolism of HDL may not be as effective as agents that stimulate its synthesis. Also, agents that act on the lipid transfer proteins would also be theoretically expected to have an effect on the process of reverse cholesterol transport. Furthermore, inhibitors of the lipid transfer proteins may elevate HDL-cholesterol, but will not be expected to increase net removal of tissue cholesterol.

The above discussion summarizes the role of HDL in reverse cholesterol transport. More recent studies have also shown that HDL can enhance the lysis of fibrin in vitro in the presence of plasminogen and its activator, thus indicating that HDL may also prevent thombosis (4). Recent evidence from my laboratory indicates that gemfibrozil (a widely used lipid-regulating agent) is the only drug known at present that stimulate the synthetic rate of apo AI and apo AII. Another related drug (bezafibrate) does not affect apo-A metabolism (5). We believe that a greater understanding of the physiology of reverse cholesterol transport and the stimulation of this process by various agents will be an important direction for future research related to the question of regression of lesions characterized by cholesterol accretion.

2 Role of Gemfibrozil in Stimulation of HDL Synthesis

These studies were carried out by my collaborators K. Saku, B. Hynd, S. Mendoza and P. Gartside (6, 7). As background, gemfibrozil has been used widely for treatment of lipid disorders. Its major effects were described by Manninen et al. in 254 patients followed for 5 years (8). These investigators reported an average reduction of plasma total triglycerides of 40%, plasma total cholesterol of 16%, reductions in LDL-cholesterol of 23%, an increase in HDL-cholesterol by 23%, and an increase in HDL to total cholesterol ratio of 45%.

We assessed the mechanism by which gemfibrozil increases plasma HDL levels and decreases triglycerides. Six patients with primary HDL deficiency associated with endogenous hypertriglyceridemia were investigated. The mean level of HDL-cholesterol was 22 mg dl^{-1} and the mean level of plasma triglycerides was 978 mg dl^{-1} at entry into the study. No patients had any secondary cause for the hyperlipidemia, none had diabetes mellitus, and all patients were within 15% of the desirable body weight. After diagnosis and evaluation, each subject was placed on a diet appropriate for the lipid abnormality and followed until they were stabilized. They were then admitted to the clinical research center where baseline studies, described below, were conducted. This was followed by treatment of each subject with the diet plus gemfibrozil in a dose of 600 mg twice a day for 2 months after which time the research studies were repeated. In the clinical center the research studies included measurements of apo AI and AII turnover using autologous radiolabeled HDL, measurements of hepatic and extrahepatic lipoprotein lipases, and characterization of VLDL and HDL. A control group of healthy volunteers was also studied.

Two months of treatment with gemfibrozil resulted in a marked reduction in triglycerides from 449 ± 87 mg dl^{-1} to 189 ± 41 mg dl^{-1}, HDL-cholesterol rose from a mean of 25.4 ± 2.3 to 33.9 ± 2.1 mg dl^{-1}. Apo AI increased in each subject, from a mean of 70.4 ± 2.7 mg dl^{-1} to 90.6 ± 3.7 mg dl^{-1}. Likewise, there was also an increase in the level of apo AII from 24.2 ± 1.6 to 32.9 ± 1.6 mg dl^{-1}. Extrahepatic lipoprotein lipase (LPL) increased significantly in these subjects. There was no significant change in the hepatic triglyceride lipase (HTgL) activities. Assessment of the VLDL apo-C content revealed a significant reduction in the ratio of apo CIII to apo CII, the inhibitor and activator of the enzyme lipoprotein lipase, respectively. These studies indicated that gemfibrozil reduced triglycerides by stimulating lipoprotein lipase and possibly by affecting the ratio of apo CIII to apo CII in the VLDL.

Results of the turnover studies revealed the following results. Compared to healthy control subjects, the low levels of apo AI and AII were found to be the result of decreased synthesis and increased fractional catabolic rate (7). Following gemfibrozil treatment, there was a significant increase in the absolute of both apo AI and apo AII synthetic rate. Apo AI synthetic rate increased from 12.07 ± 0.40 mg kg^{-1} body wt. day to 15.24 ± 0.49 mg kg^{-1} body wt. day^{-1}. Apo AII synthetic rate increased from 3.86 $\pm$ 0.25 to 5.12 ± 0.27 mg kg^{-1} body wt. day^{-1}. Analysis of the decay curves of radiolabeled HDL as well as the decay of radioactivity in apo AI or apo AII did not reveal any significant effect on the fractional catabolic rate of apo AI or apo AII. Examination of the HDL particle size by gradient gel electrophoresis revealed a significant reduction in the molecular diameter of HDL as well as the molecular weight of HDL, implying the presence in plasma, of smaller HDL particles, consistent with the view that gemfibrozil stimulated the synthesis of HDL.

3 Significance of the Mechanism by Which Gemfibrozil Increases HDL Levels

Under steady-state conditions the absolute synthetic rate is equal to the absolute catabolic rate, which is also equal to the turnover of apo AI and apo AII, major proteins of HDL. Since HDL and its A-apoproteins are involved in reverse cholesterol transport as discussed above, stimulation of this process will be expected to enhance removal of body tissue cholesterol with important implications on the mechanisms by which such agents may exert antiatherosclerotic effects.

References

1. Lipid Research Clinics Program (1984) The lipid research clinics coronary primary prevention trial results. I. Reduction in incidence of coronary heart disease. JAMA 251:351–364
2. Lipid Research Clinics Program (1984) The lipid research clinics coronary primary prevention trial results. II. The relationship of reduction of incidence of coronary heart disease to cholesterol lowering. JAMA 251:365–374
3. Levy RI, Brensike JF, Epstein SE et al. (1984) The influence of changes in lipid values induced by cholestyramine and diet on progression of coronary artery disease: Results of the NHLBI type II coronary intervention study. Circulation 69(2):325–337
4. Saku K, Ahmad M, Glas-Greenwalt P, Kashyap ML (1985) Activation of fibrinolysis by apolipoproteins of high density lipoproteins in man. Thrombosis Res 39:1–8
5. Shepherd J, Packard CJ, Stewart JM et al. (1984) Apolipoprotein A and B (Sf 100–400) metabolism during bezafibrate therapy in hypertriglyceridemic subjects. J Clin Invest 74:2164–2177
6. Saku K, Gartside PS, Hynd BA, Kashyap ML (1985) Mechanism of action on gemfibrozil on lipoprotein metabolism. J Clin Invest 75:1702–1712

7. Saku K, Gartside PS, Hynd BA, Mendoza SG, Kashyap ML (1985) Apolipoprotein AI and AII metabolism in patients with primary high density lipoprotein deficiency associated with familial hypertriglyceridemia. Metabolism 34:754−764
8. Manninen V, Malkonen M, Eisalo A, et al. (1982) Gemfibrozil in the treatment of dyslipidemia. A 5-year follow-up study. Acta Med Scand (Suppl) 668:82−87

Hemorheological Activity of Gemfibrozil in Primary Hyperlipidemias

G. CIUFFETTI, G. ORECCHINI, D. SIEPI, G. LUPATTELLI, and A. VENTURA [1]

Abbreviations: SV = serum viscosity; PV = plasma viscosity; BF = blood filterability; WBV = whole blood viscosity; TC = total cholesterol; TG = triglycerides; F = fibrinogen; A/G = albumin/globulin ratio; Hct = hematocrit; WBC = white blood cells; Pts = platelets; γ = shear rate.

1 Introduction

All the various hemorheological determinants play a central role in the genesis and clinical developement of atherosclerosis (Dintenfass 1965; Smith and Staples 1981). As reported by several AA, modifications in the hemorheological pattern may be observed in different types of hyperlipidemias. A rise in SV and PV has been chiefly related to an increase in very low density lipoprotein (Leonhardt and Klemens 1977), while a decrease in BF, an expression of erythrocyte membrane flexibility, has been chiefly related to an increase of low density lipoprotein. The more membrane cholesterol is increased, the more inflexible the erythrocyte membrane becomes (Pretolani et al. 1983). Consequently, some studies on the effects of hypolipidemic drugs on hemorheological pattern have been carried out. Clofibrate, the most extensively investigated, decreases both PV and WBV through a reduction of fibrinogen plasma content (Dormandy et al. 1974). Today, the availability of Gemfibrozil, a new fibrate agent, has led us to investigate its action on hyperlipemic patients with an altered hemorheological pattern.

2 Materials and Methods

Twenty outpatients, affected by primary hyperlipidemias and with altered hemorheological pattern, were admitted to this study: 10 type IIA (6 males, 4 females, average age 46 ± 2 years), and 10 type IV (5 males, 5 females, average age 47 ± 1 years). During the 4 weeks preceding the study, the patients received neither hypolipidemic agents nor drugs known to have an effect on the hemorheological pattern. In this period and

---

1 Institute of 2 Clinica Medica Generale e Terapia Medica, University of Perugia, Policlinico Monteluce, 06100 Perugia, Italy

Drugs Affecting Lipid Metabolism
Ed. by R. Paoletti et al.
© Springer-Verlag Berlin Heidelberg 1987

during the study all patients followed an isocaloric diet containing less than 30% of the total calories as lipids and less than 300 mg day^{-1} cholesterol with a polyunsaturated/ saturated fatty acids ratio of 2:1. The 20 patients were divided at random into two subgroups of ten and the double-blind crossover study, lasting 90 days, was organized in three phases: (1) a 15-day drug wash-out; (2) a 60-day treatment period during which each subgroup was treated with 30 days Gemfibrozil (administered in two daily doses, totalling 1.2 g) or placebo alternated with 30 days placebo or Gemfibrozil; (3) a 15-day wash-out period like the first phase. At the beginning of the study and after 15, 45, 75 and 90 days, TC, TG and hemorheological determinations (SV, PV and WBV, F, A/G, Hct, WBC and Pts, BF) were evaluated. TC and TG were determined by enzymatic methods, viscosity by rotational viscometer at a shear rate of 225 s^{-1}, BF by the method of Reid et al. (1976). At the beginning of each phase and at the end of the study all patients underwent a general medical examination and ECG and the safety parameters were evaluated. Student's t-test was used for all the statistical analysis of paired data.

3 Results

The results of this study are reported in Tables 1, 2 and 3.

Table 1. Effects of Gemfibrozil on type IIA hyperlipidemia

	Gemfibrozil	
	At the beginning	After 30 days
TC (mg%)	339.00 ± 18.00	278.00 ± 15.00[b]
TG (mg%)	140.00 ± 7.00	85.00 ± 5.00[c]
SV (cP) γ 225 s^{-1}	1.28 ± 0.01	1.28 ± 0.01
PV (cP) γ 225 s^{-1}	1.49 ± 0.02	1.48 ± 0.01
WBV (cP) γ 225 s^{-1}	4.95 ± 0.16	4.64 ± 0.19
BF (ml min^{-1})	0.92 ± 0.04	1.06 ± 0.07[a]
F (mg%)	413.00 ± 42.00	382.00 ± 26.00

[a] $p < 0.05$, [b] $p < 0.005$, [c] $p < 0.0005$.

Table 2. Effects of Gemfibrozil on type IV hyperlipidemia

	Gemfibrozil	
	At the beginning	After 30 days
TC (mg%)	219.00 ± 6.00	193.00 ± 6.00
TG (mg%)	443.00 ± 40.00	242.00 ± 25.00[b]
SV (cP) γ 225 s^{-1}	1.29 ± 0.02	1.28 ± 0.01
PV (cP) γ 225 s^{-1}	1.54 ± 0.03	1.46 ± 0.02[a]
WBV (cP) γ 225 s^{-1}	5.10 ± 0.17	4.51 ± 0.16[b]
BF (ml min^{-1})	1.01 ± 0.02	1.08 ± 0.04
F (mg%)	382.00 ± 15.00	345.00 ± 20.00[a]

[a] $p < 0.05$, [b] $p < 0.005$, [c] $p < 0.0005$.

Table 3. Effects of Gemfibrozil on hyperlipidemic patients

	Gemfibrozil	
	At the beginning	After 30 days
Type IIA		
A/G	1.29 ± 0.05	1.34 ± 0.06
Hct (%)	43.50 ± 1.00	42.90 ± 1.00
WBC (10^9 l^{-1})	5.80 ± 0.50	5.70 ± 0.50
Pts (10^9 l^{-1})	225.00 ± 11.00	235.00 ± 11.00
Type IV		
A/G	1.29 ± 0.05	1.32 ± 0.03
Hct (%)	44.20 ± 0.73	43.90 ± 0.57
WBC (10^9 l^{-1})	6.49 ± 0.39	6.33 ± 0.33
Pts (10^9 l^{-1})	244.00 ± 8.00	236.00 ± 9.00

Placebo. No important variations in mean values of lipids and hemorheological parameters were noted in either hypolipemic phenotype during the placebo administration.

Gemfibrozil. In type IIA the drug administration resulted in a significant reduction in the mean values of TC (-18%) and TG (-38%) and a significant increase in mean values of BF (+15%) (Table 1). In type IV, the drug lowered the mean values of TG (-45%), F (-10%), PV (-4%) and WBV (-12%) (Table 2). The other hemorheological determinants did not show any significant variation in either hyperlipidemic phenotype (Table 3).

4 Conclusions

The results of our randomized double-blind crossover study, although preliminary because a more complete analysis of viscosity at all the shear rates is surely needed, allow us to conclude that Gemfibrozil, besides its well-known hypolipidemic activity, is able to modify the altered hemorheological pattern of hyperlipemic patients. In type IIA the increased BF, an expression of erythrocyte membrane flexibility, could be related to the TC lowering, while in type IV the reduction of PV and WBV could be due to the parallel decrease in TG and F. We can therefore conclude that Gemfibrozil may interfere not only with the genesis and evolution of the atherosclerotic process by reducing blood lipids and fibrinogen values, but also with the degree of blood flow by reducing viscosity and increasing blood filterability. This last factor is especially important for the microcirculation flow.

Acknowledgements. The authors would like to thank Mrs. G.A. Boyd Mancinelli B.A. (hons) for her help in the translation of this paper.

References

Dintenfass L (1965) Some rheological factors in the pathogenesis of thrombosis. Lancet II:370–372

Dormandy JA, Gutteridge JMC, Hoare E, Dormandy TL (1974) Effect of clofibrate on blood viscosity in intermittent claudication. Brit Med J IV:259–263

Leonhardt HRA, Klemens UH (1977) Studies of plasma viscosity in primary hyperlipoproteinaemia. Atherosclerosis 28:29–40

Pretolani E, Battistini G, Iosa G, Salvi P, Tonti D, Zoli I (1983) Iperlipemia e filtrabilitá eritrocitaria. Ricerca Clin Lab 13 (Suppl 3):341–344

Reid HL, Barner AJ, Lock PJ, Dormandy JA, Dormandy TL (1976) A simple method for measuring erythrocyte deformability. J Clin Pathol 29:855–858

Smith EB, Staples EM (1981) Haemostatic factors in human aortic intima. Lancet I:1171–1174

Evaluation of Hypolipidemic and Other Useful Pharmacological Properties of Various Nicotinic Acid Derivatives

G. Quack[1], L. Puglisi[2], W. Schatton[1], and A. Schwaier[1]

The action of nicotinic acid (NA) on lipid metabolism was first described by Altschul et al. (1955). Since then a large number of investigators confirmed these early observations and several aspects of the mode of action of NA on lipids, lipoproteins, and atherosclerotic processes could be clarified as well.

According to Hotz (1983) NA affects atherogenic conditions by decreasing VLDL-triglycerides and LDL-cholesterol and elevating HDL-cholesterol and the apo A I/apo A II quotient. Furthermore, it could be shown that NA acts preventively on lipid accumulation in lesioned areas of the great arteries, thus reducing the extent of plaque formation. Other favorable effects are: elevation of the PG I_2/TX A_2-quotient, decrease of platelet aggregability, and reduction of excess fibrinogen in plasma, accompanied by an improvement of rheological parameters.

Although being an old drug these data suggest that NA has many properties which are desirable for modern therapeutics in this field.

On the other hand, there are some side effects which, though being mainly transient, often lead to interruption of medication. The most frequently observed side effects are: flushing, deterioration of oral glucose tolerance, elevation of transaminases and alkaline phosphatase, hyperuricemia, and rebound of free fatty acids.

These effects are mostly explicable by the high doses (about 3 g day^{-1}) which are needed for therapy with the unmodified drug, and this is related to the rapid catabolism of NA which leads to inactive metabolites. As it is known (e.g., Subissi and Murmann 1978) the blood levels (and perhaps the tissue levels as well), which have to be reached for an adequate effect, may be much lower than those obtained after administration of 1 g t.i.d. of the plain compound. Therefore, some efforts have been undertaken to circumvent this disadvantage by synthesizing slow release forms of NA. But up to now expectations have not been fulfilled. Indeed, the therapeutic effects of currently available pro-drugs of NA are most often associated with prostacyclin-mediated flushing and rebound of free fatty acids after cessation of medication.

In an attempt to meet the goal, i.e., high therapeutic efficiency and reduced side effects, several substituted phenolesters of NA were synthesized.

We first tested their hypolipidemic properties in an animal model of diet-induced hyperlipidemia.

1 Merz & Co., Abteilung für Pharmakologie, D-6000 Frankfurt/Main, FRG
2 Institute of Pharmacological Sciences, University of Milan, Milan, Italy

Drugs Affecting Lipid Metabolism
Ed. by R. Paoletti et al.
© Springer-Verlag Berlin Heidelberg 1987

Male rats were fed a hyperlipidemic diet [orotic acid-free diet according to Standerfer and Handler (1955) with the addition of 1% cholic acid and 2% cholesterol] for 7 weeks. After 4 weeks they were treated with the test compounds (100 mg kg^{-1} day^{-1}, n = 6 per group). At the end of the study total serum cholesterol (CHOL) and serum triglycerides (TG) were determined. The results of the study are shown in Fig. 1.

Under the study conditions applied L 9, L 34, L 42, L 43, and L 44 appear to be potent drugs which are able to counteract diet-induced hyperlipidemia significantly, whereas clofibrate does not have any effect at all.

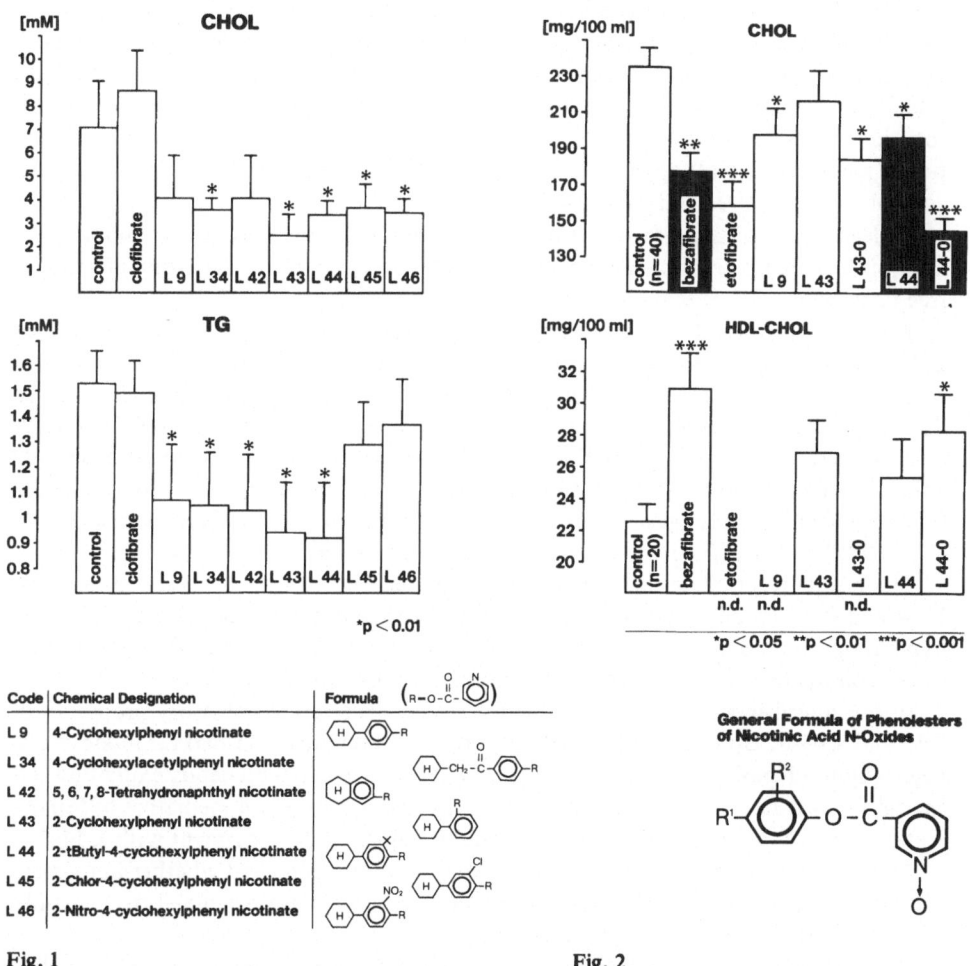

Fig. 1   **Fig. 2**

Fig. 1. Evaluation of effects on serum lipids of some phenolesters of nicotinic acid in the rat model diet-induced hyperlipidemia. Data are given as mean ± SD

Fig. 2. Evaluation of effects on serum cholesterol and HDL-cholesterol of some phenolesters of nicotinic acid and their N-oxides in the rat model diet-induced hypercholesterolemia. Data are given as mean ± SEM

Speculating that N-oxides could enhance these effects by functioning as slow release forms of NA, some phenolesters of nicotinic acid N-oxide were synthesized. Some of the N-oxides were tested together with their parent compounds and the two standard drugs etofibrate and bezafibrate in an animal model of diet-induced hypercholesterolemia. In this study female rats were fed a hypercholesterolemic diet (standard maintenance diet with the addition of 1% cholic acid and 2% cholesterol) for 4 weeks. After 2 weeks treatment commenced by including the test compounds into the diet (0.1%, corresponding to $120-150$ mg kg^{-1} day^{-1}, n = 10 per group if not otherwise stated in Fig. 2). After 4 weeks CHOL was determined; in some cases HDL-CHOL was measured additionally. In part A of the study the test compounds L 9 and L 43-0 were compared with etofibrate, in part B the NA derivatives L 43, L 44, and L 44-0 were compared with bezafibrate. Both parts were run in parallel. From the combined results outlined in Fig. 2 it is obvious that indeed, e.g., the superiority of L 44-0 over L 44 could be established. Moreover, with respect to CHOL, the effect of this N-oxide is even stronger than in the case of bezafibrate. Also, when only data derived from study B are statistically analysed, L 44-0 appears to be superior to bezafibrate (CHOL, % decrease rel. control: $-33, p < 0.01$ and $-18, p < 0.05$).

At least some of these favorable effects may be due to the long-lasting inhibition of lipolysis as it has been shown in an animal model of fast-induced hypermobilization of free fatty acids.

When we performed preclinical toxicological studies with some new NA derivatives, L 44 and L 44-0 appeared to be very well tolerated. Additionally, in a guinea pig model, we tested the potency of these drugs to induce ear reddening. Compared with NA, "flushing" was less severe. For these reasons we selected L 44 and L 44-0 for further development. Consequently, in three animal models of disease, we could show that either L 44 or L 44-0 are at least as effective as etofibrate: streptozotocin-induced hypertriglyceridemia (Puglisi et al. 1985), ethanol-induced hypertriglyceridemia (Puglisi et al. 1985) and NATH-diet induced hypercholesterolemia.

The results referred to and other not mentioned here led us to the conclusion that L 44 and its N-oxide might be useful as lipid-lowering drugs for the therapy of human hyperlipoproteinemias as well. Studies in humans have been initiated already, but results upon the hypolipidemic activity of these drugs are not yet available.

In preclinical work our interest now focuses on the anti-inflammatory properties of L44 and L44-0. Up to now we could show that L44-0 is active in two rat models: carrageen-induced paw edema and carrageen pellet-induced subcutaneous exudation. Of special interest is that in the exudative fluid the major prostanoid mediators of inflammation, i.e., $TX A_2$ and $PG E_2$ are substantially reduced after treatment with L44-0.

Due to the phenolic moiety within the compound we wondered whether lipoxygenase products like $LT C_4$ would also be affected. In Fig. 3 results derived from a study with peritoneal macrophages are presented.

Comparing these data with those reported for two other inhibitors, i.e., nordihydroguaiaretic acid (NDGA) and BW 775C (Brune et al. 1984), L44-0 at the concentration of 10^{-5} M appears to be equipotent. On the other hand, there has been seen no adverse effect with L44-0 in human volunteers so far, whereas it is reported that the two other agents are toxic (Brune et al. 1984).

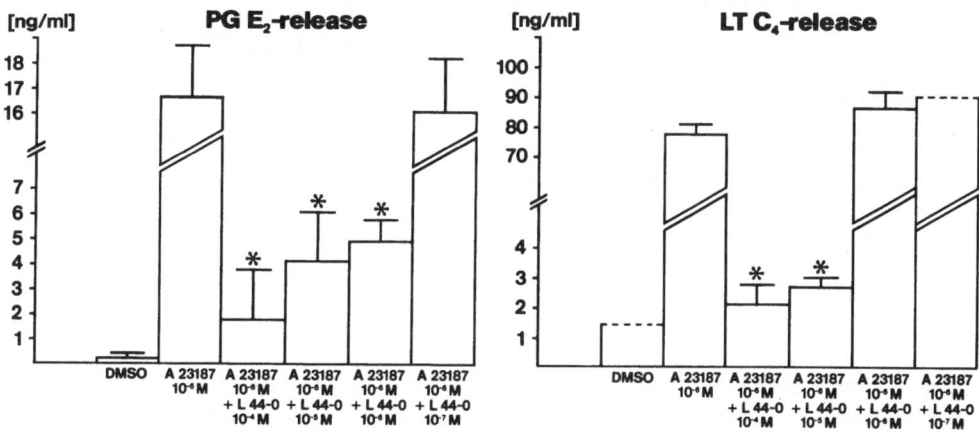

[1]BRUNE 1986, personal communication

Fig. 3. Effects of L44-0 on the A 23187-induced release of PG E$_2$ and LT C$_4$ from mouse peritoneal macrophages in vitro. Data are given as mean ± SD. (After Brune et al. 1984)

As both cyclooxygenase and lipoxygenase products are believed to be involved in the initiation and promotion of inflammation as well as atherosclerosis, phenolesters of NA such as L 44 and L 44-0 might be promising drugs in these respects, too.

References

Altschul R, Hoffer A, Stephen JD (1955) Influence of nicotinic acid on serum cholesterol in man. Arch Biochem 54:558–559

Brune K, Aehringhaus U, Peskar BA (1984) Pharmacological control of leukotriene and prostaglandin production from peritoneal macrophages. Agents Act 14:729–734

Hotz W (1983) Nicotinic acid and its derivatives: a short survey. Adv Lipid Res 20:195–217

Nath N, Wiener R, Harper AE, Elvehjem CA (1959) Diet and cholesteremia. I. Development of a diet for the study of nutritional factors affecting cholesteremia in the rat. J Nutrit 67:289–307

Puglisi L, Maggi FM, Caselli GF, Accomazzo MR, Saibene G (1985) Drugs affecting primary and secondary hypertriglyceridemia experimentally induced. In: Pozza G, Micossi P, Catapano A, Paoletti R (eds) Diet, diabetes and atherosclerosis. Raven, New York, pp 263–272

Standerfer SB, Handler P (1955) Fatty liver induced by orotic acid feeding. Proc Exp Biol Med 90:270–271

Subissi A, Murmann W (1978) Correlation between the plasma concentration of free nicotinic acid and some of its pharmacological effects in the fasted rat after an oral dose of sorbinicate and of nicotinic acid. Arzneimitt Forsch/Drug Res 28 (II):1143–1145

Pharmacokinetics and Metabolism of 2-Tert-Butyl-4-Cyclohexyl-Phenyl Nicotinate (L-44) in Rats

G. Galli[1], E. Bosisio[1], E. De Fabiani[1], M. Crestani[1], G. Quack[2], and W. Schatton[2]

This chapter is concerned with the results collected so far on the pharmacokinetics and metabolism of 2-tert-butyl-4-cyclohexyl nicotinate (L-44) (Fig. 1), a substance exhibiting hypolipidemic properties (Quack et al. 1986). For this study we utilized the compound labeled with [^{14}C] in the phenolic ring. In order to check the chemical and radiochemical purity of the labeled compound different chromatographic techniques, combined with radioactivity monitoring detectors, were employed.

L-44

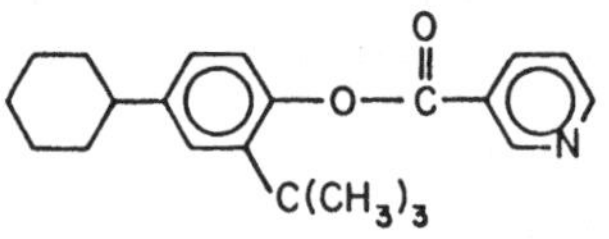

2-t-Butyl-4-cyclohexyl-phenyl nicotinate

L-44 p

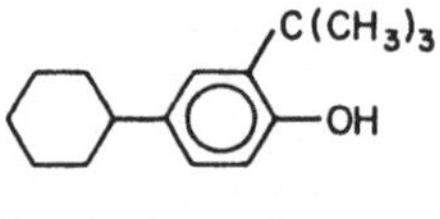

2-t-Butyl-4-cyclohexyl phenol

Fig. 1. Chemical structure of L-44 and L-44 phenol

Thin layer chromatography (TLC) radioactivity scanning (Berthold LB 2722) revealed that all the radioactivity was associated to the R_f of authentic L-44 and mass and radioactivity peaks in the traces obtained from radio-gas chromatography (Carlo Erba Model GV-Nuclear Chicago 4998 Flow Counter) and from radio-HPLC (Varian 5000 – Berthold LB 503) analysis matched perfectly.

Radioactive L-44 (specific radioactivity 26.5 μC mg^{-1}), dissolved in dimethylformamide, was administered to Sprague-Dawley Charles River rats (150–160 g) either by intravenous injection or by gastric intubation at the dosage of 1 mg kg^{-1} (body wt.).

1 Institute of Pharmacological Sciences, University of Milan, Milan, Italy
2 Merz & Co., Abteilung für Pharmakologie, D-6000 Frankfurt/Main, FRG

Drugs Affecting Lipid Metabolism
Ed. by R. Paoletti et al.
© Springer-Verlag Berlin Heidelberg 1987

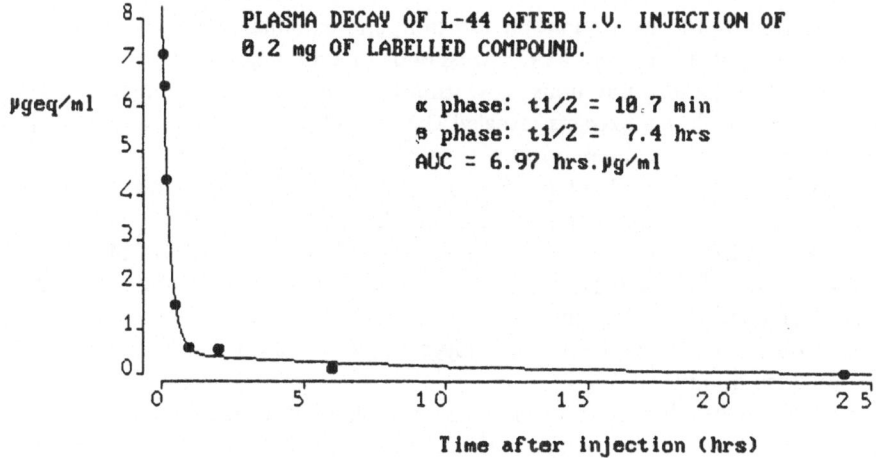

Fig. 2. Plasma decay curve after i.v. administration of [^{14}C]L-44

Blood withdrawn from the tail vein or collected after cervical dislocation at the different times after the administration was centrifuged and aliquots of plasma were used for radioactivity counting.

The plasma decay of L-44, expressed as μg Eq ml^{-1}, plotted against time after intravenous and oral administration, is shown in Figs. 2 and 3. The regression lines of the distribution (α) and of the elimination (β) phases of L-44 given intravenously were calculated according to a two-compartment open model. Half-lives of α and β phases are 10.7 min and 7.4 h, respectively, and the area under the curve (AUC) corresponds to 6.97 h μg Eq ml^{-1}.

The fitting of the curve related to plasma μg Eq ml^{-1} of L-44 after oral administration is consistent with the two-compartment model.

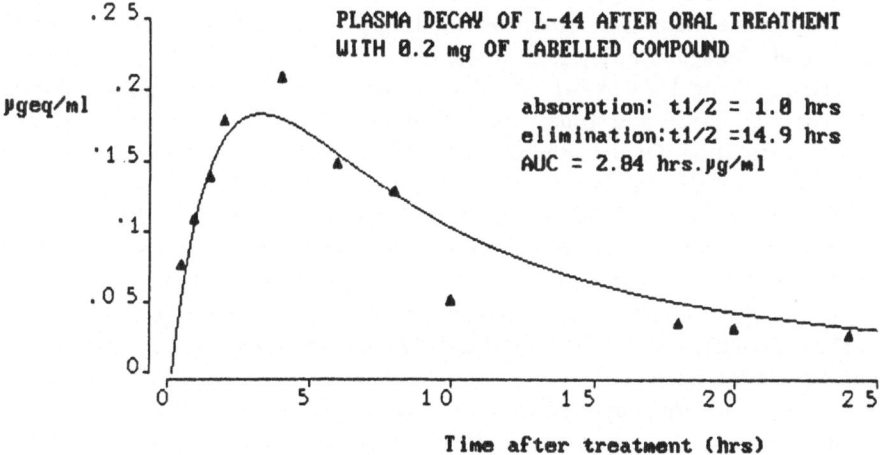

Fig. 3. Plasma decay curve after oral administration of [^{14}C]L-44

The half-lives of the absorption, distribution, and elimination phases are 1.0, 4.2, and 14.9 h, respectively, and the AUC corresponds to 2.84 h μg Eq ml^{-1}.

The ratio of oral and intravenous AUC values indicates a bioavailability of 40.8%.

TLC analysis of plasma extracts revealed that already after 2 min, following intravenous administration, all the radioactivity was completely associated to the R_f of 2-t-butyl-4-cyclohexylphenol (L-44 p) (Fig. 1), thus indicating that L-44 undergoes a very rapid hydrolysis at the ester bond. The identification of the phenolic metabolite, L-44 p, was confirmed by gas chromatography-mass spectrometry (GC-MS). A TLC radioactivity trace of extracts from plasma obtained 30 min after injection showed the appearance of a second peak at the origin of the plate. The amount of radioactivity associated to this peak increased with time, suggesting that L-44 p is further metabolized to more polar compounds.

In order to verify whether the drug behaves similarly in man, a pilot study in three human volunteers, who were given an oral dose of 1.3 g L-44 each, was performed.

The plasma concentration of L-44 and its metabolite L-44 p were measured using a GC-MS quantitative technique based on selected ion monitoring (SIM). As internal reference compounds 4-cyclohexyl-phenyl nicotinate and 4-cyclohexyl-phenol were utilized for the determination of L-44 and L-44 p, respectively.

The ratio of the intensities of the most prominent ions, arising from the compounds to be measured and from the reference substances, was plotted against the increasing amounts of the drug and its metabolite added to 1 ml plasma samples of untreated subjects. The calibration curves were linear ($r^2 > 0.995$) for both L-44 and L-44 p throughout the range tested (10–1000 ng).

The parent drug was not detected in the plasma of the volunteers. Moreover, L-44 p could be determined only in two of three subjects, indicating that also in humans the drug disappears rapidly from the systemic circulation and the metabolite production rate probably depends on many factors.

From the analysis of the plasma decay curve of L-44 p it was calculated that in one subject the $t_{1/2 \beta}$ was 1.8 h and maximal concentration of L-44 p was 14 ng ml^{-1}; in the other subject $t_{1/2 \beta}$ was 2.6 h and maximal concentration was 4 ng ml^{-1}.

To investigate the metabolic pattern of the drug [^{14}C] L-44 was administered orally to fasted male Sprague-Dawley rats, at the dose of 100 mg kg^{-1} dissolved in dimethylformamide (10 mg L-44 in 0.1 ml solvent). Animals were placed individually in metabolic cages and feces and urine were collected for 24 h, then combined and extracted with acetonitrile. Of the administered radioactivity, 75% was recovered in the feces extract and only 2% was found in urine extracts.

After evaporation, most of the residue of the feces extract (95.5% of the recovered radioactivity) was suspended in petroleum ether and partitioned on Bio-Sil A column chromatography. Seven fractions were eluted using solvent systems of increasing polarity. The qualitative composition of fractions was determined according to TLC, HPLC, and GC. The radioactivity profiles, GC, and GC-MS analysis were performed after treatment with a silylating agent.

Most of the parent drug appeared in fractions 1 and 2 (54.3% of the administered radioactivity), whereas the majority of L-44 p was eluted in the first fraction (12%). Fraction 6 (2.8%), 7 (10.6%), and the unchromatographed residue (6%) of the feces extract all contained variable amounts of L-44, L-44 p, and other metabolites. The appearance of the drug and its metabolites in different fractions is due to the heterogeneous suspension applied to the column.

Fractions 6, 7 and the unchromatographed residue were further subjected to preparative HPLC purification. Radioactive peaks were collected and aliquots of the fractions were analyzed by radio-GC and GC-MS after silylation reaction. Mass spectra of the radioactive peaks separated by gas chromatography indicate that in addition to L-44 p, two metabolites (compounds I and II), among others, are present in the feces extract in appreciable amounts. The molecular ion of the silylated compound I at m/z 392 (Fig. 4, upper panel) indicates that an additional hydroxyl group is introduced in the molecule of L-44 p. The presence of the ion at m/z 349 suggests that the hydroxyl group is probably located in the benzylic position.

The molecular weight (M^+ = 480) and the fragmentation pattern of compound II after silylation (Fig. 4, lower panel) is compatible with a metabolite of L-44 p containing two additional hydroxyl groups. Also, in this case the presence of an ion corresponding to the loss of 43 mass units from the molecular ion (m/z 437) further suggests that at least one of the hydroxyl groups is located in the benzylic position.

Table 1 summarizes the results reported up to now. The parent drug is partially eliminated without alteration in the feces (47% of the dose). L-44 p is further metabolized

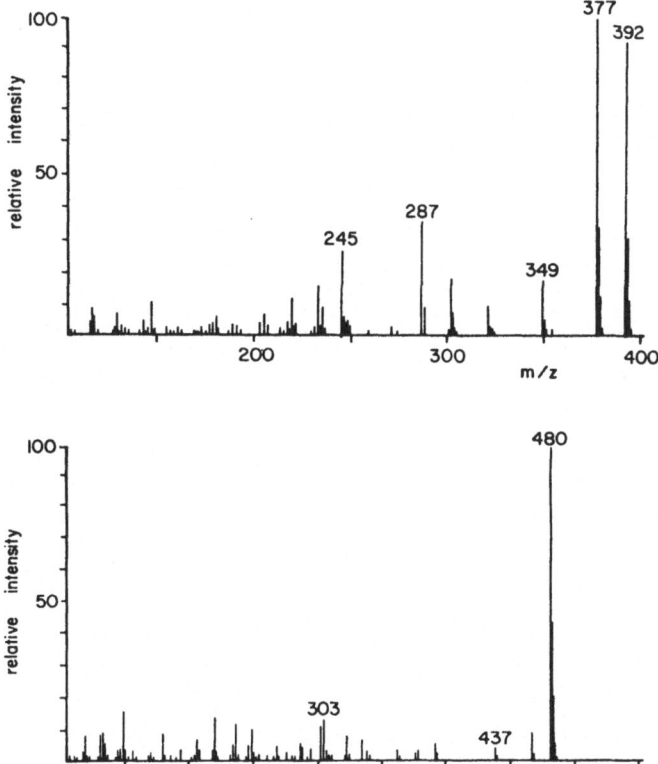

Fig. 4. Mass spectra of compound I (*upper panel*) and compound II (*lower panel*)

Table 1. Fecal radioactivity distribution after oral administration of L-44

Compound	The administered radioactivity (%)
L-44	46.9
L-44	15.8
Metabolite I	5.6
Metabolite II	2.9
Other metabolites	3.1

through subsequential hydroxylations. All combined, these metabolites account for 10–12% of the given drug. Work is in progress to determine whether the formation of hydroxylated metabolites of L-44 p is of hepatic or intestinal origin.

From the analysis of urines and bile, a novel series of metabolites was evidenced. These compounds possessed higher polarity with respect to those described previously and their presence appeared in blood samples collected 30 min after the administration. The structure of these metabolites is presently under investigation.

Reference

Quack G, Puglisi L, Schatton W, Schwaier A (1986) Evaluation of the hypolipidemic properties of various nicotinic acid derivatives in the rat. Abstr Book IXth Int Symp Drugs affecting lipid metabolism, p 67

Effects of Etofibrate on Risk Factors of Atherosclerosis

T. Seifert[1] and W. Schatton[2]

1 Hyperlipoproteinemia

For some time now the concept of the multifactorial genesis of atherosclerosis has been well accepted. Originally, it was based on the results of large epidemiological studies, which showed a powerful correlation between certain habits of life or metabolic or pathophysiological states, on the one hand, and the probability of contracting an atherosclerotic disease, on the other. The parameters of prognostic value were called risk factors. The 18-year follow-up of the Framingham study confirmed, among other factors, that serum cholesterol is a risk factor in the development of cardiovascular disease. In addition, different risk factors, when present at the same time, potentiate each other to an enormous risk. The coronary risk increases overproportionately with the blood cholesterol level, growing to a life-threatening danger in patients with familial homozygous hypercholesterolemia. An evaluation of the data of the Pooling Project by means of the Framingham HDL-risk-multipliers shows the incidence of fatal or nonfatal myocardial infarction to decrease in men at any concentration of total cholesterol by a factor of 2.5 as HDL-cholesterol increases from less than 40 mg dl^{-1} to more than 50 mg dl^{-1}. For this reason the atherogenic index LDL-cholesterol/HDL-cholesterol beats the concentration of total cholesterol in estimating the coronary risk. Hyperlipoproteinemia is not only correlated with coronary vascular disease, but also with peripheral arterial acclusion disease (PAOD). In a group of patients with symptomatic PAOD, hyperlipoproteinemia was found in 76% of the patients, which was twice as much as in subjects without PAOD. This difference was mainly due to higher levels of total and VLDL-bound triglycerides in the PAOD group (Diehm 1984). There is convincing experimental evidence that cholesterol-rich LDL is taken up by vascular smooth muscle cells, causing accumulation of cholesterol esters in the vessel wall and that HDL particles are able to transport cholesterol back from peripheral cells to the liver. But physicians are primarily interested in whether the development of atherosclerosis can be delayed or even prevented by lowering the blood cholesterol levels. Numerous large intervention trials showed indeed that primary and secondary prevention of coronary heart disease (CHD) is possible by lowering the blood cholesterol. The LRC-CPP trial was

---

1   Merz & Co., Abteilung für Klinische Forschung, D-6000 Frankfurt/Main, FRG
2   Merz & Co., Abteilung für Pharmakologie, D-6000 Frankfurt/Main, FRG

Drugs Affecting Lipid Metabolism
Ed. by R. Paoletti et al.
© Springer-Verlag Berlin Heidelberg 1987

Table 1. Changes of blood lipid and HDL concentrations under etofibrate

	N	Before treatment	After treatment	Δ (%)
Cholesterol (mg dl^{-1})	300	342	266	−22.1
Triglycerides (mg dl^{-1})	300	386	244	−36.7
HDL (mg dl^{-1})	131	40.6	48.4	+19.2

500 mg day^{-1}; 6 weeks.

carried out on 3806 hyperlipoproteinemic patients who at the beginning did not ex-hibit any symptoms of an atherosclerotic disease. In the cholestyramine-treated group there were 24% less definite fatal myocardial infarctions and 19% less definite nonfatal myocardial infarctions. Total mortality decreased by 15% if deaths from accidents and violence are excluded. In the Stockholm IHD study 555 survivors of myocardial infarc-tion were treated with a combination of clofibrate and nicotinic acid. The drug-treated group profited with 26% less total mortality and 36% less fatal myocardial infarction. Admittedly the rate of reinfarction was not reduced in this study.

In a placebo-controlled, double-blind comparison etofibrate, at a dosage of 300 mg three times a day, reduced the cholesterol concentration by 26% within 4 weeks. The initial mean value of 293 mg dl^{-1} declined to an ideal value of 220 mg dl^{-1}. At the same time, triglycerides decreased by 48%, reaching a normal level as well. The placebo did not cause any changes of these blood lipids (Altomonte et al. 1981). Similarly, favorable results can be obtained on a large scale by general practitioners. An average reduction of cholesterol by 22% and of triglycerides by 37% was achieved in 300 pa-tients with a daily dose of 500 mg etofibrate (Table 1).

During the 6-week treatment, the HDL-cholesterol rose by 19% in the evaluated subgroup of 131 patients (Schatton and Holm 1986). Within the scope of the results of the previously described intervention trials, the extent of lipid lowering under eto-fibrate is of great clinical relevance.

2 Platelet Function

Another risk factor which should be taken seriously seems to be platelet aggregability (Longenecker 1985). Ex vivo hyperreactivity as well as elevated in vivo activity of platelets have been shown in patients with ischemic heart disease, transient ischemic attacks, and peripheral arterial occlusion disease. The above mentioned risk factors cause damage of the endothelium, covering the inner side of the vessel wall. Platelets stick to these thrombogenic surfaces and thus become activated. Some risk factors, such as smoking and LDL particles, are capable of directly enhancing platelet aggre-gability. Finally, aggregating platelets further activate platelet activity. In contrast platelets, attached to the arterial vessel wall, contribute to the growth of atherosclerotic plaques at sites of endothelial lesions. They release numerous substances with chemo-tactic, mitogenic, vasoconstrictive, and proaggregatory activity. Particularly by specific binding of the platelet-derived growth factor (PDGF), arterial smooth muscle cells migrate into the intima, where they change into a proliferative state. At the same time, thrombus formation is facilitated by the vasoconstrictive action of thromboxane A_2 (TXA$_2$) and the proaggregatory effect of TXA$_2$ and ADP, deliberated from aggregating platelets. First attempts to inhibit plaque formation by antiplatelet drugs have been successful in animals and patients with intermittent claudication (Hess et al. 1985). Reduced incidence of thrombotic vessel occlusions has been obtained in patients with coronary heart disease, transient ischemic attacks, and after vascular surgery. Etofibrate exhibits a marked antiplatelet activity. This activity must be regarded as an antiatherogenic one.

3 Fibrinolytic Activity

Both plaque growth and thrombus formation are limited by fibrinolysis. For this reason reduced fibrinolytic activity can be regarded as another risk factor of atherosclerosis. If the fibrinolytic activity is diminished, as is the case in patients with atherosclerotic risk factors (Heinrich et al. 1983), fibrin withdrawal from plaques and thrombi is reduced, but fibrin incorporation is enhanced due to the increasing fibrinogen concentration. Consequences of the elevated plasma fibrinogen concentration are increased plasma viscosity and aggregation of erythrocytes, thus reducing blood fluidity and hence · blood flow. The lowered supply of oxygen and nutrients favors the formation of necrotic plaques. Up to now the causal role of reduced fibrinolytic activity for the development of vessel occlusions and symptomatic ischemic vascular disease is a hypothesis which is, however, supported by some clinical observations.

Plasma viscosity, erythrocyte aggregation, and blood viscosity were increased in patients with at least 75% stenosis of the carotid artery (Ernst 1986). A similar relation was found in patients with coronary stenosis and chest pain. Fibrinogen concentration, plasma viscosity, packed cell volume, and blood viscosity clearly increased with the extent of stenosis (Lowe et al. 1980). One intervention study showed a parallel decrease of the plasma fibrinogen concentration and improved cerebral blood flow under therapy with clofibrate.

A consistent improvement of fibrinolytic activity (measured via euglobulin lysis time, Dembinska-Kiec, personal communication), fibrinogen concentration (Spöttl and Froschauer 1979), and plasma viscosity (Pfeiffer and Tilsner 1978) was also achieved by etofibrate (Table 2).

Table 2. Effect of etofibrate on fibrinolytic activity, plasma fibrinogen concentration, and plasma viscosity

	N	Before treatment	After treatment	Δ (%)
Euglobulin lysis time (s)[a]	13	218	163	−25.2
Fibrinogen (mg dl^{-1})[b]	25	596	391	−34.4
Plasma viscosity (cP)[c]	17	2.01	1.68	−16.4

[a] 2 x 500 mg day^{-1} ; 4 weeks.
[b] 3 x 300 mg day^{-1} ; 6 months.
[c] 3 x 300 mg day^{-1} ; 6 weeks.

Table 3. Effects of etofibrate on maximal pain-free walking distance and time of pain relief in patients with arteriosclerosis obliterans

	N	Before treatment	After treatment	Δ (%)
Maximal pain-free walking distance (m)	16	286	412	+44
Time of pain relief (s)	16	93	85	− 9

2 x 500 mg day^{-1} ; 4 weeks.

As already mentioned atherosclerosis is believed to be initiated and promoted by different risk factors. Therefore, it is quite reasonable and desirable to interfere with the development of atherosclerosis at different sites. First results show this concept to be promising.

Plaque formation, artificially induced by electrical stimulation, and an atherogenic diet in rabbits could be inhibited by etofibrate (Betz and Hämmerle 1986). Furthermore, etofibrate was given to patients with symptomatic PAOD in a small pilot study. The maximal pain-free walking distance increased by 44% within 4 weeks and the time of pain relief showed some improvement (Dembinska-Kiec, pers. commun.; Table 3).

Thus, in our opinion, etofibrate is not "only" a lipid-lowering agent, but can also be regarded as a drug with antiatherogenic activity.

References

Altomonte L, Mingrone G, Negrini A, De Cunto F, Greco AV (1981) Studio a doppio cieco sull'attivita terapeutica di un nuovo farmaco ipolipemizzante: L'etofibrato. Clin Ther 96:31−38

Betz E, Hämmerle H (1986) Effect of etofibrate and its metabolites on atheromas of rabbits and on smooth muscle cell cultures. Drug Res 36:92−98

Diehm C (1984) Carbohydrate-metabolism and fat-metabolism in normal persons and patients with peripheral arterial occlusive disease − the consequences of strenuous exercise. VASA J Vasc Dis (Suppl 13):1−61

Ernst E (1986) Hämorheologie und zerebrale Insuffizienz. Therapiewoche 36:2514−2526

Heinrich D, Thilo-Körner DGS, Roka L (1983) Die Bedeutung des Gefäßendothels für die Regulation der Fibrinolyse. Drug Res 33:1375−1378

Hess H, Mietaschik A, Deichsel G (1985) Drug-induced inhibition of platelet function delays progression of peripheral occlusive arterial disease. Lancet I:415−419

Longenecker GL (1985) The platelets. Physiology and pharmacology. Academic Press, London New York Orlando

Lowe GDO, Drummond MM, Lorimer AR, Hutton I, Forbes CD, Prentice CRM, Barbenel JC (1980) Relation between extent of coronary artery disease and blood viscosity. Br Med J 673:674

Pfeiffer M, Tilsner V (1978) Der Einfluß von Etofibrat auf die Plasmaviskosität bei Hyperlipoproteinämien. Med Klin 73:60−62

Schatton W, Holm E (1986) Etofibrat bei schweren diätrefraktären Fällen von Hyperlipoproteinämie. Fortschr Med 13:281−282

Spöttl F, Forschauer J (1976) Influence of etofibrate on plasma fibrinogen and plasminogen concentrations in patients with different forms of primary hyperlipoproteinemia. Atherosclerosis 25:293−301

Hyperaggregability of Platelets in Patients with Hyperlipoproteinemia Type IIb Under Treatment with Etofibrate Retard and Acetylsalicylic Acid

B. Krüger and R. Thiemann [1]

1 Introduction

Heightened platelet function is suggested to be associated with the well-known thrombotic complications and the accelerated atherogenesis of type II hyperlipoproteinemia (Carvalho et al. 1974). With regard to the two main theories on the pathogenesis of atherosclerosis, the hypothesis of damaged endothelium and the lipid hypothesis, one has to distinguish between cause and effect. On the one side, it is not imperative that endothelial damage by rheological, metabolic, traumatic, or inflammatory causes results in atherosclerotic plaques. On the other hand, one can expect this with a much higher probability if such endothelial lesions occur in patients with hypercholesterolemia. The pathogenesis of atherosclerosis is founded upon many facts and atherosclerosis is the multifactorial result of parts of the lipid hypothesis as well as of the hypothesis of damaged endothelium.

There are five points of prophylactic and therapeutic importance with regards to atherosclerosis:

1. Reduction of mechanical, toxic, or inflammatory irritations.
2. Inhibition of platelet aggregation.
3. Inhibition of platelet-derived growth factor (PDGF).
4. Decrease of high LDL-cholesterol levels.
5. Increase of HDL-cholesterol.

The first point leads to a reduction of risk parameters like nicotin, which e.g. lowers the endothelial prostacyclin (PGI2) production, but not the production of thromboxane (TXA2) in the platelets (Hawkins 1972; Wennmalm 1982). Another risk parameter, essential hypertension, is connected with reduced production of PGI2 and prostaglandin E2 (PGE2) (Weber et al. 1979); patients with hypercholesterolemia show an increased production of TXA2 (Hirsch et al. 1981), all resulting in hyperaggregability of the platelets.

In our studies (Krüger et al. 1983) we could show that there is a connection between lipid-lowering effects in hyperaggregating patients with hyperlipidemia and normalization of the aggregability of platelets. To test the aggregability of platelets we take a

1 Medizinische Klinik der Universität Erlangen-Nürnberg, Krankenhausstraße 12, D-8520 Erlangen, FRG

Drugs Affecting Lipid Metabolism
Ed. by R. Paoletti et al.
© Springer-Verlag Berlin Heidelberg 1987

0.5 µg collagen stimulus added to 1.0 ml platelet-rich plasma (PRP). This stimulus leads to a differential aggregation of the platelets and analogously to a differential increase of light transmission through PRP, representing the actual status of aggregability. We discussed such changes of aggregability under treatment of hyperlipoproteinemia type II and assumed a causal change in the prostanoid system, especially of the arachidonate metabolites PGI2 and TXA2. To scrutinize this assumption we realized a study with patients with hypercholesterolemia and hyperaggregability with a view of the stable metabolites 6-keto-prostaglandin-F-lalpha (6k-PGF-la) and thromboxane B2 (TXB2) of the short-life arachidonate metabolites PGI2 and TXA2 under treatment with either etofibrate retard (E) or acetylsalicylic acid (ASS) as standard of a maximal antiaggregating effect.

2 Patients and Methods

Thirty outpatients with hyperlipidemia type IIb and concomitant hyperaggregability of the platelets [maximal amplitude (Ma) of the platelet aggregation test (PAT) of more than 50 mm, induction of aggregation of platelets by 0.5 µg collagen (Collagen-Suspension Horm)] were treated 4 weeks by diet and with placebo, then a further 12 weeks with either 500 mg E or 500 mg ASS daily. Controls of the plasmatic values of 6k-PGF-1a and TXB2 and of the PAT and seral values of the total cholesterol and its HDL, LDL, and VLDL fraction and of the triglycerides were made before, after 4 weeks of diet and placebo, and after 4, 8, and 12 weeks of treatment. Statistical significances are based on the Student's t-test for paired and unpaired parameters.

3 Results

The values of total cholesterol (Fig. 1) showed no difference between the two groups during the pretreatment period. In the first 4 weeks of treatment the total cholesterol values showed a statistically significant decrease under E ($p < 0.001$), but no change under ASS. This difference between the two groups remained constant over the whole period of treatment ($p < 0.001$). Accordingly, the LDL-cholesterol fraction showed a significant decrease under E ($p < 0.001$) and no change under ASS. The HDL- and the VLDL-cholesterol fractions showed in both groups no significant change over the whole 16-week period.

---→

Fig. 1A–D. 30 Patients with hyperlipidemia IIb and hyperaggregability of the platelets: 4 weeks of only diet, then further 12 weeks of diet and either 500 mg Etofibrate ret. or 500 mg Acetylsalicylic acid (ASS): A Total cholesterol in the Etofibrate ret. group (n = 15) and in the ASS group (n = 15). B Platelets aggregability under Etofibrate ret. (n = 15) or ASS (n = 15). C 6-keto-Prostaglandin-$F_{1-alpha}$, the stable metabolite of Prostacyclin, under Etofibrate ret. (n = 15) or ASS (n = 15). D Thromboxane B_2, the stable metabolite of Thromboxane A_2, under Etofibrate ret. (n = 15) or ASS (n = 15)

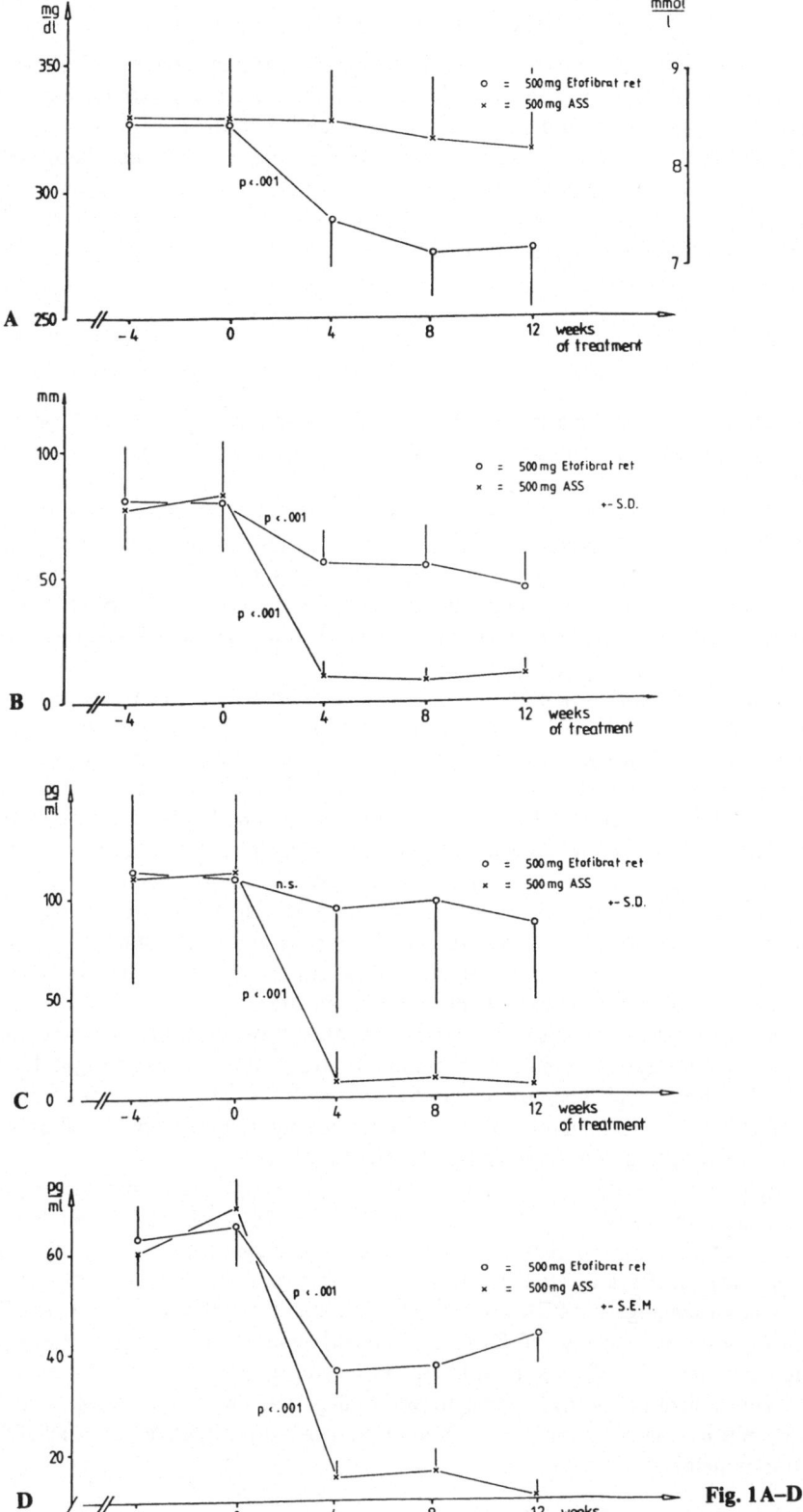

Fig. 1A–D

The maximal amplitude (Ma) of the PAT showed no difference between both groups in the pretreatment period. Then, a total inhibition of platelet aggregation was seen in the ASS group ($p < 0.001$) and a decrease of aggregability in the E group ($p < 0.001$).

The plasmatic 6k-PGF-1a showed no significant change under E and a drop nearly to zero under ASS ($p < 0.001$). The TXB2 values were similar under ASS and showed a severe decrease under E ($p < 0.001$). The corresponding ratio of 6k-PGF-1a and TXB2 increased under E and decreased under ASS.

4 Discussion

Our studies over 2 and 3 weeks treatment of hyperlipoproteinemia type II with etofibrate retard led to the assumption that there is not only a lipid-lowering effect of the substance, but also an antiaggregating effect on the hyperaggregating platelets (Krüger et al. 1983). Our data showed a wide dispersion and only tendencies according to the aggregability values. The studies of Augustin (1984) illuminated why: The 3-month period of treatment in his study showed that the first 4 weeks must be seen as an induction period. All parameters taken before such a 4-week period of treatment with etofibrate retard are interindividually widely spread, but came then to one point. The ways do not seem to be equal, but lead to the same result.

In our study with a 4-week pretreatment period and then a 12-week treatment period, it could be seen that the only dietary measures of the preperiod show no distinct effect either on the cholesterol or on the prostanoid patterns. We expected this, because all patients were well-known hyperlipidemics, and accordingly we have in Germany a nearly calculable deficiency of outpatient treatment of hyperlipidemias. Such patients are clinically trained in low-cholesterol diets, but many physicians seem to deny the necessity of further treatment by drugs, if such patients show no satisfying lowered cholesterol levels under diet. None of the patients had medical lipid-lowering or antiaggregating pretreatment for at least the last 3 months before the start of the study. We can assume that for all patients in this outpatient study, their lipids, prostanoids, and platelet-aggregation systems were in a state of balance.

Etofibrate retard showed a severe decrease of total cholesterol, LDL-cholesterol, triglycerides, platelet aggregation, and thromboxane B2 formation, and a slight decrease of the VLDL-cholesterol. The ratio of 6k-PGF-1a and TXB2 increased under etofibrate retard. ASS led to a maximal inhibition of platelet aggregation and TXB2 and 6-k-PGF-1a formation, resulting in a lower TXB2-6k-PGF-1a ratio.

The lipid-lowering effects, i.e., the normalization of aggregation parameters, must be seen as antiatherosclerotic effects. In addition, the TXB2, the stable metabolite of TXA2, one of the most potent vasoconstrictive and platelet-aggregating substances, decreases under etofibrate retard treatment and there is no influence seen on the vasodilative and antiaggregating PGI2, measured by its stable metabolite 6k-PGF-1a. These differences between decrease of TXA2 and no influence on PGI2 explain the antiaggregation potency of etofibrate retard. The antiaggregating potency of ofibrate retard leads to a normalization of the hyperaggregability in patients with hyperlipoproteinemia type IIb only by lowering the TXA2 formation. The vasoprotective endothelial PGI2 formation remains unaffected.

References

Augustin J (1984) Die Lipidhypothese. Herz Gefäße 4:201–202

Carvalho A, Colman R, Lees R (1974) Platelet function in hyperlipoproteinemia. N Engl J Med 290:434–438

Hawkins RI (1972) Smoking, platelets and thrombosis. Nature (London) 236:450–455

Hirsch PD, Campbell WB, Willerson JT, Hillis LD (1981) Prostaglandins and ischemic heart disease. Am J Med 71:1009–1014

Krüger B, Sörgel F, Kirchberg HG, Pällmann U, Geldmacher von Malinckrodt M, Lang E (1983) Einfluß von Etofibrat auf die Aggregation der Blutplättchen. Therapiewoche 33:3297–3299

Weber PC, Siess W, Scherer B (1979) Vaskuläre, thrombozytäre und renale Prostaglandine. Biochemie, Funktion und klinische Aspekte. Klin Wochenschr 57:425–435

Wennmalm A (1982) Interaction of nicotine and prostaglandins in the cardiovascular system. Prostaglandins 23:139–144

Effect of Etofibrate on Platelet Function in Hyperlipidemic Patients

G. Noseda[1], C. Fragiacomo[1], A. Radelli[1], E. Bosisio[2], E. De Fabiani[2], and G. Galli[2]

1 Introduction

Current theory holds that atherosclerotic plaque formation results from injuries which damage the histological and functional integrity of the endothelium (for review see e.g. Miller 1984). As causative factors low density lipoproteins (LDL), hemodynamic shearing at bifurcations of the arterial system, nicotine, and others have been identified in endothelial damage. Once the endothelial lining is disrupted, plasma components, especially LDL, are deposited in the denuded area followed by migration of smooth muscle cells into the subendothelial space.

Simultaneously, circulating platelets adhere and aggregate to seal off the injury, releasing thromboxane A 2 (Tx A 2) which causes further platelet clotting, vasoconstriction and other factors like the platelet-derived growth factor (PDGF), which stimulates the smooth muscle cells to proliferate into the injured site.

Hypolipidemic agents have been shown to counteract one step of this process in man by reducing LDL-cholesterol. However, there are only limited clinical data available about the effects of lipid-lowering agents on the arachidonic acid metabolism with special regard to platelets.

Etofibrate (ETO) is a well-known hypolipidemic agent with proven efficacy in various types of hyperlipoproteinemia. Furthermore, ETO is the only one with a favorable influence on arachidonic acid metabolism and platelet aggregation.

In animal models Puglisi et al. (1984) was the first to find that ETO favorably influenced the balance between Tx A 2, generated by platelets and prostacyclin I 2 (PG I 2) generated in the arterial vessels concomitant with its hypolipidemic activity. The serum concentration of Tx B 2, the metabolite of Tx A 2, was significantly reduced in rats treated with ETO. At the same time, the drug was able to stimulate the formation of antiaggregatory PG I 2 in blood vessels, measured as its metabolite 6-keto PG F 1α.

In type IIa patients ETO increased the PG I 2/Tx A 2 ratio to about 150% of pretreatment values in blood significantly (Wagner et al. 1984).

Furthermore, ETO reduced platelet aggregation in patients with enhanced platelet aggregability and cholesterol levels above 300 mg dl^{-1} (Krüger and Lang 1984). This

---

1 Ospedale Beata Vergine, CH-6850 Mendrisio, Switzerland
2 Institute of Pharmacological Sciences, University of Milan, Milan, Italy

Drugs Affecting Lipid Metabolism
Ed. by R. Paoletti et al.
© Springer-Verlag Berlin Heidelberg 1987

effect was attributed to ETO and its main metabolites, the 1,2-ethandiolmonoesters of clofibric and nicotinic acid (NA).

As in these studies the influence of ETO on either platelet aggregation or serum Tx A 2 has been measured, there is only indirect evidence that this drug inhibits platelet function in man by interfering with Tx A 2 biosynthesis or release. Therefore, we decided to investigate the effect of ETO on platelet aggregation and Tx A 2 simultaneously.

Since the transformations of arachidonic acid in platelets also include a lipoxygenase catalyzed reaction forming 12-hydroperoxyeicosatetraenoic acid (12-HPETE), which is reduced to 12-hydroxyeicosatetraenoic acid (12-HETE) with chemotactic activity toward leukocytes, we decided to measure the effect of ETO on 12-HETE formation as well.

2 Patients, Study Design, and Evaluated Parameters

In an open, randomized pilot study one group of patients received ETO (Lipo-Merz retard R), 500 mg b.i.d.; for comparison, an almost equimolar dosage with respect to NA (500 mg t.i.d.) of Xanthinol nicotinat (XAN, Complamin retard R), was given to another group of patients. Both drugs are NA derivatives, but XAN releases nicotinic· acid instantaneously, whereas ETO can be considered as a slow release form of NA liberating NA by stepwise hydrolysis via the mentioned 1,2-ethandiolmonoesters. Eleven patients (3 with type IIa, 3 with type IIb, and 5 with type IV; 8 men and 3 women; mean age 58 years) received ETO. Seven patients (1 with type IIa, 4 with type IIb, and 2 with type IV; 3 men and 4 women; mean age 65 years) were treated with XAN. After a period of controlled diet of 4 weeks the patients received therapy with XAN or ETO for 3 weeks each. The laboratory investigations included the determination of lipids (total serum cholesterol and triglycerides), determination of the platelet aggregation [in platelet-rich plasma (PRP) after stimulation with three different inducers: ADP, adrenaline, and collagen] and measurement of the concentration of Tx B 2 and 12-HETE (determined in PRP after stimulation with collagen).

3 Results

After the 3-week treatment periods the following data could be obtained:

Lipid Parameters. In the patients with initial values of total cholesterol $\geqslant 6.8$ mM $(= 260$ mg dl$^{-1})$ ETO lowered total cholesterol by approximately 12% (n = 6), whereas XAN did not induce any change (n = 5).

In the patients with initial values of $\geqslant 2.28$ mM $(= 200$ mg dl$^{-1})$ ETO lowered serum triglycerides by 32% (n = 4). In contrast, XAN increased the triglyceride levels by 5% (n = 4).

Due to the small number of cases, no statistical analysis of the lipid data was performed.

Platelet Aggregation. Using adrenaline a slight increase of the aggregation was observed in patients treated with XAN (n = 7), whereas a slight reduction was seen in the patients receiving ETO (n = 10).

The platelet aggregation stimulated with collagen was not modified under treatment with XAN (n = 7) and slightly decreased in the subjects taking ETO (n = 11).

In the case of induction of platelet aggregation with ADP, XAN did not decrease the percentage of platelet aggregation (n = 7); ETO, in contrast, diminished the aggregation (n = 11).

Tx B 2 Concentration. XAN did not change the formation of Tx B 2 (n = 7), whereas ETO lowered the formation by 15% (n = 10).

12-HETE Concentration. No change was seen, either with XAN (n = 7) or with ETO (n = 10), in the concentration of 12-HETE in PRP after collagen stimulation. In Figs. 1 and 2 the individual data for the 12-HETE and Tx B2 concentration, respectively, in PRP after stimulation with collagen are shown (concerning the prominent lines, cf. Discussion).

In summary, XAN did not show any effect on lipids, platelet aggregation, and arachidonic acid metabolites. On the other hand, ETO reduced total cholesterol and triglycerides with a concomitant decrease of platelet aggregation after ADP stimulation and a reduction of the biosynthesis release of Tx A 2 in platelets during aggregation induced by collagen.

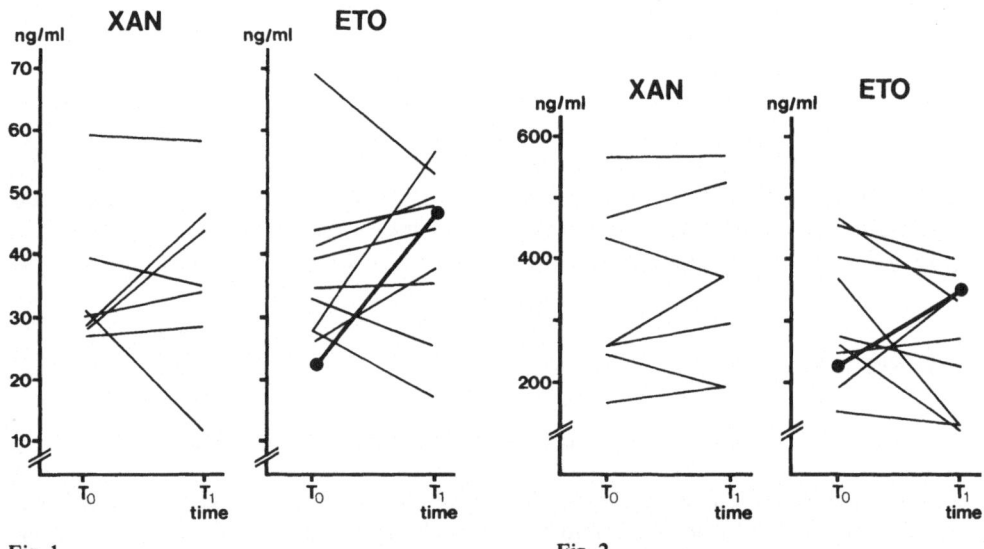

Fig. 1   Fig. 2

Fig. 1. Concentration of 12-HETE after stimulation of PRP with collagen before (T_0) and after 3 weeks of treatment (T_1)

Fig. 2. Concentration of TXB_2 after stimulation of PRP with collagen before (T_0) and after 3 weeks of treatment (T_1)

4 Discussion

With regard to the platelet aggregation and the Tx B 2 formation, our results did not reach statistical significance. This is mainly due to the limited number of patients, the inhomogeneity of the type of hyperlipoproteinemia, and the short duration of treatment.

Furthermore, after analyzing the data we had to face several problems which we would like to mention briefly as they seem to us of great importance for planning further studies.

1. At the end of the study we routinely analyzed serum samples for compliance control which has been only possible for ETO. In one case we found no trace of the hydrolysis product of ETO after the 3-week treatment period. This patient showed a strong increase in both 12-HETE (prominent line in Fig. 1) and Tx B 2 (prominent line in Fig. 2) when comparing the data of day 0 and day 21. This effect was much stronger than seen with the patients who maintained their medication. Therefore, we think that one should be cautious in relating changes of release product concentration only to the medication. Also, influences of the diet or other environmental influences have to be considered as well.

2. In this study one patient received the two medications one after the other with the 4-week wash-out period between the two treatment phases. This patient had greatly differing Tx B 2 concentrations at the two time points T 0, although no medication had been given (Fig. 3, prominent lines).

3. In several cases there were no real aggregation observable on day 0. As such a depression of aggregation is mostly seen with platelets derived from patients who are under medication with inhibitors of platelet aggregation, it is possible that some patients had taken, e.g., aspirin, for curing their headache. Other influences, e.g., diet, cannot be ruled out. Under such conditions it is obvious that the effects of the two lipid-lowering agents cannot be correctly evaluated. Taken together, on this basis it is difficult to give a quantitative estimate of drug effects. Therefore, we think that we should have more information on nonmedication-related influences on platelet meta-

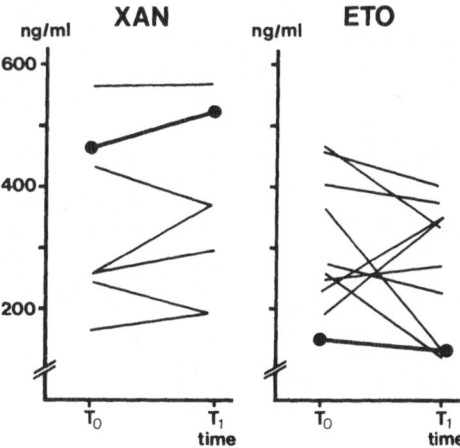

Fig. 3. Concentration of TXB_2 after stimulation of PRP with collagen before (T_0) and after 3 weeks of treatment (T_1)

bolism in normolipemic volunteers and patients with various types of hyperlipoproteinemias. Thus, dietary habits, physical activity, and stress all effect the activity state of platelets and should be controlled when investigating drug effects.

It is under consideration to conduct a more basic study on the diurnal variations of Tx B 2 and 12-HETE during the day and from one day to the other under well-defined environmental and clinical conditions. Only when it can be established that under such conditions fluctuations are minimal will we feel encouraged to do further studies with lipid-lowering agents.

Nevertheless, the data derived from this study and in consideration of the data already published, we are convinced that in patients with increased platelet sensitivity, treatment with ETO will result in a reduction of platelet aggregation via its influence on the thromboxane pathway.

References

Krüger B, Lang E (1984) Effects of two different doses of etofibrate on platelet aggregation and lipoproteins. Eur J Clin Invest 14(2), pt II, Abstr 46

Miller NE (1984) Atherosclerosis: mechanisms and approaches to therapy. Raven, New York

Puglisi L, Maggi F, Accomazzo MR, Schatton W (1984) The antiatherosclerotic property of etofibrate. Eur J Clin Invest 14(2), pt II, Abstr 44

Wagner HA, Cremer P, Creutzfeldt C, Seidel D (1984) Influence of etofibrate on prostaglandins and lipoproteins in 11 hyperlipoproteinemics. Eur J Clin Invest 14(2), pt II, Abstract 45

Importance of Phospholipids in Cholesterol-Solubilizing Capacity of High-Density Lipoproteins

G. Salvioli and R. Lugli[1]

1. Cholesterol (Ch) efflux from cells or deposits (such as the arterial wall and gallstones) is a primary goal of therapy. The removal of cholesterol requires several steps: first, it is desorbed from the membranes or from the surface of solid cholesterol; subsequently, the Ch diffuses into the unstirred water layer where several acceptors, such as albumin, lipoproteins, apolipoproteins and lecithin in monomeric form, can take it up. Phospholipids and in particular phosphatidylcholines (PC) are the physiological solubilizers of Ch and they are important constituents of high density lipoproteins (HDL). The presence of PC in the unstirred water layer is therefore important for cholesterol solubilization and transport. It is noteworthy that Ch solubility in water is higher than that of PC (10^{-8} vs 10^{-10} M) (Haberland and Reynolds 1973); consequently, Ch concentration in the unstirred water layer (UWL) is greater than that of PC. But the presence of PC in monomeric form or assembled in particles plays a key role in Ch solubilization (Rothblat and Phillips 1982). HDL bind some PC from sonicated dispersions introduced into the vein (Scherphof et al. 1978) or from mixed micelles, as in the commercial preparation Lipostabil. HDL are responsible for reverse Ch transport. Moreover, hydrophilic PC molecular species desorb more readily from membranes, since diunsaturated PC have a faster exchange rate (Robbins and Patton 1986). Recently, Nichols (1986) demonstrated that bile salts at concentrations in the range found in blood during the postprandial period bind to vesicles, reducing the stability of bilayer phospholipids and enhancing their transfer rate between the particles. Lipostabil contains dilinoleoylphosphatidylcholine and deoxycholate (molar ratio $\cong$ 1). Therefore, its infusion in man increases the availability of hydrophilic PC molecular species, above all in HDLs (Salvioli et al. 1978); moreover, it raises bile acid concentration in blood (at values close to those present in cholestasis) (Salvioli et al. 1978). Even when administered per os, it is incorporated preferentially into the HDL fraction (Zierenberg and Grundy 1982).

2. Our studies evaluated the capacity of egg lecithin (and other molecular species) alone or in association with apolipoproteins obtained from HDL to solubilize solid cholesterol monohydrate (ChM). Vesicles of PC with Ch and HDL-apolipoproteins were prepared by sonication at 4 °C (final lipid or lipid + protein concentrations were 300 mg%). The vesicles and Lipostabil were used in Ch dissolution experiments. ChM powder was used to measure the maximum Ch solubility at equilibrium (Ch-eq); a disk

1 Università di Modena, Ospedale Estense, Viale Vittorio Veneto 9, 41100 Modena, Italy

Drugs Affecting Lipid Metabolism
Ed. by R. Paoletti et al.
© Springer-Verlag Berlin Heidelberg 1987

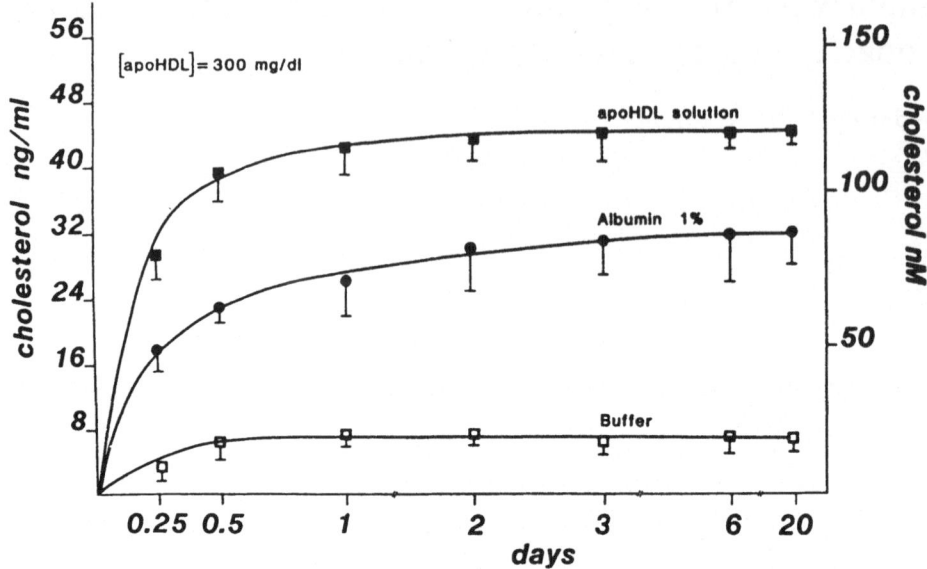

Fig. 1. Results of studies evaluating the capacity of egg lecithin

of compressed ChM with known surface was used to evaluate the initial dissolution rate (IDR). With IDR and Ch-eq values, the dissolution rate constant (K) (indicative of the general speed of the dissolution process) can be determined (Igimi and Carey 1981; Salvioli et al. 1983).

The results are reported in Fig. 1: a solution of HDL-apolipoproteins dissolves more Ch than a solution of albumin 1%; in spite of this, apoHDL is a poor Ch solubilizer and the apoHDl-Ch molar ratio at equilibrium is about 1000.

When vesicles containing egg-PC alone or with apoHDL in different proportion by weight (maintaining total solid concentration at 300 mg%), the amount of Ch solubilized at equilibrium decreased with increasing protein proportion (or decreasing PC fraction) (see Table 1). But considering the Ch/PC molar ratio in the dissolving solutions, we found values of 0.47 with PC alone and increased values when apoHDLs are present. Therefore, the capacity of PC to take up Ch rises in the presence of HDL-apolipoproteins even though the total amount of acquired Ch appears to be directly correlated to

Table 1. Values for IDR, K and Ch-eq in different solutions[a]

Solution	IDR (mmol cm^{-1} s^{-2}) x 10^{-8}	K (cm s^{-1}) x 10^4	Ch-eq (mM)
PC	5.76	0.32	1.8
PC:apoHDL 5:1	4.54	0.32	1.4
PC:apoHDL 2:1	4.23	0.39	1.1
PC:apoHDL 1:2	3.31	0.37	0.9

[a] IDR = initial dissolution rate; K = dissolution rate constant; Ch-eq = maximum cholesterol solubility. Total lipid concentration 300 mg%.

the initial amount of PC present in solution. The effects of apolipoproteins in promoting Ch efflux seems to be solubilization of PC, organizing it in to well-organized particles (DeLamatre et al. 1986). The initial dissolution rate pattern (Table 1) indicates that the presence of apoHDL reduces the speed of cholesterol solubilization. But the dissolution rate constant (K), which is greatly influenced by the amount of Ch solubilized at equilibrium, rises in their presence. In other words, apoHDLs seem able to increase the general speed of cholesterol solubilization in the presence of PC.

3. The amount of solubilized Ch at equilibrium by egg-PC, palmitoyl-oleoyl-PC, dioleoyl-PC and dilinoleoyl-PC were 1.81, 1.70, 1.65 and 1.56 mM respectively (PC concentration 300 mg% in form of vesicles). In terms of IDR the values show that egg-PC > 16:0-18:1 PC > 18:1-18:1 PC > 18:2-18:2 PC (5.76, 5.24, 5.19 and 5.02 respectively). In terms of dissolution rate constant, (evaluating the mass transfer), dilinoleoyl-PC shows the highest value of the studied molecular species of lecithin, but the differences are not great.

These results are not in accordance with the findings that PC unsaturation raises the Ch exchange rate between the vesicles (Bloj and Zilversmit 1977) and that the exchange rates of individual lecithins are positively correlated with their hydrophilicity (Robbins and Patton 1986).

4. The removal of Ch from the aqueous phase outside the cell membrane or the ChM surface requires conditions which facilitate the adsorption of Ch molecules by acceptor particles; the collision frequency between Ch molecules (desorbed from ChM) and acceptor particles play an important role. Both the particle concentration and the particle radius regulate the collision frequency with Ch molecules (Rothblat and Phillips 1982). Lipostabil, containing 18:2-18:2 PC and deoxycholate, is a micellar solution with very small particles which have a high affinity for HDLs. In general, these particles acquire free cholesterol from endogenous sources and are associated above all with HDL (Salvioli et al. 1978; Williams and Scanu 1986); in this way HDLs are enriched in polyunsaturated PC and become more able to transport cholesterol.

The maximum Ch solubilization with Lipostabil (total lipid concentration around 7 g%) is very high (14.7 mM); IDR is 0.6 (mmol cm^{-1} s^{-2}) $\times$ 10^{-7}, and the dissolution rate constant 0.6 (cm s^{-1}) $\times$ 10^4. When the solution is diluted to obtain a final concentration of 300 mg%, turbidity appears because of vesicle formation and the solubilizing capacity decreases to 1.3 mM; the dissolution rate constant becomes 0.5. Therefore, deoxycholate increases the capacity of lecithin (in a concentration of 200 mg%) to solubilize ChM.

Infusion of lecithin plus bile salts thus seems to favor the removal of cholesterol deposits and its excess in cellular membranes, as demonstrated by the reduction of circulating echinocytes in cirrhotics infused with Lipostabil (Salvioli et al. 1978).

5. In conclusion, the apolipoproteins of HDL have a low cholesterol-solubilizing capacity. This property seems dependent on the apoHDL/lecithin ratio. ApoHDLs increase cholesterol dissolution rate of PC vesicles; in fact, they impart pseudomicellar characteristics to the particles (Carey et al. 1985). Mixed micelles containing deoxycholate and polyunsaturated phosphatidylcholine (molar ratio $\cong$ 1) are available for intravenous administration; they have very high cholesterol solubilizing capacity and increase the bile acid concentration in blood. An increased availability of bile salts

enhances the spontaneous phospholipid movements around cells, facilitating the assembly of cholesterol in particles in the aqueous spaces.

References

Bloj B, Zilversmit DB (1977) Complete exchangeability of cholesterol in phosphatidylcholine/ cholesterol vesicles of different degrees of unsaturation. Biochemistry 16:3943–3948

Carey MC, Benedek GB, Donovan JM (1985) Micelles and vesicles of human apolipoproteins (APO) A-I and A-II, lecithin and bile salts: new insight from quasi elastic light scattering (QLS). In: Barbara L, et al. (eds) Advances in bile acid research. MTP, Lancaster, pp 183–189

DeLamatre J, Wolfbauer G, Phillips MC (1986) Role of apolipoproteins in cellular cholesterol efflux. Biochim Biophys Acta 875:419–428

Haberland HE, Reynolds JA (1973) Self-association of cholesterol in aqueous solution. Proc Natl Acad Sci USA 70:2313–2315

Igimi H, Carey MC (1981) Cholesterol gallstone dissolution in bile: dissolution kinetics of crystalline cholesterol with chenodeoxycholate, ursodeoxycholate, and their glycine and taurine conjugates. J Lipid Res 22:254–270

Nichols JW (1986) Low concentrations of bile salts increase the rate of spontaneous phospholipid transfer between vesicles. Biochemistry 25:4596–4601

Robbins SJ, Patton GM (1986) Separation of phospholipid molecular species by high performance liquid chromatography: potential for use in metabolic studies. J Lipid Res 27:131–139   .

Rothblat GH, Phillips MC (1982) Mechanisms of cholesterol efflux from cells. Effect of acceptor structure and concentration. J Biol Chem 257:4775–4782

Salvioli G, Rioli G, Lugli R, Salati R (1978) Membrane lipid composition of red blood cells in liver disease: regression of spur cell anemia after infusion of polyunsaturated phosphatidylcholine. GUT 10:844–850

Salvioli G, Igimi H, Carey MC (1983) Cholesterol gallstone dissolution in bile. Dissolution kinetics of chenodeoxycholate-lecithin and conjugated ursodeoxycholate-lecithin mixtures: dissimilar phase equilibria and dissolution mechanisms. J Lipid Res 24:701–720

Scherphof G, Roerdink F, Waite M, Parks J (1978) Disintegration of phosphatidylcholine liposomes in plasma as a result of interaction with high density lipoproteins. Biochim Biophys Acta 542: 296–307

Williams KJ, Scanu AM (1986) Uptake of endogenous cholesterol by a synthetic lipoprotein. Biochim Biophys Acta 875:183–194

Zierenberg O, Grundy SM (1982) Intestinal absorption of polyenephosphatidylcholine in man. J Lipid Res 23:1136–1142

Effects of Polyunsaturated Phospholipid on Cholesterol Efflux in Vitro and in Experimental Animals and Human Subjects

J. KOIZUMI, A. POSTIGLIONE, B. L. KNIGHT, A. K. SOUTAR, and G. R. THOMPSON [1]

1 Introduction

Patients with homozygous familial hypercholesterolaemia pose a difficult clinical problem, because of their poor response to dietary and drug therapy. Eleven years ago, however, we introduced the use of plasma exchange to treat this condition (Thompson et al. 1975) and we and other workers have subsequently confirmed its safety and effectiveness (Thompson 1981); treated patients now live longer than their untreated, homozygous siblings (Thompson et al. 1985).

A theoretical disadvantage of plasma exchange is that both LDL and HDL get removed; a low HDL-cholesterol is known to be associated with an increased risk of coronary heart disease. This can be overcome by perfusing plasma through affinity columns which selectively remove LDL. An alternative approach is to supplement the plasma exchange medium (plasma protein fraction, PPF, or human serum albumin) with an exogenous phospholipid (PL) emulsion, which might be expected to increase the capacity to take up cholesterol of the residual HDL after plasma exchange. There is evidence that the addition of phospholipid to albumin enhances the latter's ability to take up cholesterol from cells in vitro (Burns and Rothblat 1969) and even more so when phospholipids are first combined with apoprotein A-I, as in HDL (Stein et al. 1976).

To test the potential usefulness of adding exogenous PL to PPF to promote cholesterol efflux from tissues, polyenephosphatidylcholine (PPC) in the form of Lipostabil (5% lecithin in 4% Na deoxycholate) was incubated with cells in vitro and was also infused in vivo during plasma exchange, both in experimental animals and in man. PPC has been shown to coalesce with HDL and to increase its uptake of cholesterol by approximately 50% as compared with native HDL (Zierenberg et al. 1981).

2 Effect of Exogenous Phospholipids in Vitro

Cholesterol efflux was first studied in vitro in human monocyte macrophages isolated from normolipidaemic donors. Cells were preloaded with cholesteryl esters by incuba-

1 MRC Lipoprotein Team, Hammersmith Hospital, Ducane Road, London W12 OHS, Great Britain

Table 1. Content and efflux of [^{14}C]-cholesterol from human macrophages as a function of the presence of Lipostabil (mean ± SD from six incubations)[a]

| | Cholesterol (μg)/cell protein (mg) | |
	Cellular content	Medium
With Lipostabil	10.3 ± 2.0	6.7 ± 1.0*
Without Lipostabil	12.4 ± 2.4	4.0 ± 1.2*

[a]Human macrophages were incubated with [^{14}C]-cholesteryl oleate-labelled LDL + dextran sulphate for 16 h, washed and incubated for 6 h with 10% human serum—RPMI in the presence or absence of Lipostabil (125 μg ml^{-1}). Radioactivity in cells and medium was then assayed and expressed in relation to cell protein content.
*$p<0.02$.

ting them with LDL (d 1.019—d 1.063) which had been reconstituted with [^{14}C/-cholesteryl oleate as described by Krieger et al. (1978) and then coupled with dextran sulphate (Basu et al. 1979). The cells were subsequently incubated with 10% normal human serum in culture medium RPMI in the presence or absence of exogenous PL (Lipostabil). Results are shown in Table 1: cholesteryl ester-enriched macrophages incubated in the presence of Lipostabil show a significant increase in the amount of radioactivity released into the medium, as compared with control cells incubated in the absence of Lipostabil. However, the accompanying decrease in the cellular content of [^{14}C] was not statistically significant.

3 Effect of Exogenous Phospholipids in Vivo During Plasma Exchange

The effect of the addition of PPC to PPF during plasma exchange was studied in non-human primates and in man. Two Rhesus monkeys were used to test the safety of infusing large amounts of Lipostabil. No evidence of haemolysis or hepatic dysfunction was found when up to 2 g Lipostabil mixed with 200 ml monkey plasma was infused. Therefore 2.5—5 g Lipostabil was added to 4 litres PPF and administered to two adult males with homozygous familial hypercholesterolaemia who had been regularly undergoing plasma exchange for several years. Serum bile acid concentrations increased markedly for up to 3 h after plasma exchange, reflecting the large amount of Na deoxycholate infused, but no subjective side effects resulted from Lipostabil administration (Thompson and Jadhav 1986).

The addition of 5 g Lipostabil to PPF caused an increase in HDL-PL levels in plasma, which was still present 15 h after plasma exchange but without an increase in HDL-cholesterol, as would have been expected if cholesterol efflux from the extravascular compartment had occurred.

Although exogenous PL, in the form of Lipostabil, promoted cholesterol efflux from cells in vitro there was no evidence, over the time course of the study, that this occurred in vivo when up to 5 g were administered during plasma exchange. This may have reflected the rapid removal of exogenous PL from circulation. The possibility

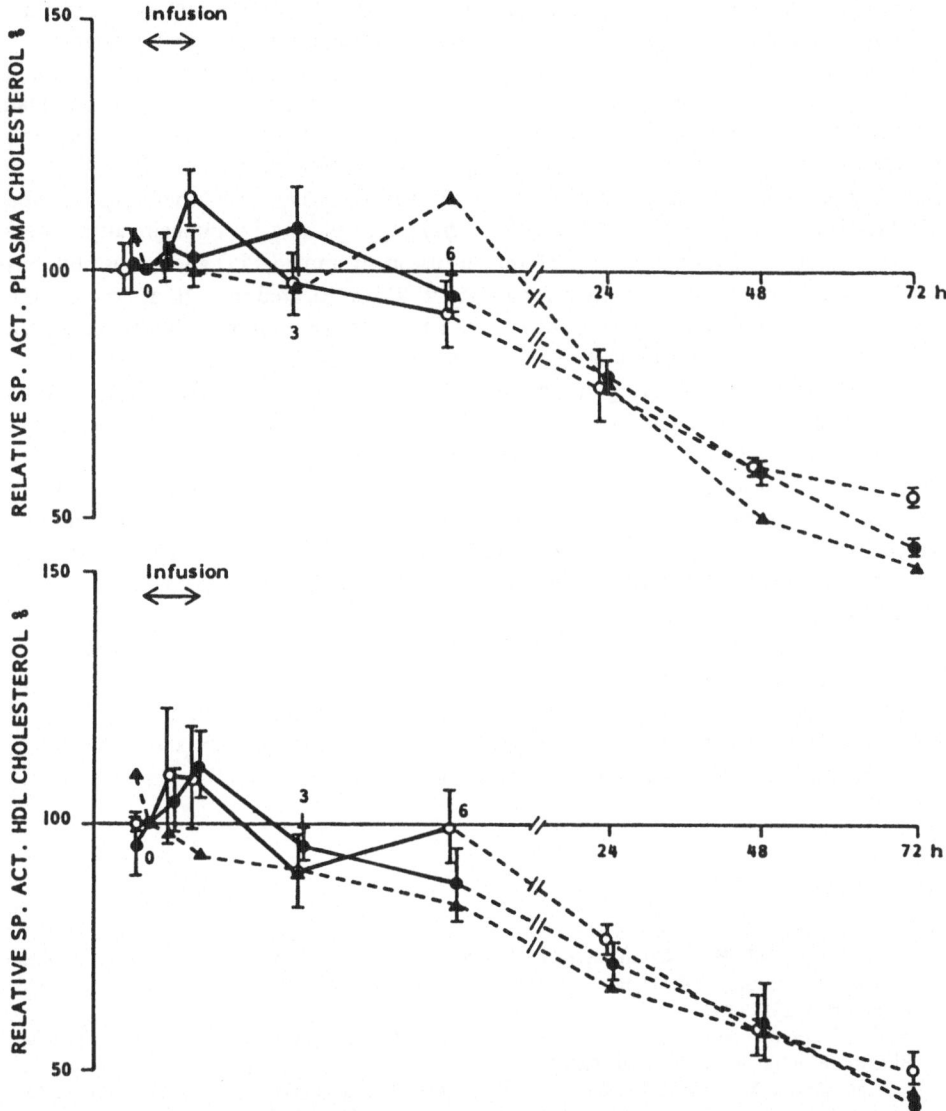

Fig. 1. Changes of relative specific acitivity in plasma and HDL-cholesterol after infusion of apoA-I/PC complex (n=3) ●; Lipostabil (n=3) ○; saline (n=2) ▲, into rabbits

that apoA-I/lecithin complexes are cleared from plasma at a slower rate, similar to that of native HDL, was therefore explored in rabbits, as described below.

4 Comparison of Lipostabil and Human apoA-I/PPC Complexes

In an attempt to label macrophages in vivo [3H]-cholesterol-labelled, acetylated LDL was prepared and injected into fasted normal rabbits, as described by Miller et al.

(1985). The specific activity of total plasma cholesterol decreased rapidly to less than 10% of its initial level within 6 h and then declined gradually in an exponential manner. Human apoA-I/PPC complexes were prepared by incubating apoA-I with Lipostabil and dialysing off the deoxycholate. The resultant particles, which were smaller than Lipostabil but slightly larger than plasma HDL, had a protein:phospholipid mass ratio of 1:2.25. 100 mg apoA-I/PC, 100 mg Lipostabil (both expressed as PL) and 26 ml saline were then injected into three groups of rabbits, 2 days after they had received [^{3}H]-cholesterol-labelled acetylated LDL. Changes in plasma PL concentrations were similar during the first 6 h after a 30-min infusion of apoA-I/PPC and Lipostabil, but 73 ± 23% of the rise immediately after apoA-I/PPC was located in HDL as against only 38 ± 9% after Lipostabil. Subsequent rises in PL levels after 6 h probably reflected the effects of re-feeding since they occurred in all three groups of animals.

Changes in PL levels during the first 6 h were accompanied by small increases in total plasma cholesterol during the first 3 h after infusion of apoA-I/PPC complexes and for 6 h after Lipostabil, but HDL cholesterol levels were virtually unchanged. As with PL presumedly non-specific rises in plasma and HDL-cholesterol occurred subsequently in all three groups. Transient increases in the specific activity of total and HDL-cholesterol were also observed, the rise in HDL-specific activity above baseline, averaging 12 ± 6% at 1 h after apoA-I/PPC complexes, had been infused and 9 ± 10% after Lipostabil (Fig. 1).

These results suggest that injected human apoA-I/PPC complexes associate to a greater extent with rabbit HDL than does Lipostabil but both seem to be cleared from plasma at a similar rate and to be roughly comparable in terms of their ability to mobilise tissue cholesterol. However, the magnitude of the latter effect and its therapeutic usefulness cannot be assessed on the basis of these experiments.

References

Basu SK, Brown MS, Ho YK, Goldstein JL (1979) Degradation of low density lipoprotein. Dextran sulfate complexes associated with deposition of cholesteryl esters in mouse macrophages. J Biol Chem 254:7141–7146

Burns CH, Rothblat GH (1969) Cholesterol excretion by tissue culture cells: effect of serum lipids. Biochim Biophys Acta 176:616–625

Krieger M, Brown MS, Faust JR, Goldstein JL (1978) Replacement of endogenous cholesteryl esters of LDL with exogenous cholesteryl linoleate. J Biol Chem 253:4093–4101

Miller NE, La Ville A, Crook D (1985) Direct evidence that reverse cholesterol transport is mediated by high density lipoprotein in the rabbit. Nature (London)314:109–111

Stein O, Vanderhoek J, Stein Y (1976) Cholesterol content and sterol synthesis in human skin fibrobalsts and rat aortic smooth muscle cells exposed to lipoprotein deficient serum and HDL/phospholipid mixture. Biochim Biophys Acta 431:347–358

Thompson GR (1981) Plasma exchange for hypercholesterolemia. Lancet i:1246–1248

Thompson GR, Jadhav AV (1986) Control of plasma HDL levels after plasmapheresis. In: Lipoprotein deficiency syndromes. Plenum, New York, pp 127–135

Thompson GR, Lowenthal R, Myant NB (1975) Plasma exchange in the management of familial hypercholesterolemia. Lancet i:1208–1211

Zierenberg O, Assmann G, Schmitz GZ, Rosseneu M (1981) Effect of PPC on cholesterol uptake by human HDL. Atherosclerosis 39:527–542

Phospholipids in Human Atherosclerosis

P. Avogaro, G. Bittolo Bon, and G. Cazzolato[1]

Despite their relevant physiological and structural role, phospholipids (PL) have not gained favour in clinical assessment of human atherosclerosis. Already in 1951, Gertler et al. recorded higher values of serum lipid phosphorus in coronary patients than in controls. Furthermore, it was shown that the serum cholesterol/lipid phosphorus is increased in the presence of coronary heart disease (CHD). The same authors suggested a "chemotype" which included the analysis of serum cholesterol, phospholipids and uric acid. A recommandation was also given to increase the values of plasma PL, thus reestablishing a normal C/PL ratio. The meaning of the recorded variations of PL plasma values has not been discussed. An increased C/PL ratio has also been recorded in type II hyperlipoproteinaemia (Peeters 1976). Increased values of ratio sphyngomyelin/phosphatidylcholine (SM/PC) was observed in types II and IIB, being normal in type IV (Peeters 1976). The lecithin/sphyngomyelin ratio is abnormally low in homozygotes with familial xanthomatosis hypercholesterolaemia (FXH) (Mills et al. 1976). In the low density lipoproteins (LDL) of FXH, an increased C/PL ratio was recorded with a decreased lecithin/sphyngomyelin ratio (Jadhav and Thompson 1979). Since this abnormal composition of LDl normalizes following plasma exchange it has been suggested that the abnormal composition of LDL in FHX is due to hypocatabolism and then to the aging of LDL (Jashav and Thompson 1979). The chemical percentage composition of the major lipoprotein classes has been studied in survivors of myocardial infarction (Avogaro et al. 1979). In both HDL_2 and HDL_3 PL and protein have significantly decreased, whereas triglycerides (TG) have increased. The chemical composition of ultracentrifugally obtained LDL (d=1040–1050) from survivors of myocardial infarction and from controls has been analyzed (Avogaro 1983). The LDL of survivors are characterized by a higher content of TG and apoB, and by a lower content of PL. From these studies it appears, therefore, that the decrease in PL content is an important feature of the deranged structure of lipoproteins in patients with clinical atherosclerosis. The meaning of the recorded variations of PL was, however, not elucidated. Recent studies have established that incubation of LDL with cultured endothelial cells (EC) converts it to a new form (EC-modified LDl) (Henriksen et al. 1981). The EC-modified LDL (EC-LDL) shows an increase in negative charge and in hydrate density; moreover, EC-LDL is characterized by extensive changes in apolipoprotein B (apoB) (Parthasarathy et al. 1985). Usually, the native LDL is recognized

---

1   Regional General Hospital, Unit for Atherosclerosis, 30100 Venice, Italy

Drugs Affecting Lipid Metabolism
Ed. by R. Paoletti et al.
© Springer-Verlag Berlin Heidelberg 1987

by receptors of fibroblasts, smooth muscle cells, endothelial cells and poorly by macrophages and doesn't produce accumulation of cholesterol esters in such cells in culture
(Goldstein et al. 1979). The EC-LDL is recognized by a specific receptor on macrophages (the acetyl LDL receptor or "scavenger" receptor) and it is taken up and degraded
three to ten times more rapidly than native LDL. It has been thought that the biological modification of LDL is due to a peroxidation from free radicals (Parthasarathy et
al. 1985). Actually, the changes are blocked by antioxidants like vitamin E and butylated hydroxytoluene (BHT) and by compounds such as EDTA which are able to che-
late iron and copper (Parthasarathy et al. 1985). The free radical-mediated peroxidation of LDL is accompanied by extensive hydrolysis of LDL phosphatidylcholine (Ptd-
CHO) to lyso-phosphatidylcholine (1-Ptd CHO) (Steinbrecher et al. 1984). Gel electrophoresis of EC-LDL shows, moreover, an almost complete loss of B-100 with the appearance of lower molecular weight bands (Steinbrecher et al. 1984). Both the hydrolysis of Ptd-CHO and the breakdown of apoB are inhibited by an inhibitor of phospholipase A_2, bromophenacyl bromite (Parthasarathy et al. 1985). It appears therefore
that (1) the LDL modification is due to free-radical peroxidation, (2) there is a phospholipid breakdown, which appears to be an obligatory element in generating EC-LDL
and (3) the breakdown is due to the action of a LDL phospholipase A_2. Phospholipids
of cellular membranes are particularly susceptible to lipid peroxidation. The lipid peroxidation of membranes induces progressive changes in membrane lipid composition,
order and fluidity. The oxidized membrane phospholipids are rapidly cleared through
an enhanced hydrolysis via phospholipase A_2 action (Sevanian and Kim 1985). Along
with this high-order removal of oxidized fatty acids a marked hydrolysis of intact
fatty acids is also observed. The extent of fatty acid release is roughly correlated with
the degree of fatty acid unsaturation.

We have recently investigated whether modified LDL is present in humans. To this
aim a procedure has been followed utilizing (1) density-gradient ultracentrifugation;
(2) ion-exchange chromatography; (3) spectrophotometric readings with two wavelengths at 280 and 250 nm respectively, for elution of proteins and of plasma peroxides and conjugated dienes; (4) HPLC double readings with two wavelengths at 205 and
234 nm respectively, for evaluation of plasma lipids and of plasma peroxides and conjugated dienes. In the eluate of the chromatographic column C, TG, PL and were determined apoprotein B on Na D SO_4 was also evaluated. Our study has demonstrated:
(1) that some modified-LDL particles (m-LDL) are present in the plasma of humans;
(2) m-LDL accounts for up to 35% of the total LDL mass; (3) m-LDL has a more electronegative charge and a hydrated density similar to that of normal LDL (n-LDL); (4)
m-LDL contains more proteins than n-LDL and is characterized by a higher FC/EC
ratio and especially by significantly lower levels of PL; (5) more conjugated dienes and
peroxides are contained in m-LDL than in n-LDL; (6) less apoB-100 is present in m-
LDL, whereas apoB aggregates appear; (7) LDL particles are more heterogeneous than
n-LDL and have a major diameter (27 vs 24 Å); (8) utilizing four monoclonal antibodies, a lower reactivity of m-LDL in the recognition sites for apoB was determined;
(9) m-LDL is poorly recognized and taken up by fibroblast LDL receptors.

The decrease of PL appears to be an index of early damage occurring at the membrane level or in the lipoprotein structure. The low PL levels occur as an effect of
Phospholipase A_2 whose intervention is aimed at reestablishing a more normal mem-

brane situation through the excision of the oxidized fatty acids. The low levels of PL and/or PUFA impair the production of prostaglandins and eicosanoids (Smith et al. 1984), thus reducing the inhibitory effects that PG_2 and eicosanoids display on the proliferation of smooth muscle cells (Smith et al. 1984).

References

Avogaro P (1983) Phospholipids and atherosclerosis. In: Avogaro P, Mancini M, Ricci G, Paoletti R (eds) Phospholipids and atherosclerosis. Raven, New York, pp 211–214

Avogaro P, Cazzolato G, Bittolo Bon G, Belussi F (1979) Levels and chemical composition of HDL_2 HDL_3 and other major lipoprotein classes in survivors of myocardial infarction. Artery 5:495–508

Gertler MM, Garn SM, White PD (1951) Young candidates for coronary heart disease. JAMA 147: 621–624

Goldstein JL, Ho YK, Basu SK, Brown MS (1979) Binding site on macrophages that mediates uptake and degradation of acetylated low density lipoprotein, producing massive cholesterol deposition. Proc Natl Acad Sci USA 76:333–377

Henriksen T, Mahoney E, Steinberg D (1981) Enhanced macrophage degradation of low density lipoprotein previously incubated with cultured endothelial cells: recognition by receptors for acetylated low density lipoproteins. Proc Natl Acad Sci USA 78:6499–6503

Jadhav AV, Thompson GR (1979) Reversible abnormalities of low density lipoprotein composition in familial hypercholesterolaemia. Eur J Clin Invest 9:63–67

Mills GL, Taylaur CE, Chapman MJ (1976) Low density lipoproteins in patients homozygous for familial hyperbetalipoproteinaemia. Clin Sci Mol Med 51:221–231

Parthasarathy S, Steinbrecher VP, Barnett J, Witzum JL, Steinberg D (1985) Essential role of phospholipase A_2 activity in endothelial cell induced modification of low density lipoprotein. Proc Natl Acad Sci USA 82:3000–3004

Peeters H (1976) The biological significance of the plasma phospholipids. In: Peeters H (ed) Phosphatidylcholine. Springer, Berlin Heidelberg New York, pp 10

Sevanian A, Kim E (1985) Phospholipase A_2 dependent release of fatty acids from peroxidized membranes. J Free Rad Biol Med 1:263–271

Smith DL, Willis AL, Mahmud I (1984) Eicosanoid effects on cell proliferation in vitro: relevance to atherosclerosis. Prostagl Leukotrien Med 16:1–10

Steinbrecher UP, Parthasarathy S, Leake DS, Witzum JL, Steinberg D (1984) Modification of low density lipoproteins by endothelial cells involves lipid peroxidation and degradation of low density lipoprotein phospholipids. Proc Natl Acad Sci USA 81:3883–3887

Effect of Polyunsaturated Phosphatidylcholine (PPC) on Plasma Lipoproteins in Patients with Type II Hyperlipoproteinemia

G. Noseda and C. Fragiacomo [1]

1 Introduction

Polyunsaturated phosphatidylcholine (PPC) is a lecithin which is extracted from soybeans and is very rich in polyunsaturated fatty acids. Earlier investigations in man have shown that PPC lowers significantly total and LDL-cholesterol (1–14). Triglycerides are also frequently (1–6, 8–10, 12), although not always, reduced (15, 16). Reports on the behaviour of HDL during treatment with PPC are contradictory (15, 17). The purpose of the present study was to test the lipid-lowering effect of PPC in doses higher than previously used.

2 Patients and Methods

I. Study

Twenty-seven patients (15 men, 12 women, age > 40 years) with hyperlipidemia type IIa (n=8) and type IIb (n=19) with LDL-cholesterol values > 200 mg dl^{-1} were investigated. During the 4 weeks preceding the study, the patients received no lipid-lowering drugs. They were all placed on a diet with a P:S ratio of $> 1,5-2$ and a cholesterol content of $< 200-300$ mg day^{-1}.

In a randomized double-blind study the subjects received during 6 weeks twice a day three capsules containing 450 mg 3-sn-poly-enylphosphatidylcholine (P0206/1/01, Nattermann, GmbH, Cologne) (14 cases, 5 with type IIa and 9 with type IIb hyperlipidemia) or placebo (13 cases, 3 with type IIa and 10 with type IIb hyperlipidemia) (Fig. 1).

The following parameters were determined: total cholesterol and triglycerides by enzymatic methods; HDL-cholesterol after precipitation of apoprotein B-containing lipoproteins with dextran sulphate-Mg^{++}; LDL and VLDL cholesterol after separation of these lipoproteins by ultracentrifugation. Apoprotein B was measured by radial immunodiffusion (Behring) and apoprotein A_1 by rocket immunoelectrophoresis.

---

1   Ospedale Beata Vergine, CH-6850 Mendrisio, Switzerland

Drugs Affecting Lipid Metabolism
Ed. by R. Paoletti et al.
© Springer-Verlag Berlin Heidelberg 1987

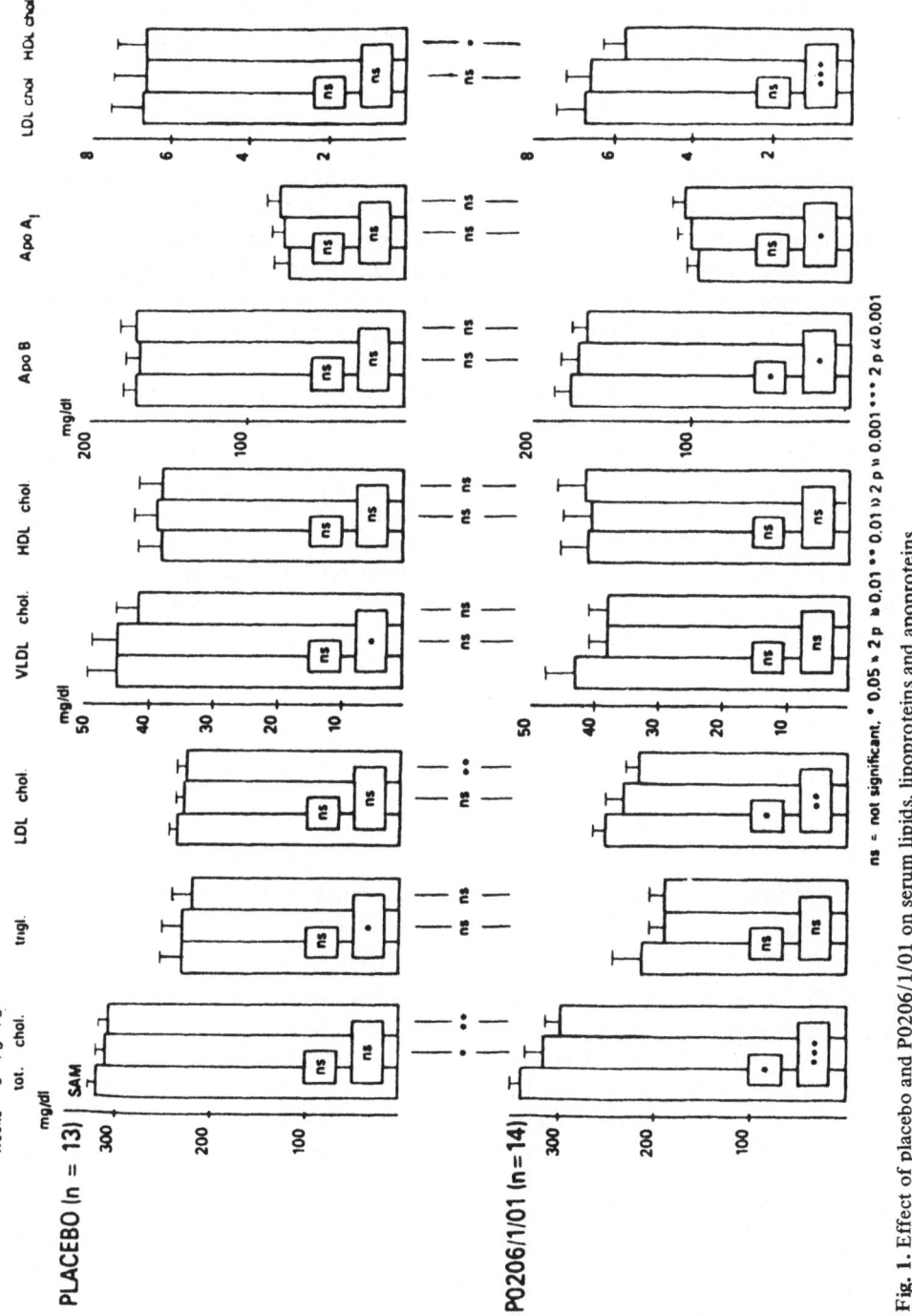

Fig. 1. Effect of placebo and PO206/1/01 on serum lipids, lipoproteins and apoproteins

ns = not significant, * 0.05 » 2 p » 0.01 ** 0.01 » 2 p » 0.001 *** 2 p « 0.001

Statistical analysis was done with the Wilcoxon Test and, to detect possible differences between the values obtained with placebo and PPC, with the paired difference U-test.

II. Study

We have commenced a second study in which we compare, versus placebo, the action of three capsules containing 450 mg PPC twice daily, to be taken in the morning and evening before meals, on the composition of the lipoprotein classes VLDL, LDL, HDL, HDL_2 and HDL_3 separated by preparative and zonal ultracentrifugation. Criteria for acceptance were: outpatients with hyperlipidemia type II, total cholesterol more than 260 mg dl^{-1}, LDL-cholesterol more than 180 mg dl^{-1}, age over 40 years.

During the 4 weeks preceding the study the patients received no lipid-lowering drugs. They were all placed on a diet rich in unsaturated fatty acids and containing less than 300 mg cholesterol/day. In a randomized double-blind study the subjects received during 12 weeks twice a day three capsules containing 450 mg PPC or placebo. Thereafter a wash-out diet period of 3 weeks terminated the study.

The measured parameters were: cholesterol, triglycerides, phospholipids, Apo-A_I, Apo-A_{II} and Apo-B in plasma; cholesterol, triglycerides, phospholipids and Apo B in VLDL and LDL; cholesterol, triglycerides, phospholipids, Apo A_I and Apo A_{II} in HDL, HDL_2 and HDL_3. Only 13 patients (8 received PPC and 5 receiving placebo) have so far concluded this second trial. Thus, our results are to be considered very preliminary.

3 Results

I. Study

Placebo: The following significant percent variations were found:

Triglycerides	after 6 weeks	−6.2% (*)
VLDL-cholesterol	after 6 weeks	−6.1% (*)

P0206/1/01: The following significant percent variations were found:

Total cholesterol	after 3 weeks	− 5.9% (*)
	after 6 weeks	−11.1% (***)
LDL-cholesterol	after 3 weeks	− 5.8% (*)
	after 6 weeks	−12.9% (**)
Apoprotein B	after 3 weeks	− 2.4% (*)
	after 6 weeks	− 5.6% (*)
Apoprotein A_I	after 6 weeks	+ 8.3% (*)
Ratio LDL chol/HDL chol after 6 weeks		−15.5% (*)

Placebo vs *P0206/1/01:* The reductions of total cholesterol after 3 and 6 weeks (*,**) LDL-cholesterol after 6 weeks (**) and the ratio LDL chol/HDL chol after 6 weeks (*) of therapy with P0206/1/01 are significant as compared to the small variations with placebo.

Table 1. II. Study (preliminary results)

	T (Weeks)	Placebo D (mg dl^{-1})	PPC D (mg dl^{-1})	P
Decrease				
HDL$_2$-chol.	6	+ 1,6	− 1,8	< 0,02
Apo A$_{II}$	12	+ 2,6	− 4,0	< 0,001
Increase				
HDL-trigl.	6	− 0,4	+ 3,0	< 0,01
Apo A$_I$	3	+ 4,5	+ 8,7	< 0,01

II. Study (Preliminary Results)

Comparing the results obtained after therapy with PPC with those obtained after placebo, only few statistical significant differences were observed (Table 1).

4 Conclusions

In a first double-blind study 27 patients with type II hyperlipidemia (8 IIa and 19 IIb) were treated as follows: 13 received placebo and 3-sn-polyenylphosphatidylcholine (PPC) (P0206/1/01, Nattermann, GmbH, Cologne) in a dose of three times 450 mg b.i.d. In all patients and also in the two subclasses of patients with type IIa and IIb hyperlipidemia, total cholesterol and LDL-cholesterol were lowered significantly by PPC. The other parameters showed only small changes. There was a downward trend in apoprotein B, triglycerides and VLDL-cholesterol, and an upward trend in apoprotein A$_I$, with virtually unchanged HDL-cholesterol. None of these variations was significant as compared to placebo.

The fall in LDL-cholesterol with unchanges HDL-cholesterol caused a statistically significant decrease in the LDL-cholesterol/HDL-cholesterol ratio, thus supporting the hypothesis of an antiatherogenic property of PPC, as demonstrated experimentally in various animals (18, 19).

For the second, still ongoing study, no final conclusions can be drawn. The most interesting result is a significant increase of Apo A$_I$ after 3 weeks of therapy with PPC as compared to placebo. This seems to confirm the already discussed antiatherogenic properties of this substance.

References

1. Nakamura H, Ebatz Y, Tonomo A, Hatz Y, Shiegematsu H, Hara T, Takeuchi I, Unretada Y (1973) Effects of EPL capsule on serum lipid level. J Med REm Clin 22:1565–1574
2. Yasugi T, Iijima M, Kinoshita T, Harada S, Matsumoto H, Sugeta K, Takeshita H, Saijo M, Shimizu T, Sassa E, Tomita M, Kobari M, Kobayashi I, Seto Y, Kuono H, Hoshino K (1973) Effects of EPL on serum lipid. Jpn J Med Treat 22:691–969
3. Sekimoto H, Nakada J, Nakai T, Nakamura K, Yamada S, Mori K (1972) Study of clinical effects of EPL for the improvement of abnormality in serum lipid in a double-blind technique. Jpn J Med Treat 5:1725–1734
4. Hevelke G, Grott E, Machalke K, Zappe R (1975) Zur Wirkung "essentieller" Phospholipide auf den Blutfettgehalt des Menschen. Folia Angiol, Suppl IV: 80–87
5. Knüchel F (1955) Untersuchungen über die medikamentöse Beeinflußbarkeit der Serumlipide bei der Arteriosklerose. Therapiewoche 5:570–574
6. Andersen P (1965) Die Behandlung von arteriosklerotischen Kreislaufstörungen und Hyperlipämien mit Phosphatidylcholin. Ther Umsch 22:614–622
7. Spigai C (1970) Einfluss der "essentiellen" Phospholipide auf den Lipidspiegel. Med Heute 19: 197–198
8. Casellas Bernat G, Riera de Barcia L, Alou M, Solivellas S, Casellas Bernat A (1975) Dislipemias y fosfatidilcolina insaturada. Clin Med 15:90–95
9. Uchida S (1968) Efficacy of EPL for lipid metabolism disorder. Shinryo Shin'yaku 5:1073–1076
10. Unger E, Walter M, Jaross W, Hanefeld M, Rossbach G (1971) Der Einfluß von essentiellen Phospholipiden auf die Fettfraktionen des Serums und der Leber. Zentralbl Pharmakol 110: 453–458
11. Uemura K (1973) Clinical experiences on polyenyl-phosphatidyl-choline (EPL). J New Remed Clin 22:1583–1597
12. Saba P, Galeone F, Salvadosini F, Pagliai E, Guidi O, Scalabrino A (1978) Effects of soybean polyunsaturated phosphatidylcholine (Lipostabil) on hyperlipoproteinemia. Curr Ther Res 4:299–306
13. Varkonyi Gy (1962) Beitrag zur therapeutischen Wirkung der oralen und intravenösen Verabreichung "essentieller" Cholinphospholipide. Z Ges Inn Med 18:830–838
14. Pupita F, Gagna G (1969) Therapie der Arteriosklerose mit "essentiellen" Phospholipiden. Med Monatsschr 23:514–515
15. Fasoli A (1982) Clinical evaluation of polyenoyl phosphatidyl-choline: effects on serum lipid and lipoprotein patterns. In: Ricci G, et al. (eds) Therapeutic selectivity and risk/benefit assessment of hypolipidemic drugs. Raven, New York, p 257
16. Blaton V, Soeteway F, Vandanne D, Declercq B, Peeters H (1976) Effect of polyunsaturated phosphatidylcholine on human types II and IV hyperlipoproteinemias. Artery 2:309
17. Dewailly P, Moulin S, Rouget JP, Sezille G, Jaillard J (1983) Effects of polyenyl-phosphatidylcholine on lipoproteins of patients with hypercholesterolemia. Eur J Clin Pharmacol (submitted)
18. Sirtori CR (1983) Phospholipids, atherosclerosis and aging. In: Avogaro P, et al. (eds) Phospholipids and atherosclerosis. Raven, New York, p 197
19. Schneider J (1982) Experimental and clinical effects of polyenoylphosphotidylcholine on erythrocytes and platelets. In: Ricci G, et al. (eds) Therapeutic selectivity and risk/benefit assessment of hypolipidemic drugs. Raven, New York, p 263

Dietary Lecithin: Metabolism, Fate, and Effects on Metabolism of Lipids and Lipoproteins

S. M. Grundy[1]

1 Introduction

The phospholipids, of which lecithin is a major component, constitute a major class of lipids in body tissues and in plasma. They play a vital role in cellular function and in transport of lipids. Since they are required for the solubilization of cholesterol, both within cells and in lipoproteins, the question naturally has arisen whether the phospholipids, or lecithin in particular, can be used in the prevention of atherosclerosis. This possibility has led investigators to feed phospholipids and to administer them intravenously with the aim of slowing down the process of atherosclerosis. Results have been conflicting. Some studies in experimental animals have suggested that the feeding of lecithin can reduce the size of atherosclerotic plaques (1,2). Several mechanisms might be responsible for this effect. There could be an increase in the activities of cholesteryl ester hydrolase in the arterial wall (3,4); there might be an activation of triglyceride lipase (5); or there could be changes in the metabolism of high density lipoproteins (HDL) (6). Further, high doses of lecithin in the diet might interfere with absorption of cholesterol, or they might alter the metabolism of triglycerides or low density lipoproteins (LDL). Because of the possible beneficial effects of dietary lecithin, our laboratory has carried out a series of studies on the metabolism of lecithin. Although these studies have not specifically examined whether the feeding of lecithin will prevent the development of atherosclerosis, they have provided new insights into the metabolism of lecithin and the effects of lecithin on the metabolism of other lipids. Our findings will be summarized in this chapter.

2 Intestinal Absorption of Lecithin

Only a limited number of studies have been carried out on the fate of dietary lecithin. Lecithin is known to be hydrolyzed by pancreatic phospholipase A_2 to lysolecithin and fatty acids in the lumen of the small intestine; both of the latter are taken up by

---

1 Departments of Internal Medicine and Biochemistry and Center for Human Nutrition, University of Texas Health Science Center at Dallas, 5323 Harry Hines Boulevard, Dallas, TX 75235, USA

Drugs Affecting Lipid Metabolism
Ed. by R. Paoletti et al.
© Springer-Verlag Berlin Heidelberg 1987

intestinal mucosal cells (7). After mucosal uptake, lysolecithin could have several fates. First, it might be reesterified with a fatty acid to produce lecithin, which can be incorporated into the surface coat of chylomicrons or become part of cell membranes (8, 9). Second, it could undergo complete lipolysis with release of fatty acids, which in turn could be incorporated into triglycerides (10—12). Finally, there could be direct absorption of lysolecithin into the portal circulation (13).

To determine how effective lecithin is absorbed when given in large quantities, Beil and Grundy (14) infused solubilized lecithin into the small intestine of seven subjects at rates of 150 mg kg^{-1} h^{-1}. Absorption was measured over a 50-cm segment of intestine by a triple-lumen technique. Over this segment, absorption of lecithin ranged from 19 to 66% (mean 36%). In two patients in whom absorption was estimated over 100 cm, the mean absorption was 55%. However, on stools collected in these patients, measurements of total lipid absorption was made by analysis of fecal lipid excretion and we found almost no malabsorption of lipid. Thus, even when intakes of lecithin are very high, it appears that lecithin absorption is almost complete. Thus, in normal individuals, there appears to be almost no malabsorption of lecithin even when intakes of lecithin are high. Our data nonetheless suggest that absorption of lecithin can occur over a long length of the intestine, and its efficiency of absorption is not as great as that for triglycerides, which are absorbed primarily in the upper intestine.

In another study, Zierenberg and Grundy (15) examined the metabolic fate of 1 g [^{3}H] [^{14}C]-labeled lecithin (dilinoleoglycerophosphocholine) in five patients after oral administration. The [^{3}H]-label was located in the choline moiety, and the [^{14}C]-label was in the two linoleic acid residues. It was found that more than 90% of both isotopes was absorbed from the intestine, again indicating that absorption of dietary lecithin is almost complete. After administration of labeled lecithin, there was a lag time of approximately 2 h before labeled lipid appeared in the blood stream. In four of the five patients, the peak of [^{14}C]-radioacitvity was reached earlier than that of [^{3}H]; the [^{14}C]-peak was between 4 and 12 h, and the [^{3}H] reached a peak between 6 and 24 h. At the peak of radioactivity, about 20% of the total dose of the [^{3}H]-label (in choline was present in the bloodstream. This represents a minimal figure for lecithin absorption. Most likely, total absorption of [^{3}H]-choline into plasma phospholipids was at least two to three times this amount. However, because of exchange processes and recycling, it is not possible to make a precise estimate.

Three possible mechanisms might explain the transfer of orally administered lecithin to plasma lecithin. First, oral lecithin could be hydrolyzed to lysolecithin in the intestine and reconverted to lecithin within the mucosa; second, oral lecithin could be hydrolyzed to glycerophosphocholine or phosphocholine in the mucosa, and these could then be resynthesized into lecithin; and third; the lecithin could be absorbed intact. In our study (15), because oral lecithin was labeled in both fatty acids and choline, it should be possible to distinguish between these three mechanisms. Certainly, if the [^{3}H]![^{14}C] ratios in oral and plasma lecithin were identical, the oral lecithin would have to be absorbed intact. Alternatively, a doubling of the ratio would indicate hydrolysis of oral lecithin to lysolecithin and resynthesis of lecithin with only one unlabeled fatty acid molecule; and finally, a complete absence of [^{14}C] from lecithin in the circulation would indicate loss of both fatty acids during absorption. In fact, when radioactivity first appeared in plasma, the [^{3}H] / [^{14}H] ratio in blood lecithin was somewhat less than twice that administered. This somewhat higher than 50% loss of $[^{14}C]$ is compatible with some, but certainly not all, lecithin being absorbed without complete hydrolysis to lysolecithin. In other words, a small portion of the lecithin could have been absorbed intact. Still, at the peak of radioactivity in plasma, the $[^3H]$ / $[^{14}C]$ ratio was essentially twice that of the ratio in the oral dose. This finding suggests that reesterification of lysolecithin is the major mechanism for absorption of dietary lecithin. The lack of a higher ratio indicates that esterification of glycerophosphocholine with two unlabeled fatty acids is not a significant mechanism for absorption of dietary lecithin. Thus, overall, the major route for lecithin entering the bloodstream is through the lysolecithin pathway. These studies, however, do not rule out the possibility that a portion of dietary lecithin is degraded completely within the intestinal mucosa so that it does not enter the body as lecithin at all.

3 Fate of Newly Absorbed Lecithin

In our studies, in which large quantities of dietary lecithin were fed (14), the appearance of lipids and lipoproteins into the plasma was examined. Since only lecithin was fed, all of the newly found lipoproteins would have to have been derived entirely from this substance. In another group of patients, safflower oil (100 mg kg^{-1} h^{-1}) was infused intraduodenally for comparison with the patients given lecithin by the intraduodenal route. Both types of fat were found to induce increases of lipoproteins of $S_f > 400$ (chylomicrons) and S_f 20–400 (VLDL). Lecithin infusions produced increases mainly in VLDL, while safflower oil produced mainly chylomicrons. Further, the chylomicrons derived from lecithin generally were smaller and had a higher phospholipid/triglyceride ratio than those induced by safflower oil. The enrichment of the VLDL fraction following the administration of oral lecithin presumably represented the occurrence of "small chylomicrons" of gut origin.

Of interest, in a parallel study in rats, the same pattern was not observed (14). In these animals, infusion of lecithin into the intestine resulted in the same large chylomicrons as observed during the infusion of safflower oil. In a previous study by Lekim (16), the oral administration of a single dose of lecithin, radiolabeled in both its fatty acid moieties, resulted in about 75% of radioactivity into lipoprotein (or chylomicron)-triglyceride.

It is interesting to speculate why infusions of identical amounts of fatty acids into the intestine with lecithin and with safflower oil produced different patterns of response in lipoproteins of humans. First, absorption of lecithin might not have been complete. However, our measurements of fecal excretion of lipids during lecithin feeding suggest that this was not the case; most of the lecithin appeared to be absorbed (14). Second, a portion of lecithin may have been degraded completely, so that only a fraction of it was available for production of plasma chylomicrons (or lipoproteins); this possibility is real, but it could not be studied. Third, the formation of smaller lipoprotein particles by lecithin feeding may have been due to a greater distribution of lecithin over the whole length of the intestine. This would result in lesser amounts of lipids entering each cell, and because of a relative "deficiency" of lipids, the resulting lipoproteins would be relatively lipid-depleted. The data from measurements of lecithin absorption are consistent with this mechanism. Finally, there might be a relatively high quantity of lecithin for a given amount of triglyceride synthesized during lecithin feeding, and this could increase the lipoprotein coat/core ratio; the result would be the formation of smaller lipoproteins.

Another question to consider is whether dietary lecithin may directly enter the HDL pool. Indeed, in our study with oral administration of labeled lecithin (15), a large amount of $[^3H]$-choline appeared rapidly in the lecithin fraction of HDL. Indeed, the specific activity of lecithin early in the study was higher in HDL than in VLDL+LDL. Seemingly, chylomicron-lecithin is shunted to HDL before it enters other lipoproteins. This observation is consistent with the hypothesis that the surface component of chylomicrons contributes to the HDL fraction (17). Another possible explanation for this finding is that the intestine itself may have produced nascent HDL containing newly absorbed dietary lecithin.

4 Effects of Dietary Lecithin on Plasma Lipoprotein Levels

In another study carried out by Kesäniemi and Grundy (18), the effects of dietary lecithin on concentrations of plasma lipoproteins were examined in patients with hypertriglyceridemia (type 4 hyperlipoproteinemia). In this study, ten patients were investigated during control periods and lecithin feeding. In the control period, 7g safflower oil were added to the diet to offset the addition of 10g lecithin in the lecithin-feeding period. Overall, the administration of lecithin did not alter plasma levels of total cholesterol, triglycerides, LDL-cholesterol, or HDL-cholesterol. In two patients, concentrations of total cholesterol and triglycerides actually rose during administration of lecithin, but in another, total cholesterol fell. Three patients had a reduction in LDL-cholesterol levels during lecithin feeding, but there was no difference for the group as a whole.

Turnover rates of VLDL-triglycerides were carried out during this study (18). On the average, compared to the control period, lecithin feeding did not produce changes in the concentrations, transport rates, or fractional catabolic rates for VLDL-triglycerides. If anything, there was a trend for lecithin feeding to cause an increase in production rates of VLDL-triglycerides, but again, this change was not significant.

5 Effects of Dietary Lecithin on Cholesterol Metabolism

The oral administration of lecithin has been reported to increase the fecal excretion of endogenous steroids, especially neutral steroids, regardless of the change in plasma cholesterol levels (19). In the study of Kesäniemi and Grundy (18), however, there were no changes in fecal outputs of neutral steroids or acidic steroids. Thus, high intakes of lecithin apparently did not affect the overall synthesis of either cholesterol or bile acids. In this study, as well as in the work of Beil and Grundy (14), there was a reduction in absorption of cholesterol. This effect has been reported by others as well (20–22). Nonetheless, the degree of reduction in cholesterol absorption was relatively small, and it probably was not enough to induce a significant decrease in the plasma LDL concentration. In this study (18), the diet was not rich in cholesterol, and our work cannot exclude the possibility that dietary lecithin would affect the plasma LDL level if the diet is high in cholesterol.

6 Conclusion

Our work (14,15,18) has provided additional insights into the metabolism of lecithin in human beings. It is clear that dietary lecithin is largely absorbed in humans, and a significant fraction of it is incorporated into the surface coat of chylomicrons. Its fatty acid components also can contribute to the triglyceride core of lipoproteins. Most of the lecithin is degraded to lysolecithin in the intestine, before being resynthesized into lecithin in the mucosa; however, we cannot rule out the possibility that a small fraction of oral lecithin is absorbed intact. After entering chylomicrons, a significant portion of newly absorbed lecithin enters the HDL fraction. Still, in our studies in patients with hypertriglyceridemia, we obtained no evidence that dietary lecithin will raise the HDL-cholesterol or lower the LDL-cholesterol levels. On the other hand, we have not carried out studies in patients with hypercholesterolemia, and our investigations do not allow us to predict whether significant changes in LDL or HDL would occur in such patients.

References

1. Adams CWM, Abdulla YH, Bayliss OB, Morgan RS (1967) Modification of aortic atheroma and fatty liver in cholesterol-fed rabbits by i.v. injection of saturated and polyunsaturated lecithin. J Pathol Bacteriol 94:77–87
2. Samochowiec L, Kadlubowska D, Rozewicka L (1976) Investigations in experimental atherosclerosis, pt 1. The effects of phosphatidylcholine (EPL) on experimental atherosclerosis in white rats. Atherosclerosis 23:305–317
3. Howard AN, Patelski J (1974) Mechanism of antiatherosclerotic action of intravenous polyunsaturated phosphatidyl choline. Scand J Clin Lab Invest 34 (Suppl):141:64–65
4. Waligora Z, Patelski J, Brown BD, Howard AN (1975) Effect of a hypercholesterolaemic diet and a single injection of polyunsaturated PC solution on the activities of lipolytic enzymes, acyl-CoA-cholesterol acyltransferase in rabbit tissues. Biochem Pharmacol 24:2263–2267
5. Blaton V, Vandamme D, Peeters H (1974) Activation of lipoprotein lipase in vitro by unsaturated phospholipids. FEBS Lett 44:185–188
6. Rosseneu M, Declercq B, Vandamme D, et al. (1979) Influence of oral polyunsaturated and saturated phospholipid treatment on the lipid composition and fatty acid profile of chimpanzee lipoproteins. Atherosclerosis 32:141–153
7. Arnesjo B, Nilsson A, Barrowman J, Borgstrom B (1969) Intestinal digestion and absorption of cholesterol and lecithin in the human. Scand J Gastroenterol 4:653–665
8. Scow RO, Stein Y, Stein O (1967) Incorporation of dietary lecithin and lysolecithin into lymph chylomicrons in the rat. J Biol Chem 242:4919–4924

9. Mansbach CM II (1977) The origin of chylomicron phosphatidylcholine in the rat. J Clin Invest 60:411–420
10. Shrivastava BK, Redgrave TG, Simmonds WJ (1967) The source of endogenous lipid in the thoracic duct lymph of fasting rats. Q J Exp Physiol 52:305–312
11. Baxter JH (1966) Origin and characteristics of endogenous lipid in thoracic duct lymph in rat. J Lipid Res 7:158–166
12. Ockner RK, Hughes FB, Isselbacher KJ (1969) Very low density lipoproteins in intestinal lymph: origin, composition, and role in lipid transport in the fasting state. J Clin Invest 48:2079–2088
13. Boucrot P (1972) Is there an entero-hepatic circulation of the bile phospholipids? Lipids 7:282–288
14. Beil FU, Grundy SM (1980) Studies on plasma lipoproteins during absorption of exogenous lecithin in man. J Lipid Res 21:525–536
15. Zierenberg O, Grundy SM (1982) Intestinal absorption of polyenephosphatidylcholine in man. J Lipid Res 23:1136–1142
16. Lekim D (1976) On the pharmacokinetics of orally applied essential phospholipids (EPL). In: Peeters H (ed) Phosphatidylcholine: Biochemical and clinical aspects of essential phospholipids. Springer, Berlin Heidelberg New York, pp 48–65
17. Tall AR, Small DM (1978) Plasma high-density lipoproteins. N Engl J Med 299:1232–1236
18. Kesäniemi YA, Grundy SM (1986) Effects of dietary polyenylphosphatidylcholine on metabolism of cholesterol and triglycerides in hypertriglyceridemic patients. Am J Clin Nutrit 43:98–107
19. Simons LA (1978) The effects of oral lecithin and clofibrate on cholesterol metabolism. Artery 4:167–182
20. Hollander D, Morgan D (1980) Effect of plant sterols, fatty acids and lecithin on cholesterol absorption in vivo in the rat. Lipids 15:395–400
21. Rampone AJ (1973) The effect of lecithin on intestinal cholesterol uptake by rat intestine in vitro. J Physiol 229:505–514
22. Rodgers JB, O'Connor PJ (1975) Effect of phosphatidylcholine on fatty acid and cholesterol absorption from mixed micellar solutions. Biochim Biophys Acta 409:192–200

Effects of Pantethine on Lipid Metabolism in Isolated Rat Hepatocytes

M. Galli Kienle, G. Cighetti, M. Del Puppo, and R. Paroni [1]

1 Introduction

Pantethine is the disulfide form of pantetheine the natural precursor of coenzyme A and of the prosthetic group of fatty acid synthetase. Interest in pantethine has increased recently as its lipid-lowering effects were demonstrated in laboratory animals (Shinomiya et al. 1980; Tomikawa et al. 1982) and in humans (Maggi et al. 1982; Avogaro et al. 1983; Gaddi et al. 1984). Since the mechanism of these effects is still unknown, we studied the modifications induced by pantethine on lipid metabolism in rat hepatocytes using either radioactive mevalonolactone (MVL) or acetate as the labeled substrate.

2 Experiments with Labeled MVL

It is known that mevalonate, both "in vivo" and "in vitro", inhibits cholesterol synthesis at the level of mevalonate formation by HMG CoA reductase (Jenke et al. 1981; Arebalo et al. 1982; Edwards et al. 1983). This effect is similar to that produced by hypercholesterolemic diets. In order to analyze the effects of pantethine under similar conditions in isolated hepatocytes (Cighetti et al. 1983), we have studied the incorporation of MVL radioactivity at a 0.5 mM concentration, which was shown by us to inhibit HMG CoA reductase via the accumulation of methyl sterols in isolated rat hepatocytes (Cighetti et al. 1986).

Pretreatment of hepatocytes with pantethine enhanced the effects of mevalonate, i.e., the incorporation of MVL radioactivity into cholesterol decreases as the concentration of pantethine increases. At the same time radioactivity associated with squalene and methyl sterols increases (Fig. 1). Accumulation of squalene and 4,4-dimethyl-5α-cholest-8-en-3β-ol was confirmed by measuring their amounts in the nonsaponifiable lipids by GC-MS (Cighetti et al. 1986) (Fig. 2).

---

1 Department of Medical Chemistry and Biochemistry, University of Milan, 20133 Milan, Italy

Drugs Affecting Lipid Metabolism
Ed. by R. Paoletti et al.
© Springer-Verlag Berlin Heidelberg 1987

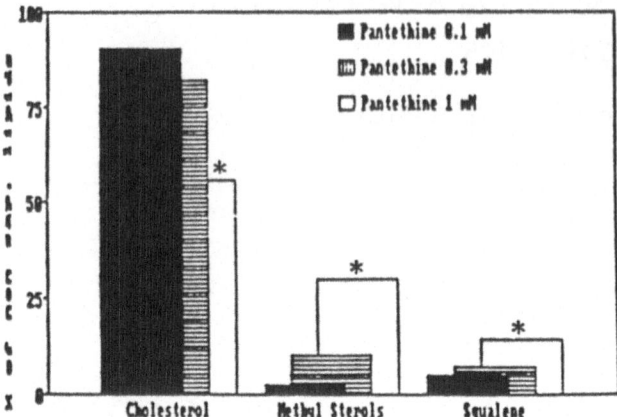

Fig. 1. Distribution of radioacitvity among cholesterol and its precursors after incubation of hepatocytes with pantethine and labeled MVL. The distribution of radioactivity was obtained from the
TLC analysis of nonsaponifiable lipids extracted from cells preincubated 1 h with pantethine and
incubated 1 h with 0.5 mM labeled MVL. Results are means obtained in duplicated incubations
with three cell preparations. * $p < 0.01$ vs controls preincubated without pantethine

Since it was previously suggested (Arebalo et al. 1982; Edwards et al. 1983) that intermediary sterol(s) accumulating after mevalonate treatment may act as modulatory
agent(s) of HMG CoA reductase, we have measured HMG CoA reductase activity in
cells treated with MVL and pantethine (Cighetti et al. 1983). The reductase activity in
microsomes prepared from cells incubated for 1 h with 0.5 mM MVL (437 ± 9 pmol
min^{-1} mg^{-1} protein; mean ± SE in three microsomal preparations) was significantly
lower than in control cells (577 ± 13). The active form was slightly modified by the
presence of MVL during the incubation. Addition of 1 mM pantethine to MVL produced an additive effect with a further decrease of the enzyme activity (284 ± 28;
$p < 0.01$ vs cells incubated only with MVL). Moreover, pantethine induced a significant
decrease ($p < 0.05$) of the active fraction of the enzyme as the expressed activity was
148 ± 35 in cells incubated with MVL plus pantethine and 237 ± 12 pmol min^{-1} mg^{-1}
protein with MVL only.

All the above described effects of pantethine were not observed when 0.01 mM MVL
was used, thus suggesting that MVL concentration is limiting for pantethine to exert
its effect.

3 Experiments with Labeled Acetate

The effects of pantethine on earlier stages of cholesterol biosynthesis and on fatty acid
synthesis were studied using labeled acetate. In this case the substrate concentration
was kept low in order to avoid modification of the endogenous pool. Incorporation of
radioactivity into cell lipids was dramatically lower in incubations carried out in the
presence of pantethine as compared to controls and this lowering effect was dependent

Table 1. Effect of pantethine on the incorporation of $[2\text{-}^{14}C]$ acetate radioactivity into CO_2 and lipids by isolated rat hepatocytes[a]

Incubation time (min)	Incorporated acetate			
	CO_2 pmol/10^7 cell		Total lipids nmol/10^7 cells	
	C	P	C	P
5	27 ± 1	35 ± 9	1.3 ± 0.10	0.47 ± 0.09*
15	210 ± 3	260 ± 30	4.8 ± 0.13	1.3 ± 0.06*
60	2800 ± 110	2000 ± 80	11.3 ± 0.29	2.6 ± 0.13**

[a]Hepatocytes (10^7/3 ml) were incubated with 1 mM pantethine (P) and without it (C). Results are means ± SE of triplicated incubations.
* $p < 0.05$; ** $p < 0.01$ vs C by Duncan's test.

on the incubation time (Table 1). On the other hand, the incorporation of radioactivity into CO_2 was only slightly modified under these conditions.

When the radioactive substrate was incubated with cells which had been preincubated for 1 h with pantethine, a concentration-dependent decrease of radioactivity incorporated into cholesterol, fatty acids, and CO_2 was observed (Fig. 3). These results suggested that preincubation with pantethine may modify the acetate pool. This was indeed the case because acetate concentration in the medium showed a dependence both on the time of incubation of cells with pantethine (not shown) and on its concentration (Fig. 3).

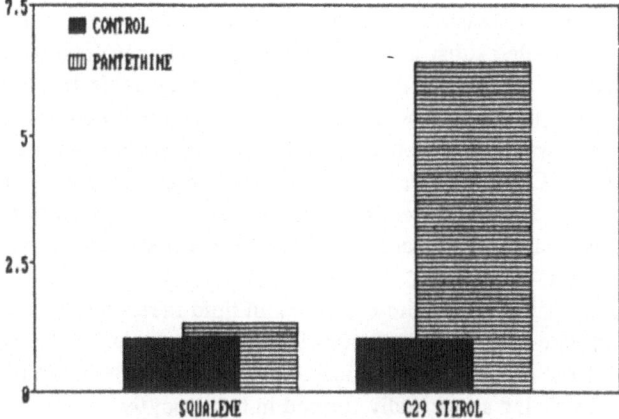

Fig. 2. Cell levels of 4,4'-dimethyl-5α-cholest-8-en-3β-ol and squalene. Results are expresses as ratios between amounts found in cells preincubated with pantethine and in control cells of the experiment described in Fig. 1

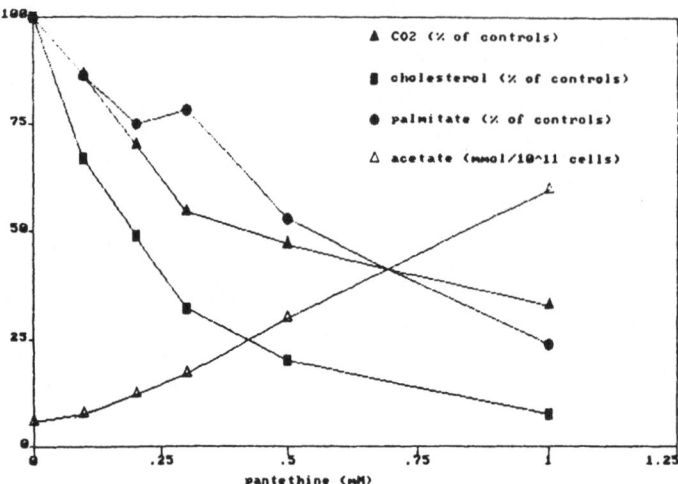

Fig. 3. Effects of pantethine on the incorporation of acetate radioactivity into CO_2, cholesterol, and fatty acids and on acetate levels. Hepatocytes ($10^7/3$ ml) were preincubated with pantethine for 1 h and incubated thereafter with labeled acetate (3.5 μCi; 58 mCi mmol^{-1}) for an additional hour. Acetate concentration was enzymatically determined in the medium of cells after 1-h preincubation. Results are means ± SE of triplicated incubations

A decrease in substrate molar radioactivity caused by the release of acetate in pantethine-treated cells may only partially account for the above observations, since the incorporation of radioactivity was far less for sterols than for CO_2 and fatty acids (Fig. 3). Moreover, in short-time incubations with labeled acetate, carried out after 1 h preincubation with pantethine, radioactivity associated with CO_2 was higher than in controls, whereas that in lipids was lower (data not shown). In order to understand how the modification of the substrate concentration affects its conversion via various pathways, incubations of cells with increasing concentrations of nonlabeled acetate mixed with a fixed amount of the labeled substrate were carried out. The results shown in Table 2 clearly indicate that the amounts of cholesterol and fatty acids formed during the incubation in cells preincubated 1 h with 1 mM pantethine are much lower than those formed in control cells with an acetate concentration of the same order of magnitude, whereas the amount of CO_2 is higher. From the double-reciprocal plots of the data, it can be calculated that pantethine treatment of the cells for 1 h results in 53 and 27% inhibition of cholesterol and fatty acid synthesis, respectively, while oxidation of acetate is almost twice that in controls.

It is not clear how pantethine brings about these changes in lipid metabolism. It is known that thioesters of pantetheine have lower affinities for the enzymes of cholesterol and fatty acid synthesis than the corresponding thioesters of CoA. It remains to be seen however if thioesters of pantetheine are actually formed in hepatocytes.

Table 2. Influence of acetate concentration on its metabolism by rat hepatocytes[a]

Acetate			Metabolites nmol/10^7 cells	
Concentration μmol/10^7 cells	Molar radioactivity[b] mCi mmol^{-1}	CO_2	Cholesterol	Acids
0.65 (C)	4.6	0.45 ± 0.1	5.1 ± 0.1	36 ± 3
1.15 (C)	2.6	1.1 ± 0.2	6.5 ± 0.7	49 ± 7
2.15 (C)	1.4	1.2 ± 0.3	6.6 ± 0.4	53 ± 4
5.15 (C)	0.58	1.7 ± 0.2	7.8 ± 0.9	68 ± 3
4.37 (P)[c]	0.69	3.0 ± 0.4	2.9 ± 0.1	36 ± 2

[a]In control samples (C) hepatocytes (10^7/3 ml) were preincubated for 1 h before the addition of acetate at the shown concentrations; in samples (P) preincubation was carried out with 1 mM pantethine and only radioactive acetate (0.06 μmol/10^7 cells) was added at the end of the preincubation. All samples were then incubated for 5 min. Results are means ± SE of triplicated incubations.

[b]Calculated from the radioactivity added (3 μCi/10^7 cells) and from the concentration of added acetate plus that found in the medium after 1-h incubation.

[c]Concentration found in the medium after 1-h preincubation plus that of radioactive acetate.

References

Arebalo RE, Tormanen CD, Hardgrave JE, Noland BJ, Scallen TJ (1982) In vivo regulation of rat liver 3-hydroxy-3-methylglutaryl coenzyme A reductase: immunotitration of the enzyme after short-term mevalonate or cholesterol feeding. Proc Natl Acad Sci USA 79:51–55

Avogaro P, Bittolo Bon G, Fusello M (1983) Effect of pantethine on lipids lipoproteins and apolipoproteins in man. Curr Ther Res 33:488–493

Cighetti G, Galli G, Galli Kienle M (1983) A simple model for studies on the regulation of cholesterol synthesis using freshly isolated hepatocytes. Eur J Biochem 133:573–578

Cighetti G, Del Puppo M, Paroni R, Galli G, Galli Kienle M (1986) Effects of pantethine on cholesterol synthesis from mevalonate in isolated rat hepatocytes. Atherosclerosis 60:67–77

Edwards PA, Lan S, Tanaka RD, Fogelman AM (1983) Mevalonolactone inhibits the rate of synthesis and enhances the rate of degradation of 3-hydroxy-3-methyl-glutaryl-coenzyme A reductase in rat hepatocytes. J Biol Chem 258:7272–7275

Gaddi A, Descovich GC, Noseda G, Fragiacomo C, Colombo L, Craveri A, Montanari G, Sirtori CR (1984) Controlled evaluation of pantethine a natural hypolipidemic compound in patients with different forms of hyperlipoproteinemia. Atherosclerosis 50:73–83

Jenke HS, Löwel M, Berndt J (1981) In vivo effect of cholesterol feeding on the short term regulation of hepatic hydroxymethylglutaryl coenzyme A reductase during the diurnal cycle. J Biol Chem 256:9622–9625

Maggi GC, Donati C, Criscuoli G (1982) Pantethine a physiological lipomodulating agent in the treatment of hyperlipidemias. Curr Ther Res 32:380–386

Shinomiya M, Matsuoka N, Shirai K, Morisaki N, Sasaki N, Murano S, Saito Y, Kumagai A (1980) Effect of pantethine on cholesterol ester metabolism in rat arterial wall. Atherosclerosis 36:75–80

Tomikawa M, Nagayasn T, Towara K, Kameda K, Abiko Y (1982) Effect of pantethine on lipoprotein profiles and HDL subfractions in experimentally hypercholesterolemic rabbits. Atherosclerosis 41:267–277

Effects of Pantethine on Experimental Atherosclerosis in Rabbits

M. R. Malinow, B. Upson, M. Axthelm, and P. McLaughlin [1]

1 Introduction

Panthetine is the disulfide dimer of panthetheine, the amide conjugate of pantothenic acid and cysteamine, which occurs naturally as a portion of coenzyme A and of acyl-carrier proteins (Wittwer et al. 1985). When pantethine was given to hyperlipoproteinemic patients, it lowered plasma cholesterol levels (Maioli et al. 1984; Cattin et al. 1985) and elevated high density lipoprotein (HDL) cholesterol levels, especially the HDL_2 fraction with a consequent increase in the HDL_2/HDL_3 ratio (Maioli et al. 1984). The effects on normolipidemic patients were less marked (Murai et al. 1983). In rabbits fed 0.5% cholesterol, pantethine also reduced total cholesterol levels and produced minor decreases in aortic atherosclerosis (Carrara et al. 1984); larger decreases were reported in a study involving a smaller number of animals and the effects were accentuated by combining the intake of pantethine with probucol (Tawara et al. 1986). We have investigated the effects of pantethine and of an ester derivative (tetra[3-(3-pyridinemethoxycarbonyl) propionyl]- pantethine tetratartrate, MG 28362) in rabbits receiving food additioned with cholesterol at the 0.2 or 0.25% level. Only results obtained with pantethine will be reported here.

2 Material and Methods

Sixty male New Zealand white rabbits weighing approximately 2.5 kg were kept in individual cages and were offered Purina Chow pellets covered with 4.2% butter, as well as 0.2% cholesterol (Bio Serv Inc., Frenchtown, NJ) for 1 month and 0.25% cholesterol thereafter. Food and water were offered ad libitum during the first month; subsequently, the food was adjusted to the estimated intake. The animals were assigned randomly to three groups of 20 rabbits each with the following additions to the diet: group I, none (controls); group II, 1.3% pantethine; group III, 1.1% MG 28362. The test drugs were suspended in egg white ($35\,\text{g}\,\text{k}^{-1}$ of chow), beaten, spread

---

1 Oregon Regional Research Primate Center, 505 NW 185th Ave., Beaverton, OR 97005; and Oregon Health Sciences University, 3181 SW Sam Jackson Rd., Portland OR 97201, USA

Drugs Affecting Lipid Metabolism
Ed. by R. Paoletti et al.
© Springer-Verlag Berlin Heidelberg 1987

over the chow pellets, and air dried. The control diet contained egg white similarly applied. The diets were prepared in 10 kg batches and kept refrigerated until used.

Blood was obtained from an ear vein 1 and 2 months after the beginning of the observation, and by cardiac puncture under anesthesia before termination, for cholesterol determination (cf. infra). The animals were weighed monthly. Acceptance of food was estimated visually during the last month of the experiment. At the end of 16 weeks, the animals were killed with intravenous Surital R (thiamytal sodium, Parke-Davis, Morris Plains, NJ). The thorax and abdomen were opened, and the liver, spleen, and kidneys were removed, blotted, weighed, and appropriate samples — together with the heart after removal — were fixed in 10% formalin, and stored for future study. The aorta was removed, opened longitudinally, washed briefly with water, fixed in 10% formalin, and stained with Sudan IV. The percent surface involvement with sudanophilia was measured "blindly" in the first 10 cm and in the remaining aorta, to be called for convenience thoracic and abdominal aorta, respectively, using a grid stencil method modified from Howard (1979).

Blood was mixed with EDTA and the plasma was separated by centrifugation. Cholesterol was determined in an alcohol-acetone extract by the $Fe Cl_3$ method of Rudel and Morris (1973); HDL-cholesterol was determined at termination after precipitation of low density (LDL) and very low density lipoproteins (VLDL) with heparin-Mn_2 (Albers et al. 1978). Since no attempts were made to validate the method against ultracentrifugation in cholesterol-fed rabbits, HDL-cholesterol findings should be considered approximations.

Group means were tested by Student's t-test adjusted by Cochran analysis of independent variables when variances were unequal (Snedecor and Cochran 1967). The integrated average of plasma cholesterol (months 1 through 4) was correlated with the percent aortic intimal sudanophilia using the Statistical Program for Social Sciences (SPSS).

3 Results

3.1 Body and Organ Weights; Food Acceptance (see Figs. 1 and 2)

One animal in the control group and two in the pantethine group (χ^2 nor significant) died during the course of the observation and their data were deleted from the experiment. Body weights increased similarly in the two groups of animals. Food acceptance (not shown) was occasionally reduced in seven animals of group 1 (control) and in one animal of group II (pantethine) (χ^2 not significant). The relative weights of liver, spleen, and kidneys were similar in both groups of animals.

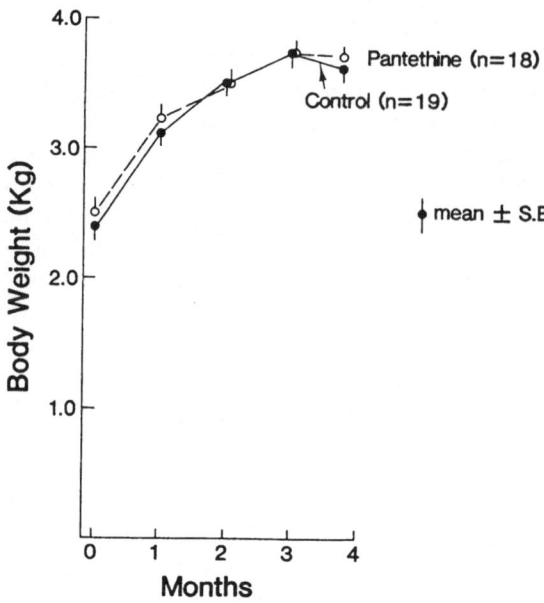

Fig. 1. Body weights of rabbits during the experimental period

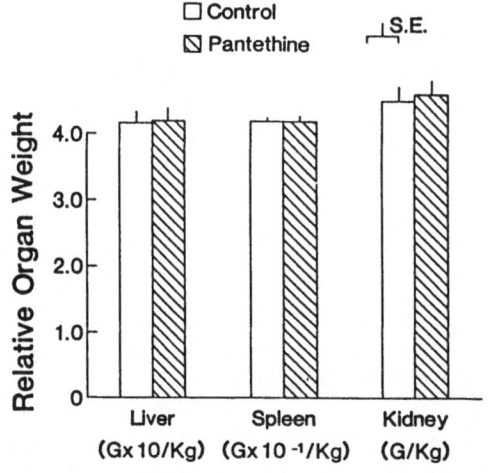

Fig. 2. Relative weights of selected organs in rabbits

3.2 Plasma Cholesterol (see Fig. 3)

Plasma cholesterol was greatly elevated in the control animals; pantethine reduced hypercholesterolemia by about 50%; the differences were significant ($p < 0.01$) at each time interval. The levels of HDL-cholesterol at termination (not shown) were similar in both groups of animals, i.e., 36 ± 5 and 39 ± 7 (mg dl^{-1}) in groups 1 and II, respectively (p = not significant).

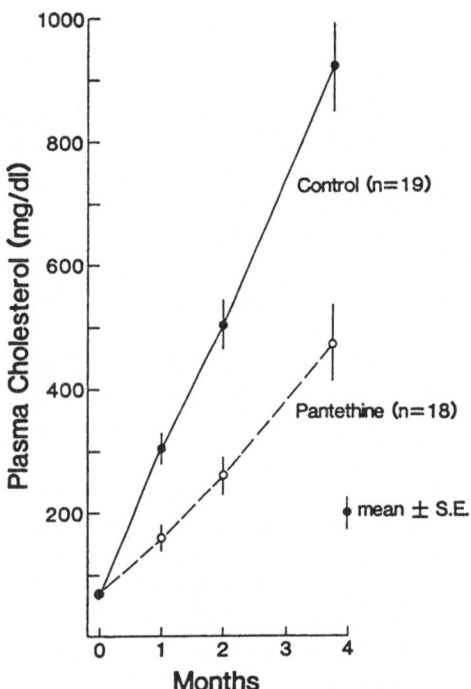

Fig. 3. Plasma cholesterol levels in rabbits. The differences are significant ($p < 0.01$) at each interval into the experiment. The initial values are derived from data reported elsewhere (Malinow et al. 1980)

3.3 Aortic Intimal Sudanophilia (see Fig. 4)

Intimal sudanophilia was reduced by pantethine from 43.1 ± 5.3 to $15.7 \pm 3.8\%$, from 27.9 ± 4.6 to $6.8 \pm 2.0\%$, and from 39.2 ± 4.7 to $12.9 \pm 3.1\%$ in the thoracic, abdominal, and total aorta, respectively ($p < 0.01$ for each contrast). These data indicate decreases of 64, 76 and 67%, respectively, from the involvement present in control rabbits.

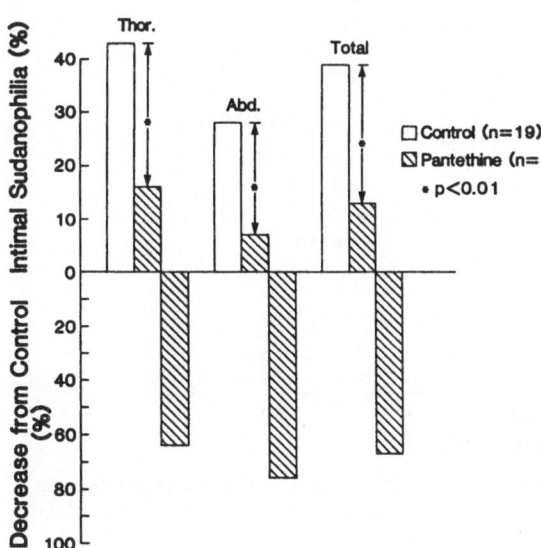

Fig. 4. Intimal sudanophilia in the thoracic (*Thor.*), abdominal (*Abd.*), and total aorta of rabbits.
(See Methods for definition of aortic segments)

4 Discussion

Our results demonstrate that pantethine réduced plasma cholesterol without changing HDL-cholesterol levels in cholesterol-fed rabbits; this was not due to reduced food intake with a consequent lower body weight. The drug also markedly decreased aortic intimal sudanophilia, which mainly corresponds to the extent of fatty streaks and fatty-fibrous plaques in the vessel wall. The findings are similar to those observed by Carrara et al. (1984) and Tawara et al. (1986), although the antiatherosclerosis effects were more marked in our experiment, perhaps due to the smaller content of cholesterol in the food.

Figure 5 shows that there is a positive correlation between total aortic intimal sudanophilia and the level of plasma cholesterol averaged throughout the observation (r=0.751; $p<0.01$). The index of determination ($r^2=0.56$) suggests that the average plasma cholesterol, including the hypocholesterolemic effects of pantethine, *explains* about 56% of the variance of aortic sudanophilia. The remaining variance may be related to undetermined arterial wall factors involved in vascular reactivity to choles-terolemia (Malinow et al. 1958, 1976). Additionally, other reported effects of pantethine could be involved, namely, removal of fatty acids through increased oxida-tion (Miyagawa 1976 quoted by Farina et al. 1980); increased levels of HDL_2 with a higher HDL_2/HDL_3 ratio (Maioli et al. 1984); stimulation of acid cholesterol esterase activity in the arterial wall (Shinomiya et al. 1980), and prevention of LDL peroxida-tion (Bon et al. 1985). Future studies are needed to determine the contribution of these mechanisms to the antiatherogenic effects of pantethine.

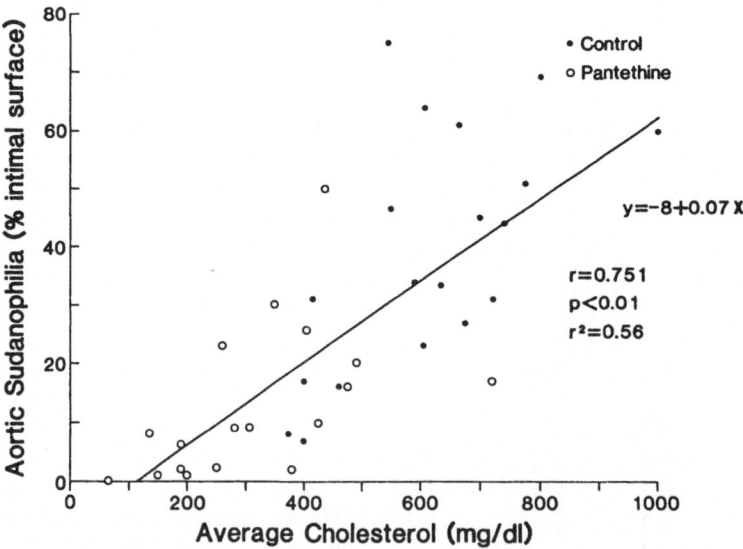

Fig. 5. Correlation between percent intimal sudanophilia in the total aorta vs the integrated plasma cholesterol levels from months 1 through 4; r, Pearson's index of correlation; r^2, index of deter-mination

5 Summary

Rabbits were fed Purina rabbit chow additioned with butter and 0.2 or 0.25% cholesterol, without or with 1.3% pantethine for 16 weeks. Pantethine reduced cholesterolemia by about 50% and aortic intimal sudanophilia by about 70%. No signs of toxicity were detected.

Acknowledgements. The work described in this article, Publication No. 1487 of the Oregon Regional Primate Research Center, was supported with grants RR-00163 and HL-16587 from the National Institutes of Health and by funds from Maggioni-Winthrop S.P.A., Milan, Italy. We are thankful to Dr. Mario Riva, Dr. Franco Ravenna, and Dr. Carlo Donati for the generous supply of pantethine and MG 28362.

References

Albers JJ, Warnick GR, Wiebe D, King P, Steiner P, Smith L, Breckenridge C, Chow A, Kuba K, Weidman S (1978) Multi-laboratory comparison of 3 heparin-MN2+ precipitation procedures for estimating cholesterol in high-density lipoprotein. Clin Chem 24:853–856

Bon GB, Cazzolate G, Zago S, Avogaro P (1985) Effects of pantethine on in vitro peroxidation of low density lipoproteins. Atherosclerosis 57:99–106

Carrara P, Matturri L, Galbussera M, Lovati MR, Franceschini G, Sirtori CR (1984) Pantethine reduces plasma cholesterol and the severity of arterial lesions in experimental hypercholesterolemic rabbits. Atherosclerosis 53:255–264

Cattin L, Da Col PG, Fonda M, Mameli MG, Pilotto L, Vanuzzo D, Feruglio GA (1985) Treatment of hypercholesterolemia with pantethine and fenofibrate; an open randomized study of 43 subjects. Curr Ther Res 38:386–394

Farina R, Lovati MR, Raucci G, Sirtori CR (1980) Effects of pantethine on different models of experimental hyperlipidemia in rodents: a comparison with clofibrate. Pharmacol Res Commun 14:499–510

Howard CF, Jr. (1979) Aortic atherosclerosis in normal and spontaneously diabetic Macaca nigra. Atherosclerosis 33:479–493

Maioli M, Pacifico A, Cherchi GM (1984) Effect of pantethine on the subfractions of HDL in dyslipidemic patients. Curr Ther Res 35:307–311

Malinow MR, Pellegrino AA, Ramos EH (1958) Chemical and anatomical correlations in cholesterol-fed rabbits. Acta Physiol Lat Am 8:37–46

Malinow MR, McLaughlin P, Papworth L, Kaito HK, Lewis L, McNulty WP (1976) A model for therapeutic intervention on established coronary atherosclerosis in a nonhuman primate. Adv Exp Med Biol 67:3–31

Malinow MR, McLaughlin P, Stafford C, Kohler GO, Livingston AL (1980) Alfalfa saponins and alfalfa seeds: dietary effects in cholesterol-fed rabbits. Atherosclerosis 37:433–438

Murai A, Miyahara T, Tanaka T, Sako Y, Nishimura N, Kameyama M (1983) The effect of pantethine on lipid and lipoprotein abnormalities in survivors of cerebral infarction. Artery 12:234–243

Rudel LL, Morris MD (1973) Determination of cholesterol using opthalaldehyde. J Lipid Res 14:364–366

Shinomiya M, Matsuoka N, Shirai K, Morisaki N, Sasaki N, Murano S, Saito Y, Kumagai A (1980) Effect of pantethine on cholesterol ester metabolism in rat arterial wall. Atherosclerosis 36:75–80

Snedecor GW, Cochran WG (1967) Statistical methods, 6th edn. State Univ Press, Ames, Iowa, pp 114–116

Tawara K, Ishihara M, Ogawa H, Tomikawa M (1986) Effect of probucol, pantethine and their combinations on serum lipoprotein metabolism and on the incidence of atheromatous lesions in the rabbit. Jpn J Pharmacol 41:211–222
Wittwer CT, Gahl WA, Butler JDeB, Zatz M, Thoene JG (1985) Metabolism of pantethine in cystinosis. J Clin Invest 76:1665–1672

Modified LDL in Humans: Effect of Pantethine

G. Bittolo Bon, G. Cazzolato, and P. Avogaro [1]

1 Introduction

Previous studies have established that low density lipoproteins (LDL) incubated with malondialdehyde (MDA) and oxidized LDL undergo extensive modification in structure and are specifically recognized by the acetyl-LDL receptor of the macrophages (Fogelman et al. 1980; Henriksen et al. 1981; Steinbrecher et al. 1985). Oxidatively modified LDL are also potentially toxic to endothelial cells (Hessler et al. 1983). The addition of pantethine [D-bis(N-pantothenil-B-aminoethyl)-disulfide] in vitro during the incubation of LDL with O_2 prevents the increase of the electrophoretic mobility, the increase of MDA content and the appearance of apoB aggregates in oxidized LDL (Bittolo Bon et al. 1985). The aim of the study was to assess the effect of treatment with pantethine in modifying the LDL peroxidation in vitro. Moreover, the effect of the drug on an LDL subfraction was studied, which was by ion exchange chromatography, that according to our data represents on aliquot of human LDL peroxidized in vivo.

2 Methods

LDL was isolated by density gradient ultracentrifugation from freshly drawn plasma from eight patients before and after 2 weeks of therapy (pantethine 600 mg b.i.d). Chemical analyses and the determination of lipid peroxides (Yagi 1982) were done in freshly prepared LDL and in aliquote exposed to peroxidative stress by dialysis for 24 h at room temperature against phosphate buffer saline continually streamed with O_2. The study of the effect of pantethine on the more electronegative subfraction of LDL present in vivo was done in eight patients affected by coronary atherosclerosis. Fasting blood was collected before and after 2 weeks of therapy (doses as above) in test tubes containing EDTA (1 mg ml^{-1}), reduced glutathione (0.5 mg ml^{-1}), chloramphenicol (0.05 mg ml^{-1}), d-alpha-tocopherol (50 μg ml^{-1})

[1] Regional General Hospital, Unit for Atherosclerosis, 30100 Venice, Italy

Drugs Affecting Lipid Metabolism
Ed. by R. Paoletti et al.
© Springer-Verlag Berlin Heidelberg 1987

and the inhibitor of phospholipase A2 4-bromophenacylbromide (40 μM). LDL was isolated by density gradient ultracentrifugation and dialyzed 24 h under an N_2 stream in the dark at 4°C. The LDL was then chromatographed on DEAE CL68 and eluted with a linear gradient from 0 to 0.3 M NaCl. The effluent was monitored at 280 nm for proteins and at 254 nm for fatty acid conjugated dienes (Pryor and Castle 1984). Two ml fractions were collected for chemical analyses and for estimation by HPLC of diene-conjugated fatty acids of cholesteryl esters (CE-DC), triglycerides (TG-DC) and diene conjugate products (DCP) (Cawood et al. 1983).

3 Results

Freshly prepared LDL from untreated patients showed a lipid peroxide content of 0.29 ± 0.09 (SE) nM MDA/M LDL cholesterol. The incubation with O_2 increased the amount of MDA in LDL nine times (Fig. 1). Following therapy with pantethine the MDA content of LDL was only slightly lower than under basal conditions. In such LDL, however, the lipid peroxidation by O_2 was significantly reduced (Fig. 1). In both treated and untreated patients of the second group the elution by ion exchange chromatography allowed the identification of an LDL subfraction with a more electronegative change than the LDL bulk. This subfraction, operatively called modified LDL (mLDL), was characterized by the presence of a distinct peak at 254 nm and by a higher percentage content of protein, a lower CE/FC ratio and by an extremely lower content of phospholipids (PL) than the LDL bulk. Moreover, mLDL had a

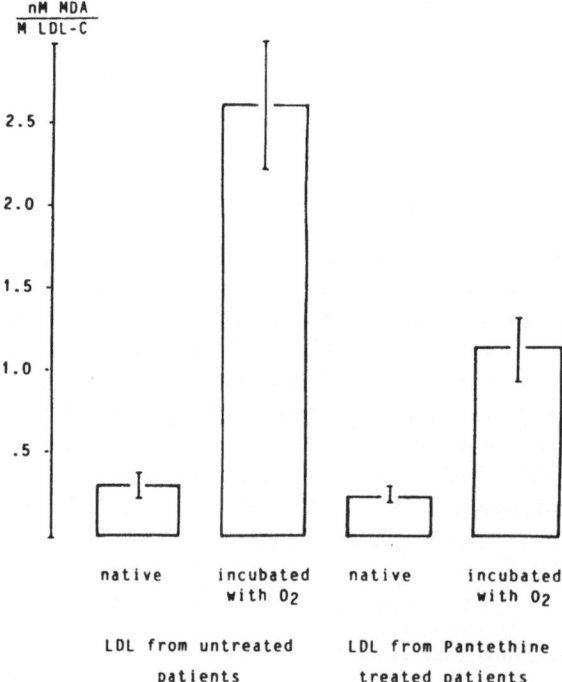

Fig. 1. Lipid peroxide content in native LDL and after 24 h incubation with O_2 before and after therapy with pantethine (values are expressed as mean ± SE)

Table 1. Mass value, percent contribution to total LDL, chemical composition, values of conjugated dienes and peroxides in native LDL (nLDL) and modified LDL (mLDL)

	Basal		After pantethine	
	nLDL	mLDL	nLDL	mLDL
Mass (mg dl⁻¹)	300 ± 70	106 ± 55	314 ± 65	63 ± 26*
% of total LDL[a]	74 ± 12	26 ± 12	83 ± 8***	17 ± 8***
CE %	45 ± 3	48 ± 3	46 ± 4	47 ± 8
FC	12 ± 1	15 ± 2*	10 ± 1	14 ± 4*
TG	4 ± 1	4 ± 2	5 ± 2	3 ± 2
PL	23 ± 2	9 ± 3**	23 ± 2	10 ± 3**
P	16 ± 2	24 ± 3**	16 ± 3	26 ± 7**
CE/FC	4.0 ± 0.7	3.2 ± 0.7**	4.6 ± 0.8	3.7 ± 1.3**
CE-DC	1.6 ± 0.5	4.8 ± 1.7**	0.7 ± 0.5*	4.6 ± 1.4**
TG-DC	13 ± 4	16 ± 8	10 ± 3	14 ± 5
DCP	0.3 ± 0.1	1.6 ± 0.8**	0.3 ± 0.1	1.9 ± 0.5**

* $P < 0.05$; ** $P < 0.01$ vs nLDL; *** $P < 0.01$ vs basal values.
[a] CE = cholesteryl esters; FC = free cholesterol; TG = triglycerides; PL = phospholipids; CE-DC, TG-DC = conjugated dienes of cholesterol and triglycerides; DCP = conjugated diene products; CE-DC, TG-DC and DCP = given as integrated area units/lipid concentration expressed in nmol l⁻¹; P = proteins.

higher content of CE-DC, TG-DC and DCP. The relative and absolute concentration of plasma mLDL was significantly reduced after pantethine (Table 1).

4 Discussion

The possibility that modification of LDL may play a role in the pathogenesis of atherosclerosis increased the interest focussed on LDL peroxidation and on its prevention by anti-oxidant compounds (Steinberg 1986). Pantethine is a natural compound with a plasma lipid-lowering action (Avogaro et al. 1983). The drug furthermore shows a protective effect against endothelial lesions in cholesterol-fed animals, which seems mostly independent from the cholesterol-lowering action (Carrara et al. 1984), and is effective in inhibiting the LDL peroxidation in vitro (Bittolo Bon et al. 1985). This report stresses that LDL isolated from the plasma of patients treated with conventional doses of pantethine is resistant to peroxidation in vitro. Moreover, the drug induces a significant reduction of plasma mLDL. Because of their electric charge and the high content of fatty acids CD and DCP, we believe this subfraction is the in vivo peroxidized aliquot of human plasma LDL.

Although there is no evidence and we lack an assay of the drug, pantethine is lipophilic and probably binds LDL. Thus, it would offer protection for LDL against in vivo peroxidative processes. If oxidative-induced modification of LDL is of relevance in atherogenesis, then pantethine may have antiatherogenic properties either lowering plasma lipids or inhibiting lipid peroxidation.

References

Avogaro P, Bittolo Bon G, Fusello M (1983) Effect of pantethine on lipids, lipoproteins and apolipoproteins in man. Curr Ther Res 33:488

Bittolo Bon G, Cazzolato G, Zago S, Avogaro P (1985) Effects of pantethine on in vitro-peroxidation of low density lipoproteins. Atherosclerosis 57:99–106

Carrara P, Matturri L, Galbusera M, Lovati MR, Franceschini G, Sirtori CR (1984) Pantethine reduces plasma cholesterol and the severity of arterial lesions in experimental hypercholesterolemic rabbits. Atherosclerosis 53:255

Cawood P, Wickens DG, Inversen SA, Braganza JM, Dormandy TL (1983). The nature of diene conjugation in human serum, bile and duodenal fluid. FEBS Lett 162:239–243

Fogelman AM, Shechter I, Seager J, Hokom M, Childs JS, Edwards PA (1980) Malondialdehyde alteration of low-density lipoproteins leads to cholesterol ester accumulation in human monocyte-derived macrophages. Proc Natl Acad Sci USA 77:2214–2218

Henriksen T, Mahoney E, Steinberg D (1981) Enhanced macrophage degradation of low-density lipoprotein previously incubated with cultured endothelial cells; recognition by receptors for acetylated low density lipoproteins. Proc Natl Acad Sci USA 78:6499–6503

Hessler JR, Morel DW, Lewis JL, Chisolm GM (1983) Lipoprotein oxidation and lipoprotein-induced cytotoxicity. Arteriosclerosis 3:215–222

Pryor WA, Castle L (1984) Chemical methods for detection of lipid hydroperoxides. In: Packer L (ed) Methods in enzymology 105: Oxygen radicals in biological systems. Academic Press, London New York Orlando, pp 293–299

Steinberg D (1986) Studies on the mechanism of action of probucol. Am J Cardiol 57:16H–21H

Steinbrecher UP, Parthsarathy S, Leake DS, Witzum JL, Steinberg D (1985) Modification of low-density lipoproteins by endothelial cells involves lipid peroxidation and degradation of low-density lipoprotein phospholipids. Proc Natl Acad Sci USA 81:3883–3887

Yagi K (1982) Assay for serum lipid peroxide level and its clinical significance. In: Yagi K (ed) Lipid peroxides in biology and medicine. Academic Press, London, New York, pp 223–242

Changes in Platelet Membrane Lipid Composition Following Pantethine Administration

D. Prisco, P. G. Rogasi, and G. G. Neri Serneri[1]

1 Introduction

Hyperlipoproteinemias, and in particular type IIa hyperlipoproteinemia, have been found to be one of the main risk factors for the development of atherosclerosis and its thromboembolic complications.

In an attempt to modulate or control lipid changes associated with the disease and possibly to prevent or to reverse the evolution of atherosclerosis, a number of anti-lipemic drugs have been tested and some have found clinical application. Unfortunately, the clinical use of these drugs has not been completely successful because of side effects, sometimes serious, which cast doubts on the feasibility of their prolonged use. These difficulties led to the study of some natural compounds which could interfere with lipid metabolism by physiological mechanisms.

One of these substances is pantethine, the oxidation form of pantethine, a component of the synthetic pathway of coenzyme A. In the last years the clinical use of pantethine in patients affected by hyperlipoproteinemias was reported to induce a reduction of plasma cholesterol and triglycerides with an increase of HDL-cholesterol (Hiramatsu et al. 1981; Avogaro et al. 1983).

A restricted number of investigations were performed on the effects of pantethine treatment on platelet functions, in particular on platelet aggregation (Hiramatsu et al. 1981; Prisco et al. 1984). A reduction of adenosine diphosphate-induced platelet aggregation (Hiramatsu et al. 1981), collagen-induced platelet aggregation (Prisco et al. 1984) and thromboxane A2 platelet production (Prisco et al. 1984) was observed after oral treatment with pantethine in hyperlipoproteinemic patients. Some of these effects of pantethine on platelets could be mediated by its effects on platelet lipids. In fact, membrane platelet lipids play an important role in platelet functions related to hemostasis and thrombosis so that alterations in their composition can result in platelet functional changes (Goodnight et al. 1982).

The aim of this study was to examine the effects of pantethine oral treatment on plasma and platelet lipid composition in a group of type IIa hyperlipoproteinemic patients.

---

1 Clinica Medica I dell'Università, Viale Morgagni 85, 50134 Florence, Italy

Drugs Affecting Lipid Metabolism
Ed. by R. Paoletti et al.
© Springer-Verlag Berlin Heidelberg 1987

2 Material and Methods

2.1 Subjects Investigated

We studied 14 patients (10 males and 4 females; 21 to 52 years) affected by type IIa hyperlipoproteinemia according to the Fredrickson-Lees classification. All had at least one first-degree relative with hypercholesterolemia. Patients had a total plasma cholesterol concentration above 7.5 mmol l^{-1} and a normal plasma triglyceride concentration. None had diabetes.

2.2 Experimental Procedure

The study was based on the crossed administration of the active drug (Pantetina, Maggioni Farmaceutici SpA, Milan, Italy), 1200 mg a day, orally or a placebo for 28 days according to a single-blind design. After a 28-day wash-out period the treatment was crossed. Throughout the study patients did not modify their dietary habits.

2.3 Methods

After the preparation of washed platelets and platelet-poor plasma, samples were extracted and analyzed for their lipid composition according to procedures previously described (Prisco et al. 1986).

2.4 Statistical Analysis

To express the results we considered determinations obtained at the end of active drug and placebo treatment. Student's t-test for paired data was employed for statistical evaluation.

3 Results

A decrease of cholesterol (placebo: 8.2 ± 1.1 mmol l^{-1}; pantethine: 7.0 ± 1.5 mmol l^{-1}, $p < 0.005$) and total phospholipid (placebo: 3.4 ± 0.8 mmol l^{-1}; pantethine: 2.8 ± 0.9 mmol l^{-1}, $p < 0.01$) content in plasma was found after pantethine treatment. Moreover, a slight but significant decrease in the sphyngomyelin/phosphatidylcholine ratio was found after pantethine treatment (placebo: 0.33 ± 0.06; pantethine: 0.29 ± 0.05, $p < 0.05$).

In the three major plasma phosphoglyceride fractions we could observe, after pantethine treatment, a reduction of saturated fatty acids (Fig. 1).

After pantethine treatment we observed a reduction both of platelet cholesterol and phospholipids (Fig. 2) without any change of their ratio.

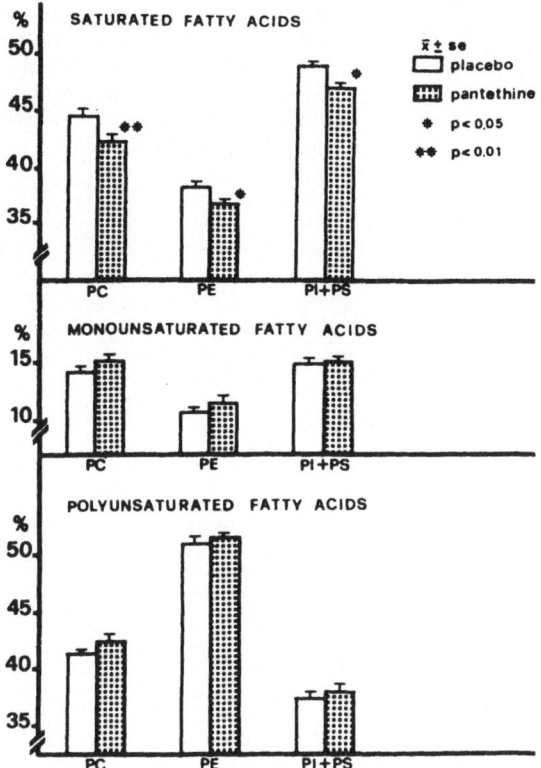

Fig. 1. Fatty acid composition of plasma phospholipids after placebo and pantethine treatment (*PC* phosphatidylcholine; *PE* phosphatidylethanolamine; *PI+PS* phosphatidylinositol + phosphatidylserine)

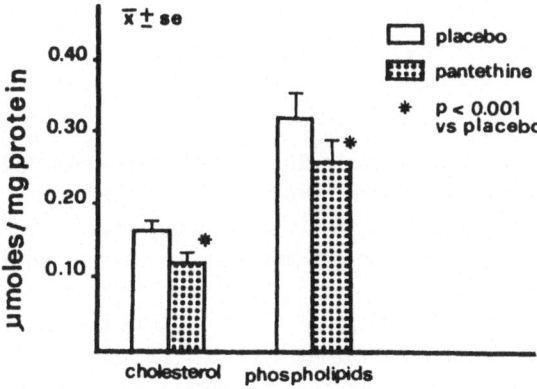

Fig. 2. Cholesterol and phospholipid content in platelets after placebo and pantethine treatment

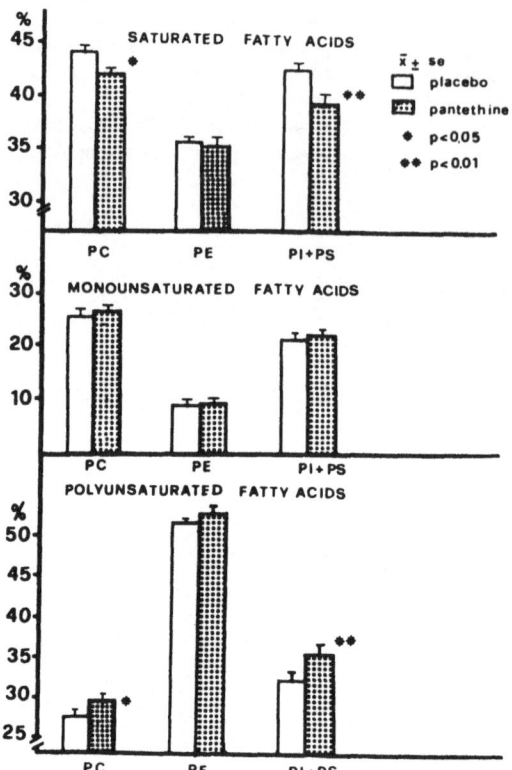

Fig. 3. Fatty acid composition of platelet phospholipids after placebo and pantethine treatment (*PC* phosphatidylcholine; *PE* phosphatidylethanolamine; *PI+PS* phosphatidylinositol + phosphatidylserine)

The analysis of fatty acids esterified in the three main platelet phosphoglyceride fractions showed slight but significant variations after pantethine treatment (Fig. 3): a decrease of saturated fatty acids and an increase of polyunsaturated.

4 Discussion

In IIa hyperlipoproteinemia elevated concentrations both of total cholesterol and total phospholipids in platelets in comparison to controls were reported (Shattil et al. 1977) and were found even in our laboratory. The results of our study demonstrate that oral treatment with pantethine in IIa hyperlipoproteinemia can interfere with lipid metabolism and can modify not only plasma but also platelet lipid composition.

The effects on platelet lipids are most likely not specific for these cells but rather secondary to changes in the plasma lipid composition. Extrinsic factors, mainly plasma lipids, play a major role in the regulation of platelet lipid pattern and, in particular, platelets can incorporate both cholesterol and phospholipids from plasma (Shattil et al. 1975; Joist et al. 1976).

In conclusion, pantethine is able to reduce plasma cholesterol and to increase the ratio of polyunsaturated to saturated fatty acids. Accordingly, pantethine affects the composition of platelet lipids and this effect seems to also have functional implications because membrane platelet lipids play a major role in some platelet functions such as platelet aggregation and thromboxane A2 synthesis (Goodnight et al. 1982). Moreover, pantethine, in addition to its well-documented action on plasma cholesterol and triglycerides, can favourably affect two determinants of plasma lipid fluidity: the polyunsaturated/saturated fatty acid ratio and the sphyngomyelin/phosphatidyl-choline ratio. These variations can influence the fluidity of cell membranes (McMurchie and Raison 1979) and can enhance the lipoprotein clearance rate (Morrisset et al. 1977).

Therefore, these results support once more a favourable activity of pantethine in the treatment of type IIa hyperlipoproteinemia.

References

Avogaro P, Bittolo Bon G, Fusello M (1983) Effect of pantethine on lipids, lipoproteins and apolipoproteins in man. Curr Ther Res 33:488–493

Goodnight SH, Jr, Harris WS, Connor WE, Illingworth DR (1982) Polyunsaturated fatty acids, hyperlipidemia and thrombosis; a review. Arteriosclerosis 2:87–113

Hiramatsu K, Nozaki H, Arimori S (1981) Influence of pantethine on platelet volume, microviscosity, lipid composition and functions in diabetes mellitus with hyperlipidemia. Tokai J Exp Clin Med 6:49–57

Joist JHG, Dolezel G, Lloyd JV, Mustard JF (1976) Phospholipid transfer between plasma and platelets in vitro. Blood 48:199–211

McMurchie EJ and Raison JK (1979) Membrane lipid fluidity and its effect on the activation energy of membrane associated enzymes. Biochim Biophys Acta 554:364–374

Morrisset JD, Pownall HJ, Jackson RL, Sogura R, Gotto AM, Taunton OD (1977) Effects of polyunsaturated and saturated fat diets on the chemical composition and thermotropic properties of human lipoproteins. In: Kuhnau WH, Holman RT (eds) Polyunsaturated fatty acids. Am Oil Chem Soc, Champaign, Ill, pp 139–161

Prisco D, Rogasi PG, Matucci M, Costanzo G, Gensini GF (1984) Effect of pantethine treatment on platelet aggregation and thromboxane A2 production. Curr Ther Res 35:700–706

Prisco D, Rogasi PG, Matucci M, Abbate R, Gensini GF, Neri Serneri GG (1986) Increased thromboxane A2 generation and altered membrane fatty acid composition in platelets from patients with active angina pectoris. Thromb Res 44:101–112

Shattil SJ, Anaya-Galindo R, Bennet JS (1975) Platelet hypersensitivity induced by cholesterol incorporation. J Clin Invest 55:636–643

Shattil SJ, Bennet JS, Colman RW (1977) Abnormalities of cholesterol-phospholipid composition in platelets and low-density lipoproteins of human hyperbetalipoproteinemia. J Lab Clin Med 89:341–353

Pantethine Treatment
in Type III Hyperlipoproteinemia

G. Franceschini, P. Apebe, G. Gianfranceschi, A. Gaddi, M. Sirtori, and C. R. Sirtori [1]

Abbreviations: arg=arginine; cys=cysteine; HDL=high density lipoproteins; IEF=isoelectric focusing; LDL=low density lipoproteins; TC=total cholesterol; TG=triglycerides; VLDL=very low density lipoproteins.

1 Introduction

Type III hyperlipoproteinemia is a rare familial disorder of plasma lipoprotein metabolism, characterized by increased plasma cholesterol and triglyceride levels and by the presence of a broad β-band on paper and agarose gel electrophoresis (Brown et al. 1983). Patients with type III hyperlipoproteinemia develop accelerated atherosclerosis, often leading to death for coronary artery disease.

The primary genetic defect in type III hyperlipoproteinemia is represented by the presence of an abnormal apolipoprotein E (apo E) in the lipoproteins of the affected subjects (Mahley 1983). Apo E is a major constituent of human chylomicron and very low density lipoprotein (VLDL) remnants, and plays a key role in lipoprotein metabolism, permitting the hepatic uptake of remnant particles through the interaction with a specific cell surface receptor (Mahley 1983). This interaction is impaired in type III patients, who show a delayed plasma clearance of remnants (Gregg et al. 1981) with accumulation of apo E and cholesterol-rich VLDL with β-mobility.

The apo-E system in man consists of three major isoforms identical, except at two substitution sites, in positions 112 and 158 of the amino acid sequence (Mahley et al. 1984). In the E3 isoform, cysteine occurs at position 112 and arginine at 158. In the more acidic E2 isoform, arg_{158} is substituted with cysteine, whereas in the more basic E4 isoform, cys_{112} is substituted with arginine. Several other rare apo-E mutants have been identified (Mahley et al. 1984) with variable receptor-binding capacity. The characterization of these mutants led to the identification of the receptor-binding domain in the apo-E sequence. In particular, the arginine residue in position 158 has been shown to be essential for a correct interaction between apo E and its receptor (Mahley 1983).

1 Center E. Grossi Paoletti, Institute of Pharmacological Sciences, University of Milano and Clinica Medica II, University of Bologna, Italy

Drugs Affecting Lipid Metabolism
Ed. by R. Paoletti et al.
© Springer-Verlag Berlin Heidelberg 1987

The three major apo-E isoforms, coded by independent alleles at one single gene locus, lead to the appearance in the population of six phenotypes, three homozygotes, E2/2, E3/3, E4/4, and three heterozygotes, i.e. E2/3, E3/4, E2/4 (Zannis and Breslow 1982). Type III patients are all homozygotes E2/2, lacking the normal E3 isoform (Utermann et al. 1975). The absence of the positive charge at position 158 in the apo-E2 isoform, due to the arg→cys substitution, does not allow the interaction of apo E2 with the receptor, resulting in a defective in vitro binding (Weisgraber et al. 1982) and in a delayed in vivo clearance (Gregg et al. 1981). The in vitro binding of apo E2 to the receptor can be restored by treatment of the isoprotein with cysteamine (Weisgraber et al. 1982), a small thiol compound forming a mixed disulphide with cysteine, and converting it to a lysine analogue. Cysteamine treatment, although potentially useful for type III patients, is not advisable, because of the pulmonary toxicity of the compound (Corden et al. 1981).

Pantethine, a natural compound, part of coenzyme A, has been proven to lower plasma lipid levels in several forms of hyperlipidemia (Gaddi et al. 1984). Pantethine (P) is converted in vivo, through the action of pantetheinase, to pantothenic acid and cysteamine (Dupré et al. 1970). By this mechanism, P depletes cultured cystinotic fibroblasts of intracellular cystine (Butler and Zatz 1983). We postulated that P might be useful in the treatment of type III patients, by converting in vivo the non-functional apo-E2 isoform to an apo-E3 analogue, with receptor-binding capacity.

2 Methods

Seven type III hyperlipoproteinemic patients (5 males, 2 females), aged between 38 and 55 years, were selected for the study. All subjects had a VLDL-cholesterol/plasma triglyceride (TG) ratio > 0.3, β-VLDL on agarose gel electrophoresis and an apo E2/2 phenotype (Brown et al. 1983). The patients were treated with a fibrate derivative (Fenofibrate or Bezafibrate) and maintained their therapy throughout the study. P was administered at a daily dose of 400 mg t.i.d. for a period of 6 months.

Cholesterol and TG, both in plasma and separated lipoproteins, were determined by enzymatic methods (Bucolo and David 1973; Röschlau et al. 1974). Lipoproteins were separated by ultracentrifugation, according to the NIH protocol (Fredrickson et al. 1968); high density lipoproteins (HDL) were obtained after selective precipitation of apo B containing lipoproteins with dextran-$MgCl_2$ (Warnick et al. 1982). VLDL were delipidated by chloroform:methanol (2:1) and apo-E isoforms separated by analytical isoelectric focussing (IEF) on polyacrylamide gel rods (0.5 x 7 cm), using pH 5–7 Ampholines. IEF gels, after staining, were scanned by an Elvi Seroskope, previously standardized with appropriate reference apolipoproteins.

3 Results

Pantethine was well tolerated by the selected patients, who all completed the study. During the fibrate treatment, significant changes in plasma lipid and lipoprotein levels were noted in all the examined patients (Table 1). Plasma total cholesterol (TC) and TG levels decreased by 29 and 51% respectively; VLDL-cholesterol (VLDL-C) was markedly reduced by fibrate treatment (-52%), whereas cholesterol associated with low density lipoproteins (LDL-C) decreased by 12%. HDL-cholesterol (HDL-C) increased significantly (+16%). Despite these marked changes, the plasma lipid levels were completely normalized by fibrate therapy only in two of the seven examined patients. Furthermore, the composition of VLDL, abnormally enriched in cholesterol (Brown et al. 1983), was not modified, the VLDL-C/VLDL-TG ratio decreasing from 0.51 ± 0.12 to 0.49 ± 0.09 (p = n.s.); the apo-E isoform distribution was also unchanged by fibrate treatment.

The addition of P to the fibrate therapy led to a further improvement of the lipid-lipoprotein pattern. After 6 months of combined treatment, the plasma lipid levels were fully normalized in five of the seven treated patients. Cholesterol and TG decreased by 49 and 71% respectively, compared to the pre-treatment values, and by 28 and 41%, compared to fibrate alone (Table 1). These changes were due again to a marked decrease of VLDL-C (-80 and -57%, compared to basal and fibrate values) and to a less significant reduction of LDL-C (-33 and 24% respectively). By contrast, HDL-C levels were markedly increased (+32 and +13%). VLDL composition was slightly modified, the VLDL-C/VLDL-TG ratio decreasing from 0.49 ± 0.09 to 0.44 ± 0.08.

The apo-E isoforms were separated by analytical isoelectric focussing from VLDL, isolated before and after 6 months of fibrate+P therapy. In one patient, we observed the appearance of a new protein band, focussing in the same position of the apo-E3 isoform (Fig. 1), lacking in all the examined patients.

Table 1. Plasma lipid and lipoprotein levels in type III patients treated with fibrate and fibrate + pantethine

	Basal	Fibrate	Fibrate + pantethine
		mg dl^{-1}	
Total cholesterol	388.7 ± 87.7	281.5 ± 47.5[a]	203.8 ± 40.0[b,c]
Triglycerides	516.1 ± 136.4	254.5 ± 110.8[a]	148.8 ± 33.2[b,c]
VLDL-C	178.4 ± 46.5	85.7 ± 33.3[a]	36.8 ± 21.7[b,c]
LDL-C	169.2 ± 53.5	147.5 ± 29.4	112.4 ± 33.5[a]
HDL-C	41.5 ± 16.2	48.2 ± 15.0[a]	54.7 ± 13.4[b,c]

n = 7; mean ± SD.

[a] $p < 0.02$. [b] $p < 0.005$ vs basal. [c] $p < 0.05$ vs fibrate.

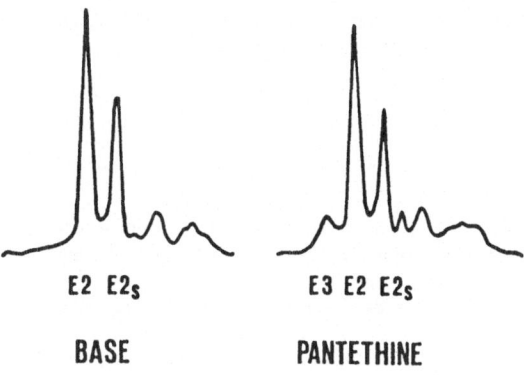

Fig. 1. Scans of IEF gels (apo-E region) from an E2/E2 homozygous patient before and after pantethine treatment

E2 E2$_S$ E3 E2 E2$_S$

BASE   **PANTETHINE**

4 Discussion

Type III hyperlipoproteinemia is generally associated with severe hypercholesterolemia and hypertriglyceridemia, both sensitive to diet and drug therapy (Brown et al. 1983). Clofibrate and its analogues are particularly effective in reducing plasma lipid levels in type III patients and have been widely considered as drugs of choice for this metabolic disorder. Fibrates generally act by improving the catabolism of TG-rich lipoproteins through the stimulation of lipoprotein lipase (Sirtori 1981), the key enzyme in this process. By this mechanism, fibrates can normalize the delayed clearance of VLDL, and particularly of VLDL remnants, observed in type III patients.

Fibrate treatment, however, cannot correct the primary defect in type III hyperlipidemia, i.e. the presence of a non-functional apo E (Mahley 1983). The accumulation of VLDL remnants in the plasma of type III patients is, in fact, due to the inability of the apo-E2 isoform, the only apo E present in these patients (Utermann et al. 1975), to interact with the apo-E receptor. This interaction can be restored in vitro by treatment of type III VLDL with cysteamine, converting the E2 isoform to a functional E3 analogue (Weisgraber et al. 1982). Based on these in vitro experiments, cysteamine has been recently proposed for the treatment of type III patients (Fisher and Gahl 1982). Major drawbacks for the therapeutic use of cysteamine arise from its disagreeable taste and odor, and high toxicity (Corden et al. 1981).

P is a natural compound, converted in vivo to cysteamine through the action of pantetheinase (Dupré et al. 1970). It has been widely used for the treatment of several forms of hyperlipidemia (Gaddi et al. 1984), being extremely well tolerated. We tested the hypothesis that P could be effective in type III hyperlipoproteinemia, by in vivo conversion of the non-functional apo-E2 isoform to an apo-E3 analogue with receptor-binding capacity. The reported findings indicate that, especially when added to a standard fibrate therapy, P is very effective in improving the altered lipid-lipoprotein pattern of type III patients. The mechanism responsible for this hypolipidemic effect is at present unknown. However, the appearance in the VLDL of some type III patients, of an apo E3-like protein band, may be indicative of an activity of P on the primary defect responsible for the disease. In view of these results, P may represent a safe and effective tool for the treatment of type III patients.

Acknowledgement. Supported in part by the Consiglio Nazionale delle Ricerche of Italy (Progetto Finalizzato Ingegneria Genetica e Basi Molecolari delle Malattie Ereditarie).

References

Brown MS, Goldstein JL, Fredrickson DS (1983) Familial type III hyperlipoproteinemia. In: Stanbury JB, Wyngaarden JB, Fredrickson DS, Goldstein JL, Brown MS (eds) The metabolic basis of inherited disease. Mc Graw-Hill, New York, p 655

Bucolo G, David H (1973) Quantitative determination of serum triglycerides by use of enzyme. Clin Chem 19:476–483

Butler JD, Zatz M (1983) Pantethine depletes cystinotic fibroblasts of cystine. J Pediat 5:796–798

Corden BJ, Schulman JD, Schneider JA, Thoene JG (1981) Adverse reactions to oral cysteamine use in nephropathic cystinosis. Dev Pharmacol Ther 3:25–32

Dupré S, Graziani MT, Rosei MA, Fabi A, Grosso ED (1970) The enzymatic breakdown of pantethine to pantothenic acid and cysteamine. Eur J Biochem 16:571–578

Fisher EA, Gahl WA (1982) Cysteamine in treatment of type III hyperlipidaemia? Lancet ii: 1131–1132

Fredrickson DS, Levy RI, Lindgren FT (1968) A comparison of heritable abnormal lipoproteins patterns as defined by two different techniques. J Clin Invest 47:2446–2451

Gaddi A, Descovich GC, Noseda G et al. (1984) Controlled evaluation of pantethine, a natural hypolipidemic compound, in patients with different forms of hyperlipoproteinemia. Atherosclerosis 50:73–83

Gregg RE, Zech LA, Schaefer EJ, Brewer HJ, Jr (1981) Type III hyperlipoproteinemia: defective metabolism of an abnormal apolipoprotein E. Science 211:584–586

Mahley RW (1983) Apolipoprotein E and cholesterol metabolism. Klin Wochenschr 61:225–232

Mahley RW, Innerarity TL, Rall SC, Weisgraber KH (1984) Plasma lipoproteins: apolipoprotein structure and function. J Lipid Res 25:1277–1294

Röschlau P, Bernt E, Gruber W (1984) Enzymatische Bestimmung des Gesamt-Cholesterins im Serum. Z Klin Chem Biochem 12:402–407

Sirtori CR (1981) Mechanism of action of clofibrate. Lancet i: 1362

Utermann G, Jaeschke M, Menzel J (1975) Familial hyperlipoproteinemia type III: deficiency of a specific apolipoprotein (apo E III) in the very low density lipoproteins. FEBS Lett 56:352–355

Warnick GR, Benderson J, Albers JJ (1982) Dextran sulfate precipitation procedure for quantitation of high density lipoprotein. Clin Chem 28:1379–1388

Weisgraber KM, Innerarity TL, Mahley RE (1982) Abnormal lipoprotein receptor-binding activity of the human E apoprotein due to cysteine-arginine interchange at a single site. J Biol Chem 257:2518–2521

Zannis VI, Breslow JL (1982) Apolipoprotein E. Mol Cell Biochem 42:3–20

Subject Index